Energy-Efficient Communication Networks and Systems

Editor

Josip Lorincz

MDPI

Basel • Beijing • Wuhan • Barcelona • Belgrade • Novi Sad • Cluj • Manchester

Editor
Josip Lorincz
Faculty of electrical engineering,
mechanical engineering and
naval architecture
University of Split
Split
Croatia

Editorial Office
MDPI
St. Alban-Anlage 66
4052 Basel, Switzerland

This is a reprint of articles from the Special Issue published online in the open access journal *Sensors* (ISSN 1424-8220) (available at: https://www.mdpi.com/journal/sensors/special_issues/EECNS).

For citation purposes, cite each article independently as indicated on the article page online and as indicated below:

Lastname, A.A.; Lastname, B.B. Article Title. *Journal Name* **Year**, *Volume Number*, Page Range.

ISBN 978-3-0365-8966-4 (Hbk)
ISBN 978-3-0365-8967-1 (PDF)
doi.org/10.3390/books978-3-0365-8967-1

Contents

About the Editor

Josip Lorincz

Josip Lorincz received a B.Sc. (M.S. equivalent) and a Ph.D. degree in telecommunications engineering and computer science from the University of Split, Croatia, in 2002 and 2010, respectively. In 2003, he joined the Faculty of electrical engineering, mechanical engineering, and naval architecture, at the University of Split, Croatia, where he currently works as an associate professor. In the academic year 2009/2010, he was a visiting researcher at the Department of Electronics, informatics, and bioengineering of the Politecnico di Milano, Italy. As a project leader or researcher, he participated in more than thirty scientific and professional projects funded by the EU, public or private sector. He is the founder and chair of the Symposium on Green Networking and Computing, organized in the frame of the International Conference on Software, Telecommunications and Computer Networks (SoftCOM). He also serves as the technical program committee member for many international scientific conferences and as a reviewer for top scientific journals. He was a guest editor in different scientific journals and an evaluator of international scientific projects. His current research interests include energy-efficient wireless and wired networks, optimization in telecommunications, advanced design, management and analyses of sensor networks, and evolution of routing protocols. He has authored more than 60 research papers published in different scientific conferences and journals. He is a senior IEEE member, senior ACM member, associate member of the Croatian Academy of Engineering, and the first president of the Croatian ACM chapter. Since 2004, he has owned Cisco CCNA, CCAI, and BCMSN certificates. In 2012, he received an award for scientific work from his home institution, FESB, and in 2013, he was awarded as an outstanding young researcher by the Croatian Academy of Engineering. In 2023, from his home University of Split, he received a science award for present achievements in science.

Preface

Information and communication systems are integral parts of modern society, facilitating the exchange of information and connecting people across the globe. However, the usage of information and communication systems comes with significant environmental costs, which are the consequence of the two causes. The first one is the high energy consumption required for the operation of information and communication systems, and the second one is related to the carbon emissions that occur during their operation. As a consequence, the energy efficiency of information and communication systems has become an increasingly important topic, as the need to reduce energy consumption and minimize carbon emissions has become a critical priority.

One of the most significant challenges in improving the energy efficiency of information and communication systems and consequently reducing carbon emissions is the sheer volume of data that needs to be processed and transmitted. As the volume of data continues to grow due to the proliferation of new services and applications and the increase in the number of network users, the energy required for processing and transmitting this data also increases. This has led to a continuous increase in the energy consumption of data centers and communication networks, which are responsible for storing and transmitting data. This growth in energy consumption has resulted in various negative economic impacts related to the large operational expenditures (OPEX) of communication networks and system operators and environmental impacts related to the pollution caused by increased carbon emissions. Hence, there is a constant need for the development of solutions that can reduce energy consumption while maintaining high performance, quality of service, and reliability of communication networks and systems.

Although the contribution of communication networks and systems to energy consumption and carbon emissions on a global level cannot be completely nullified, in order to stop or even reverse the increasing trend of energy consumption and carbon emissions contributed by communication networks and systems, these contributions should be maximally decreased. To achieve this, improved or completely new strategies and approaches in terms of the development and operation of communication networks and systems must be devised and practically implemented. Hence, energy-efficient operation of communication networks and systems should be envisioned on all Open System Interconnection (OIS) layers. This can be realized through developing more energy-efficient communication protocols, designing more energy-efficient hardware components, envisioning novel algorithms that will minimize the number of active components at any given time, reusing them in a better way, and exploiting renewable energy sources in powering communication networks and systems. To accomplish this challenging task, a joint effort of different participants coming from governments, standardization organizations, academia, and industry must take place. Hence, energy efficiency has an important role in ensuring the sustainable development of communication networks and systems, and it represents a topic of increasing interest and importance in modern society.

Despite such astonishing interests in improving the energy efficiency of communication networks and systems perceived during the last decade, the research on energy-efficient communication networks and systems on different levels and in many fields demands enhanced or completely new solutions. Also, versatile basic or highly sophisticated problems are still unsolved or are even in their infancy. Hence, improving the energy efficiency of communication networks and systems are and will proceed in being an actual economic, social, industrial, and particularly research challenge. This book, entitled "*Energy-efficient Communication Networks and Systems*", is dedicated to all aspects of the research and development related to solving such challenges. The reprint provides an overview of the latest research and technologies in the field of energy-efficient communication networks and systems. It contains the accepted scientific papers gathered in the "*Special Issue on Energy-efficient Communication Networks and Systems*" organized for the Sensors journal which is published by MDPI (Multidisciplinary Digital Publishing Institute). Twelve high-quality articles have been collected and reproduced in this book, demonstrating significant achievements in the field to which this reprint is dedicated. Among published scientific papers, one paper is editorial, one paper is a review type of paper, and the remaining ten works are research articles. Published papers are consolidated in this reprint as self-contained peer-reviewed scientific works.

(1). The editorial paper "*Lorincz J.; Klarin Z.; Begusic D.; Advances in Improving Energy Efficiency of Fiber–Wireless Access Networks: A Comprehensive Overview, Sensors 2023, 23, 2239.*", gives a comprehensive survey of recent research on approaches that contribute to energy efficiency (EE) improvements of Fiber-Wireless (FiWi) access networks. The presented EE analyses are performed for different types of FiWi networks, including the radio-and-fiber (R&F) networks, the radio-over-fiber networks (RoF), the FiWi networks based on multi-access edge computing (MEC), and the software-defined network (SDN)-based FiWi networks. For the R&F networks, energy conservation techniques and research studies related to the optical and wireless domains were presented, as well as related works that deal with the improvement of FiWi networks' EE through the cooperation of techniques in wireless and optical domains. Furthermore, two basic RoF techniques, the digital-RoF (D–RoF) and analog-Rof (A–RoF), were elaborated in the context of EE, and an overview of research studies in the field of improving the EE of D–RoF and A–RoF systems was given. Additionally, the cloud–radio access network (C-RAN) architecture was reviewed through the prism of energy consumption with the presentation of current research efforts related to the improvement of the C-RAN energy efficiency. The MEC-based FiWi networks, which introduce cloud computing at the edge of the mobile network, were further presented, and articles dedicated to the mechanisms and concepts for the optimization of the MEC FIWi network's energy consumption were highlighted. Finally, flexible SDN FiWi networks that offer high scalability and ease of management were presented, with an emphasis on research related to energy conservation techniques implemented in such networks. The last part of the paper also discusses future directions for improving the EE in the FiWi networks.

(2). In addition, the second review paper, "*Depasquale E.-V.; Davoli F.; Rajput H.; Dynamics of Research into Modeling the Power Consumption of Virtual Entities Used in the Telco Cloud, Sensors 2023, 23, 255.*" gives analytical and a graphical survey of the literature over the period 2010–2020, on the measurement of power consumption and developed power models of virtual entities (virtual machines (VMs) or containers) implemented in the telecommunication (telco) operators cloud. The paper presents a thorough analysis of the dynamics of research related to the virtual entities (VEs) implementation challenges, approaches, pitfalls, fallacies, and research gaps with respect to the predictive modeling and supporting measurements of individual VEs power consumption that is relevant to the telco cloud. Research dynamics is characterized through a publication frequency analysis, performed based on the application of a novel developed method that is unique in its ability to parse research literature. Through the visual aids and cross-cutting themes, the authors in the paper provide a thorough characterization of the problems, approaches, developments, formal methods, pitfalls, fallacies, and research gaps that characterize the research space of predictive modeling and measurements of individual telco VEs power consumption. The presented survey can serve as a reference in the selection of the most appropriate power consumption model of VEs implemented in the telco clouds.

(3). The third published paper, "*Lorincz J.; Klarin Z.; How Trend of Increasing Data Volume Affects the Energy Efficiency of 5G Networks, Sensors 2022, 22, 255.*" analyses the impact of the expected increase of data volumes (DVs) through the 2020s, on the energy efficiency (EE) of the fifth generation (5G) radio access network (RAN) by using standardized data and coverage EE metrics. An analysis was performed for five different macro and small 5G base stations (BSs) implementation and operation scenarios and for rural, urban, dense-urban, and indoor-hotspot device density classes (areas). The results of analyses reveal a strong influence of increasing DV trends on standardized data and coverage EE metrics of 5G heterogeneous networks (HetNets). For every device density class characterized with increased DVs, elaboration on the process of achieving the best and worst combination of data and coverage EE metrics for each of the analyzed 5G BSs deployment and operation approaches have been performed. This elaboration is further extended on the analyses of the impact of 5G RAN instant power consumption and 5G RAN yearly energy consumption on values of standardized EE metrics. The presented analyses can serve as a reference in the selection of the most appropriate 5G BS deployment and operation approach, which will simultaneously ensure the transfer of permanently increasing DVs in a specific device density class and the highest possible levels of data and coverage EE metrics.

(4). In the fourth paper *"Lorincz J.; Ramljak I.; Begušic D.; Analysis of the Impact of Detection Threshold Adjustments and Noise Uncertainty on Energy Detection Performance in MIMO-OFDM Cognitive Radio Systems, Sensors 2022, 22, 631."*, the efficiency of spectrum sensing performed with the energy detection (ED) method realized through the square-law combining (SLC) of the received signals at secondary users (SUs) has been analyzed. The analyses take into account the detection threshold (DT) adjustments performed according to noise uncertainty (NU) variations in multiple-input multiple-output (MIMO)—orthogonal frequency division multiplexing (OFDM) communication systems. The mathematical expression of the main parameters used for the evaluation of the ED performance as a local spectrum sensing technique employing SLC in MIMO-OFDM CR systems has been introduced. In addition, the algorithm for simulating the ED method in versatile operating environments characterized by the influence of distinct levels of NU and performed with dynamic DT (DDT) adjustments has been presented. The analysis of ED sensing efficiency has been performed through extensive simulations, which indicates how different working parameters, including the number of samples used in the ED process, the transmit powers of the primary user (PU), the DDT and NU factors, the probabilities of false alarm, and the signal-to-noise (SNR) level impact the probability of the detection of PU signals in MIMO-OFDM CR systems.

(5). To improve energy management, in the fifth paper, *"Marques D.; Senna C.; Luís M.; Forwarding in Energy-Constrained Wireless Information Centric Networks, Sensors 2022, 22, 1438."*, authors propose an efficient forwarding scheme in energy-constrained wireless information-centric networks (ICNs). Analyzed ICNs are composed of a large number of sensors allocated across the smart city. To achieve the stated goal dedicated to the improvement of sensor nodes' energy management, authors, among other parameters, consider the different types of sensor devices, their internal energy consumption, and the network context. The proposed forwarding strategy extends and adapts concepts of ICNs through the implementation of packet domain analysis, neighborhood evaluation, and sensor node sleeping and waking strategies. In order to consistently address sensor mobility and to improve the quality of content delivery, the proposed solution takes advantage of the neighborhood awareness of the moments that indicate when to listen and forward data packets. The evaluation of the proposed strategy is performed by simulation with real datasets obtained based on real urban mobility. The results show that the proposed data forwarding strategy enables significant improvements in the network content availability, overall energy savings, and the network lifetime.

(6). In the paper *"Liu Y.; Li C.; Li J.; Feng L.; Joint User Scheduling and Hybrid Beamforming Design for Massive MIMO LEO Satellite Multigroup Multicast Communication Systems, Sensors 2022, 22, 6858."*, authors investigated the robust joint user scheduling and hybrid beamforming design scheme that can maximize the energy efficiency (EE) of the massive multiple-input multiple-output (MIMO) low Earth orbit (LEO) satellite multigroup multicast communication system. To solve the stated problem, the authors first adopted the hierarchical clustering algorithm to group users and then applied the semidefinite programming (SDP) algorithm and the concave-convex process (CCCP) framework to tackle the optimization of user scheduling and hybrid beamforming design. To obtain the digital and analog satellite antenna beamforming matrix in a hybrid beamformer, the alternative optimization algorithm based on the majorization-minimization framework (MM-AltOpt) is also proposed. Numerical simulation results show that the energy efficiency of the proposed joint user scheduling and beamforming algorithm is higher than that of the traditional decoupling algorithms.

(7). To improve the power efficiency (PE) of the fifth generation (5G) heterogeneous network, in the seventh paper "*Osama M.; El Ramly S.; Abdelhamid B.; Binary PSO with Classification Trees Algorithm for Enhancing Power Efficiency in 5G Networks, Sensors 2022, 22, 8570.*", an approach based on switching on and off the redundant small cells (SCs) using machine learning (ML) techniques is proposed. The proposed scheme needs to ensure a reduction in the energy consumption of the radio part of the 5G network, while the quality of service (QoS) for every user must be satisfied. The proposed approach is based on a linearly increasing inertia weight–binary particle swarm optimization (IW-BPSO) algorithm for SC on/off switching. Moreover, a soft frequency reuse (SFR) algorithm is proposed using the classification trees (CTs) approach to alleviate the interference and elevate the system throughput. The obtained results show that the proposed algorithms, in terms of energy efficiency improvements, outperform the other conventional energy-saving algorithms for 5G networks. The proposed algorithms reduce the power consumption of the network and the interference among the SCs while ensuring improvements in the total throughput and the PE of the system.

(8). Next published paper, "*Han D.; So J.; EnergyEfficient Resource Allocation Based on Deep Q-Network in V2V Communications, Sensors 2023, 23, 1295.*", utilizes the artificial intelligence concept of a deep Q-network (DQN) to select the transmit resource blocks and transmit power of vehicles in the vehicle-to-vehicle (V2V) network. The goal of the proposed concept is to maximize the sum rate of the vehicle-to-infrastructure (V2I) and V2V communication links while reducing the power consumption and latency of those links. The exploited DQN concept also utilizes the channel state information, the signal-to-interference-plus-noise ratio (SINR) of V2I and V2V links, and the latency constraints of vehicles to find the optimal resource allocation scheme. The proposed DQN-based resource allocation scheme ensures energy-efficient transmissions that satisfy the latency constraints for the V2V links while reducing the interference of the V2V network to the V2I network. The performance of the proposed scheme has been evaluated in terms of the sum rate of the V2X network, the average power consumption of the V2V links, and the average outage probability of the V2V links using a real case study. The simulation results show that the proposed scheme greatly reduces the transmit power of the V2V links, especially when compared to the transmit power of the V2V links in conventional reinforcement learning-based resource allocation schemes.

(9). In the paper "*Rau, F.; Soto, I.; Zabala-Blanco, D.; Azurdia-Meza, C.; Ijaz, M.; Ekpo, S.; Gutierrez, S.; A Novel Traffic Prediction Method Using Machine Learning for Energy Efficiency in Service Provider Networks. Sensors 2023, 23, 4997.*" The authors propose a systematic approach for solving complex prediction problems with a focus on energy efficiency. The approach involves the usage of neural networks, specifically recurrent and sequential networks, as the main tool for prediction. In order to test the proposed methodology, a case study was conducted to address the issue of energy efficiency in the data centers of telecommunications service providers. The case study involved comparing four neural network types, more specifically, recurrent neural networks (RNNs), long short-term memory

(LSTM), gated recurrent units (GRUs), and online sequential extreme learning machine (OS-ELM), for determining the best network in terms of network traffic prediction accuracy and computational time. The results show that OS-ELM outperformed the other networks in both accuracy and computational efficiency. The simulation was applied to real traffic data and showed significant energy savings potential in a single day, which offers a real solution for energy efficiency and energy savings that can be applied not only to the core part but also to the aggregation networks. The obtained results confirmed that utilizing a neural network as the primary tool for prediction enables high accuracy and adaptability to different data types of data center networks.

(10). In the paper "*Kinman, G.; Žilić, Ž.; Purnell, D.; Scheduling Sparse LEO Satellite Transmissions for Remote Water Level Monitoring. Sensors 2023, 23, 5581.*", authors have presented and evaluated the transmission scheduling for emerging sparse low earth orbit (LEO) satellite services suitable for IoT (Internet of Things) applications. More specifically, the authors explore the use of LEO satellite links in the long-term monitoring of water levels across remote areas. Since emerging sparse LEO satellite constellations maintain a sporadic connection to the ground station(s), the monitoring sensor's transmissions to the satellite need to be scheduled for satellite overfly periods. Thus, for remote sensing, energy consumption optimization is critical, and this motivates authors to develop a learning approach for scheduling the transmission times for the monitoring sensors. A detailed probabilistic energy consumption model was developed in order to evaluate the proposed online learning scheme for predicting transmission periods. The proposed online learning-based approach combines Monte Carlo and modified k-armed bandit approaches in order to produce an inexpensive scheme that applies to scheduling any LEO satellite transmissions. The proposed learning approach is inexpensive computationally, learns in small increments in a modest number of training epochs, and is interpretable, unlike most modern machine learning approaches. Results presented in the paper demonstrate the ability to adapt scheduling transmission time, which results in 20-fold energy saving of the transmission energy. The presented study applies to a wide range of IoT applications in areas with no existing wireless coverages.

(11). The paper "*Sterz, A.; Klose, R.; Sommer, M.; Höchst, J.; Link, J.; Simon, B.; Klein, A.; Hollick, M.; Freisleben, B.; Energy-Efficient Decentralized Broadcasting in Wireless Multi-Hop Networks. Sensors 2023, 23, 7419.*" presents a novel multi-hop data broadcasting protocol named BTP (Broadcast Tree Protocol). The proposed protocol uses a game-theoretical model to construct a spanning tree in a decentralized manner. The goal of the developed spanning tree model is to minimize the transmission power of each node and consequently minimize the total energy consumption of a network. The authors in the paper integrate three algorithms capable of inhibiting the creation of graph cycles into the design of BTP. Although based on a game-theoretical model, the proposed BTP neither requires information exchange between distant nodes nor time synchronization during its operation, while BTP inhibits graph cycles effectively. Authors, through simulations, evaluate BTP with respect to various aspects, compare BTP to other algorithms from the literature, and investigate the scalability of BTP. The proposed protocol is evaluated in Matlab and NS-3 simulations and through real-world implementation on a testbed of 75 Raspberry Pis. The evaluation conducted shows that the proposed protocol can achieve significant energy reduction when compared to a simple broadcast protocol in real-world experiments.

(12). Last published paper, "*Carbajal Ipenza, S.J.; Masiero, B.S.; Efficient Sigma–Delta Sensor Array Beamforming. Sensors 2023, 23, 7577.*" analyses beamforming algorithm implementation for sensors based on sigma–delta modulators ($\Sigma\Delta$Ms), which are widely used in consumer, industrial, automotive and medical applications such as micro-electro-mechanical systems (MEMSs) or digital microphones. Although $\Sigma\Delta$Ms have become a cost-effective and convenient way to deliver data to digital processors, the ΣΔMs output a pulse-density modulated (PDM) bitstream signal, which is the reason why sensors require either built-in or external high-order decimation filters to demodulate the PDM signal to a baseband multi-bit pulse-code modulated (PCM) signal. Because of this extra circuit requirement, the implementation of sensor array algorithms, such as beamforming in embedded systems (where the processing resources are critical) or in very large-scale integration (VLSI) circuits (where the power and surface usage are crucial), becomes especially expensive as a large number of parallel decimation filters are required. Thus, the article proposes a novel architecture for beamforming algorithm implementation that fuses delay and decimation operations based on maximally flat (MAXFLAT) filters to make array processing more affordable. As proof of concept, the paper presents an implementation example of a delay-and-sum (DAS) beamformer at given spatial and frequency requirements using a novel proposed approach. The presented results show that in comparison with the most efficient state-of-the-art beamformer architectures, the proposed architecture requires significantly lower storage and computational resources, which contributes to the improvement of network energy efficiency.

The 40 different authors coming from academia and industry have contributed to this book. Their contributions provide valuable insights into the latest research and technologies in the area of energy-efficient communication networks and systems. The reprint also includes case studies and examples of the usage of energy-efficient communication technologies in practice, providing readers with practical insights into the implementation of these technologies. The reprint is intended for researchers, engineers, and students who are interested in the design and implementation of energy-efficient communication networks and systems. Hence, the reprint gives insights and solutions for a range of problems in the field of obtaining energy-efficient communication systems and networks, and it lays the basis for solving new challenges toward achieving future advances.

The reprint editor would like to thank all authors who have submitted their articles and all reviewers for their valuable work dedicated to giving an expert review for submitted papers. Moreover, the reprint editor is grateful to all persons involved in the edition of this reprint for their invaluable support, including the *Sensors* Journal Section managing editor Mr. Winston Wang, and other editorial team members involved in editing the papers submitted to *Sensors* journal *"Special Issue on Energy-efficient communication networks and systems"*.

The reprint editor sincerely hopes that this reprint will be a valuable source of information presenting the recent advances in different fields related to greening and improving energy efficiency and sustainability of those communication networks and systems particularly addressed in this reprint.

Josip Lorincz
Editor

sensors

MDPI

Editorial

Advances in Improving Energy Efficiency of Fiber–Wireless Access Networks: A Comprehensive Overview

Josip Lorincz [1,*], Zvonimir Klarin [2] and Dinko Begusic [1]

1 Faculty of Electrical Engineering, Mechanical Engineering and Naval Architecture (FESB), University of Split, R. Boskovica 32, 21000 Split, Croatia
2 University of Applied Sciences Sibenik, Trg Andrije Hebranga 11, 22000 Sibenik, Croatia
* Correspondence: josip.lorincz@fesb.hr

Abstract: Due to the growing impact of the information and communications technology (ICT) sector on electricity usage and greenhouse gas emissions, telecommunication networks require new solutions which will enable the improvement of the energy efficiency of networks. Access networks, which are responsible for the last mile of connectivity and also for one of the largest shares in network energy consumption, are viable candidates for the implementation of new protocols, models and methods which will contribute to the reduction of the energy consumption of such networks. Among the different types of access networks, hybrid fiber–wireless (FiWi) networks are a type of network that combines the capacity and reliability of optical networks with the flexibility and availability of wireless networks, and as such, FiWi networks have begun to be extensively used in modern access networks. However, due to the advent of high-bandwidth applications and Internet of Things networks, the increased energy consumption of FiWi networks has become one of the most concerning challenges required to be addressed. This paper provides a comprehensive overview of the progress in approaches for improving the energy efficiency (EE) of different types of FiWi networks, which include the radio-and-fiber (R&F) networks, the radio-over-fiber networks (RoF), the FiWi networks based on multi-access edge computing (MEC) and the software-defined network (SDN)-based FiWi networks. It also discusses future directions for improving the EE in the FiWi networks.

Keywords: radio; access; networks; fiber; D-RAN; C-RAN; Mobile fronthaul; RoF; digitized RoF; analog RoF; NGFI; FiWi; NG–PON; energy efficiency; SDN; NFV; MEC

Citation: Lorincz, J.; Klarin, Z.; Begusic, D. Advances in Improving Energy Efficiency of Fiber–Wireless Access Networks: A Comprehensive Overview. *Sensors* **2023**, *23*, 2239. https://doi.org/10.3390/s23042239

Academic Editor: Peter Chong

Received: 22 January 2023
Accepted: 14 February 2023
Published: 16 February 2023

1. Introduction

Climate change manifested by global warming caused by large amounts of greenhouse gas (GHG) emissions presents a serious issue that affects today's society. This issue is contributed to by increased energy demand for powering today's Information and Communication Technology (ICT) systems. This increased energy demand for ICT systems additionally poses an economic issue due to the increasing cost and consumption of energy. Therefore, improving the energy efficiency (EE) of ICT systems has become an important consideration, since addressing this issue can contribute to the global reduction of GHG emissions and operating expenditure (OPEX) costs of service providers and network owners.

Despite the ability to reduce GHG emissions in some real-life scenarios (e.g., transportation reduction and document dematerialization), ICT is becoming a major contributor to global energy consumption [1]. The EE in telecommunication networks, as one of the significant contributors to the energy consumption of the overall ICT industry, is attracting global attention due to its large contribution to increasing OPEX costs and carbon dioxide emissions. Depending on the usage scenario, it is projected that communication technology can contribute to an increase in global electricity usage between 8% and 51% by 2030 [2]. Today's communication networks are dominated by wireless technologies (e.g., mobile cellular or wireless local/personal/access networks), and wireless traffic is expected to increase dramatically with the full deployment of the fifth-generation (5G) mobile network [3]. This is a consequence of the basic aim of the 5G technology, which is envisioned to support a massive number of connected devices through the Internet of Things (IoT) concept. This will result in usage scenarios where communication technology will be incorporated into every aspect of today's society. To support an enormous number of devices, the 5G is designed as a heterogeneous network (HetNet) with the ultra-dense deployment of small base stations (BSs) operating in sub-6 GHz and millimeter-wave (mmWave) frequency bands and employing massive multiple-input multiple-output (mMIMO) transmission technologies. Hence, an increasing number of connected devices with a constant demand for higher data rates will support new services and applications, requiring an increase in the number of supporting network infrastructure elements, which negatively affects the EE of mobile networks (MNs). Radio access networks (RAN) and particularly BSs are the largest, in terms of number and power consumption, individual elements of the RAN, and consequently they add a significant contribution to the overall energy consumption of telecommunication networks. Therefore, the EE of RAN should be a key concern in the design and deployment of future MNs [4].

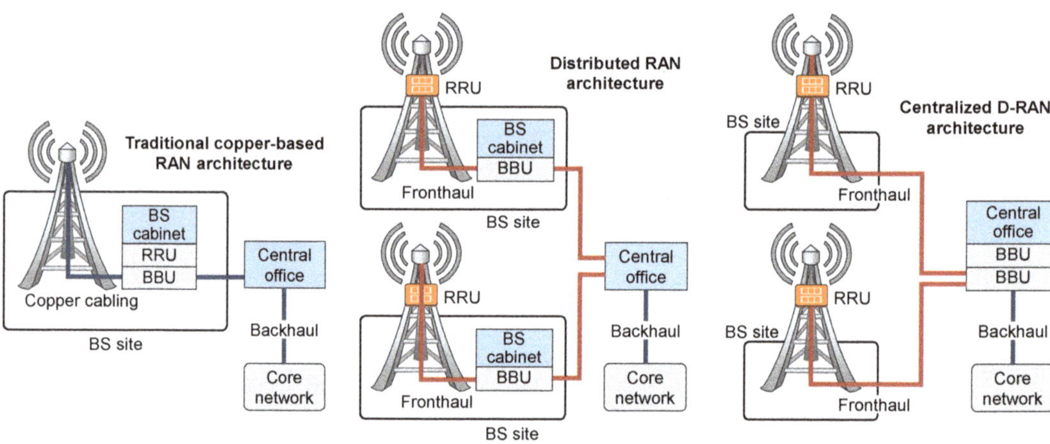

Figure 1. Comparison of traditional RAN and D-RAN architecture.

Due to the abovementioned necessity for satisfying implementation demands related to the simultaneous serving of higher numbers of users, offering larger data rates or ensuring a better quality of service, mobile network operators (MNOs) in past decades upgraded their radio access network (RAN) architectures. One of the main RAN architecture upgrades is related to the replacement of obsolete and overhauled traditional waveguide and coaxial-based BSs architecture (Figure 1) with those based on fiber–optic systems. Such architecture, through exploiting wired fiber–optic links for connecting physically spaced parts of BSs, establishes a completely new concept of data transmission in the RANs known as Distributed-RAN (D-RAN) (Figure 1). Introduced for third-generation (3G) mobile networks, the D-RAN architecture is characterized by modular BS design. This design decouples BS hardware into a remote radio unit (RRU) dedicated to the transmission and reception of wireless signals and a Baseband Unit (BBU) dedicated to baseband digital signal processing (Figure 1). Such D-RAN architecture, also known as Fiber to the Antenna (FTTA) architecture, creates two physically separated BS components that are connected with optical fiber at the location of the BS site, or up to a few hundred meters away.

The introduction of such modular BS design with fiber–optic cabling between BBU and RRU provides numerous advantages which primarily include higher bandwidth, reduced transmission losses and lower sensitivity to electromagnetic interference and noise.

However, one additional positive aspect of such architecture is reducing the increased energy consumption of network equipment, since RRU can exploit the concept of natural air cooling, which reduces the energy needed for cooling overall BSs composed of BBU and RRU in the same cabinet (Figure 1).

Additionally, the introduction of optical communication between RRU and BBU in D-RAN becomes the basis for the development of the newest generation of RAN architectures, which are based on centralized wireless access networks known as Cloud-RAN (C-RAN). In the C-RAN, decupled BS RRU and BBU components are placed at different locations which can be up to tens of kilometers away (Figure 1). Locations are connected with fiber–optic technology which establishes a new RAN entity defined as a mobile fronthaul network (MFN). Although the realization of MFN based on the C-RAN architecture was initially introduced for fourth-generation (4G) mobile networks, the superior bandwidth capabilities and network scalability that C-RAN provides set the basis for the full exploitation of the C-RAN concept in the realization of the MFN of 5G networks and future sixth-generation (6G) mobile networks. Although most of the modern BSs currently use D-RAN architecture, the C-RAN can be viewed as the architectural evolution of the D-RAN and the C-RAN architecture and will be essential for the full practical deployment of the 5G and the future 6G RANs.

Besides implementation in mobile cellular networks, the integration of fiber optics and wireless technology in access networks through the combination of fiber and wireless technologies, also known as the fiber–wireless (FiWi) concept, presents an alternative cost-effective solution to the realization of access networks. FiWi access networks integrate the high capacity of optical fiber networks with the coverage and flexibility of wireless networks. They create a robust infrastructure for the development and deployment of current and future applications and services and they are considered to be the most promising option for the realization of next-generation access networks [5]. By using optical networks as the backhaul and wireless networks as the front-end, FiWi networks aim to provide straightforward access for users [6]. These networks are expected to meet the demands of future access networks, such as providing high bandwidth, reliability, low cost and flexibility [7]. Additionally, it has been reported that access networks account for approximately 70% of energy usage in the ICT industry, and as a result of an increasing number of communication devices and bandwidth rates, this percentage is likely to increase in the future [6]. Therefore, to decrease overall network energy consumption, it is crucial to develop energy-efficient architectures of access networks, and architecture based on the FiWi concept is seen as a promising contributor to this goal.

In a FiWi network, both radio-over-fiber (RoF) and radio-and-fiber (R&F) technologies are utilized. Radio-over-fiber (RoF) constitutes a popular communication system design that addresses the increasing bandwidth demand and enables optical and wireless integration in modern C-RAN systems. The radio-and-fiber (R&F) network, in contrast, is an approach based on connecting distributed radio transmitters (such as wireless local area network (WLAN) access points (APs)) and centralized WLAN controllers with fiber–optic networks. R&F networks have the advantage of building wireless local area network (WLAN)-based FiWi networks. These advantages are reflected in the possible coverage of larger areas without the limitations related to the optical back-end size, while the RoF network has a limited reach of deployed fiber imposed by the propagation delay in fiber. Using the same infrastructure for both wired and wireless services, FiWi networks can merge the traditionally separate optical and wireless access networks, thus leading to potential cost savings [8].

The high performance requirements of future 5G networks, including extremely low latency, uninterrupted user experience and high throughput and reliability, will necessitate the use of localized services that are within RANs located closer to mobile subscribers.

The multi-access edge computing (MEC) paradigm is exploited as a solution for offering services closer to mobile subscribers. Through telecom operators, information technology (IT) and cloud computing, the MEC aims to bring cloud services to the users directly from the network edge [9]. However, the concept of MEC over FiWi networks must address not only the 5G network realization challenges, but it must also address unique challenges related to the integration of the MEC concept with existing wired and wireless infrastructures, and managing resources effectively in the context of backhaul and RAN coordination [9]. Furthermore, a software-defined network (SDN) architecture which uses a centralized controller to manage networks is a revolutionary approach that has the potential to greatly improve the control and management of FiWi access networks, and thus increase the EE of the entire FiWi network [10]. Therefore, in this paper, EE aspects of the FiWi networks with particular emphases on the RoF, R&F, MEC and SDN technologies were overviewed.

The paper is further structured as follows: the overview of the FiWi access networks is presented in Section 2. In Section 3, the EE challenges of R&F networks are discussed. The energy-efficient RoF solutions are elaborated in Section 4, with an emphasis on the impact of the C-RAN network architecture on the EE. The EE solutions for MEC-based FiWi access networks are reviewed in Section 5. In Section 6, the SDN-based energy-saving schemes in FiWi networks are elaborated. Section 7 discusses the results of performed analyses dedicated to improving FiWi networks EE. Finally, conclusion remarks related to the review of EE methods for FiWi networks are given in Section 8.

2. Fiber–Wireless Broadband Access Network

Merging optical fiber and wireless technologies in broadband access networks by leveraging the advantages of both technologies presents a viable solution to support the increasing bandwidth demands of future applications [11,12]. The concept of FiWi access networks is based on the seamless integration of the mobility and coverage offered by wireless networks and the high bandwidth and better stability offered by optical technology. In comparison with traditional access networks, FiWi access networks can be considered as an alternative or transitional solution that provides high data rates more cost-effectively with the desired quality of service (QoS).

2.1. Architecture of FiWi Networks

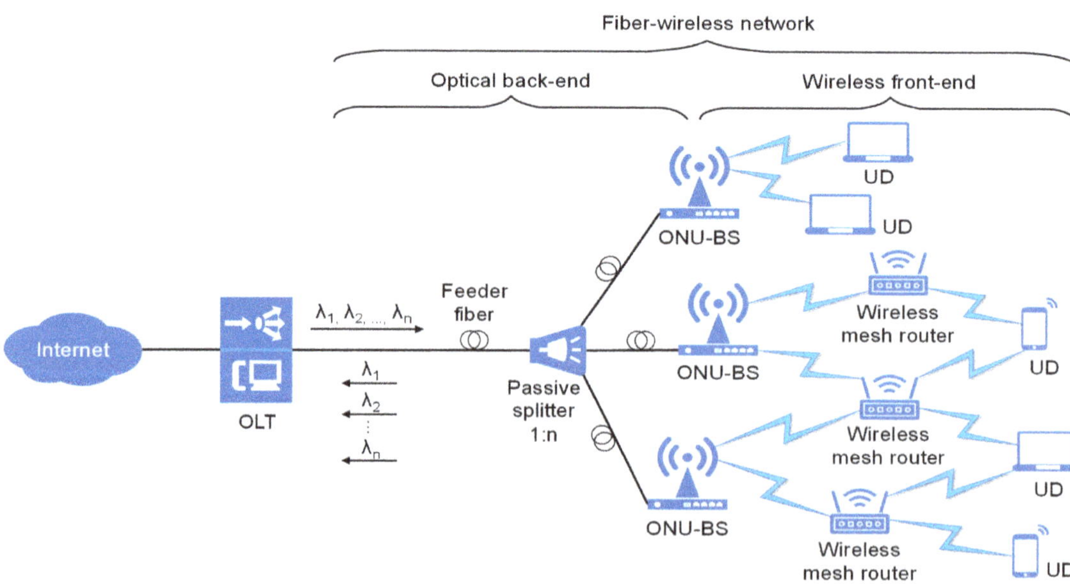

Figure 2. Illustration of R&F integration between optical and wireless networks.

The most common architecture of FiWi access networks comprises optical and wireless network domains. Ethernet passive optical networks (EPONs) or gigabit passive optical networks (GPONs) are commonly used in the optical segment of the network. In the wireless domain, a wireless point-to-point or mesh network (WMN) is operated according to the IEEE 802.11, WiMAX and LPWAN standards [11] and specific routing protocols [13], and cellular mobile network (4G, 5G) low-power wireless access network (LPWAN) IoT standards (LoRa, Sigfox, NB-IoT, etc.) in the wireless segment of the network [14]. The generic architecture of the FiWi access network is shown in Figure 2 [15]. The optical transport network between the optical line terminal (OLT) unit and the optical network unit (ONU) can reach up to 100 km. In the downstream direction, an OLT connects the FiWi access network to the core network, while in the upstream direction it is responsible for scheduling resources toward the ONUs located near mobile users.

The allocation of resources by OLTs is usually dynamic. It involves resource planning in terms of allocating time slots for data transmission to a single ONU device using time-division multiplexing (TDM) or assigning a single wavelength using the wavelength-division multiplexing (WDM) mechanisms. Such dynamic bandwidth allocation (DBA) is a key component of the TDM–PON mechanism that prevents collisions and improves bandwidth utilization through a polling scheme based on determining the ONU bandwidth requirements [16]. Connecting the OLT device with several ONU devices is realized through a shared optical fiber, at the ends of which the passive optical splitters are installed (Figure 2). Such architecture enables point-to-multipoint topology between single ONU and multiple OLT devices.

It is also possible to combine these two solutions using time and wavelength division multiplexing (TWDM), which is characteristic of Next-Generation Passive Optical Network 2 (NG–PON2) and 100G-Ethernet Passive Optical Network (100G-EPON). The TWDM–PON uses the same physical network architecture as TDMA–PON (Figure 2). However, the OLT in TWDM–PON requires multiple transceivers that operate at different wavelengths and the ONUs are equipped with either a tunable transceiver (NG–PON2) or multiple transceivers (100G-EPON) to match specific OLT wavelengths in upstream transmission. The OLT's transceiver sends packets to all ONUs on the same wavelength in the downstream direction, and each ONU determines whether to receive or ignore the packet that is based on its type and destination address. In order to allow the ONUs to access the shared upstream channel, the OLT employs dynamic wavelength and bandwidth allocation (DWBA) scheduling. Depending on the size of its queue, this system allocates a specific time slot and wavelength for transmission to each ONU.

In classical PON networks, ONU devices terminate at user locations, where it is possible to offer the service to end users exclusively through optical media. However, in FiWi networks, ONU devices are deployed in the characteristic locations (Figure 2). These characteristic locations are planned according to the needs of the spatial coverage of individual BSs or mesh portal points (MPPs) which are integrated with ONU devices. Integrated ONU-BS or ONU-MPP devices represent the interface between the optical and wireless segments of the FiWi access network (Figure 2).

The transmission of optical signals in FiWi access networks can be implemented by RoF and R&F technology. Generally, both RoF and R&F technologies can be deployed in FiWi access networks; however, R&F technology is a very good solution for high-range FiWi networks based on the WLAN standard [14]. In comparison with RoF, the R&F approach also has certain shortcomings in terms of coordination between optical and wireless domains. These shortcomings are related to ensuring QoS and service viability [17]. The R&F concept requires the integration of wireless and optical segments on the physical and media access control (MAC) layers, which then requires the translation of protocols on ONU and BS interfaces. This translation introduces additional cost and complexity in the realization of such networks [17].

The application of FiWi networks is particularly interesting in sensor networks which represent the basis for the realization of IoT concepts in the fields such as telemedicine, smart agriculture, Industry 4.0, smart cities, smart buildings, autonomous driving, etc. This will result in the rapid enlargement of machine-to-machine (M2M) communications, which are characterized by a massive connectivity of communicating nodes (sensors) and transmissions of smaller-size data. Improvements to the FiWi network that could facilitate application within the IoT concept include the use of the millimeter-wave communication band (mmWave) for the transmission at high data rates over shorter distances and the concepts of network function virtualization (NFV) and SDN, which could significantly affect the reduction of capital expenditure (CAPEX) and OPEX of the network [18].

2.2. Challenges in Realization of the FiWi Networks

Although the RoF network exploits the advantages of both the optical distribution network and wireless mobile network, the main challenge in the realization of the RoF concept is ensuring appropriate MAC protocol functionality. The merging of wireless and optical network segments has an impact on RoF network functionality, which is primarily reflected in the occurrence of additional propagation delays among different network segments. This can cause the expiration of certain timeouts of wireless MAC protocols that can significantly reduce network performance [14]. For example, when using the IEEE 802.11 standard, a distributed coordination function (DCF) of the MAC technique, the additional propagation delay has a large impact on the performance of the FiWi networks. There are different ways to solve this problem within the RoF concept; however, all of them are solutions that balance the the length of the optical cable and the network bandwidth [14].

One of the possible solutions related to the elimination of the MAC coordination problem is the implementation of R&F technology (Figure 2). While, in the RoF approach, the optical cable is utilized as a medium for transmitting analog signals and management and access control to optical and wireless media is centralized in the central office (CO), in R&F technology, access control and management of optical and wireless media are separated and two different MAC protocols for each used network domain [14]. Therefore, in R&F networks, the optical and wireless networks are combined to create a single, integrated network. In general, R&F networks use distinct MAC protocols in each part of the network and therefore client access control is handled separately [5]. This means that traffic generated solely from wireless communication does not need to be transmitted through the optical network, as in the case of the RoF system. Distributed MAC protocols, such as IEEE 802.11, can avoid the additional propagation delays caused by fiber optic cables, which can negatively impact their performance. This feature allows for greater flexibility in the length of the deployed fiber optic cables and also increases the system's resilience, since local wireless traffic can still be served even if connectivity with the optical segment is lost [19].

3. Energy Efficiency Analyses of Radio-and-Fiber Networks

Although flexible and robust FiWi networks have emerged as an attractive solution for the realization of today's modern access networks, the low utilization of the optical network elements and high overhead in packet data communication raise an issue of optimizing the EE of FiWi networks. The utilization issue is a consequence of the data traffic variations wherein, during low or no utilization periods, active elements of FiWi networks consume energy as in the periods of moderate or high utilization. Therefore, improving the EE of FiWi networks is a non-negligible challenge and an active research area [6].

Over the years, many power-saving (PS) software techniques have been adopted to increase the EE of hybrid FiWi networks. These mechanisms differ depending on whether the PS technique is implemented in an optical or a wireless domain. Most research studies are focused on increasing the EE of these two domains separately (for the optical back-end and wireless front-end of the network). However, it has been shown that in terms of EE of hybrid FiWi networks, cooperation between PS mechanisms of both domains achieves better results [20]. In this section, the recent optical back-end and wireless front-end PS techniques are first reviewed, and then joint cooperative PS techniques between these two domains are analyzed. Table 1 summarizes related work on the improvement of EE of R&F networks.

3.1. Power-Sawing Techniques in the Optical Domain of FiWi Networks

The most popular optical back-end technology currently used in the realization of the FiWi network is the passive optical network (PON). The PON is a cost-effective point-to-multipoint network access architecture. In the PON network, OLT is allocated in the CO and via a passive optical splitter connected with numerous ONUs which are allocated near users (Figure 2) [21]. Due to the constant growth in the number of end users, the largest part of energy consumption in telecommunications falls on access networks [22]. Additionally, in optical access networks, 90% of the energy is consumed for powering ONUs and, therefore, the energy-conservation techniques for ONU devices are of great importance [23]. Moreover, in the PONs, the OLT is continuously occupied with transmitting and receiving data, making it unsuitable for energy-saving strategies. On the other hand, the ONU's transmitter is idle for the majority of the time and it is appropriate for the implementation of energy-saving strategies [16].

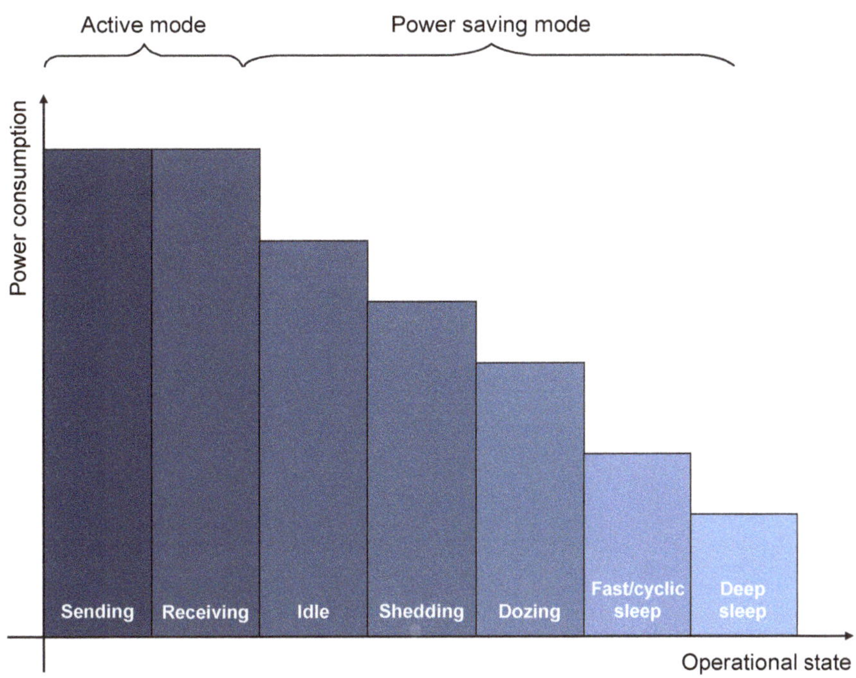

Figure 3. ONU power consumption under different active and energy-saving states.

Reducing ONU device power consumption can be achieved primarily through different PS operation modes prescribed by the ITU-T standard [24] (Figure 3). These PS modes of operation are characterized according to the operational state of the ONU receiver and transmitter. Accordingly, the ONU operation state can be in one of the following energy-saving modes: ONU power shedding, ONU dozing or ONU (fast and deep) sleeping modes (Figure 3). The power-shedding mode conserves energy by switching off only unnecessary ONU features, while keeping the receiver and transmitter fully operational. Thus, in the power-shedding mode, minimal power savings are ensured (Figure 3), while the best possible performance is maintained.

Table 1. Summary of related works on PS methods in R&F networks.

	Reference	PS Method	Summary of Contributions Related to the Improvement of FiWi Network Energy Efficiency
ONU sleeping mechanisms	[25]	ONU sleep mode with ALR	Hybrid PS technique that includes adaptive link rate control and sleep functions.
	[26]	WSM	A PS mode that combines the doze and cyclic sleep modes into a single mode.
	[27]	ONU sleep mode and WSM	Comparative performance analysis of fast/cyclic and watchful sleep modes
	[28]	WSM operation mode with DBA	Performance of watchful sleep mode that utilizes the dynamic bandwidth allocation.
	[29]	AWSM for UNUs	An introduction of adaptive watchful sleep mode
	[30]	Load-adaptive ONU sleeping scheme	Load-adaptive ONU PS mechanism that adjusts the number of sleeping ONUs based on the overall load on the network.
	[16]	PS mechanism based on OSC, GDBA and TSC components	A PS method based on the SIEPON standard.
	[31]	Decentralized PS mechanism based on ONU queue manager, TRx controller, sleep manager, OLT queue manager and GDBA components	Decentralized PS solution based on the SIEPON standard.
	[32]	Optimization of sleep interval using ANN	Determination of optimal fast/cyclic sleep interval for energy-efficient XG-PON. The ANN model is used to estimate the optimal sleep interval values.
Wireless power-saving techniques	[33,34]	Integer linear programming (ILP) optimization model and heuristic algorithm	An approach for finding the most efficient way to save energy in wireless access networks using heuristics.
	[35]	Transmit power scaling and on/off switching	Extensive studies on the impact of changing the transmit power and turning the BSs on and off on the instantaneous power consumption of macro BSs. Real-world measurements are used from a range of different macro BSs to develop linear power consumption models.
	[36]	Adaptive PSM	An adaptive PS method in wireless networks that are based on the IEEE 802.11 standard.
	[37]	Scheduled PSM based on a time-slicing mechanism	A PS method based on a time-slicing mechanism in a multi-traffic environment with high background traffic.
	[38]	PSM based on the execution of the heuristic algorithm	A generic power management model according to which the wake-up scheduling mechanism is controlled by the AP. Proposes two heuristic algorithms to address the downlink scheduling optimization problem, identifying the importance of tuning the length of the beacon interval in order to conserve energy and reduce delay.
	[39]	C-PSM	A centralized PS mechanism that improves the EE of wireless clients in an 802.11 infrastructure network.
	[40]	SAPSM	A PS method that uses a ML classifier to assign priorities to applications, where applications classified as high-priority can switch to active mode, while traffic classified as low-priority is optimized for EE.

Table 1. *Cont.*

	Reference	PS Method	Summary of Contributions Related to the Improvement of FiWi Network Energy Efficiency
Wireless power-saving techniques	[41]	A ML method of identifying and categorizing network traffic	A ML-based approach for optimizing power saving in WLANs by classifying network traffic based on contextual factors, and adjusting the listen interval accordingly.
	[42]	Overview of PS methods	Overview of the power supply system parameters for powering the BS sites with renewable energy sources. Approaches for reducing telecom operator energy and CAPEX based on different air-conditioning systems for BS sites.
	[43]	Overview of PS methods	Overview of the renewable energy sources for powering base station sites. Comparison of the EE among hybrid systems that use multiple renewable energy sources and systems that use a single renewable energy source.
	[44]	Save energy and maximize connectivity (SEMC) algorithm	A generic algorithm for ad-hoc wireless networks that conserves energy and maintains good connectivity through adjusting transmission range and choosing a transmission time based on data rates, which results in reduced transmission power and energy savings.
Cooperating optical and wireless techniques	[21]	ONU sleep mode with PSM	A method for determining the optimal sleep period and behavior for optical ONUs for improving throughput and reducing energy consumption.
	[45]	ONU sleep mode with radio interface standby	A wireless–optical topology reconfiguration scheme that enables integrated energy saving through reconfiguration of the optical topology using ONU sleep mode and the wireless topology using radio interface standby.
	[46]	ONU sleep mode with PSM and adaptive PSM	A method for controlling the ONU sleep period based on the energy control mechanism of wireless stations.
	[47]	ONU sleep mode with powering off radio interfaces	An ONU sleep algorithm for dynamic scheduling of the power states of ONUs based on their traffic profile and load thresholds. An algorithm for dynamic radios turning off in order to reconfigure the wireless topology by dynamically controlling the power states of radios.
	[6]	ONU sleep with adaptive frame aggregation and load transfer mechanism	Proposed adaptive frame aggregation mechanism that optimizes energy consumption by adjusting frame lengths based on channel quality. Proposed the delay-aware load transfer mechanism that maximizes ONU sleep time and ensures reliable service transmission by allocating traffic load based on QoS requirements.
	[48]	ONU sleep mode with PSM and DBA	A PS scheme that coordinates power-saving modes for wireless stations, APs and ONUs, in order to reduce energy consumption.
	[20]	TDMA mechanism between ONU and wireless station and between OLT and ONU with DBA	A technique that aims to reduce delays and improve EE by organizing the system into clusters of ONUs and using an equal partition approach. Using this approach the ONUs in the back-end and the wireless stations in the front-end are active only during certain timeslots in the TDMA cycle.

Table 1. *Cont.*

	Reference	PS Method	Summary of Contributions Related to the Improvement of FiWi Network Energy Efficiency
Cooperating optical and wireless techniques	[49]	Load transfer region sleep mechanism between ONU and wireless stations	A collaborative sleep mechanism that uses load transfer to determine which nodes should sleep and adjusts routes for affected services based on service priority.
	[50]	Genetic algorithm, teaching-learning-based optimization, spiral update positioning and encircling prey mechanism	Several different ONU placement optimization algorithms are compared in extensive simulations.

In ONU dozing mode, the ONU transmitter is powered off for a considerable amount of time, while the ONU receiver maintains an active state at all times. This corresponds to even lower instantaneous power consumption of the ONU dozing mode when compared with the power consumption of the power-shedding mode (Figure 3). The ONU sleeping modes are, according to [24], further divided into two subgroups: deep sleep and fast sleep (also known as cyclic sleep) (Figure 3). The ONU sleeping modes are characterized by powering off both the ONU receiver and the transmitter for a considerable amount of time. When using ONU sleeping mechanisms, it is necessary to make a trade-off between energy savings and network performance. Due to powering off the receiver and the transmitter during the entire PS mode period, during the ONU deep sleep mode maximum energy conservation can be achieved at the cost of significant performance degradation. On the other hand, during the ONU fast/cyclic sleep mode, the transmitter and the receiver alternate between on and off periods, forming sleep cycles, which results in lower energy savings when compared with the energy savings of deep sleep mode (Figure 3).

Another PS method in PONs is the implementation of the adaptive link rate (ALR) concept. It is based on energy reduction that can be achieved by adapting different transmission rates between optical devices [51]. In optical access networks where multiple data transmission rates are available (e.g., GPON and G-EPON), reducing transmission rates in low-traffic periods can contribute to the reduction of the energy consumption of the optical unit and this reduction improves network EE. Moreover, these PS techniques can be combined with ONU sleep modes for even greater energy savings. The authors in ref. [25] proposed a hybrid scheme that combines the ONU sleep mode with ALR mechanisms in order to improve the EE of 10G-EPON systems (Table 1). Such a hybrid PS scheme activates the ONU sleep mode in the absence of network traffic, while in the presence of network traffic, the ALR function for adapting the downlink data rates is activated based on the levels of traffic load. It was shown in ref. [25] that such a hybrid PS approach can contribute to significant energy consumption reductions in the optical part of the network.

In addition to the aforementioned PS modes, a newer PS scheme called watchful sleep mode (WSM) has emerged and is also included in all major PON standards [52–54]. The WSM acts as a unified solution that combines cyclic sleep and doze PS modes. It eliminates the need for mode negotiation between optical units and maximizes the ONU's EE. It is expected that WSM will be used as the only PS mode for PONs in future (Table 1) [26]. In ref. [27], the authors conducted a performance comparison between PS fast/cyclic sleep modes and a WSM for a GPON ONU and concluded that both approaches offer similar energy savings. However, the WSM performed better regarding state transition latency (Table 1). The simulation results in ref. [28] showed a decrease in downstream and upstream transmission delays with a significant energy savings when WSM was implemented in combination with dynamic bandwidth allocation (DBA) in a 10-gigabit-capable passive optical network (XG–PON) (Table 1). In addition, the authors in ref. [29] recently proposed a new energy-efficient scheme called adaptive watchful sleep mode (AWSM). The AWSM increases the energy savings of the standard WSM scheme by minimizing the

ONU receiver's active (on) time during the watch state (Table 1). In ref. [30], a PS strategy was proposed for edge-enhanced metro FiWi networks using matching game theory. The proposed method optimizes the number of sleeping ONUs depending on network traffic (Table 1) and thus improves network EE.

Furthermore, the IEEE service interoperability in the Ethernet passive optical networks (SIEPON) work group (SIEPON 1904.1) standardized two additional PS mechanisms, namely, transmitter (Tx) and transceiver (TRx) sleep modes for EPON systems [55]. Similar to the doze PS mode of the ITU-T standard, in the Tx sleep mode, ONU transmitter subsystems can enter the sleep (PS) mode, while the receiver components remain fully operational. In contrast, the TRx sleep mode is equivalent to the ITU-T fast/cyclic sleep mode and this PS approach enables entering the sleep (PS) state for both the transmitter and receiver subsystem of the ONU. Compared with the Tx sleep mode, the TRx sleep mode is more energy efficient at the expense of the increased delay. In both the Tx and the TRx sleep modes, the OLT is responsible for controlling the ONU sleep intervals, with the main task of determining the right sleep interval that can satisfy both the delay and the energy-saving requirements [56].

To optimize Tx sleep interval wile satisfying QoS in TDM PON systems, the authors in ref. [16] proposed a SIPEON-based energy-saving scheme (Table 1). The proposed scheme is based on adding new components to the conventional PON architecture, namely the ONU sleep controller (OSC), the green DBA (GDBA) mechanism, and the Tx sleep controller (TSC). The OSC and GDBA components are a part of the OLT hardware architecture and they are used to calculate the ONU Tx sleep periods based on received report messages. The TSC component, as part of the ONU, is responsible for ONU transmitter control and for monitoring the incoming traffic. The TSC enables the exit and entering of ONU in sleep mode according to instantaneous traffic intensity. The simulation results presented in ref. [16] showed that this energy-saving scheme can decrease the Tx power consumption of ONU while satisfying desired SIEPON and ITU-T QoS performance requirements.

In contrast to the centralized sleep mechanism introduced in the SIPEON standard, the authors in ref. [31] proposed a decentralized energy-saving solution that is also based on the SIPEON standard (Table 1). In the PS solution proposed in ref. [31], instead of the OLT unit, the Tx or TRx sleep mode is initiated by the ONU. To achieve this, the ONU hardware architecture is enriched with new components known as the ONU queue manager and TRx controller, while the OLT hardware architecture is enriched with the sleep manager, OLT queue manager and GDBA mechanism. The ONU Tx sleep duration is calculated by the TRx controller and reported to the OLT. Then the sleep manager calculates the sleep intervals of the ONU receiver and, based on these intervals, the type of sleep mode is determined. The obtained simulation results showed that significant energy consumption reductions could be achieved in the optical segment of the network while satisfying QoS requirements.

In the past, the fast/cyclic sleep mechanism of the ONU has been extensively researched to reduce energy consumption in XG-PON. However, due to the emergence of new types of network traffic with stringent demands, further improvements in sleep time interval selection are required. Hence, the paper [32] proposed the use of an artificial neural network (ANN) to enable the ONU to determine the optimal sleep time interval values by learning from past experiences. The M/G/1 queueing system was used for theoretical analysis prior to simulation, and the ANN was trained and tested for the XG-PON network to make optimal sleep time interval decisions. The results indicate that as the network load increases, sleep time interval decreases for both methods. The ANN network records a wider range of sleep time interval values than the theoretical values. As a result, these findings will enable network operators to determine the optimal sleep time interval values at the current network conditions with more flexibility.

3.2. Power Saving Techniques in the Wireless Domain of FiWi Networks

Besides the optical domain, different PS techniques have been developed in the wireless domain of FiWi networks. Since the highest consumers of energy in the radio part of the network are wireless network (front-end) devices (e.g., BSs, APs, LPWAN gateways, etc.), different solutions have been proposed for improving the EE of the wireless segment of FiWi networks. Most of the proposed solutions are based on dynamic control of the activity state of wireless network devices in accordance with traffic variations. Besides ONTs, wireless network devices can also be, at a certain moment, in an active or in a sleep state. Whether it is transmitting or being idle in the active state, wireless network devices consume significantly more energy compared with the sleep (partially or fully turn-off) state [57]. Hence, different PS approaches have been presented in refs. [33–35,58] as techniques for wireless network station on- and off-activity adjustments depending on the data load.

Moreover, a mechanism called an adaptive power saving mechanism (PSM) was introduced in ref. [36], which, instead of inefficient fixed wake-up intervals, uses an adjustment constant in order to adjust the wake-up intervals of wireless network devices in a more efficient manner (Table 1). Similar to ONU sleep techniques, using PSM and adaptive PSM, the wireless network station needs to buffer the incoming data, which consequently causes additional transmission delay [18].

To address these issues associated with initial PSM, the authors in ref. [37] presented a scheduled access point (AP)-centric PSM protocol based on a time-slicing mechanism. The proposed protocol enables the improvement of energy-efficient scheduling of AP activity at the expense of minimal delay in a multi-traffic environment with heavy background traffic (Table 1). The authors in ref. [38] developed a generic power management model (GPMM), according to which the wake-up scheduling mechanism is controlled by the AP. Additionally, to address the downlink scheduling optimization problem and to achieve more energy savings, in ref. [38] two heuristic algorithms for optimal AP activity adjustments were proposed (Table 1). Furthermore, the authors identified the importance of tuning the length of the beacon interval in order to conserve energy and reduce delay. Since both of the aforementioned scheduling techniques are computationally demanding, the authors in ref. [39] proposed a centralized power-saving mode (C-PSM). In order to reduce latency and increase EE, the C-PSM approach uses traffic pattern statistics to calculate the optimal AP listening intervals, beacon intervals and congestion window size (Table 1).

Effective management of WLAN power consumption on smartphones can have a significant impact on energy consumption. The authors in ref. [40] have shown through experiments that the WLAN power management process on various smartphones is autonomous and occurs entirely at the driver level. However, the limitation of driver-level implementations is that essential power management decisions can only be made by monitoring packets at the MAC layer, which disables distinguishing between applications. As a result, each application has an equal chance to consume more energy and determining which applications can impact WLAN power management is crucial. To solve this problem, the authors have introduced a smart adaptive power save mode (SAPSM). SAPSM uses a machine learning (ML) classifier to assign a priority label to each application. Only applications that have high-priority can affect the client's behavior to switch to active mode, while traffic with low-priority is optimized for EE. It is shown that the SAPSM implementation on an Android smartphone device significantly improves EE under typical usage scenarios.

As an extension of the SAPSM approach, the authors in ref. [41] have proposed a new classification method of network traffic using ML classifiers to optimize WLAN power saving. The approach utilizes the contextual degrees of traffic interaction in the background for ML classifier applications. The output traffic is then classified to optimize context-aware listen interval PSMs. The study evaluates the performance of several ML classifiers using a real-world dataset of several smartphone applications that enable the reflection of various types of network interactions and behaviors.

In addition to the aforementioned PS methods in the wireless domain of the FiWi network, methods for powering the locations of wireless network devices (AP and BS) using renewable energy sources have also been proposed [42,43]. These methods represent an effective way to save energy and reduce OPEX for network owners [42]. Due to the lack of ability for BSs to access the electricity grid, having limited access to daily power supply or simply having large expenses for consumed electricity, mobile network operators have a significant interest in using renewable energy sources for powering remote BS sites. Therefore, using renewable energy sources can reduce OPEX and improve the energy efficiency of the wireless domain in FiWi networks.

Furthermore, the authors in ref. [44] proposed an algorithm that aims to conserve energy and maintain connectivity in mobile ad-hoc networks (MANETs). The algorithm aims to conserve energy in nodes that have limited battery life and maintain connectivity between nodes, which is crucial for route discovery. The proposed algorithm is a generic algorithm that can be applied in various situations and operates at layer 2 of the Open Systems Interconnection (OSI) model, making it independent of routing algorithms. The simulation results showed that the proposed algorithm significantly reduces energy consumption while maintaining good connectivity over time.

3.3. Cooperating Power-Saving Techniques in the Optical and Wireless Domain of FiWi Networks

Since the FiWi network consists of optical and wireless domains that are realized with two different technologies with versatile PS mechanisms, cooperation between these two PS mechanisms is found to be a desirable option. It was shown in refs. [21,46] that unsynchronized PS mechanisms between optical and wireless network domains lead to the degradation of the performance and EE of the FiWi network. To reduce such degradations, the authors in ref. [21] proposed a method that increases ONU EE and reduces wireless network stations' latency without sacrificing throughput in a FiWi network. The proposed cooperative PS method simultaneously executes the ONU sleep mechanism in the optical domain and the PSM mechanism in the wireless domain of the network (Table 1).

For the realization of joint EE improvement in both domains of the FiWi access network, another interactive PS method based on the wireless–optical topology reconfiguration (WOTR) technique is proposed in ref. [45] (Table 1). The proposed reconfiguration scheme uses the ONU sleep method in combination with the method which puts the radio interface (RI) in the standby state. Implementation of this method is realized through two interactive modules: one for the optical back-end and one for the wireless front-end. Through simulations, the authors in ref. [45] demonstrated significant energy consumption reduction with negligible network throughput degradation when compared with optical-only PS schemes.

Due to the issues related to the increased latency and degraded EE caused by the simultaneous use of unsynchronized PS schemes in the optical and wireless domains of the FiWi network, the authors in ref. [46] proposed a cooperative ONU sleep scheme that dynamically adjusts the ONU sleep period according to the conditions in the wireless domain (Table 1). By proposing an integrated EE scheme that jointly schedules both ONU and radio sleep states, the authors in ref. [47] designed an energy-efficient FiWi network in which power consumption optimization was based on the developed heuristic algorithms (Table 1). The proposed network design uses an ONU sleep mechanism that dynamically schedules ONU sleep periods and a "Radios Off" algorithm for wireless topology reconfiguration. The QoS is ensured by employing wireless rerouting. Moreover, the frame aggregation scheme introduced in WLANs for boosting the overall performance of the wireless front-end was proven in ref. [59] to be an effective approach for improving the EE of the FiWi networks. In order to reduce the power consumption of the FiWi networks, authors in ref. [6] present an adaptive frame aggregation and load-transfer scheme. The proposed scheme jointly maximizes the EE and ONU sleeping periods of the FiWi IoT networks, realized as PON in the optical domain and wireless mesh network (WMN) in the wireless domain (Table 1).

The authors in ref. [48] proposed a cooperative PS scheme called the energy-conservation scheme for FiWi networks (ECO-FiWi) that synchronously deploys wireless front-end and optical back-end PS techniques and integrates them into a dynamic bandwidth allocation (DBA) procedure by leveraging the time division multiple-access (TDMA) operations (Table 1).

Table 2. Summary of the related works on PS methods in RoF networks.

References	PS Method	Summary of Contributions Related to the Improvement of FiWi Network Energy Efficiency
[60]	Number and directivity of antennas and antenna position optimization	Method for improving EE of a DAS by increasing the number of antennas and optimizing the antenna position.
[61]	Antenna unit output power optimization	Comparison of the EE of optimized narrowband single-service and broadband multi-service DAS solutions.
[62]	Antenna position optimization and the selection of the optimal number of antennas	Investigation of the EE of the RoF DAS technologies by measuring the power consumption of 802.11 APs and smartphones. Proved the existence of the optimal number of distributed antennas for a given indoor environment topology.
[63]	Selection of the optimal number of antennas with data frame aggregation mechanisms	A method for evaluating and optimizing energy consumption in 802.11n RoF DAS systems. Proved the existence of an optimal number of distributed antennas for a given scenario, and that the data frame aggregation mechanisms can further improve EE.
[64]	Centralized RoF architectures using dedicated RoF links for each cell	An EE model for RoF networks confirmed better network EE when designed with small cells and when the energy usage of the remote units surpasses a certain threshold.
[65]	Analyses of the impact of E/O/E conversion, number of services and wireless network capacity on EE of the RoF links	Confirmed that E/O/E losses degrade EE on an optical link using the A–RoF technique. Confirmed that the D–RoF link shows the degradation of EE at higher Nyquist zones due to RF signal reconstruction. Confirmed that the wireless bandwidth can improve the EE of both, the A–RoF and D–RoF connections and that the amount of energy savings in the presence of multiple services depends on the specific wireless environment.

The study found that the use of TDMA in FiWi networks significantly improves the energy efficiency while maintaining delay performance, with the feature of increasing energy savings proportionally with traffic load. Another cooperative mechanism called the delay-controlled and energy-efficient clustered (DEC) TDMA scheme was presented for the FiWi network in ref. [20] (Table 1). This is a cooperative scheme that jointly considers the EE of the OLT, ONU and wireless network devices by leveraging the TDMA technique. Moreover, to further decrease delay and energy consumption, the DBA mechanism was used to allocate data transmission slots between wireless network devices and ONU, and between ONU and OLT. It was shown that the EE of the DEC TDMA FiWi network was equivalent to ECO-FiWi, but in terms of transmission delay, DEC TDMA outperformed the ECO-FiWi scheme.

In order to achieve energy conservation and ensure satisfactory QoS, a collaborative sleep technique between optical and wireless nodes was developed in ref. [49] (Table 1). This technique consists of two sub-techniques. The first one is a technique based on load balancing for achieving optimal planning of transmission route selection, and the second one is a cooperative sleep technique based on load transfer using priority-based rerouting. The simulation results showed that this technique can achieve high throughput and EE with low latency for high-priority services. In ref. [50], the authors proposed a framework that simultaneously focuses on ensuring the EE and survivability of the FiWi networks using optimization algorithms for ONU placement (Table 1). Several different optimization algorithms were compared in extensive simulations, namely, genetic algorithm, teaching–learning-based optimization, spiral update positioning and encircling prey mechanism. The authors concluded that the genetic algorithm provides the best performance regarding EE, with an average performance in terms of the FiWi network survivability.

4. Energy Efficiency Analyses of Radio-over-Fiber Networks

The implementation of the FiWi networks using the RoF technology is based on exploiting optical fiber as a means for the direct transport of radio frequency (RF) signals among one or more remote access (antenna) units (RAU) and network operator central office (CO) (Figure 4). The CO is responsible for managing access to both the optical and wireless domains [18].

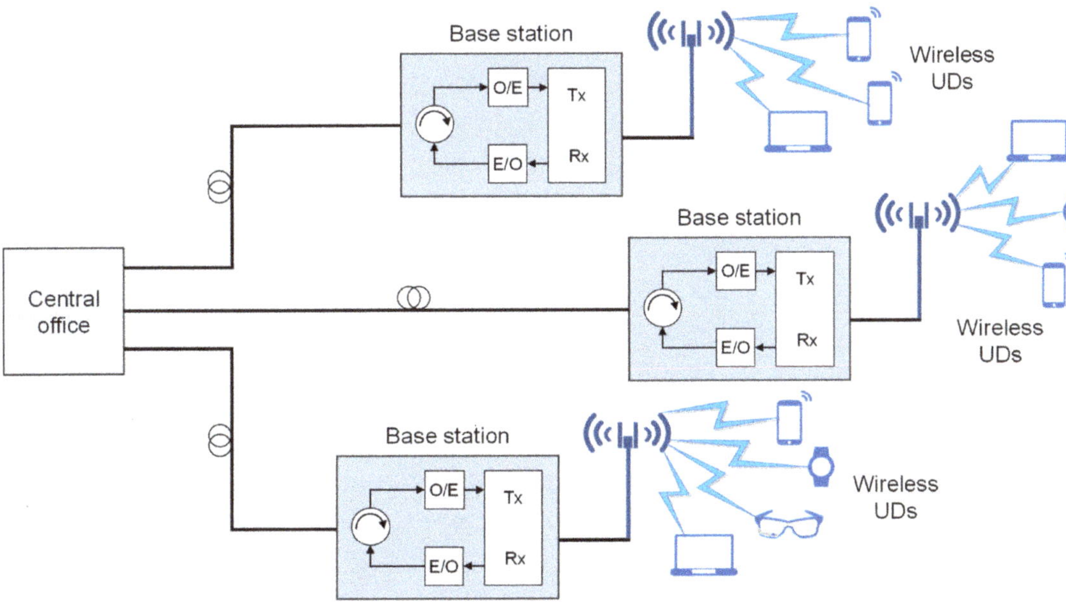

Figure 4. Generic RoF architecture.

The optical modulation of the RF signal is realized through electrical-to-optical (E/O) conversion, which is required for transmitting an RF signal over an optical–fiber link. Compared with traditional coaxial-based RAN systems, the RoF network paradigm offers several advantages. Those advantages are related to the high bandwidth of fiber links with high data rates and low signal attenuation. This allows greater transport distances and better resistance to electromagnetic noise and thereby increases the flexibility of the allocation of network devices, reduces BS size due to the BS design with RRUs and BBUs, increases installation flexibility, allowing multiple operators to share the same RoF network and improves the EE through dynamic resource allocation realized using the centralization of baseband functions at CO premises [66].

Depending on how the RF signal is transmitted through an optical–fiber link, there are two basic RoF communication techniques known as digitized RoF (D–RoF) and analog RoF (A–RoF) [67]. In the A–RoF system (Figure 5), the transmission of the RF signal is carried in the analog domain, where analog RF signal is used for light modulation of the optical signal transmitted in the fiber. On the other hand, in the D–RoF system (Figure 6), the D–RoF signal is digitized prior to transmission over an optical communication link.

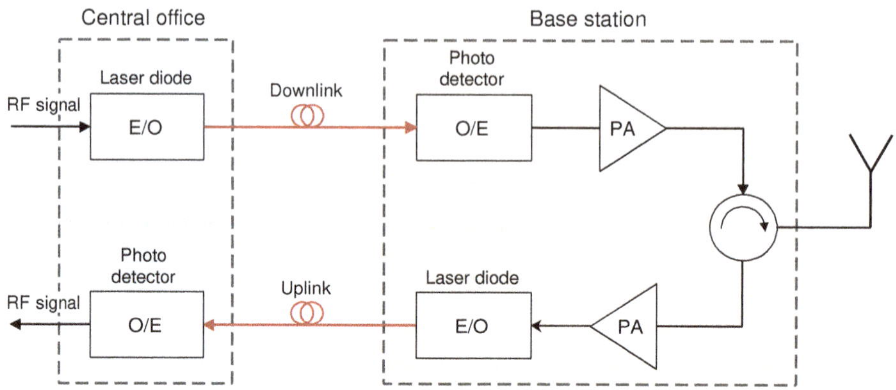

Figure 5. General A–RoF optical link architecture.

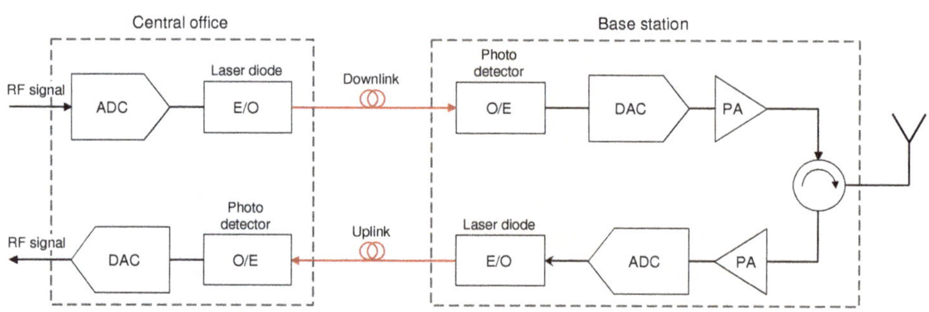

Figure 6. General D–RoF optical link architecture.

For example, the MFH link between the RRU and the BBU designed for the LTE advanced (LTE-A) C-RAN architecture based on a common public radio interface (CPRI protocol is a prominent representative of the D–RoF. The main concept of such an MFH solution based on the CPRI interface is to use a digitized version of the baseband signal before being transmitted over the optical–fiber link. Although the CPRI-based fronthauling solution has proven to be effective for previous mobile network generations (i.e., 2G, 3G and 4G), the D–RoF approach has one major issue. Namely, the CPRI has shown to be bandwidth-inefficient, as it uses a constant data rate for transmitting the signals [68]. Hence, this bandwidth inefficiency can become critical in 5G networks using carrier aggregation (CA) and mMIMO techniques wherein high bandwidth and capacity requirements are expected. Moreover, the CPRI does not support flexible rerouting that allows automatic RRU switching to another BBU and the data rate is dependent on a number of antennas which can also be critical in 5G mMIMO implementations [69].

Presumably, the D–RoF-based solutions may have been sufficient for the early stages of the 5G network, but as 5G reaches its full potential, digitized optical MFH transmission based on the conventional CPRI protocol will generally not be sufficient [70]. Analog–optical transmission, commonly referred to as the A–RoF, presents a traditional RoF technique that has promising potential to overcome D–RoF's CPRI limitations (Figure 5). Analog modulation of the optical signal with no prior digitization addresses bandwidth limitations, making it more suitable for high-bandwidth networks. On the other hand, due to the characteristics of the fiber link, transmitted A–RoF signals can experience numerous link impairments such as attenuation, chromatic dispersion and fiber nonlinearities [71]. Depending on the carrier frequency used for modulation of the optical signal, two common A–RoF techniques are usually considered, i.e., the radio frequency-over-fiber (RFoF) and the intermediate frequency-over-fiber (IFoF). The main difference between the RFoF and the IFoF techniques is in the carrier frequency used for modulation of the optical signal for transmission over a fiber–optical link. In the case of the RFoF technique, the analog signal directly modulates the optical signal, while in the IFoF technique, the signal of intermediate frequency is used for optical signal modulation.

4.1. Approaches for Improving Energy Efficiency in General RoF Networks

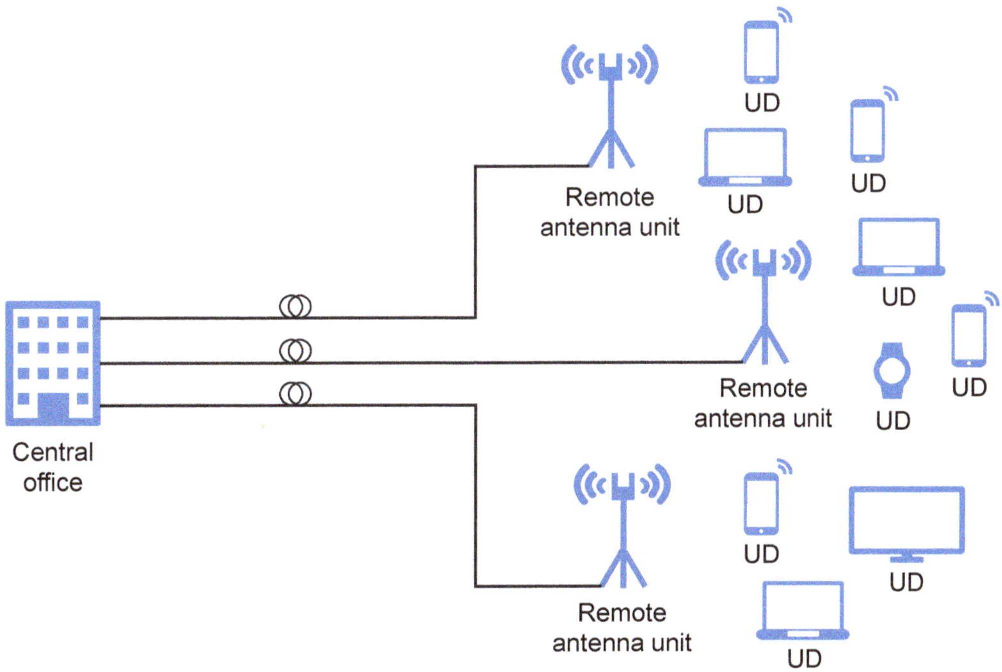

Figure 7. RoF DAS architecture.

As the RoF integration and mobile broadband access techniques have been developed, studies focused on improving the EE of the RoF networks have been performed. An overview of research activities related to the EE improvements of the ROF-based systems is presented in Table 2. In ref. [60], it was shown that RoF-distributed antenna systems (DAS) are the most appropriate architecture for the deployment of high-capacity wireless communication systems (Figure 7). The authors in ref. [60] demonstrated that, for an active DAS system that utilizes optical fibers, there is a specific number of antennas that results in the highest EE when the power consumption of both the radio and optical components is considered (Table 2). In contrast to traditional WLANs, the RoF DAS architecture uses multiple remote antenna units instead of individual APs while maintaining all signal processing functions at a CO (Figure 7). This allows for a reduction in the complexity and power consumption of the remote antenna units [63].

One way to optimize the RoF DAS performance is to minimize the power consumed at each remote antenna unit, while providing wireless coverage to a specific area. In ref. [61], the authors proposed a solution to decrease energy usage in the RoF DASs by adjusting the RF output power at the antenna units for optimal efficiency (Table 2). They found that a narrow-band DAS is more energy efficient than a wide-band DAS, due to the improved efficiency of the power amplifiers, for a single wireless service. However, when more than two services are needed, a multi-service broadband DAS can be more energy efficient than multiple narrow-band DASs when used for offering individual wireless services.

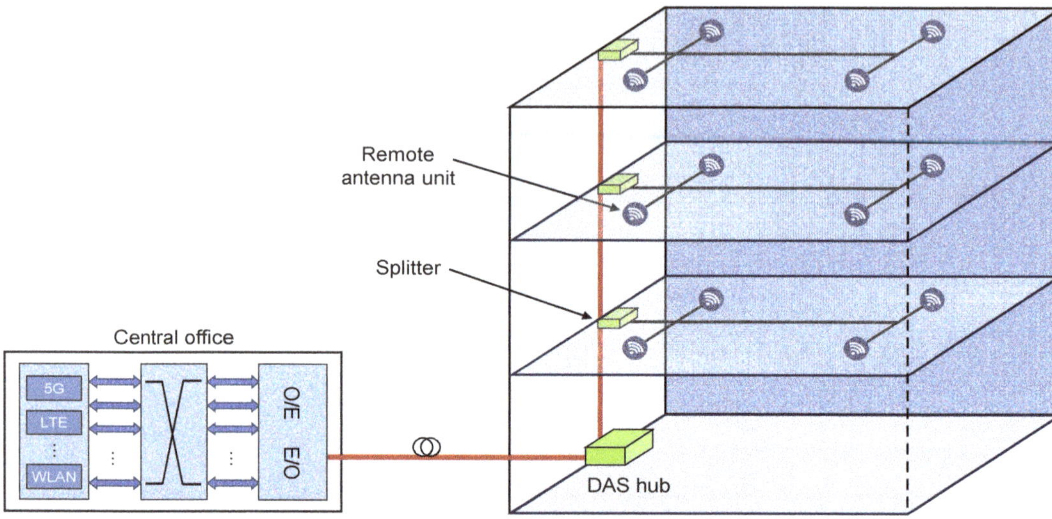

Figure 8. An example of indoor DAS implementation.

It was shown in ref. [62] that DAS architecture can enhance the coverage and performance of wireless communications within indoor RoF DAS deployments (Figure 8). In a specific study, the authors suggested a method to determine the location of distributed antennas to optimize network capacity based on the probability of non-uniform user presence (Table 2). They also used power consumption measurements of smartphones and 802.11 APs to calculate the EE of a DAS that uses radio-over-fiber (RoF) technology. For the given indoor environment topology, the simulation results indicated that there was a specific number of distributed antennas that maximized the EE of the system.

In ref. [63], the authors studied the EE of IEEE 802.11n-based RoF DAS architectures (Figure 7) and developed a method for evaluating and optimizing the energy consumption in these systems (Table 2). They found that there was an optimal number of distributed antennas for specific AP implementation scenarios, and that the aggregation mechanisms of the IEEE 802.11n standard could further improve the EE in the RoF DAS. Additionally, they demonstrated that MAC protocol data-unit-aggregation techniques are more effective in providing higher end-to-end throughput and greater EE than MAC service data-unit-aggregation schemes in the IEEE 802.11n RoF DAS.

The authors in ref. [64] conducted a study on the EE of indoor networks, which provide high-speed mobile access to end users using hybrid RoF technology (Table 2). Using a validated EE model, they found that while individual RoF links may not be as energy efficient as traditional baseband-over-fiber links, the RoF networks could be more energy efficient when carefully designed with small cells and when the energy usage of the remote units is above a certain level.

In ref. [65], for an indoor network, a theoretical evaluation model was presented in order to evaluate the effect of wireless bandwidth, multiple services and loss due to electrical–optical–electrical (E/O/E) conversion on the EE of the optical links in A–RoF- and D–RoF-based networks (Table 2). It was shown that E/O/E loss had a large impact on the EE of the optical link when the A–RoF transmission technique was used. On the other hand, it was shown that the D–RoF link was less susceptible to E/O/E loss. However, the D–RoF link showed a degradation of the EE due to RF signal reconstruction at higher Nyquist zones. Furthermore, it was shown that the EE could be improved on both the

A–RoF and the D–RoF connections by increasing the wireless bandwidth. It was also shown that in the presence of multiple services, additional energy savings depend on the wireless environment.

4.2. Approaches for Improving the Energy Efficiency of Cloud Radio Access Networks

4.2.1. Energy-Saving Potential of Cloud Radio Access Network Architecture

The usage of optical fiber instead of coaxial cabling for connecting BBU and RRU locations in the D-RAN network architecture constitutes the basis for the realization of next-generation cellular mobile networks (Figure 1) [72]. Due to communication between RRUs and BBUs realized over fiber links, such systems can achieve communication over longer distances than traditional BSs, having collocated RRU and BBU in the same cabinet or having the connection between RRU and BBU using coaxial cables or waveguides at the BS site (Figure 1).

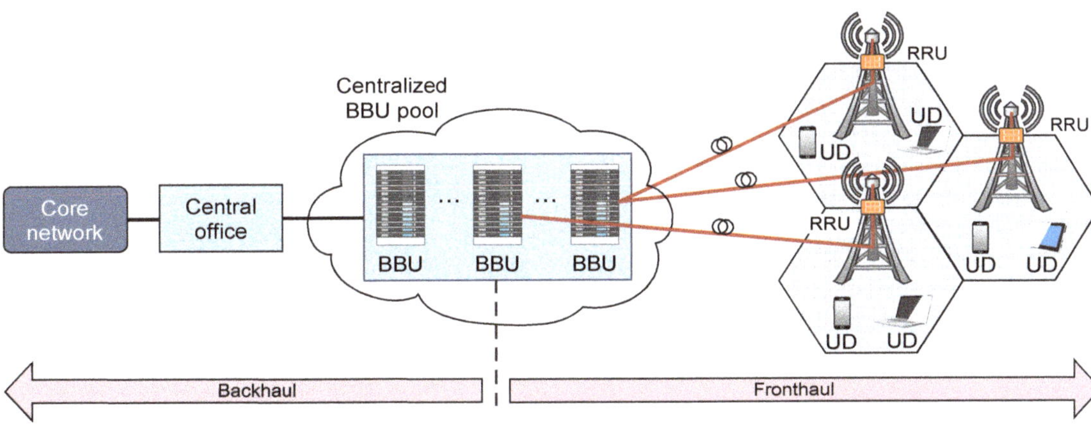

Figure 9. C-RAN architecture.

The D-RAN concept is particularly suitable for the realization of RAN as a HetNet with the large number of distributed small cells realized as elements of DAS. Small-cell communication systems realized through DAS show potential for increasing the capacity and data rate of RANs. However, such RAN communication systems must be simple and have low power consumption. To meet these requirements, the RAN architecture is evolved from D-RAN to cloud-based RAN (C-RAN). The C-RAN architecture involves separating the BBU of traditional cell sites from remote RRUs and allocating the BBUs in a central cloud location (Figure 9). This allows complex and power-inefficient traditional macro BS sites to be simplified. Simplification means the allocation of just RRUs at the BS site (Figure 9). This approach reduces BS site power consumption due to RRU natural air conditioning and decreases BS site maintenance costs due to the lower amount of components installed at the BS site [73].

Although three communication standards in D-RAN systems have been developed (i.e., Open Base Station Architecture Initiative (OBSAI), CPRI and Open Radio equipment Interface (ORI)) for digital–optical interface communication between RRU and BBU [74], the CPRI has become the most common D–RoF protocol, which is also predominantly used in C-RAN networks (4G/5G C-RAN). It defines ten different options in terms of transmission bit rates, which range from 0.61 Gbit/s for Option 1 to 24.33 Gbit/s for Option 10 [75].

Table 3. Summary of the related works on PS methods in C-RAN networks.

References	PS Method	Summary of Contributions Related to the Improvement of FiWi Network Energy Efficiency
[76]	CnR algorithm	A method for estimating the resource utilization rate of BBUs. Additionally, the CnR algorithm to save energy in the BBU pool is presented, and it is shown that the proposed algorithm is effective at decreasing energy consumption in the BBU pool and overall system.
[77]	Wake-on-RRU and wake-on-BBU approach	An approach that uses WoL packets sent by the RRU to wake up BBUs and an approach that uses WoL packets to wake up BBUs sent by the controller in the BBU pool.
[78]	Dynamic resource provisioning (DRP) algorithm	A dynamic resource-allocation algorithm to select active RAUs and consolidate virtual machines onto computing units in order to minimize energy consumption in C-RANs. In order to achieve this goal, the proposed algorithm uses a context-aware scheme to minimize the number of virtual machine migrations.
[79]	Graph partitioning algorithm and rejoining algorithm	A scheme for associating BBUs and RRUs based on graph partitioning and rejoining in order to minimize power consumption.
[80]	Power control algorithm	A power control algorithm based on mobility prediction for improving the EE of 5G H-CRAN.
[81]	H-CRAN energy-efficient radio resource management (HERM) algorithm	The HERM algorithm to solve the network EE optimization problem. The results showed that the developed algorithm significantly improves the EE of H-CRAN.
[82]	MIMO–RoF system	An adaptive RoF system for next-generation C-RANs that takes into account energy consumption, capacity per wavelength and distribution range.
[83]	Particle-swarm optimization (PSO), quantum PSO (QPSO) and genetic algorithm (GA) approaches	The optimal number of virtual machines that maximize the EE of C-RAN.
[84]	Heuristic-Assisted Deep Reinforcement Learning (HA-DRL) BBU aggregation scheme	An aggregation scheme for BBU based on HA-DRL that ensures both energy efficiency and guaranteed QoS.
[85]	Double Deep Q Network (DDQN) resource allocation framework	Framework based on DDQN resource allocation method that maximizes the overall EE in C-RAN.

The C-RAN architecture (Figure 9) was introduced in 2010 by China mobile as a centralized solution to better support the requirements of future HetNets [86]. Since traditional RAN systems suffered from numerous challenges related to increased CAPEX and OPEX with spectral and energy inefficiency, in the cloud centralized solution has become a necessity in designing future RAN systems for next-generation mobile networks. The C-RAN thus presents a logical evolution step from the D-RAN (and the centralized D-RAN), where baseband processing is separated from the cell site and is performed in the remote CO (Figure 9). In comparison with the centralized D-RAN (Figure 1), the C-RAN architecture enables complete RAN architecture to be based on the cloud-centralized management and control paradigm (Figure 9). In remote CO, multiple BBUs can be aggregated in a BBU pool that can utilize the computational resources of multiple BSs (Figure 9). The RRUs are connected to the BBU pool through C-RAN mobile fronthaul (MFH) realized with high-speed and low-latency optical links [87]. The distance between RRUs and the BBUs physically located in the cloud can be up to hundreds of kilometers away. With such an approach, centralized operation and management of overall RAN can be ensured [88]. This BBU centralization can significantly reduce mobile operator OPEX, since multiple BBUs aggregated in the same central equipment room (CO) in the cloud can share the same resources (e.g., power supply, air-conditioning) and thus improve the EE of the network [89].

Furthermore, in C-RAN network architecture, RRU design and functionality are much simpler, and this further contributes to the reduction of BS site power consumption and maintenance costs. Additionally, the C-RAN has an architecture suitable for the implementation of cooperative techniques such as coordinated multi-point (CoMP) processing technology. The cooperation of multiple BBUs in the large cloud BBU pool using CoMP technology enables the sharing of different system information in the cloud. This can improve the spectral and EE of the RAN and also can contribute to alleviating inter-cell interference (ICI) of the densely deployed small cells [86].

Therefore, through the implementation of the cloud-computing paradigm, the C-RAN architecture is able to aggregate multiple BS resources into a central BBU pool in the cloud. According to ref. [90], with such centralized architecture, it is possible to optimize multi-cell cooperation processing, which results in more energy-efficient operation of RAN than those having decentralized BBUs in the cloud. Therefore, full implementation of the C-RAN architecture can bring considerable potential energy savings to mobile network operators in the future.

4.2.2. Techniques for Improving Energy Efficiency in Cloud Radio Access Networks

Since energy efficiency has become one of the main concerns when designing RAN, some initial research analyses have been published with respect to improving the EE of C-RAN networks and they are summarized in Table 3. In the case of the latest 5G mobile cellular networks, the heterogeneous C-RAN (H-CRAN) architecture is composed of a small number of macro BSs and the large number of small BSs (micro, pico, femto) in combination with algorithms for the effective allocation of radio resources in centralized and integrated BBU pools is considered as a promising approach for minimizing network energy consumption. However, to support the expected increase in the number of UDs, wireless services and applications in the future, the number of network elements (the RRUs on the front-end side and the BBUs on the back-end side) will also increase, and the need for improving the RAN's EE will remain.

One additional issue with C-RAN architecture is that all of its BBUs are always active, even when user traffic is low. This leads to the high energy consumption of the BBU pool. To address this issue, the authors in ref. [76] proposed a method for estimating BBUs' resource-utilization rate. The method takes into account the data rate requirement, the number of mobile UDs, the RRU bandwidth and the transmission power between the RRU and mobile UDs (Table 3). They also presented the combine and remove (CnR) algorithm for deciding when to switch BBUs off and on. The proposed algorithms were developed with the aim of maximizing the number of sleeping BBUs while maintaining the QoS. The simulation results showed that when compared with traditional RANs, the proposed scheme can save the energy consumption of the BBU pool and the overall RAN system.

Table 4. Summary of the related works on PS methods in networks based on the MEC concept.

References	PS Method	Summary of Contributions Related to the Improvement of FiWi Network Energy Efficiency
[9]	Unified resource management scheme	The realization of the FiWi network with MEC led to a significant reduction in power consumption and an increase in the battery life of edge devices.
[91]	Unified resource management scheme and cloudlet-aware DBA algorithms	A resource management scheme that takes into account the use of cloudlets and incorporates offloading tasks into the FiWi DBA process. The proposed management scheme could significantly reduce the amount of energy used by edge devices and extend their battery life significantly.
[92]	Priority-based task offloading and caching (PrO) method	The proposed PrO scheme efficiently manages tasks by caching, offloading and performing local computing while preserving the priority order, which resulted in reduced delay and energy consumption.
[93]	ACCO and GT-CCO PS methods	Proposed a two cloud–MEC collaborative computation offloading mechanisms. Using a combination of MEC and centralized remote cloud services resulted in significantly lower energy consumption compared with solutions without a centralized cloud.
[94]	GT-CCO PS method	Proposed a FiWi network architecture that enables the coexistence of centralized cloud and MEC in the IoT applications. A game-theoretic collaborative computation offloading scheme was proposed as a solution for improving energy efficiency and handling a large number of mobile devices effectively.
[7]	ISA-CCO PS method	Confirmed that the proposed ISA-CCO solution is more effective than previously proposed ACCO and GT-CCO in terms of reducing energy consumption and improving processing response time on mobile devices.
[95]	TSGO PS method	Energy-efficient offloading strategy for MEC-enhanced FiWi. The three EE benchmarks to evaluate EE mechanisms for MEC FiWi were proposed and it was confirmed that the proposed strategy could significantly decrease energy consumption in MEC-enhanced FiWi networks.

In the C-RAN, the BBU pool serves a large number of RRU units (Figure 9) and by effectively coordinating the BBUs, network performance in terms of power consumption can be optimized. In order to minimize power consumption caused by communication overhead between the BBUs and RRUs, the authors in ref. [79] presented a scheme for the association of the BBU and the RRU based on graph partitioning and rejoining (Table 3). By using a partition and rejoin scheme, the authors assigned the BBUs to RRUs based on both the individual resource requirements of the RRUs and their communication with each other. The simulation results showed that the algorithm proposed in ref. [79] could reduce power consumption in the BBU pool with linear computational complexity.

In ref. [80], the authors presented a power-control algorithm that utilizes mobility prediction to improve EE in the 5G H-CRAN (Table 3). The proposed algorithm predicts the movement of user equipment in vehicular mobility situations and performs RRU switching based on the prediction results. The authors proposed an RRU-switching approach using Markov mobility prediction and a gradient-optimized transmission power method. The simulation results indicated that the proposed algorithm performed better regarding EE compared with existing RRU-switching algorithms.

Furthermore, the authors in ref. [81] developed a radio resource management method for energy-efficient H-CRAN (Table 3). The EE of H-CRAN was studied using an energy consumption model that includes all networking devices. Using this model, the authors developed an energy-efficient radio resource management algorithm to improve the network EE of the H-CRAN.

The authors in ref. [82] proposed an adaptive RoF system for reducing energy consumption while maintaining the required transmission rate the next-generation C-RAN systems (Table 3). The proposed system is based on the 2x2 MIMO–RoF model that employs coherent optical orthogonal frequency division multiplexing (CO-OFDM) technology. Through extensive numerical analyses, the authors verified that the proposed system could achieve significantly lower energy consumption with a high spectrum efficiency.

Additionally, the proliferation of the NFV technique has enabled the adoption of virtual machines which execute BBU functions in a way such that multiple virtual machines can run on a single, generic server at the BBU pool (Figure 9). By increasing the number of deployed RRUs and active virtual machines within the server, the authors in ref. [83] emphasized the need for solving the problem of increased power consumption. Therefore, the authors in ref. [83] proposed a power model that maximizes the network energy efficiency through the estimation of the optimal number of active virtual machines within the BBU server (Table 3).

To address the unsustainable growth of energy consumption caused by the increased mobile traffic, the C-RAN architecture separates the BBU from the RRH in the BSs and consolidates the BBUs into a BBU pool to allow low-utilized BBUs to enter sleep mode during decreased network activity, thereby reducing energy consumption. However, when a BBU is in sleep mode, the RRHs connected to it must be switched to another BBU, which can impact the QoS for those RRHs. To address this issue, the authors in ref. [84] have proposed a deep reinforcement learning (DRL) based BBU aggregation scheme that ensures both minimal energy usage of BBU and minimal migration of RRH traffic at the same time. Furthermore, several heuristic algorithms are introduced to assist with the DRL training. The proposed heuristic-assisted DRL (HA-DRL) approach is evaluated numerically and is found that the proposed approach outperforms the benchmarks by achieving the lowest cost for all scenarios. Additionally, the authors have found that the DRL agent is able to achieve better results by trading BBU power consumption for RRH traffic migration.

The utilization of C-RAN has proven to be effective in improving network performance. This gain is due to the smart management of RRHs with regard to power consumption and on/off operation modes. However, conventional resource allocation techniques maximize network efficiency without taking into account the overhead of RRH switching in adjacent time intervals. Therefore, in ref. [85], the authors aim to optimize EE while adhering to per-RRH transmission power and user data rate constraints. To achieve this, authors have formulated the EE problem as a Markov decision process (MDP) and have implemented DRL techniques to gain cumulative EE rewards. Simulation results showed that a proposed double deep Q network (DDQN)-based framework outperforms traditional approaches due to its ability to consider future effects of actions, and the ability to overcome the issue of action overestimation. This leads to a significant improvement in EE compared to benchmarks.

The presented analyses of related works indicate that many challenges related to improving EE in FiWI networks based on the C-RAN concept have not been addressed. It can be observed that further research activities related to the development of advanced architectures, signalizations, protocols and scheduling algorithms must be performed in order to make such FiWi networks more energy efficient.

5. Energy Efficiency Analyses of FiWi Networks Based on Multi-Access Edge Computing

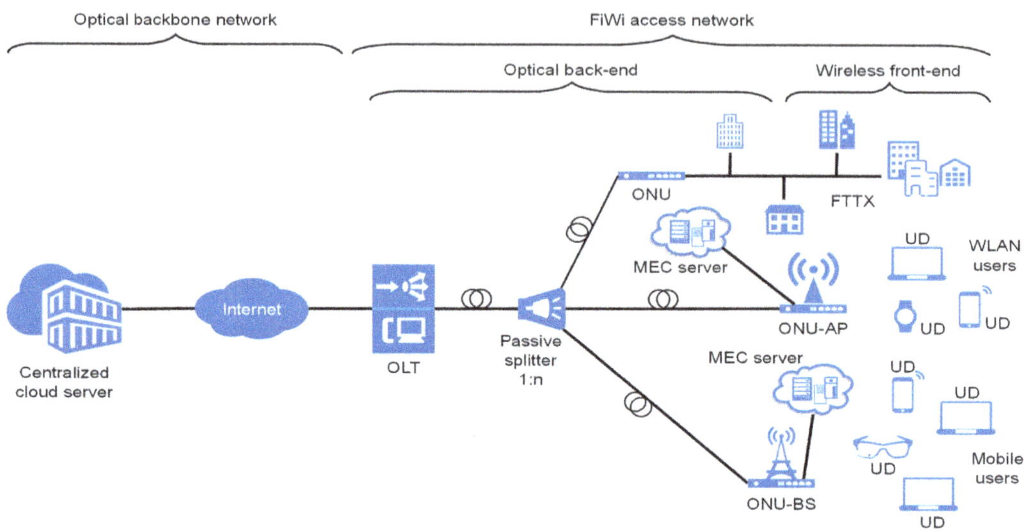

Figure 10. The generic architecture of a CM-FiWi access network.

The MEC concept introduces the cloud computing processes into FiWi networks, which are closer to mobile end UDs, more specifically at the mobile network edge. By exploiting an approach based on processing part of the user traffic in the cloud servers located at the edge of the mobile network, the transmission delay in the network is reduced and central servers at the cloud are less loaded. Figure 10 shows the generic network architecture that encompasses MEC and a centralized cloud paradigm over the FiWi access network. Such a network is referred to as a cloud–MEC FiWi broadband access network (CM-FiWi).

The architecture presented in Figure 10 consists of a long-reach optical backbone network connected to a centralized cloud and a standard FiWi access network supporting different wireless access technologies such as LTE, 5G or WLAN. One or more MEC servers are connected near the UDs of the fixed-access network or integrated with the ONU with the AP (ONU-AP) node or the ONU with base station (ONU-BS) node via a direct optical connection (Figure 10). This connection enables cloud services at the edge of the network, which can comprise the access of some wireless local/personal networks or cellular mobile networks (Figure 10). In such a FiWi network based on the MEC concept, ONU devices can also retain their traditional role, i.e., provide fixed services to users via the fiber-to-the-x (FTTx) concept (Figure 10).

This technology and architecture are key to the evolution towards full installation of the 5G networks and especially 6G networks, since it enables the transformation of mobile networks towards a programmable platform that meets the mobile cellular network requirements for increased bandwidth, lower delay and better scalability and configurability [96]. As one of the key technologies for enabling the full potential of 5G networks [97], MEC creates a pathway for the practical implementation of applications that require extremely low latencies with high reliability. Some examples of such applications include tactile internet, augmented reality/virtual reality, connected cars and mission-critical IoT systems. Moreover, the FiWi networks based on the MEC concept can reduce the overall operator CAPEX and OPEX through the exploitation of the existing infrastructure and through the implementation of integrated resource-management mechanisms.

The authors in ref. [9] reviewed the challenges and possible design scenarios for implementation of MEC-enabled hybrid FiWi networks with various RAN technologies

(WLAN, 4G, LTE-A and HetNets) (Table 4). Additionally, using the TDMA scheduling resource-management scheme, the Ethernet-based FiWi network was further inspected regarding delay performance, battery life of edge UDs and response-time efficiency. The results showed that the MEC over FiWi can significantly reduce power consumption and extend the battery life of edge UDs (Figure 10). Furthermore, the same group of authors proposed, in ref. [91], a cloudlet-aware resource-management scheme that decreases the delay of offloading tasks and extends the battery life of edge UDs (Table 4). This scheme incorporates offloading the FiWi dynamic bandwidth allocation process and was designed using two TDMA layers to improve network performance. Analysis showed that the proposed solution, which incorporates cloudlets into the MEC FiWi networks, could lead to a significant reduction in energy usage for edge UDs and extend their battery life by several hours. The proposed architecture and resource-management strategy could be a useful solution for the implementation of MEC in future technologies such as the 5G tactile internet.

For the computational offloading scheme that involves moving intensive computing tasks to a cloud located at the edge of the mobile network is shown to be beneficial in improving EE and reducing latency for UDs in the mobile network [92]. Hence, the authors in ref. [92] proposed a priority-based offloading model that takes into account offloading and caching optimization that is combined with a local computation policy. The study shows that the proposed model has a substantial impact on decreasing both, the delay and energy usage in a cellular network.

Two collaborative computation offloading schemes using the CM-FiWi architecture, namely, an approximation collaborative computation offloading scheme (ACCO) and a game-theoretic collaborative computation offloading scheme (GT-CCO), were studied in ref. [93] (Table 4). The simulation results showed that using both MEC and centralized cloud services resulted in notably better energy efficiency of the network compared with the MEC schemes without centralized cloud offloading. Furthermore, in ref. [94], a generic FiWi architecture with a combination of centralized cloud and distributed MEC for IoT connectivity was presented (Table 4). The problem of collaborative computation offloading for the IoT over FiWi was addressed through the GT-CCO scheme. The numerical results showed that the proposed scheme was energy efficient and able to effectively handle a large number of mobile devices.

Due to limitations of mobile UDs, e.g., reduced computing resources, memory capabilities and limited battery capacity, offloading compute-intensive tasks to the MEC server or a remote cloud server emerged as a viable and promising solution for today's computation and delay-sensitive applications [7]. This offloading scheme, called cloud–MEC collaborative computation offloading (CMCCO), takes advantage of both types of cloud services, i.e., centralized remote cloud service and a decentralized MEC service as two complementary technologies. The authors in ref. [7] proposed an energy-aware collaborative computation offloading (EA-CCO) system which can perform various computing tasks in the CM-FiWi network (Table 4). An iterative searching algorithm for collaborative computation offloading scheme (ISA-CCO) was developed in order to decrease task offloading overhead by taking into account residual battery rate, transmit power allocation and the scaling of computing resources. The simulation results showed that the proposed ISA-CCO solution achieved superior results compared with the aforementioned ACCO and GT-CCO paradigms in terms of energy consumption and processing response time of the mobile UDs.

Due to the removal of unnecessary data traffic from optical backbone networks, the MEC-enhanced FiWi network presents a viable choice for practical implementation in cases where resource-intensive and delay-sensitive mobile applications will be used.

However, the assumption that the FiWi infrastructure offers unlimited and free resources is not realistic and, hence, the energy-efficient offloading techniques must be considered in the realization of MEC-enhanced FiWi networks. Motivated by the need to develop energy-efficient offloading mechanisms for the MEC-enhanced FiWi networks, the authors in ref. [95] presented the two-layer Stackelberg game offloading (TSGO) strategy (Table 4). The FiWi layer and mobile edge computation offloading layer of this strategy are responsible for performing bandwidth allocation and offloading decisions, respectively. In addition, three EE benchmarks were proposed to evaluate the performance of the proposed strategy. The simulation results showed that the TSGO strategy supports the concept of green communications and could effectively decrease the energy usage of the MEC-enhanced FiWi networks.

6. Implementation of SDN-Based Energy Conservation Concepts in the FiWi Networks

In recent years, traditional access networks have become difficult to manage due to the constant increase in the number of users and network devices, which is followed by the increased complexity of the network structure. Compared with traditional communication networks, networks exploiting the SDN concept introduce flexible, dynamic and programmable networking concepts. The implementation of SDN results in easier network management and improved overall network performance [10]. The SDN allows for the virtualization of network functions (NFV) using applications that run on top of the SDN controller. These applications use a programming language that simplifies the rapid deployment of new services and capabilities. The SDN and its NFV applications can be used to manage the PON devices of different standards in a coordinated and dynamic manner [98].

Table 5. Summary of the related works on PS methods in SDN-based networks.

References	PS Method	Summary of Contributions Related to the Improvement of FiWi Network Energy Efficiency
[99]	SDN control mechanism through OpenFlow protocol	Confirmed that SDN-based control architecture has the potential to reduce energy consumption in the FiWi access network.
[100]	Enhanced standard PON devices with OpenFlow SDN technology and SD controller	An adaptive SD ONU PS mechanism that uses enhanced standard PON devices with advanced SDN capabilities. The simulation results showed that the proposed scheme could increase the EE while still guaranteeing the QoS requirements in a TDMA–PON system.
[98]	SD TWDM–PON architecture with OpenFlow technology	Development of the architecture that uses SD orchestration to coordinate wavelength/link speed deployment and to improve EE by adapting the link rate or activity state of the OLT/ONU transceivers during periods of low traffic while still maintaining the required QoS.
[101]	SDN-based 5G EPON architecture	Proposed an open control layer SDN-based framework that aims to minimize energy consumption in EPON while avoiding adding additional packet delay.
[102]	Controllers in OpenFlow technology	Energy-saving scheme for a FiWi access network that combines the OpenFlow technology in the SDN.
[103]	EEWA scheme	Proposed the EE scheme that significantly optimizes energy usage and workload allocation in a network combining the neighbor edge servers, local edge servers and the remote cloud. Proposed a path priority selection method to decrease the probability of network blocking and to improve the use of available spectrum.

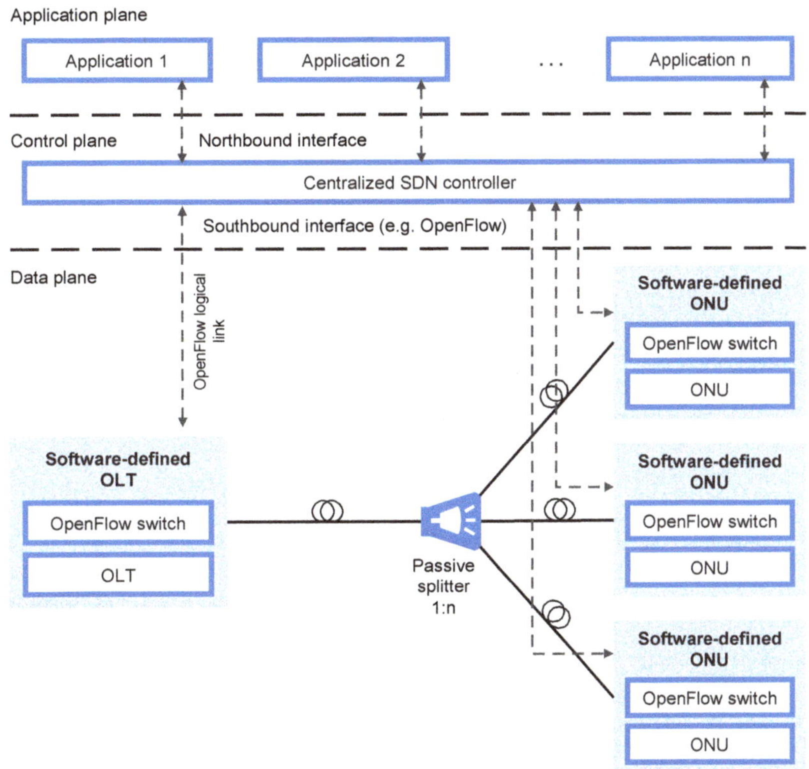

Figure 11. Software-defined PON architecture.

According to Figure 11, which presents a visualization of the SDN PON architecture, the separation of the data and control planes are the main features of SDN. Such separation enables centralized network management which is executed in the centralized controller. Two controller interfaces, namely, a northbound and southbound interface, are defined within the SDN architecture. They enable the centralized controller to communicate with applications and network equipment (Figure 11). The northbound interface is used to communicate with the application layer, which enables the application of SDN control management functions. On the other hand, the southbound interface is used for communication between a centralized controller located within the control plane and the data plane (Figure 11). The data plane consists of network equipment and it is responsible for packet-forwarding decisions. OpenFlow is a commonly used protocol at the southbound interface (Figure 11). It provides access to the data plane and is considered the enabler of the SDN concept [104].

Due to the constant network infrastructure growth, the power consumption of access networks has become one of the main concerns for companies that own larger networks and telecom operators. The SDN-based approach has emerged as a viable and effective solution for reducing power consumption of access networks [10]. Table 1 summarizes the related works on PS methods in SDN-based networks. The authors in ref. [99] showed through simulation that the proposed SDN-based energy management scheme in the EPON system can reduce the overall energy consumption of optical access networks (Table 5).

Although PON is considered an energy-efficient fixed network access technology, due to its mass deployment worldwide, PON systems need to be even more energy efficient to meet today's green policy requirements. Due to modern consensual requirements related to ensuring more sustainable and energy-efficient operation of access networks, PON devices need to be enhanced in order to more easily adapt to such an environment. Having this as a goal, the authors in ref. [100] proposed an energy-conservation scheme through adaptive SDN-based TDMA–PON system architecture (Table 5). They introduced software-defined (SD)–, OLT– and SD–ONU architecture and employed an energy-saving scheme using an OpenFlow-based SD controller (Figure 11). To enable PS operation, the SD–OLT and SD–ONU devices were enhanced with SD agents connected to the SD controller via the OpenFlow protocol. The approach uses flow tables for packet classification and forwarding and an energy-saving table to store SD–ONU receiver sleeping periods. The transceiver of the SD–ONU was decoupled and thus the independent working state of the transmitter and receiver was enabled in such an architecture. Hence, depending on the traffic characteristics, the energy conservation of the SD–ONU was orchestrated via the SD controller through the assignment of the SD–ONU receiver's sleep period and the SD–ONU transmitter's wake-up threshold. The simulation results performed in low-traffic conditions showed a significant reduction in energy consumption, while maintaining QoS requirements.

In addition, an energy-efficient SD optical access architecture based on TWDM–PON was proposed in ref. [98] (Table 5). In this architecture, the authors enhanced existing fiber–optic network equipment with SD capabilities. SD orchestration was proposed to manage the use of different wavelengths and link speeds in the network. Through this management, saving energy could be achieved by turning off or decreasing the data rate of the OLT/ONU transceivers when traffic was low, while still maintaining the required quality of service. The authors in ref. [98] also evaluated the performance and EE of the proposed system in various scenarios which consider factors such as packet delay, link capacity and data loss in the network. The simulation results showed that various trade-offs should be considered when evaluating the EE of the system.

In order to minimize power consumption without adding additional delay in the 5G EPONs, the authors in ref. [101] proposed a framework based on the SDN and NFV paradigms (Table 5). The authors proposed an SDN 5G EPON network design that reduces the complexity of managing and operating the network, while improving the usage of network resources. The suggested solution extends the multipoint MAC control sub-layer in the OLT and divides its responsibilities between the OLT and the SDN controller. The management and control functions of the OLT and RAN were moved to the SDN controller, while the other functional components remained in the OLT and were integrated with the OpenFlow switch. Therefore, the functions that operate on longer time scales were transferred to the SDN controller, while those related to shorter time scales were kept in the OLT and RAN. This design allowed for the integration of EPON and RAN energy conservation techniques and additionally enabled the minimization of CAPEX costs. Nevertheless, the authors in ref. [101] emphasized the need for further evaluation of the proposed solution in order to clarify levels of energy savings.

The authors in ref. [102] proposed the PS scheme for FiWi networks based on an SDN approach (Table 5). The scheme is based on the OpenFlow protocol and a newly introduced controller that manages ONU's working and sleeping states based on the traffic flow threshold. According to the simulation results, such a proposed centralized energy-saving approach based on SDN has a significant advantage. These advantages are mainly due to the central management provided by the introduced controllers. The controllers enable fair distribution of available resources, which results in greater energy savings. The authors demonstrated that the proposed SDN-based energy-saving scheme in the FiWi networks is effective and has practical value.

To achieve energy-efficient computing in a specific environment such as cloud-edge FiWi networks, it is crucial to coordinate the actions of edge servers and cloud servers in order to reduce energy consumption. Therefore, the authors in ref. [103] presented a solution for reducing energy consumption in such an environment by improving the cooperation between edge servers and cloud servers (Table 5). The proposed solution, called energy-efficient workload allocation (EEWA), was tested on an SDN testbed to demonstrate its feasibility. In addition, simulations were conducted to find the best possible outcomes for a set of task requests. The simulation results indicated that the EEWA scheme significantly decreased the network blocking probability and the average energy consumption in edge-cloud FiWi networks.

While implementation of SDN in RAN shows potential for improving EE of the FiWi networks, much work remains to be done within the areas of algorithms, standardization, architectures and interfaces.

7. Discussion

The development of new architectures, protocols and algorithms for FiWi networks has been a hot topic in the telecommunications industry in recent years. These networks combine the high speed and reliability of fiber optic cables with the convenience and accessibility of wireless technology, providing a cost-effective solution for delivering broadband services to a wide range of customers. Research on FiWi networking involves the integration of optical fiber and wireless broadband access technologies. FiWi broadband access networks resulting from the integration of optical fiber and wireless technologies can utilize both RoF and R&F hybrid networking technologies.

The carbon footprint of ICT infrastructure is gaining increasing public attention and raises concerns related to global warming. This has led to increased research into developing energy-efficient solutions and research dedicated to improving the EE of the FiWi networks is not an exception. The use of fiber optic cables, which transmit data using light signals, allows FWI networks to operate with lower power requirements than traditional copper wire or coaxial cable networks. This not only helps to reduce the energy costs of operating FWI networks, but also has environmental benefits, as it reduces the carbon footprint of telecommunications infrastructure.

In addition to the energy-efficient nature of fiber optic cables, FiWi networks can also incorporate other energy-saving technologies. In previous analyses presented for the RoF, R&F, MEC and SDN-based FiWi networks, it has been shown that advanced power-management schemes can be used to dynamically completely or partially turn off components in the FiWi network to reduce the power consumption of certain components when they are not in use.

While FiWi networks offer significant energy-saving potential, it is important for telecommunications companies to carefully consider the energy consumption of all aspects of FiWi networks, which includes the wireless components and the fixed network infrastructure required to support them. By adopting energy-efficient technologies and practices, service providers and telecom operators can significantly help to reduce the overall energy consumption of their FiWi networks and contribute to a more sustainable future.

Energy efficiency in C-RAN has become an increasingly important topic in recent years as the demand for mobile data services has continually grown. The C-RAN architecture, in which the baseband processing is centralized and the RF processing is distributed, has been proposed as a solution for improving energy efficiency in cellular networks. One of the main benefits of C-RAN is the ability to centralize baseband processing. By centralizing the baseband processing, the power consumption of C-RAN is reduced, since the centralized baseband unit (BBU) can serve more RRUs and thus can be designed to be more energy efficient in comparison with BBUs serving individual RRUs. Centralization also allows for the sharing of resources between BSs, which further contributes to reducing RAN power consumption. Furthermore, the C-RAN enables the dynamic power management of RRUs and BBUs by allowing them to sleep when the traffic is low, and wake up when the traffic demand increases. These BBU and RRU sleeping cycles contribute to the reduction of energy consumption.

Additionally, recent research studies proposed a wide range of methods for reducing energy consumption in the C-RAN. These methods include techniques for optimizing the use of resources in the BBUs, strategies for managing the sleep state of the BBUs, resource utilization algorithms for increased energy savings, approaches for associating

BBUs and RRUs, power management algorithms of the RRUs and the use of NFV to execute the functions for improving the EE of BBUs using virtual machines. Therefore, C-RAN architecture has the potential to significantly improve the energy efficiency of cellular networks through centralization, fronthaul compression, dynamic power management of radio resources and the use of the SDN and NFV. However, careful planning is required to ensure that the high bandwidth requirements of the fronthaul links can be met and to avoid adding complexity to the network.

In the presented analyses, it is shown that EE is an important concern in the design and operation of the FiWi networks due to the fact that the energy consumption of these networks can be a significant portion of the total OPEX of network operators. One way to improve the EE of FiWi networks is to use MEC technology. The MEC allows the deployment of computing and storage resources closer to the edge of the network and thus closer to the users. This can reduce the energy consumption of the network by reducing the amount of data that needs to be transmitted over long-distance fiber–optic links between remote cloud servers and users located at the edge of the wireless network, since more data processing can be carried out locally at the edge servers. Optical fibers are already recognized for their EE and, implemented with MEC technology, can additionally diminish unneeded data transfer in optical networks. However, energy efficiency in the MEC-based FiWi networks is still largely unexplored and it should be a highly prioritized subject of future research.

In addition, energy-efficient techniques for MEC-enabled FiWi networks include power-aware resource management schemes. These schemes optimize the allocation of resources to the fiber–optic and wireless links based on the traffic demand and the available energy. This optimization can be performed in combination with collaborative computation offloading schemes, which computationally redirect demanding tasks from the central remote server in the cloud to a server located closer to the users at the edge of the network. These concepts have the potential to contribute to the full proliferation of the green communications paradigm and to significantly decrease energy consumption in the FiWi networks.

One advantage of using the SDN in FiWi networks is the ability to dynamically control and optimize network resources. With the SDN concept, network administrators can use a centralized controller to manage and configure network devices, which allows the simple allocation of resources and the adjustment of network configurations as needed. Such an SDN concept can also help to improve the efficiency and performance of FiWi networks, particularly in dynamic or high-traffic-demand environments. Because network configurations can be easily adjusted using the centralized controller, it is easier to add or remove devices and services as needed. This will enable telecommunication providers to satisfy the changing needs of their customers and adaptation to new technologies and trends. Overall, the use of SDN in the FiWi networks can help to improve the efficiency, performance and flexibility of these networks, making them a more attractive option for telecommunications providers and their customers.

In conclusion, FiWi networks offer a cost-effective and energy-efficient solution for delivering broadband services to a wide range of customers. By leveraging the energy-saving potential of fiber optic cables and incorporating other energy-efficient techniques and strategies, telecommunications providers can help to reduce the environmental impact of their networks and contribute to a more sustainable future. The use of energy-efficient techniques presented in this survey can significantly improve the EE of FiWi networks, thus reducing their operating costs and carbon footprints.

8. Conclusions

This paper presented a comprehensive survey of recent research on approaches that contribute to energy efficiency improvements of FiWi access networks. Emphasis was given to the extensive literature review of various power-saving techniques and energy-efficient models that are dedicated to the improvement of FiWi network energy efficiency. The presented review covered energy efficiency analyses of different types of FiWi networks which include the R&F FiWi networks, the RoF FiWi networks, the MEC-based FiWi networks and the FiWi networks based on the SDN concept. For the R&F networks, energy conservation techniques and research studies related to the optical and wireless domains were presented, as well as related works that deal with the improvement of FiWi networks' energy efficiency through the cooperation of techniques in wireless and optical domains. Furthermore, two basic RoF techniques, the D–RoF and A–RoF, were elaborated in the context of energy efficiency, and an overview of research studies in the field of improving the energy efficiency of D–RoF and A–RoF systems was given. Additionally, the C-RAN architecture was reviewed through the prism of energy consumption with the presentation of current research efforts related to the improvement of the C-RAN energy efficiency. The MEC-based FiWi networks, which introduce cloud computing at the edge of the mobile network, were further presented, and articles dedicated to the mechanisms and concepts for the optimization of the MEC FIWi network's energy consumption were highlighted. Finally, flexible SDN FiWi networks that offer high scalability and ease of management of the FiWI networks were presented, with an emphasis on research related to energy conservation techniques implemented in such networks. The literature suggests that there are many areas in which the energy efficiency of the FiWi networks can be enhanced. Overall, the overview presented in this work showed that EE is one of the major concerns in the FiWi networks and further intensive research attempts should be carried out in the endeavor of improving the energy efficiency of the FiWi networks.

Author Contributions: Conceptualization, J.L.; methodology, Z.K. and J.L.; validation, J.L. and D.B.; formal analysis, J.L. and Z.K.; investigation, Z.K. and J.L.; writing–original draft preparation, J.L. and Z.K.; writing–review and editing, J.L. and Z.K.; visualization, J.L. and Z.K.; supervision, J.L. and D.B. All authors have read and agreed to the published version of the manuscript.

Conflicts of Interest: The authors declare no conflict of interest.

Abbreviations

100G-EPON	100 Gbit/s Ethernet Passive Optical Network
10G-EPON	10 Gbit/s Ethernet Passive Optical Network
3G	3rd Generation Mobile Network
4G	4th Generation Mobile Network
5G	5th Generation Mobile Network
6G	6th Generation Mobile Network
ACCO	Approximation Collaborative Computation Offloading
ALR	Adaptive Link Rate
ANN	Artificial Neural Network
AP	Access Point
A–RoF	Analog Radio-over-Fiber
AWSM	Adaptive Watchful Sleep Mode
BBU	Baseband Unit
BS	Base Station
CA	Carrier Aggregation
CAPEX	Capital Expenditure
CMCCO	Cloud–MEC Collaborative Computation Offloading
CM-FiWi	Cloud–MEC FiWi
CnR	Combine and Remove
CO	Central Office
CoMP	Coordinated Multi-Point
CO-OFDM	Coherent Optical Orthogonal Frequency Division Multiplexing
CPRI	Common Public Radio Interface
C-PSM	Centralized Power-Saving Mode
C-RAN	Cloud Radio Access Network
DAS	Distributed Antenna System

DBA	Dynamic Bandwidth Allocation
DCF	Distributed Coordination Function
DDQN	Double Deep Q Network
DEC	Delay-Controlled and Energy-Efficient Clustered
D-RAN	Distributed Radio Access Network
DRL	Deep Reinforcement Learning
D–RoF	Digitized Radio-over-Fiber
DRP	Dynamic Resource Provisioning
DWBA	Dynamic Wavelength and Bandwidth Allocation
E/O	Electrical-to-Optical
E/O/E	Electrical-Optical-Electrical
EA-CCO	Energy-Aware Collaborative Computation Offloading
ECO-FiWi	Energy Conservation Scheme for FiWi Networks
EE	Energy Efficiency
EEWA	Energy-Efficient Workload Allocation
EPON	Ethernet Passive Optical Network
FiWi	Fiber–Wireless
FTTA	Fiber to the Antenna
FTTx	Fiber to the x
GA	Genetic Algorithm
GDBA	Green Dynamic Bandwidth Allocation
G-EPON	Gigabit Ethernet Passive Optical Network
GHG	Greenhouse Gas
GPMM	Generic Power Management Model
GPON	Gigabit Passive Optical Networks
GT-CCO	Game-Theoretic Collaborative Computation Offloading Scheme
HA-DRL	Heuristic-Assisted Deep Reinforcement Learning
H-CRAN	Heterogeneous Cloud Radio Access Network
HERM	H-CRAN Energy-Efficient Radio Resource Management
HetNet	Heterogeneous Network
ICI	Inter-Cell Interference
ICT	Information And Communications Technology
IFoF	Intermediate Frequency-over-Fiber
ILP	Integer Linear Programming
IoT	Internet Of Things
ISA-CCO	Iterative Searching Algorithm for Collaborative Computation Offloading
IT	Information Technology
LoRa	Long Range
LPWAN	Low-Power Wireless Access Network
LTE-A	Long-Term Evolution Advanced
M2M	Machine-to-Machine
MAC	Media Access Control
MANET	Mobile Ad Hoc Network
MDP	Markov Decision Process
MEC	Multi-Access Edge Computing
MFH	Mobile Fronthaul
MFN	Mobile Fronthaul Network
ML	Machine Learning
mMIMO	Massive Multiple-Input Multiple-Output
mmWave	Millimeter Wave
MNO	Mobile Network Operator
MN	Mobile Network
MPP	Mesh Portal Point
NB-IoT	Narrowband Internet of Things
NFV	Network Function Virtualization
NG–PON2	Next-Generation Passive Optical Network 2
OBSAI	Open Base Station Architecture Initiative
OLT	Optical Line Terminal

ONU	Optical Network Unit
ONU-AP	Optical Network Unit with Access Point
ONU-BS	Optical Network Unit with Base Station
OPEX	Operating Expenditure
ORI	Open Radio Equipment Interface
OSC	ONU Sleep Controller
OSI	Open Systems Interconnection
PON	Passive Optical Network
PrO	Priority-based task offloading and caching
PS	Power-Saving
PSM	Power Saving Mechanism
PSO	Particle Swarm Optimization
QoS	Quality Of Service
QPSO	Quantum Particle Swarm Optimization
R&F	Radio-and-Fiber
RAN	Radio Access Network
RAU	Remote Access (Antenna) Units
RF	Radio Frequency
RFoF	Radio Frequency-over-Fiber
RI	Radio Interface
RoF	Radio-over-Fiber
RRU	Remote Radio Unit
SAPSM	Smart Adaptive Power Save Mode
SD	Software-Defined
SDN	Software-Defined Network
SEMC	Save Energy and Maximize Connectivity
SIEPON	Service Interoperability in the Ethernet Passive Optical Networks
TDM	Time-Division Multiplexing
TDMA	Time Division Multiple Access
TRx	Transceiver
TSC	Tx Sleep Controller
TSGO	Two-Layer Stackelberg Game Offloading
TWDM	Time And Wavelength Division Multiplexing
Tx	Transmitter
WDM	Wavelength-Division Multiplexing
WLAN	Wireless Local Area Network
WMN	Wireless Mesh Network
WOTR	Wireless–optical Topology Reconfiguration
WSM	Watchful Sleep Mode

References

1. The Climate Group. SMART 2020: Enabling the Low Carbon Economy in the Information Age. 2008, A Report by The Climate Group on Behalf of the Global eSustainability Initiative (GeSI). Available online: https://www.compromisorse.com/upload/estudios/000/36/smart2020.pdf (accessed on 13 February 2023).
2. Andrae, A.S.G.; Edler, T. On Global Electricity Usage of Communication Technology: Trends to 2030. *Challenges* **2015**, *6*, 117–157. [CrossRef]
3. Jejdling, F. *Ericsson Mobility Report*; Ericsson: Stockholm, Sweden, 2021.
4. Deruyck, M.; Vereecken, W.; Tanghe, E.; Joseph, W.; Pickavet, M.; Martens, L.; Demeester, P. Power consumption in wireless access network. In Proceedings of the 2010 European Wireless Conference (EW), Lucca, Italy, 12–15 April 2010; pp. 924–931.
5. Maier, M.; Ghazisaidi, N.; Reisslein, M. The Audacity of Fiber-Wireless (FiWi) Networks. In *Lecture Notes of the Institute for Computer Sciences, Social Informatics and Telecommunications Engineering*; Springer: Berlin/Heidelberg, Germany, 2008; Volume 6, pp. 16–35.
6. Zhang, H.; Hu, Y.; Wang, R.; Li, Z.; Zhang, P.; Xu, R. Energy-Efficient Frame Aggregation Scheme in IoT Over Fiber-Wireless Networks. *IEEE Internet Things J.* **2021**, *8*, 10779–10791. [CrossRef]
7. He, C.; Wang, R.; Tan, R.W.Z. Energy-Aware Collaborative Computation Offloading over Mobile Edge Computation Empowered Fiber-Wireless Access Networks. *IEEE Access* **2020**, *8*, 24662–24674. [CrossRef]
8. Ghazisaidi, N.; Maier, M. Fiber-wireless (FiWi) access networks: Challenges and opportunities. *IEEE Netw.* **2011**, *25*, 36–42. [CrossRef]

9. Rimal, B.P.; Van, D.P.; Maier, M. Mobile Edge Computing Empowered Fiber-Wireless Access Networks in the 5G Era. *IEEE Commun. Mag.* **2017**, *55*, 192–200. [CrossRef]
10. Rawat, D.B.; Reddy, S.R. Software Defined Networking Architecture, Security and Energy Efficiency: A Survey. *IEEE Commun. Surv. Tutor.* **2017**, *19*, 325–346. [CrossRef]
11. Lorincz, J.; Udovičić, G.; Begušić, D. Implementation of HFR/WLAN network. In Proceedings of the 13th International Conference on Software, Telecommunications and Computer Networks (SoftCOM 2005), Split, Croatia, 22–24 September 2005; pp. 269–274.
12. Ghazisaidi, N.; Maier, M.; Assi, C.M. Fiber-wireless (FiWi) access networks: A survey. *IEEE Commun. Mag.* **2009**, *47*, 160–167. [CrossRef]
13. Lorincz, J.; Ukic, N.; Begusic, D. Throughput Comparison of AODV-UU and DSR-UU Protocol Implementations in Multi-hop Static Environments. In Proceedings of the 9th International Conference on Telecommunications (ConTel 2007), Zagreb, Croatia, 13–15 June 2007.
14. Maier, M.; Ghazisaidi, N. *FiWi Access Networks*; Cambridge University Press: Cambridge, UK, 2011.
15. Beyranvand, H.; Levesque, M.; Maier, M.; Salehi, J.A.; Verikoukis, C.; Tipper, D. Toward 5G: FiWi Enhanced LTE-A HetNets with Reliable Low-Latency Fiber Backhaul Sharing and WiFi Offloading. *IEEE/ACM Trans. Netw.* **2016**, *25*, 690–707. [CrossRef]
16. Nikoukar, A.; Hwang, I.-S.; Wang, C.-J.; Ab-Rahman, M.S.; Liem, A.T. A SIEPON based transmitter sleep mode energy-efficient mechanism in EPON. *Opt. Fiber Technol.* **2015**, *23*, 78–89. [CrossRef]
17. Jia, Z.; Yu, J.; Ellinas, G.; Chang, G. Key Enabling Technologies for Optical–Wireless Networks: Optical Millimeter-Wave Generation, Wavelength Reuse, and Architecture. *J. Light. Technol.* **2007**, *25*, 3452–3471. [CrossRef]
18. Liu, J.; Guo, H.; Nishiyama, H.; Ujikawa, H.; Suzuki, K.; Kato, N. New Perspectives on Future Smart FiWi Networks: Scalability, Reliability, and Energy Efficiency. *IEEE Commun. Surv. Tutor.* **2016**, *18*, 1045–1072. [CrossRef]
19. Tsagklas, T.; Pavlidou, F.N. A survey on radio-and-fiber FiWi network architectures. *J. Sel. Areas Telecommun. (JSAT)* **2011**, 18–24.
20. Mishra, V.; Upadhyay, R.; Bhatt, U.R.; Kumar, A. DEC TDMA: A delay controlled and energy efficient clustered TDMA mechanism for FiWi access network. *Optik* **2020**, *225*, 164921. [CrossRef]
21. Togashi, K.; Nishiyama, H.; Kato, N.; Ujikawa, H.; Suzuki, K.-I.; Yoshimoto, N. On the effect of cooperation between power saving mechanisms in WLANs and PONs. In Proceedings of the 2013 IEEE International Conference on Communications (ICC), Budapest, Hungary, 9–13 June 2013; pp. 6225–6229.
22. Lange, C.; Kosiankowski, D.; Weidmann, R.; Gladisch, A. Energy Consumption of Telecommunication Networks and Related Improvement Options. *IEEE J. Sel. Top. Quantum Electron.* **2010**, *17*, 285–295. [CrossRef]
23. Dixit, A.; Lambert, S.; Lannoo, B.; Colle, D.; Pickavet, M.; Demeester, P. Towards energy efficiency in optical access networks [Invited]. In Proceedings of the 2013 IEEE International Conference on Advanced Networks and Telecommunications Systems (ANTS), Kattankulathur, India, 15–18 December 2013; pp. 1–6.
24. ITU-T, Series G, Supplement 45 (05/2009). Transmission Systems and Media, Digital Systems and Networks; GPON Power Conservation. Series G: Transmission Systems and Media, Digital Systems and Networks. International Telecommunication Union, 2009; pp. 1–98. Available online: https://www.itu.int/itu-t/recommendations/index.aspx?ser=G (accessed on 13 February 2023).
25. Kubo, R.; Kani, J.-I.; Ujikawa, H.; Sakamoto, T.; Fujimoto, Y.; Yoshimoto, N.; Hadama, H. Study and Demonstration of Sleep and Adaptive Link Rate Control Mechanisms for Energy Efficient 10G-EPON. *J. Opt. Commun. Netw.* **2010**, *2*, 716–729. [CrossRef]
26. Hirafuji, R.O.C.; da Cunha, K.B.; Campelo, D.R.; Dhaini, A.R.; Khotimsky, D.A. The watchful sleep mode: A new standard for energy efficiency in future access networks. *IEEE Commun. Mag.* **2015**, *53*, 150–157. [CrossRef]
27. Butt, R.A.; Idrus, S.M.; Zulkifli, N. Comparative analysis of cyclic and watchful sleep modes for GPON. In Proceedings of the 2016 IEEE 6th International Conference on Photonics (ICP), Sarawak, Malaysia, 14–16 March 2016; pp. 1–3.
28. Zin, A.M.; Idrus, S.M.; Ramli, A.; Butt, R.A.; Atan, F.M.; Ismail, N.A. Performance Evaluation of XG-PON with DBA Based-Watchful Sleep Mode. In Proceedings of the 2018 IEEE 7th International Conference on Photonics (ICP), Kedah, Malaysia, 9–11 April 2018; pp. 1–3.
29. Butt, R.A.; Akhunzada, A.; Faheem, M.; Raza, B. Enhanced Energy Savings with Adaptive Watchful Sleep Mode for Next Generation Passive Optical Network. *Energies* **2022**, *15*, 1639. [CrossRef]
30. Yin, S.; Chu, Y.; Yang, C.; Zhang, Z.; Huang, S. Load-adaptive energy-saving strategy based on matching game in edge-enhanced metro FiWi. *Opt. Fiber Technol.* **2022**, *68*, 102762. [CrossRef]
31. Hwang, I.-S.; Nikoukar, A.; Su, Y.-M.; Liem, A.T. Decentralized SIEPON-Based ONU-Initiated Tx/TRx Energy-Efficiency Mechanism in EPON. *J. Opt. Commun. Netw.* **2016**, *8*, 238. [CrossRef]
32. Zin, A.M.; Idrus, S.M.; Ismail, N.A.; Ramli, A.; Atan, F.M. Determination of optimized sleep interval for 10 gigabit-passive optical network using learning intelligence. *Int. J. Electr. Comput. Eng.* **2022**, *12*, 2663–2671.
33. Lorincz, J.; Bogarelli, M.; Capone, A.; Begušić, D. Heuristic approach for optimized energy savings in wireless access networks. In Proceedings of the 18th International Conference on Software, Telecommunications and Computer Networks (SoftCOM 2010), Dubrovnik, Croatia, 23–25 September 2010; pp. 60–65.
34. Lorincz, J.; Capone, A.; Begušić, D. Heuristic Algorithms for Optimization of Energy Consumption in Wireless Access Networks. *KSII Trans. Internet Inf. Syst.* **2011**, *5*, 626–648. [CrossRef]
35. Lorincz, J.; Matijevic, T.; Petrovic, G. On interdependence among transmit and consumed power of macro base station technologies. *Comput. Commun.* **2014**, *50*, 10–28. [CrossRef]

36. Pack, S.; Choi, Y. An adaptive power saving mechanism in IEEE 802.11 wireless IP networks. *J. Commun. Netw.* **2005**, *7*, 126–134. [CrossRef]
37. He, Y.; Yuan, R.; Ma, X.; Li, J.; Wang, C. Scheduled PSM for Minimizing Energy in Wireless LANs. In Proceedings of the 2007 IEEE International Conference on Network Protocols, Beijing, China, 16–19 October 2007; pp. 154–163.
38. Lee, J.; Rosenberg, C.; Chong, E.K.P. Energy Efficient Schedulers in Wireless Networks: Design and Optimization. *Mob. Netw. Appl.* **2006**, *11*, 377–389. [CrossRef]
39. Xie, Y.; Luo, X.; Chang, R.K.C. Centralized PSM: An AP-centric power saving Mode for 802.11 infrastructure networks. In Proceedings of the 2009 IEEE Sarnoff Symposium, Princeton, NJ, USA, 30 March–1 April 2009; pp. 1–5.
40. Pyles, A.J.; Qi, X.; Zhou, G.; Keally, M.; Liu, X. SAPSM: Smart Adaptive 802.11 PSM for Smartphones. In Proceedings of the 2012 ACM Conference on Ubiquitous Computing, UbiComp '12, Pittsburgh, PA, USA, 5–8 September 2012; pp. 11–20.
41. Saeed, A.; Kolberg, M. Towards Optimizing WLANs Power Saving: Novel Context-Aware Network Traffic Classification Based on a Machine Learning Approach. *IEEE Access* **2019**, *7*, 3122–3135. [CrossRef]
42. Lorincz, J.; Bule, I.; Kapov, M. Performance Analyses of Renewable and Fuel Power Supply Systems for Different Base Station Sites. *Energies* **2014**, *7*, 7816–7846. [CrossRef]
43. Lorincz, J.; Bule, I. Renewable energy sources for power supply of base station sites. *Int. J. Bus. Data Commun. Netw. (IJBDCN)* **2013**, *9*, 53–74. [CrossRef]
44. Tolba, F.D.; Magoni, D.; Lorenz, P. Saving Energy and Maximizing Connectivity by Adapting Transmission Range in 802.11g MANETs. *J. Commun. Softw. Syst.* **2007**, *3*, 81. [CrossRef]
45. Liu, Y.; Guo, L.; Yang, J. Energy-efficient topology reconfiguration in green Fiber-Wireless (FiWi) access network. In Proceedings of the 2013 22nd Wireless and Optical Communication Conference, Chongqing, China, 16–18 May 2013; pp. 555–559.
46. Nishiyama, H.; Togashi, K.; Kawamoto, Y.; Kato, N. A Cooperative ONU Sleep Method for Reducing Latency and Energy Consumption of STA in Smart-FiWi Networks. *IEEE Trans. Parallel Distrib. Syst.* **2015**, *26*, 2621–2629. [CrossRef]
47. Han, P.; Guo, L.; Liu, Y.; Hou, J.; Han, X. Joint Wireless and Optical Power States Scheduling for Green Multi-Radio Fiber-Wireless Access Network. *J. Light. Technol.* **2016**, *34*, 2610–2623. [CrossRef]
48. Van, D.P.; Rimal, B.P.; Maier, M.L.; Valcarenghi, M. ECO-FiWi: An Energy Conservation Scheme for Integrated Fiber-Wireless Access Networks. *IEEE Trans. Wirel. Commun.* **2016**, *15*, 3979–3994. [CrossRef]
49. Peng, X.; Xu, S.; Guo, S.; Qiu, X.; Li, S.; Yu, B. Collaborative Sleep Mechanism between Cross-domain Nodes in FiWi Network Based on Load Balancing and QoS Awareness. In Proceedings of the 2019 IFIP/IEEE Symposium on Integrated Network and Service Management (IM), Washington, DC, USA, 8–12 April 2019; pp. 270–278.
50. Chouhan, N.; Bhatt, U.R.; Upadhyay, R. An optimization framework for FiWi access network: Comprehensive solution for green and survivable deployment. *Opt. Fiber Technol.* **2019**, *53*, 102002. [CrossRef]
51. Dhaini, A.R.; Ho, P.-H.; Shen, G. Toward green next-generation passive optical networks. *IEEE Commun. Mag.* **2011**, *49*, 94–101. [CrossRef]
52. ITU-T, G.984.3 (01/2014). Gigabit-capable Passive Optical Networks (G-PON): Transmission Convergence Layer Specification. International Telecommunications Union—Telecommunication Standardization Sector (ITU-T). 2014; pp. 1–170. Available online: https://www.itu.int/rec/dologin_pub.asp?lang=f&id=T-REC-D.170-201005-I!Sup1!PDF-E&type=items (accessed on 13 February 2023).
53. ITU-T, G.987.3 (01/2014). 10-Gigabit-Capable Passive Optical Networks (XG-PON): Transmission Convergence (TC) Layer Specification. International Telecommunications Union—Telecommunication Standardization Sector (ITU-T). 2014; pp. 1–146. Available online: https://www.itu.int/ITU-T/itr-eg/files/resolution146.pdf (accessed on 13 February 2023).
54. ITU-T, G.989.3 (05/2021). 40-Gigabit-Capable Passive Optical Networks (NG-PON2): Transmission Convergence Layer Specification. International Telecommunications Union—Telecommunication Standardization Sector (ITU-T). 2021; pp. 1–280. Available online: https://www.itu.int/rec/T-REC-G.989.3 (accessed on 13 February 2023).
55. *IEEE Std 1904.1-2013*; Standard for Service Interoperability in Ethernet Passive Optical Networks (SIEPON). IEEE Communications Society: New York, NY, USA, 2013; pp. 1–834. [CrossRef]
56. Mohammed, A.F.Y.; Newaz, S.H.S.; Sankar, D.S.; Ahsan, S.; Um, T.-W. A Green Converged TWDM-PON and 5G HetNet Catering Applications Demanding Low Latency. *Opt. Fiber Technol.* **2020**, *58*, 102261. [CrossRef]
57. Mishra, V.; Upadhyay, R.; Bhatt, U.R. A Review of Recent Energy-Efficient Mechanisms for Fiber-Wireless (FiWi) Access Network. In *Progress in Advanced Computing and Intelligent Engineering*; Springer: Singapore, 2018; pp. 539–545.
58. *IEEE 802.11*; Wireless LAN Medium Access Control (MAC) and Physical Layer (PHY) Specifications. IEEE Standards Association: Piscataway Township, NJ, USA, 2012.
59. Alaslani, M.; Showail, A.; Shihada, B. Green frame aggregation scheme for Wi-Fi networks. In Proceedings of the 2015 IEEE 16th International Conference on High Performance Switching and Routing (HPSR), Budapest, Hungary, 1–4 July 2015; pp. 1–6.
60. Popov, M.; Peinado, D.; Nilson, M.; Västberg, A.; Sjolund, T. Green distributed antenna systems: Optimized design and upper bound for energy efficiency. In Proceedings of the 21st International Conference on Software, Telecommunications and Computer Networks (SoftCOM), Split, Croatia, 18–20 September 2013; pp. 1–5.
61. Crisp, M.; Penty, R.V.; White, I.H.; Bell, A. Wideband Radio over Fiber Distributed Antenna Systems for Energy Efficient In-Building Wireless Communications. In Proceedings of the 2010 IEEE 71st Vehicular Technology Conference, Taipei, Taiwan, 16–19 May 2010; pp. 1–5.

62. Josse, Y.; Fracasso, B.; Castignani, G.; Montavont, N. Energy-efficient deployment of distributed antenna systems with radio-over-fiber links. In Proceedings of the 2012 IEEE Online Conference on Green Communications (GreenCom), Online, 26–28 September 2012; pp. 7–13.
63. Deronne, S.; Lucarz, F.; Moeyaert, V.; Fracasso, B.; Bette, S. Energy efficiency analysis of aggregation mechanisms in IEEE 802.11n radio-over-fiber-based distributed antenna systems. *Photonic Netw. Commun.* **2015**, *30*, 96–107. [CrossRef]
64. Gowda, A.S.; Dhaini, A.R.; Kazovsky, L.G.; Yang, H.; Abraha, S.T.; Ng'Oma, A. Towards Green Optical/Wireless In-Building Networks: Radio-Over-Fiber. *J. Light. Technol.* **2014**, *32*, 3545–3556. [CrossRef]
65. Gowda, A.S.; Yang, H.; Abraha, S.T.; Ng'Oma, A.; Dhaini, A.R.; Kazovsky, L.G.; Albeyoglu, K.M. Energy consumption of indoor radio-over-fiber distribution links: Experimental findings. In Proceedings of the 2014 IEEE Global Communications Conference, Sydney, Australia, 8–12 December 2014; pp. 2612–2617.
66. Beas, J.; Castanon, G.; Aldaya, I.; Aragon-Zavala, A.; Campuzano, G. Millimeter-Wave Frequency Radio over Fiber Systems: A Survey. *IEEE Commun. Surv. Tutor.* **2013**, *15*, 1593–1619. [CrossRef]
67. ITU-T, Series G, Supplement 55. Radio-Over-Fiber (RoF) Tech-Nologies and Their Applications. July 2015. Available online: https://www.itu.int/rec/T-REC-G.Sup55/en (accessed on 30 September 2021).
68. de la Oliva, A.; Hernandez, J.A.; Larrabeiti, D.; Azcorra, A. An overview of the CPRI specification and its application to C-RAN-based LTE scenarios. *IEEE Commun. Mag.* **2016**, *54*, 152–159. [CrossRef]
69. Chih-Lin, I.; Yuan, Y.; Huang, J.; Ma, S.; Cui, C.; Duan, R. Rethink fronthaul for soft RAN. *IEEE Commun. Mag.* **2015**, *53*, 82–88.
70. Apostolopoulos, D.; Giannoulis, G.; Argyris, N.; Iliadis, N.; Kanta, K.; Avramopoulos, H. Analog radio-over-fiber solutions in support of 5G. In Proceedings of the International Conference on Optical Network Design and Modeling (ONDM), Dublin, Ireland, 14–17 May 2018; pp. 266–271.
71. Asha, D.S. A comprehensive review of Millimeter wave based radio over fiber for 5G front haul transmissions. *Indian J. Sci. Technol.* **2021**, *14*, 86–100.
72. Pliatsios, D.; Sarigiannidis, P.; Goudos, S.; Karagiannidis, G.K. Realizing 5G vision through Cloud RAN: Technologies, challenges, and trends. *EURASIP J. Wirel. Commun. Netw.* **2018**, *136*, 1–15. [CrossRef]
73. Zhang, J.; Xiao, Y.; Li, H.; Ji, Y. Performance Analysis of Optical Mobile Fronthaul for Cloud Radio Access Networks. *J. Phys. Conf. Ser.* **2017**, *910*, 012053. [CrossRef]
74. Nahas, M.; Saadani, A.; Charles, J.; El-Bazzal, Z. Base stations evolution: Toward 4G technology. In Proceedings of the 19th International Conference on Telecommunications (ICT), Jounieh, Lebanon, 23–25 April 2012; pp. 1–6.
75. CPRI Cooperation. Common Public Radio Interface (CPRI); Interface Specification (V7.0). Tech. Rep., Ericsson AB, Huawei Technologies Co. Ltd., NEC Corporation, Alcatel Lucent, and Nokia Networks, October 2015. Available online: http://www.cpri.info/downloads/CPRI_v_7_0_2015-10-09.pdf (accessed on 13 February 2023).
76. Guo, H.; Wang, K.; Ji, H.; Leung, V.C.M. Energy saving in C-RAN based on BBU switching scheme. In Proceedings of the 2016 IEEE 5th International Conference on Network Infrastructure and Digital Content, IC-NIDC 2016, Beijing, China, 23–25 September 2016; pp. 44–49.
77. Bluemm, C.; Zhang, Y.; Alvarez, P.; Ruffini, M.; DaSilva, L.A. Dynamic energy savings in Cloud-RAN: An experimental assessment and implementation. In Proceedings of the 2017 IEEE International Conference on Communications Workshops (ICC Workshops), Paris, France, 21–25 May 2017; pp. 791–796.
78. Yu, N.; Song, Z.; Du, H.; Huang, H.; Jia, X. Dynamic Resource Provisioning for Energy Efficient Cloud Radio Access Networks. *IEEE Trans. Cloud Comput.* **2017**, *7*, 964–974. [CrossRef]
79. Sahu, B.J.R.; Dash, S.; Saxena, N.; Roy, A. Energy-Efficient BBU Allocation for Green C-RAN. *IEEE Commun. Lett.* **2017**, *21*, 1637–1640. [CrossRef]
80. Park, H.; Lim, Y. Energy-Effective Power Control Algorithm with Mobility Prediction for 5G Heterogeneous Cloud Radio Access Network. *Sensors* **2018**, *18*, 2904. [CrossRef]
81. Liu, Q.; Han, T.; Ansari, N.; Wu, G. On Designing Energy-Efficient Heterogeneous Cloud Radio Access Networks. *IEEE Trans. Green Commun. Netw.* **2018**, *2*, 721–734. [CrossRef]
82. Wang, B.; Peng, L.; Ho, P.-H. Energy-efficient radio-over-fiber system for next-generation cloud radio access networks. *EURASIP J. Wirel. Commun. Netw.* **2019**, *2019*, 118. [CrossRef]
83. Alhumaima, R.S.; Ahmed, R.K.; Al-Raweshidy, H.S. Maximizing the Energy Efficiency of Virtualized C-RAN via Optimizing the Number of Virtual Machines. *IEEE Trans. Green Commun. Netw.* **2018**, *2*, 992–1001. [CrossRef]
84. Zhu, M.; Gu, J.; Shen, T.; Shi, C.; Ren, X. Energy-Efficient and QoS Guaranteed BBU Aggregation in CRAN Based on Heuristic-Assisted Deep Reinforcement Learning. *J. Light. Technol.* **2022**, *40*, 575–587. [CrossRef]
85. Iqbal, A.; Tham, M.L.; Chang, Y.C. Double Deep Q-Network-Based Energy-Efficient Resource Allocation in Cloud Radio Access Network. *IEEE Access* **2021**, *9*, 20440–20449. [CrossRef]
86. China Mobile Research Institute. C-RAN: The Road Towards Green RAN. White Paper, 2010. Available online: https://silo.tips/download/c-ran-the-road-towards-green-ran (accessed on 13 February 2023).
87. Checko, A.; Christiansen, H.L.; Yan, Y.; Scolari, L.; Kardaras, G.; Berger, M.S.; Dittmann, L. Cloud RAN for Mobile Networks—A Technology Overview. *IEEE Commun. Surv. Tutor.* **2014**, *17*, 405–426. [CrossRef]

88. Niu, H.; Li, C.; Papathanassiou, A.; Wu, G. RAN architecture options and performance for 5G network evolution. In Proceedings of the IEEE Wireless Communications and Networking Conference Workshops (WCNCW), Istanbul, Turkey, 6–9 April 2014; pp. 294–298.
89. Chih-Lin, I.; Li, H.; Korhonen, J.; Huang, J.; Han, L. RAN Revolution With NGFI (xhaul) for 5G. *J. Light. Technol.* **2017**, *36*, 541–550.
90. Wu, J.; Zhang, Z.; Hong, Y.; Wen, Y. Cloud radio access network (C-RAN): A primer. *IEEE Netw.* **2015**, *29*, 35–41. [CrossRef]
91. Rimal, B.P.; Van, D.P.; Maier, M. Cloudlet Enhanced Fiber-Wireless Access Networks for Mobile-Edge Computing. *IEEE Trans. Wirel. Commun.* **2017**, *16*, 3601–3618. [CrossRef]
92. Nur, F.N.; Islam, S.; Moon, N.N.; Karim, A.; Azam, S.; Shanmugam, B. Priority-Based Offloading and Caching in Mobile Edge Cloud. *J. Commun. Softw. Syst.* **2019**, *15*, 193–201. [CrossRef]
93. Guo, H.; Liu, J. Collaborative Computation Offloading for Multiaccess Edge Computing Over Fiber–Wireless Networks. *IEEE Trans. Veh. Technol.* **2018**, *67*, 4514–4526. [CrossRef]
94. Guo, H.; Liu, J.; Qin, H. Collaborative Mobile Edge Computation Offloading for IoT over Fiber-Wireless Networks. *IEEE Netw.* **2018**, *32*, 66–71. [CrossRef]
95. Yang, B.; Wu, D.; Wang, H.; Gao, Y.; Wang, R. Two-Layer Stackelberg Game-Based Offloading Strategy for Mobile Edge Computing Enhanced FiWi Access Networks. *IEEE Trans. Green Commun. Netw.* **2020**, *5*, 457–470. [CrossRef]
96. Tornatore, M.; Chang, G.-K.; Ellinas, G. *Fiber-Wireless Convergence in Next-Generation Communication Networks;* Springer: Berlin/Heidelberg, Germany, 2017.
97. ETSI ISG. Mobile-Edge Computing—A Key Technology Towards 5G; White Paper, 2015; no. 11. Available online: https://www.etsi.org/images/files/etsiwhitepapers/etsi_wp11_mec_a_key_technology_towards_5g.pdf (accessed on 13 February 2023).
98. Pakpahan, A.F.; Hwang, I.-S.; Liem, A.T. Enabling Agile Software-Defined and NFV based Energy-Efficient Operations in TWDM-PON. In Proceedings of the 2019 7th International Conference on Cyber and IT Service Management (CITSM), Jakarta, Indonesia, 6–8 November 2019; Volume 7, pp. 1–7.
99. Yan, B.; Zhou, J.; Wu, J.; Zhao, Y. Poster: SDN based energy management system for optical access network. In Proceedings of the 9th International Conference on Communications and Networking in China, Maoming, China, 14–16 August 2014; pp. 658–659.
100. Pakpahan, A.F.; Hwang, I.-S. Adaptive ONU Energy-Saving via Software-Defined Mechanisms in TDMA-PON. In Proceedings of the 2018 Tenth International Conference on Ubiquitous and Future Networks (ICUFN), Prague, Czech Republic, 3–6 July 2018; pp. 137–142.
101. Khalili, H.; Khodashenas, P.S.; Rincon, D.; Siddiqui, S.; Piney, J.R.; Sallent, S. Design Considerations for an Energy-Aware SDN-Based Architecture in 5G EPON Nodes. In Proceedings of the 2018 20th International Conference on Transparent Optical Networks (ICTON), Bucharest, Romania, 1–5 July 2018; pp. 1–4.
102. Li, S.; Hu, Y.; Guo, Z.; Shou, G. SDN-based energy-saving scheme for fiber-wireless network. In Proceedings of the 11th International Conference on Wireless Communications, Networking and Mobile Computing (WiCOM 2015), Abu Dhabi, United Arab Emirates, 19–21 October 2015; pp. 1–4.
103. Wang, S.; Chen, B.; Liang, R.; Liu, L.; Chen, H.; Gao, M.; Wu, J.; Ju, W.; Ho, P.-H. Energy-efficient workload allocation in edge-cloud fiber-wireless networks. *Opt. Express* **2022**, *30*, 44186. [CrossRef]
104. Open Networking Foundation (ONF). OpenFlow Switch Specification. March 2015. Available online: https://opennetworking.org/wp-content/uploads/2014/10/openflow-switch-v1.5.1.pdf (accessed on 5 December 2022).

sensors
MDPI

Review

Dynamics of Research into Modeling the Power Consumption of Virtual Entities Used in the Telco Cloud

Etienne-Victor Depasquale [1,*], Franco Davoli [2] and Humaira Rajput [3]

1 Department of Communications and Computer Engineering, University of Malta, MSD 2080 Msida, Malta
2 The National Laboratory of Smart and Secure Networks (S2N), Italian National Consortium for Telecommunications (CNIT), Department of Electrical, Electronic and Telecommunications Engineering and Naval Architecture (DITEN), University of Genoa, 16145 Genoa, Italy
3 Department of Electrical, Electronic and Telecommunications Engineering and Naval Architecture (DITEN), University of Genoa, 16145 Genoa, Italy
* Correspondence: edepa@ieee.org

Abstract: This article is a graphical, analytical survey of the literature, over the period 2010–2020, on the measurement of power consumption and relevant power models of virtual entities as they apply to the telco cloud. We present a novel review method, that summarizes the ***dynamics*** as well as the results of the research. Our method lends insight into trends, research gaps, fallacies and pitfalls. Notably, we identify limitations of the widely used linear models and the progression towards Artificial Intelligence/Machine Learning techniques as a means of dealing with the seven major dimensions of variability: workload type; computer virtualization agents; system architecture and resources; concurrent, co-hosted virtualized entities; approaches towards the attribution of power consumption to virtual entities; frequency; and temperature.

Keywords: virtualization; power consumption; power models; power meters; energy-aware algorithms

Citation: Depasquale, E.-V.; Davoli, F.; Rajput, H. Dynamics of Research into Modeling the Power Consumption of Virtual Entities Used in the Telco Cloud. *Sensors* **2023**, *23*, 255. https://doi.org/10.3390/s23010255

Academic Editor: Josip Lorincz

Received: 31 October 2022
Revised: 19 December 2022
Accepted: 21 December 2022
Published: 26 December 2022

1. Introduction

Several surveys of the results (see Section 1.6, "Related Surveys") of research into modeling the power consumption of virtual entities (VEs, i.e., virtual machines (VMs) or containers) have been written. In this work, our contribution lies in ***a thorough analysis of the dynamics of research itself: the challenges, the approaches, the pitfalls, the fallacies, and the research gaps, without neglecting the fruits of the research.*** Our intended audience is the prospective researcher, seeking to understand the dynamics of research into the predictive modeling and supporting measurements of power consumption by individual VEs relevant to *the telco cloud*. Dynamics are characterized through a thorough frequency analysis, which we conduct **through the application of a novel method we have developed** [1] that is unique in its ability to ***parse*** research literature. Through the visual aids we provide, and our observations through cross-cutting themes, a prospective researcher obtains a thorough characterization of the problems, approaches, developments, formal methods, pitfalls, fallacies and research gaps that characterize this research space.

Among the themes that our survey has brought us to identify, we have pointed out that all the problem categories we identified touch one or more of a set of seven main variables that may affect power consumption by virtual entities and the ensuing model representations: workload type, characteristics of the virtualization agent (VM or container), host machine resources and architecture, temperature, operating frequency, attribution of a fraction of consumed power to individual VEs and the mutual influence of concurrent VEs.

Among the major pitfalls that emerged from our thematic analysis, we highlight here the misconception of the Data Plane Development Kit's (DPDK) power efficiency (commonly misportrayed as a power hog), the often-unacknowledged limitations of the widely used linear models, the problematic use of benchmarks in model validation, the failure to precisely identify the physical contexts of some experimental research, the influence of synthetic workload generators on measurements and the sometimes-overlooked relevance of processor organization on power consumption measurements. We have also pointed out the unavoidable need to precisely identify the scope and limitations of models and the fallacy of the quest for a "universal" power model.

The research gaps we identify are four: (a) modeling of containers' power consumption; (b) the effect of overcommitment on power efficiency; (c) investigation and classification of DPDK applications; and (d) modeling of power consumption by virtualized I/O (a challenge which is starting to receive some attention).

1.1. *Why Is Research Needed?*

Precise measurement of a VE's power consumption is difficult since measurements of its host's power consumption cannot be related directly to it. Hardware power meters are incapable of measuring the power consumption of individual VEs co-hosted on a physical machine. Moreover, power consumption of a VE varies with its hosting machine. Therefore, for VEs, accurate measurement is predicated upon precise ***modeling*** of energy- and/or power consumption.

A general prerequisite to devising energy- and/or power-efficient ***operations*** is accuracy in power and energy ***measurements***. With specific regard to VEs, it is also essential for billing in multi-tenant environments, so that the Infrastructure Provider (IPr) can charge customers the fair amount for the resources (including energy) they consume. Before proceeding to the scope of our survey, we pose three salient questions that frame our work.

1.2. *Why Should Energy Consumption Be a Topic of Interest?*

Energy efficiency in the Internet (and in computing and telecommunication networks in general) has become a significant problem, which has received increasing attention since the early years around 2000 (see, e.g., [2–4] and references therein), starting from cloud computing infrastructures, and then extending to mobile and fixed networks. Indeed, it has been shown that the smaller data centers, within which telecommunications points of presence (PoPs) may be classified, represent around 95% of the United States' data center energy use [5]. Furthermore, this use is comparatively inefficient when compared with that of the hyperscale server farms (the remaining 5%).

1.3. *What Is the Underlying Cause of Increased Energy Consumption?*

Traffic growth is the primary cause of increased energy consumption. Table 1 shows the consistency with which Cisco's Visual Networking Index (VNI) has been predicting heavy growth in traffic exchanged over the access network by both businesses and consumers with:

- endpoints over managed networks;
- endpoints over unmanaged networks ("Internet traffic").

Table 1. Compound Annual Growth Rate (CAGR) reported in Cisco's VNI over four consecutive years.

Period	Fixed Internet Traffic	Managed IP Traffic	Mobile Data
2014–2019 [6]	23	13	57
2015–2020 [7]	21	11	53
2016–2021 [8]	23	13	46
2017–2022 [9]	26	11	46

(Note that the figures refer to compound annual growth rate (CAGR); they do ***not*** refer to the percentage share of total traffic.).

The key observation lies in the realization that, year after year, significant (heavy, in the case of mobile data) growth is persistently predicted. This observation is corroborated by several other researchers, with perspectives varying from traffic at the access segment to traffic in transit between Internet Service Providers (ISPs) [10–12]. Cisco [6] and Sandvine [13,14] identify "video traffic" and "real-time entertainment" as the drivers of this growth. A later edition of Sandvine's Global Internet Phenomena report [15] dedicates its executive summary exclusively to video traffic; the report shows video as consuming 60% of downstream traffic—a further 2% increase over 2018. Alcatel-Lucent Bell Labs [11] observed that growth in the metro-core due to video traffic exceeds growth in video traffic crossing the long-haul core. Clearly, it is widely recognized that video is the prime driver of this growth in traffic and attendant energy consumption and research has been addressing this problem [16–28].

1.4. How Is Energy Consumption Being Tackled?

From these (referenced) works, the development of cache architectures emerges as an important approach to controlling energy consumption, but other mechanisms exist and yet more are emerging within the general thrust towards "future networks" (ITU-T Y.3001 [29]). Networks are evolving into flexible and programmable "softwarized" virtualized infrastructures, and strongly integrated paradigms, such as 5G [30,31], rewrite the research agenda. We may thus distinguish between radical and reformist approaches:

- Reformist approaches seek to improve the caching of content and are characterized by their investigation of the reduction of the length of the path between the source and destination of IP traffic.
- Radical approaches employ the dynamic and reactive control afforded by the softwarized, virtualized infrastructures. Radical methods are enabled by standardized architectures (e.g., [32,33]) that equip the control plane with uniform interfaces that exploit extant and emerging green capabilities.

Indeed, ***the impact of virtualization technologies on power consumption in public telecommunication networks (PTNs) is still unclear***. There is a general belief that Network Functions Virtualization (NFV) should result in reduced energy consumption, owing to a consolidation of resources and increased flexibility in turning unused hardware (HW) on and off as needed. However, it is also true that "the massive introduction of general-purpose HW enabled by NFV would tend to increase power requests with respect to specialized HW solutions" [30]. Therefore, there is a need to operate power-aware management and control mechanisms in these environments. At the same time, it is necessary to limit the complexity of these mechanisms and the level of human intervention therein, to keep Operational Expenditures (OPEX) within reasonable limits. One approach to understanding this impact consists of comparative analyses of the implementations of infrastructure, with and without virtualization. This approach is taken in [34], where the evolved packet core (EPC) is studied. This work shows that the virtualized implementation is indeed less energy efficient. Unfortunately, the scope of virtualization and containerization within the converged wireless and wireline infrastructure is very broad and consideration of a single "use-case" [35] cannot be generalized to an overall statement. We therefore note that the scope for our survey needs an operational context which we suggest in the following.

1.5. The Scope of This Survey

1.5.1. Telco Cloud: The Operations Context

We suggest that the telco cloud is our network-operations context. "Telco cloud" is an evolving notion that evokes a number of common terms in attempts to describe it. Virtualization, software-defined networking (SDN), automation and orchestration are four such terms. Other prominent terms are edge computing, containerization, microservices and resilient infrastructure [36]. We suggest three key observations that organize these terms into a coherent image of the telco cloud.

1. **The telco cloud is, fundamentally, a hybrid cloud:**
    a. Self-sourced virtualization and containerization;
    b. Out-sourced (public cloud) containerization.

The complementary collaboration of the PTN operator's (PTNO) network, computing and storage infrastructure, with that of global providers of infrastructure and applications, is manifested well in [37]. A distributed cloud infrastructure operates at network (transport and interconnect) junctions. It includes (cloud) infrastructure owned and operated by the PTNO, by public cloud providers and by enterprises which consume their joint service.

2. **The telco cloud serves both internal and external clients** [36]:
    a. Internal use can suggestively be referred to as the ***IT Cloud*** [38]. This consists of applications specific to PTNOs: operational support systems (OSS) and business support systems (BSS), as well as more general applications, such as customer relationship management (CRM);
    b. External uses are growing organically on the basis of use cases seeded by ETSI [35] and the 5GPPP [39].
3. **The service-based architecture (SBA) of the 5G Core is a good fit with cloud-native computing**. Containerization is distinctively central to cloud-native computing [40,41]. The Cloud Native Computing Foundation explicitly identifies containers as components of the approach to the concept of cloud native computing [42]. There is a clear drive towards the use of containers in lieu of virtual machines as the operating environment for network functions [40], and the 5G Core's SBA provides a clear scope for employing containers.

We end this subsection by indicating the real estate where the VEs in our context may be deployed:

1. **Datacenters**: here, the real estate referred to consists of points of presence (PoPs) such as metro-core PoPs at the near edge;
2. **Softwarized and virtualized networks**: here, we refer to points of presence such as central offices (COs) and sites even deeper into the edge, such as remote radio head (RRH) sites and roadside cabinets.

1.5.2. Identifying the Models in the Scope of This Study

We survey **predictive** energy and power **models**, as well as **measurements** that facilitate the qualitative and/or quantitative prediction, of consumption by individual VEs relevant to the telco cloud. A simple interpretation of the rationale that drove our selection is that we have sought works that **measure real-time power consumption by VEs** and/or model **real-time power consumption by VEs**. The object of measurements and modeling is strictly the VE.

Nonetheless, the devil is in the details and therefore the details of this simple rationale must be worked out. One important, finer point regards the VEs themselves. There are software technologies, which we shall elaborate upon in later sections, which are **functionally critical** to VEs. Works that measure, and/or model, such technologies' power consumption are in scope. A justification of this claim on scope is not hard. Since power consumption is a scalar quantity, the reduction of power consumption of a component of a VE translates into the reduction of power consumption by the VE. In fairness, the translation is not direct (1:1). A generalization of Amdahl's law comes to mind: improvement in a component, measurable by some metric, is attenuated by the ratio of that component's use (measurable by that metric) to the system's (the VE's) use (measurable by the same metric). However, we can safely summarize the finer point we alluded to at the start of this paragraph as follows. Research that studies the measurement and modeling of power consumption by a component of a VE is in scope. We clarify this by example. For instance, we would:

1. **Include** the Data Plane Development Kit (DPDK) [43], as it serves the critical function of networking (VEs that serve as virtual network functions (VNFs));
2. **Include** a comparative study that measures power consumption by a VE using two different implementations of input/output virtualization technology, say: SR-IOV (single-root IO virtualization) and paravirtualization;

3. **Exclude** a comparative study that measures power consumption by various network adapter (or network interface card (NIC)) architectures, **unless** it reveals the impact of these architectures on VEs' power consumption.

Further detail emerges from the "real-time" requirement. This term is a reflection of the need for **application agnosticism**. Power meters that follow from such measurements and models may be used regardless of whether instantaneous or statistical readings of power are required. Studies (on measurements and models) meet this requirement by satisfying the following criteria.

1. Predictors:
    a. Must be of fine temporal granularity;
    b. Must be updated with the regularity of the temporal granularity;
    c. Must enable prediction of power consumption at the same temporal granularity.
2. Workload: Only resource-specific constraints are considered. That is, in the course of testing using, say, workloads that are processor-intensive (hence the workload is specific to the processing resource), no other constraints are allowed in works included in this study. In particular, models must not constrain the stochasticity of the workload.

Our last detail regards workload. We observe that the universal power model is a fallacy and the principal reason for this is that the interaction between workload and architecture cannot be pinned down indefinitely. This does not mean that modeling is a fruitless endeavor. It simply means that validity constraints must be placed on the model in terms of workload and architecture. Therefore, we do not exclude modeling and measurement because of its workload-scope or architectural scope. We do, however, observe that such models are pitfalls for those who apply them without knowledge of such limits.

We now proceed to present some recent, related surveys, highlighting their methodology and analytical approach. Against this background, we summarize the novelty of this survey.

1.6. Related Surveys

In recent years, much research has modeled the power consumption of servers in data centers and cloud environments in general. Several surveys in this regard cover different aspects of modeling and energy efficiency approaches for servers and virtual entities.

An analysis of power models the from micro- to macro-level is presented in [44]. This survey covers different aspects and levels of both hardware- and software-centric modeling techniques. Researchers studied models based on computing resources (CPU, memory, storage and I/O), system architecture (such as single or multiple cores), the availability of Graphics Processing Units (GPUs), system/network components, operating systems and virtualization environments. They categorize the existing models at different layers moving from architecture level modeling to the level of power models for whole data centers. However, the survey did not focus greatly on power modeling techniques that consider the effect of different virtual entities.

According to the categorization in [44], power models depend on different organizational contexts, including the power consumed by system components, running applications, and/or the execution strategy of processes. These metrics, however, can derive additive component-based, regression-based or machine-learning-based power models. Additive models usually present an aggregated view of server power consumption, which could be based on different resources (such as CPU, ram, I/O, disk), or the disaggregate static and dynamic power consumption of the server. Regression-based models are mostly based on the relation of power to the dynamic evolution of some measured system parameters. Power modeling using machine learning techniques is an advanced research area which can be further classified as supervised, unsupervised, reinforcement and evolutionary learning. Furthermore, at a much higher level, such as that of data center environments as a whole, power models for groups of servers, datacenter networks, conditioning and cooling systems were studied. The survey also analyzed the modeling methods at operating system (OS) and virtualization level, and for data-intensive, communication-intensive and general applications. At the end, the researchers compare the power models against their complexity, effectiveness, application and use-cases.

The survey carried out in [45] analyzes the power models based on their modeling approaches. It claims that power modeling methods can be divided into two main themes: analytical models and formula-learned models. In the former, parameters affecting the system power are known to the researcher; however, the weight for each parameter needs to be determined. On the contrary, in high-level formula-learned or machine-learning-based modeling, no prior knowledge of the system is required and the model is developed from scratch using the provided data set. The survey further describes commonly used machine learning techniques, and later focuses on the use and effectiveness of neural networks in power modeling. According to this survey, the approach to model power for any server may vary in several aspects. These different characteristics are summarized as (i) ***degree of autonomy***, i.e., to which level the power model is dependent on external hardware; (ii) ***level of granularity***, which describes the depth of logical (core, thread, etc.) or physical (device specific, system level, etc.) levels to which the model can precisely estimate the power; (iii) ***methodology***, describing the selection of method, which could be simulation, analytical modeling or data-training; (iv) ***simplicity***, which can be assessed by the number of variables, method selection and models' computing overheads; (v) ***portability***, as most of the models developed are effective for some specific architecture, workload or environment, and their generalization is still a question; (vi) ***accuracy***, which defines the measurement accuracy or estimation precision of power for any model; and (vii) ***power meter*** (as most studies use external power meters as a ground truth for their modeling, their accuracy is also a major concern for developing a more efficient model).

The survey in [46] adopts a slightly different analytical approach and evaluates selected existing power modeling techniques in a unified environment. Comparative analysis has been performed for twenty-four different software power models and measurement methods, with nine different benchmarks under a single experimental environment. It evaluates the existing software power measurement techniques and models for different applications, benchmarks, systems configuration, server architecture and for their estimation errors. The authors claim that most of the software-based power models use system performance metrics provided by the operating system, or performance monitoring counters provided by hardware sub-systems of the server. Software power models considered in this study are categorized in three types as single variable CPU-based, multi-variable CPU-based, and single-variable throughput-based. The result of this unified experimental setup shows that power models based on a support vector machine (SVM) and interpolation techniques show the least error for different resource-intensive applications, whereas lasso regression with 30 variables was found to be the worst power model with the highest error. Furthermore, the modeling techniques are mainly divided into two categories; linear and non-linear, where each category is further classified based on its derivation from mathematical modeling or machine learning techniques.

Another survey [47], with a more limited scope, focuses on the power modeling of servers in the cloud while considering the complexity stemming from the diversity of host hardware platforms, virtualization environment and workload. It presents the usability, applicability and limitations of the studied approaches, and also presents the analysis of the traditional existing and emerging modeling techniques. It reviews the existing power modeling methods at three stages: data acquisition methods, power consumption models and power modeling methods for servers, VMs and containers. According to this study, the collection of data for power models could be based on: (a) instruments such as external power meters, (b) dedicated acquisition systems which are generally products customized by developers for a specific hardware, (c) simulation-based, and (d) software monitoring tools. The latter constitute a widely used method that is based on system indicators and sensors. For the second stage—i.e., the power modeling—either hardware-centric, virtualization-centric or application-centric power modeling schemes can be used. Each type has its own drawback and usability; hence, the domain is still an open research area for further generalization and measurement precision. The selection of method, which is considered as the third step of the modeling in [47], is characterized as power-modeling-based on: (a) empirical parameterization, (b) function regression, (c) machine learning, and (d) evolutionary algorithms. The survey, after analyzing different power models, concluded that functional regression and machine learning methods yield high accuracy when provided with large enough data sets and clear power behavior of servers. This survey is more similar to ours; however, we distinguish further by providing statistical analysis of existing approaches, used tools, methods and measurement metrics. Moreover, our survey focuses on the modeling of power for individual virtual entities and on the effect of the virtualization environment on server and network functions' power behavior.

In general, the surveys mentioned above present the literature on server models irrespective of the operational environment. One of the major trends, not only in data centers but also in networking, is the presence of virtual entities and the adoption of virtualization environments. In contrast, our survey concentrates on the power models of virtualized entities, and focuses mainly on components and parameters that could affect their power consumption. It further provides a statistical analysis of included research works with respect to their data acquisition methods, measurement tools and modeling methods, to give researchers a bird's-eye view of the existing literature through which one can identify the approaches carried out in different studies.

1.7. Organization of This Survey

The survey is organized as follows:

1. In [1], we describe our method: a problem–approach–development (PAD) triad, which, to the best of our knowledge, we are the first to use to identify research dynamics. We have delegated the description of the method to a separate paper, to resolve the difficulty of elaborating fully on the method without distracting attention from the results which we have obtained and which we document separately in this review.
2. Section 2 presents the detailed results.
3. Section 3 is our analysis of the results. There, we give a qualitative assessment through themes which emerged as we organized the data. We have classified these themes as "state of the art", "fallacies" and "pitfalls", to suggest guidance and warnings which we were able to glean from others' experiences.
4. Section 4 concludes by attempting to encapsulate the insights we have gained through this work.
5. Appendix A illustrates the use of structural coding on a sample of the corpus on a popular research area: predictive models of renewable energy consumption in the radio access network.
6. Abbreviations lists and expands the acronyms used in this paper.

We complement this paper with an online repository (https://github.com/humaira-salam/PowerMeasurementAndModelingRawData, accessed: 23 December 2022), that carries our raw data.

2. Survey Results: A Digest of Challenges, Approaches and Developments

2.1. A Taxonomy of the Problem Space

As our parsing of the literature proceeded (our method is described in [1]), we observed that the scope of this survey is relatively narrow and the problems in our set are not fully independent of one another. Rather, the problems diverge from one another only as ***aspects*** (we could also say that they are ***derivatives***) ***of the core challenge of modeling the power consumption of virtualized entities***. Each RU (research unit) (The RU, or unit of research, is "a publication (excluding surveys) in conference proceedings and journals" that "ha[s] three common manifest properties", i.e., problem(s), approach(es) and development(s) [1]) is rooted in this core challenge, but the derivative problems (our *Problem categories* (P-categories) and their members) addressed differ from one RU to another. Figure 1 is an illustration of a simple organization of the challenges which have been tackled in the literature and shows their frequency of presence in RUs. The organization gives prominence to how challenges have been perceived:

1. One group regards the concern with obtaining an understanding of the dependency of power consumption on some genre of artifacts. Categories P1,2 and P9–11 are in this group;
2. The other group regards the concern with how to predict power consumption. Categories P3–8 and P12 are in this group.

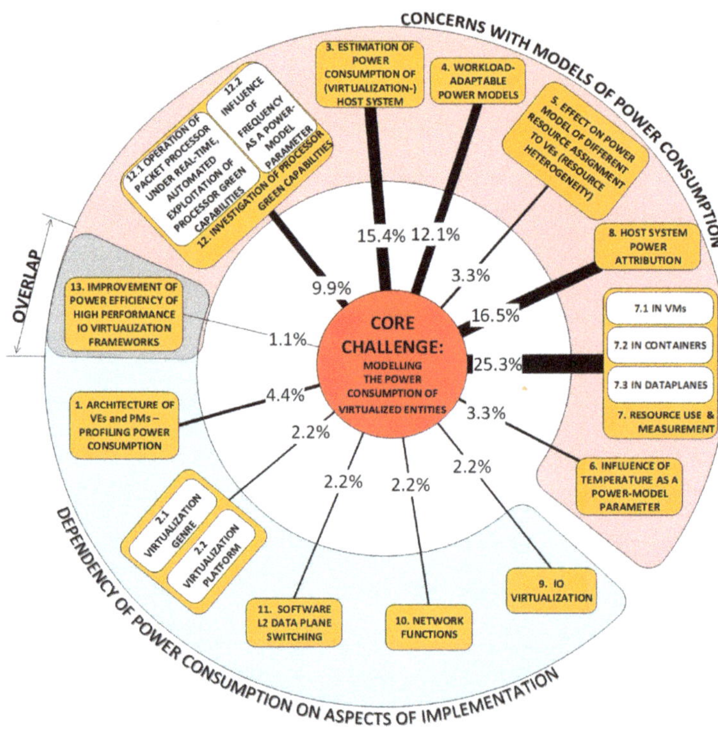

Figure 1. The core challenge and its derivatives; research interest R_{P_k} is shown in line thickness and as percentage.

We now proceed to describe the categories in more detail. Each description is preceded by a list of references to works that tackle a challenge in the problem category.

2.1.1. Problem Category P1: Host System Hardware Architecture Perspective: Dependency of VE Power Consumption on Host System Architecture [48–51]

Challenges in problem category P1 address the impact of specific architectural attributes of the host system on the power consumption of VEs. They relate to changes in power consumption (the behavior) as major attributes of architecture and system-level designs are adjusted, inserted or removed. Insertions and removals are coarse configurative actions such as enabling or disabling; adjustments consist of progressive modifications such as adding increments of a resource. Examples of attributes which have been tackled include multiple processor cores, processor frequency scaling, Non-Uniform Memory Access (NUMA) and hardware threads (e.g., Intel Hyper-Threading). For network functions, the importance of knowledge about the power efficiency of NUMA and multiple-core architectures has the added relevance of these architectures' relationship to ***determinism*** [37]. We dwell further on the underlying premise of ***hard partitioning*** in our consideration of the impact of the high-performance data plane on power efficiency (see Section 3.3.2).

Research that investigates the dependency of power consumption on architecture is ***exploratory, charting*** work. It attempts to provide a framework for detailed modeling through the discovery of broad relationships. Problems in this category arise with developments in architecture and system-level design. For example, while [49] is a comparatively old work that tackles architecture, [50] is newer and finds scope for research in system software's exploitation of NUMA. One recent, highly significant scope is that of the use of ***domain-specific architectures (DSAs)***. Researchers are exploring specialized hardware in the quest for the improvement of the energy–performance–cost ratio, and will investigate energy efficiency in the process of their research. As domain-specific cores are mixed with general-purpose cores, many architectures will be investigated from each of the three pinnings: energy consumption, performance and cost. A particularly relevant set of DSAs regards real-time packet processing by computer systems hosting NFs at intermediate nodes (INs) at the network edge. Concern lies with expediting the common tasks, such as sending/receiving packets and processing headers. SR-IOV is a good example (see, for example [52], and its inclusion in [48]), but software-only solutions, such as poll-mode drivers, may also help to cut through the many middlemen characteristics of general-purpose operating systems [48,53]. Introduced to serve the perspective of performance, it is now necessary to understand their impact on power efficiency. Therefrom, it is necessary to understand how to ***control*** their power consumption. We suggest that "***profiling***", the term chosen in [49,54], is a helpful descriptor of this kind of research. Just as a profile produces an external boundary within which to fill detail, so does this kind of research provide a framework through which modeling work is facilitated and within which modeling work provides details of power consumption relationships.

2.1.2. P2: Impact of Alternative Virtualization Genres and Virtualization Platforms on VE Power Consumption [55,56]

Here, a behavior that consumes power is investigated across different ***implementations*** of a ***system concept***. We have observed investigation of two different system concepts: (a) virtualization genres and (b) virtualization platforms. There are three members of the virtualization genre group: containers, para-virtualization and hardware-assisted virtualization. In the virtualization platform group, examples include Xen, Hyper-V, Kernel Virtual Machine (KVM), Docker and Linux Containers (LXC). Research questions typical of category P2 seek to control the scope of experimentation through exercise of specific resources, e.g., per-host networking using emulated switches (software switches) [55,56], processor-bound and memory-intensive processing [54].

We consider genres and platforms as sub-categories of the same overarching problem category. Namely, this is system-level exploration that attempts to establish generalizations about an uncharted space. Like problems in category P1, new problems in this category arise with fresh alternative virtualization genres and platforms. However, here the scope of investigation is broader than with works classified under P1. Unlike P1, where specific architectural aspects (e.g., NUMA, hardware threads) are explored, the perspective taken here is a concern with the impact of the choice of implementation of a system.

2.1.3. P3: Estimation of Power Consumption of (Virtualization-) Host System [57–69]

Measurement of a single server's power consumption through the use of an external power meter is a trivial task. However, at the scale of cloud datacenters, it is a logistical burden. In addition, travel to the datacenter's site may be burdensome. Furthermore, service availability would be reduced by the process of attaching a physical power meter to the hardware in the virtualization platform, e.g., between the server's power inlet and the outlet in the racking cabinet's power distribution unit (PDU) (naturally, availability would only be affected in cases that do not integrate (management and) measurement facilities within the PDU).

The alternative is the deployment of software power meters. In the scope of this survey, the cases we consider are meters that attempt to predict host power consumption on ***the basis of activity in the VE***. This challenge is tackled, for example, in [58–60,62,67,70]. These works then proceed to tackle the problem of attribution of system power to the guest VEs. Indeed, inclusion within the scope of both challenges (modeling power consumption of VEs and that of the host system power) seems to significantly enhance the usefulness of such research, with relatively less effort.

Host power consumption may be predicted in terms of VE resource utilization, or in terms of ***a simple characterization of the VEs' workload***. The use of simple workload characterization as a predictor requires knowledge of workload parameters such as the number of processes, number of threads, web interactions per second and network interface utilization. Enokido's and Takizawa's work [67,68,71] is noteworthy in its consistency in modeling in these terms but other variants of this approach have been found: (a) web interactions per second [65] and (b) number of VMs running processor- and/or network-intensive workloads [66].

To contrast: works such as [57,72–76] are not included within this category, notwithstanding their development of models for the prediction of host system power consumption. In these works, host system models were developed as part of the scope of the challenge of modeling virtualized entities. Therefrom, the challenge of system power attribution (problem category P8) was tackled to proceed to guests' power models.

2.1.4. P4: Dependency of Power Model on Workload [56,62,67–69,71,75–79]

This category regards the perceived dependency of a VE's power consumption model on the tasks it is processing. While it is intuitive to expect power consumption to depend on the workload, it seems far less intuitive to expect the model to depend on the workload. If this dependency is detected, the problem of model formation must undertake this aspect of investigation. Two different, major approaches towards achieving ***adaptability*** of the model to the workload have been observed:

1. ***Adaptation during run-time:*** the selected mode of instrumentation may not be suited to a generalizable, closed-form relationship between inputs and power consumption. In this case, model parameters must be re-trained online. This approach is therefore of the operating-time, or run-time, kind;
2. ***Off-line adaptation:*** a larger set of inputs may need to be identified to comprehensively characterize the variation of power consumption with the workload. This approach is therefore of the design-time, or off-line, kind.

We conclude this part with a note about two descriptors of the workload: homogeneous and specific. The term "homogeneous" is encountered in the literature to refer to the case where host system deployments within scope are subjected to a single workload. The term seems to originate in warehouse scale computing (WSC). Conclusions drawn from this kind of workload have drawn criticism as the results, while significant by virtue of the mass of WSC, are not generalizable. The other term—"specific"—identifies a single application; for example, a member of the Standard Performance Evaluation Corporation (SPEC) CPU2006 suite [80]. This term is used to indicate that models tested under such a workload are application-dependent and are valid only within a limited range of this dimension of variability (i.e., the "workload" dimension, see the treatment of the seven dimensions of variability).

2.1.5. P5: Dependency of VE's Power Consumption and Power Model on VE's Resource Configuration (Heterogeneity) [62,73,78]

This category regards the perceived dependency of a VE's power consumption and/or the dependency of its power consumption model on (a) the physical host's resource configuration and (b) the individual VE's resource allocation. Research here is concerned with two cases of very practical problems: the impact on power consumption of (a) the differences between hosting machines/containers and (b) the differences between virtual machines. We have observed that occurrences of research that undertake this challenge tackle it as an adjoint to another focus, not as the research's primary objective.

1. ***Physical host configuration:*** Host machines in a cloud datacenter may be expected to come in a limited variety of types, principally differing in resource capacities such as the number of processor packages per server, cores per processor package, amount of RAM per server, spread in storage device sizes, etc. Processor power consumption is notably variable, even within a single family of processors. Indeed, specialization in optimized power consumption within a family of processors is a part of the study carried out in [34] within the context of an edge cluster for use in NFV. As a VE migrates from one processor within a family to a processor of a different specialization, its power model will change.
2. ***Individual VE's resource allocation:*** The power consumed by a VE varies with the allocation of resources to (i.e., in use by) a VE, which can be dynamically varied. The number of virtual cores assigned to a VE is a notable example, see, e.g., [57,61,72–74]. Moreover, VEs are commonly offered in sizes, e.g., small, medium and large, where allocation varies within all the major resource categories, demanding prediction of power consumption matched to the size of the purchased VE.

2.1.6. P6: Impact of Temperature and/or Frequency on Models That Predict VEs' Power Consumption [60,81,82]

This category regards the challenge of the inclusion of processor package temperature in models of power consumption. Works that tackle this challenge are concerned with detailed models of power consumption. Here, the interest lies in obtaining models that incorporate dependence on hyper-parametric attributes such as temperature.

2.1.7. P7: Loading the VE's Resources and Measuring Resource Use [56–59,61,62,65,68,69,71,73,75–79,81–87]

This category regards the use of computing resources and the measurement of such use by VEs. Interest stems from the role of resources ***as predictors in modeling***. The researcher is firstly concerned with ***loading*** (i.e., effecting the use of) resources. What means ***within the operating context of the VE*** can be used to load a resource? Should it be loaded in isolation (using synthetic loads) or should it be loaded using representative (realistic) workloads? Once these problems have been addressed, the concern with the ***measurement*** of resource use arises. The problem here consists of identifying the means that quantify resource use made by the loading.

2.1.8. P8: Attribution of Host System Power Consumption to Individual VEs [57–59,61,62,73,75–77,79,81,82,84,86]

The attribution of host system power to individual VEs is a fundamental problem in proceeding from the directly measurable (host system power consumption) to the indirectly measurable (individual VEs' power consumption). Direct measurement of host system power is possible (e.g., at the wall outlet, or through voltage rail in-line resistors), and such empirical evidence can be used as a ground truth and compared with power consumption inferred through modeling. How, then (and herein lies the problem), can this consumption be attributed to the host's individual guests (the VEs)?

Within the host system, power consumption may be divided into ***idle (static), active (dynamic)*** and ***overhead***.

1. Idle (static) power consumption:
    a. Power consumption while idle is not attributable to any VE at all, as this consumption arises out of the electronic behavior of semiconductor material, not of computation, communication or storage;

   b. Nonetheless, this power consumption must be accounted for and different approaches have been followed. For example, the physical machine's idle power is attributed to individual VEs in fractions equal to the ratio of each VE's virtual CPUs (vCPUs) count to the total complement of vCPUs active on the physical machine [57,72–74].

2. Active (dynamic) power consumption:
   a. The active component can be linked to a particular VE;
   b. This includes active power consumption in peripherals, e.g., network interface cards/adapters (NICs) and mass storage devices.
3. Overhead, e.g.:
   a. Operation of heat dissipating units (fans) to prevent thermal runaway;
   b. Losses in the power supply.

A "top" (host)—"down" (guest) approach to attribution has been observed.

1. Decide on what host system power consumption is within the scope of the study and how to divide it. The problem of attribution of the above three causes may be summarized as follows:
   a. Is idle power attributed to the VEs or is it attributed to the host/a privileged guest?
   b. Is consumption by peripherals within the scope? How will this be attributed?
   c. Are overheads modeled or is correlation with other sources of power consumption going to account for them?
2. Select a set of performance metrics that are correlated to a VE's power consumption.
3. Select a model that maps a VE's performance metrics to its power consumption.
4. This fourth consideration is tackled only by those researchers who investigate the ***adaptability*** of the attribution obtained through steps 1, 2 and 3. Does the obtained attribution adapt well to concurrent, co-hosted VEs? That is: if concurrent, co-hosted VEs were to be investigated, would the division, metrics and model still result in accurate prediction?

2.1.9. P9: Implementation of Virtual I/O; P10: Implementation of Network Functions; P11: Implementation of Software Layer 2 (L2) Data Plane Switching [48,51,53]

These three categories are introduced together, since elements from the respective categories are commonly implemented ***as a set*** for the purpose of the realization of the virtualization of network functions. Here, researchers seek comparative statements and/or broad correlations (e.g., independent, positive, negative) between the workload (often in terms of packet rate and size) and power consumption, across implementations of the same type. As was observed for categories P1 and P2, researchers seek a profile of the power characteristics of implementations. It may be helpful to repeat that by "profile", we understand that characteristics sought here are not of the detailed form of closed-form expressions. Examples of elements from the respective categories are:

1. Virtual I/O (P10): virtio [88] and DPDK poll-mode drivers (PMDs) [89];
2. Network functions (P11): Bro (now Zeek) [90] and Snort [91];
3. Software layer 2 data plane switching (P12): Open Virtual Switch (OvS) [92] and VALE [93].

Problems in each of these three categories merit separate classification as they have been tackled separately in the literature. For example, in [53], a number of components are investigated: three different implementations of software virtual switch (P12), two different I/O virtualization devices and two different implementations of the same network function (intrusion detection system (IDS)). In [51], power consumption by packet transmission under DPDK is investigated under the condition of enforcement of (a) the network adapter's affinity to NUMA nodes and (b) DPDK process pinning to processor cores.

2.1.10. P12: Investigation of Processor Green Capabilities [50,60,67,81–83,85,87,94]

Works in this category investigate the low-power idle (LPI) and adaptive rate (AR) operation of a processor as a means of reducing power consumption. The challenge is broad enough to permit a sub-categorization into (a) those works that investigate the influence of frequency as a power-model parameter [81,82] and (b) other works that address improved, real-time governance of LPI and/or AR [83,87,94] to minimize the power consumed to process a load.

2.1.11. P13: Improvement of Power Efficiency of High-Performance IO Virtualization Frameworks [87]

A separate classification was set up for [87] as this work represents an evolution of those classified under P9. This work extends beyond profiling and suggests use of low-power instructions as the means to balance performance and power efficiency.

2.2. A Taxonomy of Approaches

Figure 2 illustrates the taxonomy we use to structure the approaches detected in research work. Line thickness and percentage values represent the ***utility*** of the specific approach. Utility is best understood within the context of all the observed triads in a literature corpus. For any specific approach, this may be used to solve a variety of problems and its application may result in a variety of developments. One may therefore think of the approach as a ***nexus***, or a point of confluence through which many researchers pass as they attempt to solve problems. Thereafter, researchers diverge radially outward from this point of confluence towards some achievement (some development). A suggestive image is to think of the approach as the center of a star, but spokes converge onto it from problems and diverge away from it onto developments. When the count of these triads (each composed of two radial lines, or dyads) is divided by the sum of all such counts for all approaches, we obtain a metric: a normalized quantity obtained within the context of all approaches. See our method [1] for the formal statement of ***utility***.

We now proceed to describe the categories within the context of the taxonomy. Each approach-category's description is preceded by a list of references to works, each one of which uses a component within that category's set.

2.2.1. Analytical Foundations

This group of categories regards the theory and hypotheses that comprise the essential abstractions at the basis of scientific research.

A1: Power attribution principle [48,53,57–59,61,62,69,73,75–77,79,81,82,84,86]: When ***host system*** power is measured, whether at the wall outlet or at one or more of the power supply's output lines, there is the problem of attributing the measurement to the logical divisions (VEs) of the host computer system. The attribution of system power to modeled entities starts with a decision on which power consumption is within scope (see Section 2.1.8). Next, power in scope is attributed to (burdened on) one or more entity. For example: will idle power consumption be attributed to the host system or will it be attributed to the VEs?

A2: Modeling bias [57–64,72,75–77,79,81,83–85,87,94]: Researchers approach the problem of developing a model under some bias which conditions their final outcome. This bias is manifest in researchers' selection of a particular type of regression to apply to their data. We note that this same observation is carried in [95]. Here, our purpose is solely to draw attention to what we have observed as researchers' **modus operandi** without analyzing their choice of approach.

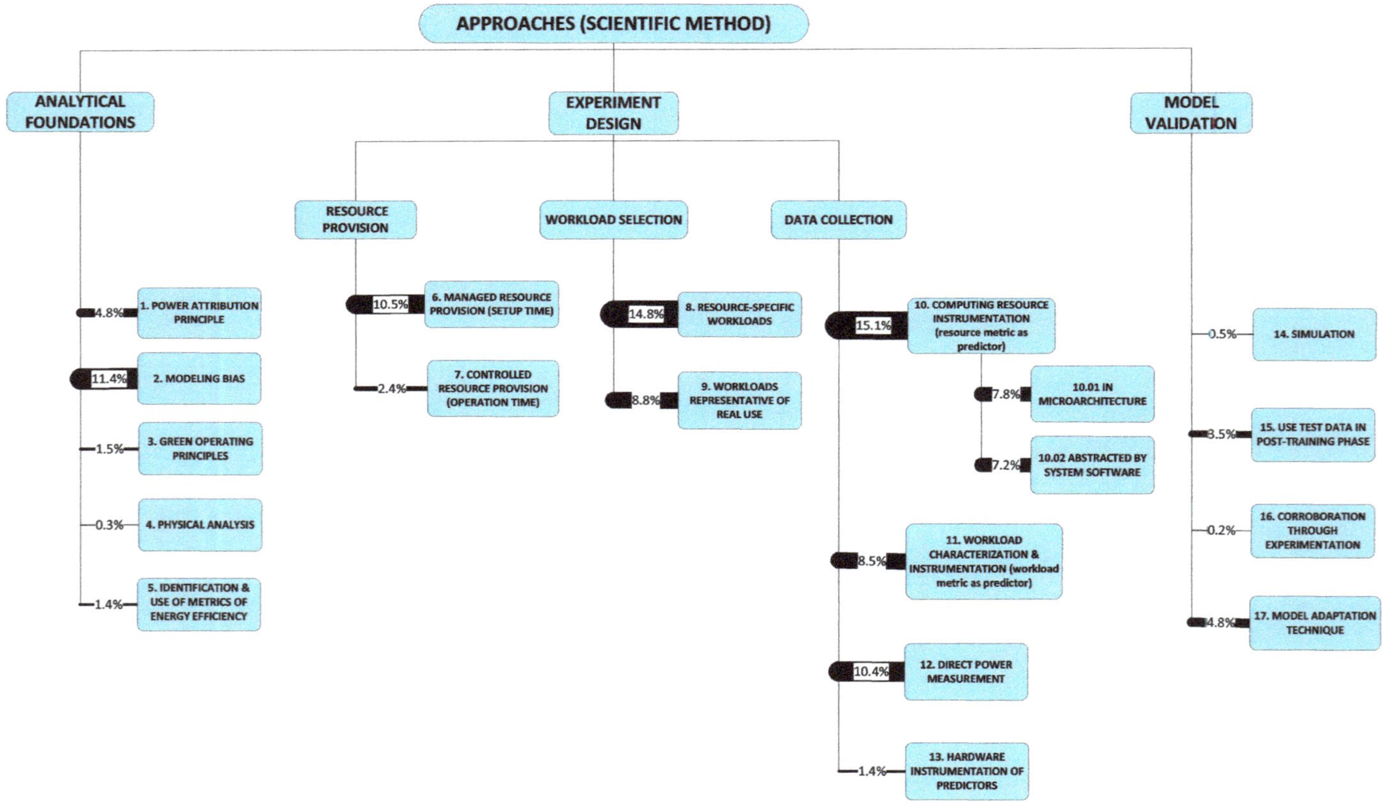

Figure 2. A taxonomy of approaches; approach utility U_{A_k} is shown in line thickness and as percentage.

A3: Green operating principles [83,87,94]: Works in this category weave radical approaches to power efficiency into their developments. For example, instead of conventional scheduling, ***run-to-completion*** [96] is exploited to obtain dedicated (or, at least, very sparsely shared) resources for the processing of packets. This approach is further nuanced by the real-time control of adaptive rates and sleep depth. In one particular case [87], the novel concept of a low-power instruction instead of transitions to/from low-power idle (sleep) states is used.

A4: Physical analysis [60]: This category regards approaches rooted in the physical properties of (semiconductor) material in the consumption of energy. Only one work [60] was found fitting this category. However, another two that used this approach to study the power consumption of physical entities (and therefore was outside the scope of this study, which is concerned with VEs) were found and they are described next to illustrate the approach better. In [70], a study implicitly applies Dennard's law in the process of obtaining weights that scale a processor sub-unit's contribution to power consumption. In [97], the physical cause of power consumption in metal-oxide-semiconductor (MOS) material is examined and used as the basis for modeling equations.

A5: Identification and use of metrics of energy efficiency [49,50,65,85]: The relationship between system architecture and power consumption can be investigated through the identification of the relevant metrics of energy efficiency. For example, an easily recognizable metric, albeit rather broad in possible interpretations, is the J/b (joule/bit). The use of such metrics encourages joint consideration of function and power consumption.

2.2.2. Experiment Design

The practical, hands-on aspects of the empirical process are the product of a (probably cyclical) design phase, concerned with a number of stewarding activities pertinent to test subjects and ancillary objects in the testing scenario, instrumentation, inputs and outputs. The activities include selection, configuration, interconnection, initiation, observation and termination. We have identified several examples of such activities within our research scope and grouped them under ***resource provision*** (categories A6, A7), ***workload selection*** (A8, A9) and ***data collection*** (A10–A13). We describe these categories next. Admittedly, the activities referred to (i.e., selection, configuration, etc.) have broad meanings; therefore, in the course of describing the categories, references to the activities are emphasized by bold, italicized text.

A6: Managed resource provision [48–51,53,55,56,60,66–68,73,81–83] ***(selection, configuration)*:** This concerns the provision of resource capacity either to a specific VE (the guest system) or to the physical entity (the host system) hosting the VEs. Within the empirical process, the techniques in this category provide the means to observe the effect on the power consumption of managed changes in resource provision. Examples include the (manual) pre-***configuration*** of:

- The frequency of operation of processor cores [48,50,81];
- Core affinity [51,55] and hardware-thread affinity [68];
- The network interface data rate capacity capping [66].

These techniques are executed as part of the process of ***selection*** of the operating parameters, i.e., setting up experimentation, ***before*** operations start. This qualification is necessary to distinguish from such approaches as may change the operating, run-time context.

A7: Controlled resource provision [78,83,85,87] ***(configuration)*:** Provision of resources may change during the running of an experiment (rather than before it starts). Approaches in this category include the automated adjustment of:

- Processor frequency (also known as performance state, or P-state) [78,83,85,87];
- The depth of processor sleep (also known as low-power idle state, or C-state) [83];
- The number of hardware threads [78];
- The time spent running a low-power instruction [87].

These techniques are approaches to solving the problem of full-throttle operation. Without a guided operation of adjustments such as those listed above, operation of the processor may quite reasonably be likened to a multi-assembly-line manufacturing plant that operates line machinery whether there are goods to produce or not.

A8: Resource-specific workload, A9: Representative workloads *(selection, configuration, interconnection)*: These two categories regard the ***workload selection*** stage, within experiment design in the scientific method. The workload comprises the ***inputs*** referred to earlier; inputs must be ***interconnected*** to the system under test, and this is often not a trivial task. In our thematic analysis in Section 3, we identify ***workload type*** as one of the seven dimensions of the variability of power models. The influence of workload type on the model obtained is evident in the attention paid by researchers to their selection of workload type. We can distinguish two broad categories of type.

Resource-specific workloads (A8) [48–51,53,55,56,59–61,63,64,66,67,69,71,75–77,84,86] are applied to investigate the impact of the utilization of specific resources on power consumption. Such ***synthetic*** workloads are applied ***(interconnected)*** to a machine (whether virtual or physical) to reduce (as much as possible) the scope of power-consuming resources to a targeted set. Resource-specific workloads are most commonly used in exploratory work, to gain an understanding of the behavior of a resource's power consumption. We refer to this approach as resource-specific workloads, synthetic workloads or resource isolation.

Representative workloads (A9) [48,56–58,60,62,65,68,69,73,75,78,79,81–85,87] may be used as complementary with, or alternative to, resource-specific (synthetic) workloads. A notable complementary use is made in the testing (post-training) phase of model development, when representative workloads are used to validate a model (obtained using synthetic workloads). They may also be used in an entirely alternative approach to synthetic loading, to support development of application-agnostic models. Representative workloads lead to training data that incorporates variation in utilization of more than one resource at a time; hence providing at least limited application agnosticism.

The next four categories (A10–A13) regard the ***data collection*** stage, within experiment design in the scientific method. Categories A8 and A9 regard the ***selection*** of workload type. The approaches described here regard measuring ***how much*** of a resource is being used, or workload is being applied, and how long to apply the workload to obtain statistically valid results ***(initiation and termination)***. Categories A10, A11 and A13 regard instrumentation ***(observation)*** of those variables considered (by the researchers concerned) to be reliable predictors of power consumption.

Resource instrumentation in microarchitecture and system software (A10) [48,49,56,58,59,61–64,69,73,75–79,81–87] includes approaches that measure resource use. These measurements are then used to predict power consumption. We make a somewhat weak distinction, for reasons we shall refer to, between instrumentation of microarchitecture and instrumentation by system software. The former regards parts of the processor interface that address the processor's infrastructure for monitoring, event counters (e.g., instructions retired, last-level-cache (LLC) misses and translation lookaside buffer misses) and, more recently, power counters (e.g., Intel's Running Average Power Limit (RAPL)) [98]. On the other hand, system software's instrumentation is carried out through intermediary system software and includes, most notably, processor utilization. We have seen references to these two categories as "hw counters" and "sw counters", respectively [99]. The distinction is weak since system software is increasingly exposing microarchitecture instrumentation (consider, for example, Linux's ***perf*** tool). This reduces the need to directly access hardware registers and blurs the separation between what is abstracted and what is concrete, raw, hardware data. Nonetheless, through our understanding of the data used, we have been able to separate the approaches into two sub-categories.

Category A11 regards the use of a simple characterization of workload as a predictor of power consumption [48,49,53,64,65,67,68,71,78,83,85,87]. This is notably different from approaches in category A10, which are concerned with resource use as a predictor of power consumption. Other examples (apart from those given earlier) of workload metrics as inputs are the "number of processes" [71] (from the same process image), "transmission rate" [68] and millions of instructions per second (MIPS) [64].

Characterizations may need to be sharper. For example, since packet network traffic arrival is known to often have the properties of a Batch Markov Arrival (stochastic) Process (BMAP), this is an operating constraint (***selection, configuration*** and ***interconnection*** are all ingredient activities here) adopted in the approach of several works studying power efficiency of network functions [83,87,94,100].

Category A12 refers to the measurement of (host) power consumption [48–50,53,55–59,61–64,67,68,71,73,76,79,84] ***(observation)*** which is most usually measured at the wall outlet or at the power supply inlet. More granular approaches are desirable, and indeed we do find cases [62] that attempt measurement at the power supply output. The principal drawback of such granular techniques is not (principally, at least) construction of intermediary hardware (e.g., riser boards or line resistors) but the difficulty in attributing power drawn through any single dc voltage output (or group thereof) to particular consumers. With the advent of RAPL and certain guarantees on its accuracy, the need for direct power measurement has been, at least partially, avoided.

Category A13 regards the use of hardware sensors to obtain inputs and/or parameters for the power model [60,81,82,87]. These include:

1. Voltage sensors (processor supply voltage);
2. Temperature sensors (processor package and memory temperatures);
3. Fan speed sensors (processor and chassis fans);
4. Wall-clock time measurement.

Some of these variables are used in models that predict power consumption while accounting for the effect of the drift of temperature and automated supply voltage adjustment (in dynamic voltage and frequency scaling—DVFS).

2.2.3. Model Validation

Model validation is a multi-faceted endeavor and this is reflected in the approaches we have detected. The approaches range across the candidates that would typically be considered: simulation (A14) [87], use of test data (A15) [57,62–64,77,85] and corroboration through experimentation (A16) [83,87]. We skip elaborating on these categories as they are either self-evident (A15) or because they are too rarely used to permit general commentary. However, to these three categories we add a fourth (A17), namely, the model adaptation technique, which we describe below and explain why it fits within this branch of the taxonomy.

Model adaptation technique (A17) [57–59,61,62,67–69,71,73,75,76,78,79,81,82,84,86]: This refers to the approach(es) taken (if any) to develop an adaptable model or modeling system. Here, ***adaptability*** refers to the fitness for use which the model exhibits under changes in one or more of the seven dimensions of variability that will be defined in detail in Section 3.1.1 below. Model adaptability is essential for practical virtualization, where changes in, for example, the number of co-hosted, concurrent VEs, or in workload type, are commonplace. Here, we list the major approaches taken towards producing adaptable models. Since these approaches emerge in the context of validating a model's accuracy under some limited range of the seven-dimensional space of operating conditions, we classify this category of approaches as an aspect of model validation.

1. ***Adaptation to change in the number of co-hosted, concurrent VEs*** is widely achieved through the ***time-division multiplexing*** of event counters [59,62,69,79,86], RAPL counters and CPU utilization [61,73]. This approach enables the use of such metrics as predictors of dynamic (active) power consumption, by apportioning counts to VEs in accordance with the time during which the VEs were active.
2. ***Adaptation to uncorrelated causes of power consumption*** can be achieved through ***additional predictors*** [82] to follow causes of power consumption that do not correlate well with counters within the current set. This case reflects itself as poor accuracy in

predicted power consumption. Although counter-based models are reported to fit a variety of processor- and memory-intensive workloads well, it may be necessary to account for unanticipated activity through the approach of adding previously unused counters.

3. ***It may not be possible to fit a single model with parameters known* a priori**, to the whole range of inputs within the scope of study, notwithstanding the diversity of predictors employed in this pursuit. The following adaptive techniques have been found in the literature.
    a. ***Dataset partitioning***, where the dataset is of the form { {predictors}, response } and is used in [77] to match the best model out of a set of models to the current, actual operation. An individual model in the set is associated with a single node in a decision tree and the node is selected according to features pertinent to the current, actual operation. A simpler, but conceptually similar, approach is taken in [58]. A number of models are devised and model-selection features are limited to the number of active VMs and a coarse grading of CPU utilization.
    b. ***Modeling on demand*** is the term we use to succinctly refer to the fourth adaptation class of techniques:
        i. One early example of this approach is found in [75], where the dependence of the model on the workload has been addressed through ***online training***, whenever prediction accuracy of the extant models falls out of a range of tolerance. The rationale adopted is that if model adaptation to such an "unseen" case is limited to parametric tuning, then a modeling system might be able to construct a model while VEs are in operation;
        ii. A broader perspective is found in [78]. An automated system for profiling containerized applications is described and demonstrated. Containerized applications are profiled from three perspectives: computing resources consumed, energy consumed and performance. In this case, the rationale is that energy consumption can be optimized by the determination of a frequency-and-hardware-threads host configuration that meets performance requirements. Thus, starting from central functional requirements (performance requirements), operating conditions are determined that minimize energy consumption. This approach is capable of meeting the challenges posed by heterogeneous host hardware and application (workload) diversity, at the cost of analytical modeling. Indeed, characteristic curves can be derived but causes underlying observed behaviors remain unaddressed.

2.3. A taxonomy of Developments

Developments fall cleanly into one of two groups: (a) models of power consumption and (b) observations on dependencies of power consumption. The first group (D1–D10) includes developments that ***predict power consumption*** over a sub-space of the seven-dimensional space of operating conditions. The second group (D11–D18) includes developments that are oriented towards the ***correlation of power consumption*** with aspects of system integration. As the taxonomy is rather broad, we present it in three parts:

1. Figure 3: top-level fork into models and dependencies;
2. Figure 4: the taxonomy of models;
3. Figure 5: the taxonomy of dependencies

D1–D4: We first present four categories of developments that concern ***models of host-system power consumption*** characterized by the condition where ***workload is processed by VEs***:

- D1: linear regression models;
- D2: non-linear regression models;
- D3: machine-learnt models;
- D4: models of local mass storage.

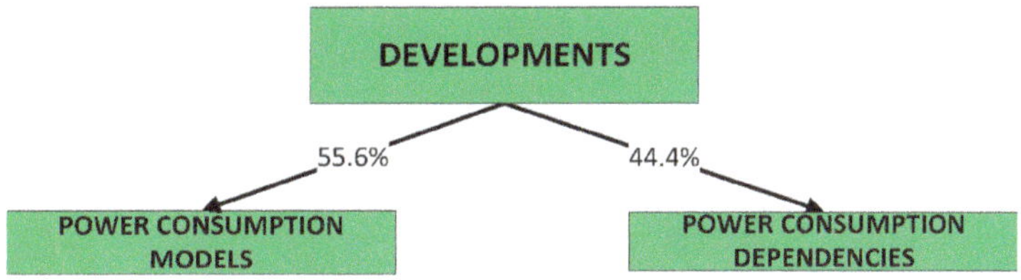

Figure 3. A top-level division of the developments, with frequency of occurrence shown as a percentage.

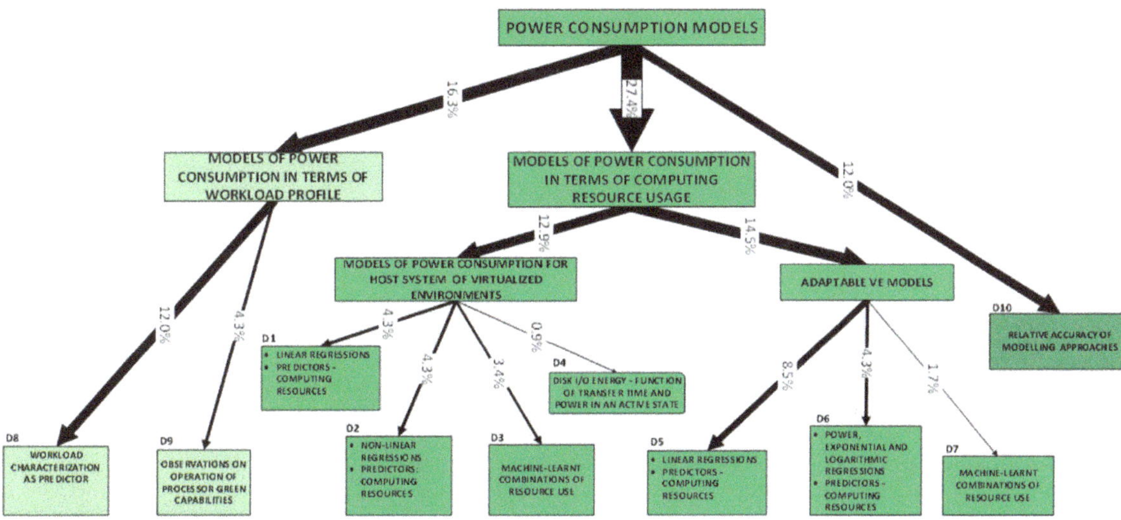

Figure 4. A taxonomy of power models, with frequency of occurrence shown in line thickness and as percentage.

This contrasts with the scope of developments referred to under categories D5, D6 and D7, where ***models of VE power consumption*** are presented.

Models in these categories are interesting from the perspective of analyses of ***sets of hosted VEs*** that seek to identify operating conditions of optimal host power efficiency. As predictors, such analyses use instrumentation that measures resources used by the VEs. Categories D1, D2 and D3 all predict power consumption in terms of resource use but differ in the type of model produced.

- D1 regards models of power consumption through linear combinations of scalar predictors [57–59,69,75]. The scalar predictors are resource usage metrics.
- D2 regards polynomial or simple mathematical powers of resource use (the scalar predictors) [57,60,61,76,81].
- D3 regards models that employ machine learning (e.g., Gaussian Mixture, Support Vector Machine, Neural Networks) [62,77,79].

Category D4 regards the models of power consumed by mass storage local to the host system [86]. These models attempt to predict power consumption in terms of activity metrics such as the total amount of time spent in a known state (in terms of power consumption, e.g., active/idle), the rate of data exchange (MB/s or input/output operations per second) and the mode of operation (sequential/random and read/write). In the context of the approximations observed in the development of these models, their accuracy cannot be fair.

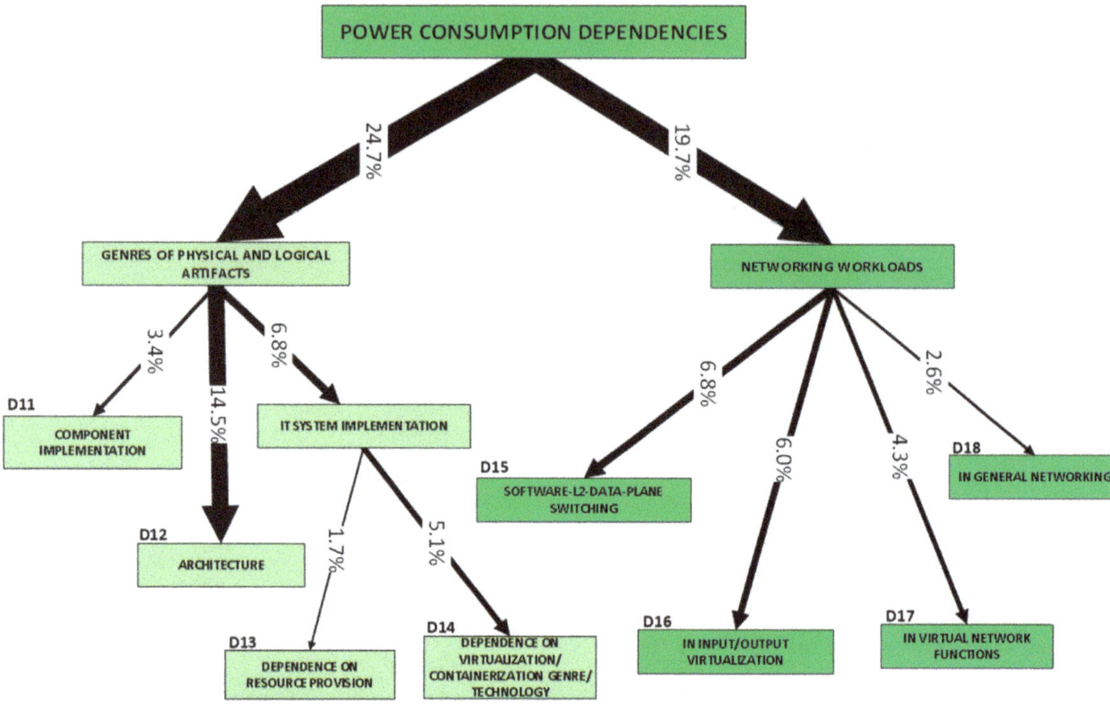

Figure 5. A taxonomy of power dependencies, with frequency of occurrence shown in line thickness and as a percentage.

2.3.1. Models of Power Consumption

Adaptable VE models (D5–D7): Developments within these categories consist of ***adaptable*** models of the virtualized entity's power consumption and have two important characteristics in common:

1. They are adaptable to a variable number of concurrent, co-hosted (active on the same host system) VEs (we refer to the latter scope of variability as the seventh dimension of variability);
2. The predictors are the measured amount of computing resources used by the VEs.

Models can be distinguished by the predictors they use, workloads employed and modeling approach:

1. ***Predictors*** (of VE power consumption) are obtained from system software's instrumentation, e.g., CPU utilization (see approach categories A10–A13), and from microarchitecture instrumentation, e.g., LLC misses (again, see approach categories A10–A13);
2. ***Workloads*** used to obtain the model (this restricts the range of workloads within which the model is valid) may be:
    a. Specific workloads: the most restrictive, as they relate to a particular test load;
    b. Synthetic workloads: less restrictive than specific but limited to the exercise of one resource, typically the CPU;
    c. Combinational workloads: still less restrictive, involving the exercise of a number of resources of the host system (e.g., SPEC CPU benchmarks may be both processor and memory intensive);
    d. Representative workloads (e.g., TPC-W [101]) produce models that are readily associated with use cases.

3. The modeling approach may be:
    a. Linear regression (category D5) [58,59,69,73,75,81,82,84,86];
    b. Power (integer- and non-integer powers), exponential and logarithmic regressions (category D6) [61,66,76];
    c. Machine-learnt combinations of resource use (category D7) [62,79].

Models of power consumption that use workload profile as predictors (D8, D9): Categories D8 and D9 group developments from (two) sets of RUs that predict power consumption of hosts and/or VEs through (the measurement of) some characteristic of the submitted workload. This contrasts with RUs in categories D1–D7, where prediction is obtained through (the measurement of) some computing resource (processor and/or memory and/or I/O). Most developments in this category are obtained through the abstraction of hardware by one or more model parameters that express power consumption under case-specific conditions of operation. Some of these abstractions are identified in the descriptions of these two categories.

Developments in D8 [63–68,71,78,83,85,87,94] use:

1. Processing load (number of processes, millions of instructions per second (MIPS), etc.) pertaining to a specific application, as predictors of ***host system*** power [63–67,71];
2. Packets per second, through an intrusion detection system implemented in a VNF [85];
3. Transcoded frames per second, through a transcoder implemented in a containerized network function (CNF), and inferred images per second, also in a CNF [78];
4. Average network transmission rate, as a predictor of host system power [68];
5. Statistics of a Batch Markov Arrival Process (BMAP) (packet traffic) as a means of the prediction of power consumption by a VNF [83,94].

Hardware is abstracted through the measurement of power consumption at some operating point (a specific operation is being carried out), or a change in power consumption over some operating range. Examples follow:

1. In [65], where energy efficiency of an interactive web service is studied, the operating point is an entire VM running the TPC-W benchmark [101].
2. In [67,71], the operating points are the host's power consumption when (a) idle, (b) one core is active (processing load) and (c) maximum, with all cores active. Furthermore, use is made of the step increment in consumption corresponding to the activation of each additional core. Cores are activated when they are utilized by VEs.
3. In [68], the operating point is the power consumption when co-hosted VEs are transferring a file to a client computer. An ***affine*** relationship between the host's power consumption and its transmission rate (transmissions originate on hosted VEs) is found.
4. In [64], an operating range is used: the increase in power consumption that corresponds to an increase in MIPS on the VEs.
5. In [83], the operating point is the power consumption when a VE running on a single processor core is switching packets at the maximum rate for a given performance state.
6. In [85], fifty-four (54) different features of network traffic are input to an artificial neural network that selects the operating frequency that optimizes power consumption.

With one exception, none of the works in the above list uncovers the hood to peer at the processor's internals (to obtain predictors of power consumption). The exception is [83]; yet even in this case, the performance monitoring counters are not used as direct predictors of power consumption, but to obtain (a) the timing information necessary for a queueing model and (b) the operating state (ACPI (Advanced Configuration and Power Interface [102]) P- and C-states).

Developments in D9 [87] are set within the models branch of the taxonomy. These developments may be considered as useful observations on ***the operation of processors' green capabilities***. Examples of these observations (all from [87]) include:

1. Low-power instructions might be a better candidate than low-power idle to save power under higher link utilization;

2. Operation in full ACPI P-state, operation with low-power instructions on idle detection and operation with low-power idle on idle detection are (a) in ascending order of latency to return to active processing and (b) in descending order of power consumption;
3. A processor utilization threshold below which low packet latency is guaranteed under BMAP traffic arrival is 80%.

By "green capabilities", we refer to a broader range of microarchitectural aspects than the by-now-conventional adaptive rate and low-power idle operation. While these latter two remain at the center of attention, there is also the means of low-power instructions [87] that has been successfully employed to improve power efficiency. Notwithstanding the origin of these observations in modeling work, it may be argued that they might also be classified within the dependencies branch of the taxonomy. We have chosen the models branch, but as further studies add to the body of knowledge on how to operate processor green capabilities, this category's position in the taxonomy might need to be changed.

Relative accuracy of modeling approaches (D10) [60,62,63,77]: Developments presented under category D10 are comparisons of the relative accuracy of alternative modeling methods with respect to conventional polynomial (including linear), power, exponential and logarithmic regression. These developments have been found within works that show models classified under category D9. The purpose is to qualify and quantify improvements of machine-learnt models with respect to conventional regression models.

2.3.2. Dependencies of Power Consumption

This parent node of the taxonomy is divided into two child nodes that are not strictly mutually exclusive. For example, the software data plane is considered in works under D15. Clearly, the software data plane is a logical artifact and might be included within a child node of "physical and logical artifacts", or directly thereunder as a leaf node (i.e., as a category). The choice of separation of D15–D18 and the inclusion under the parent node "networking workloads" was taken for two principal reasons. Firstly, the recurrence of investigation of power consumption's dependency on networking workload merits attention through separation. Secondly, as this survey caters for an audience with an interest in softwarized networking, an emphasis on the power dependency on networking workload seems justified.

Knowledge about the dependency of power consumption on specific hardware (D11); the dependency of power consumption on architecture (D12); the dependency of power consumption on resource provisioning (D13); and the dependency of power consumption on virtualization genre and technology (D14): Categories D11–D14 are grouped into a set of works that obtain the sense of the correlation (positive/negative/none) between power consumption and some genre of artifacts:

1. Specific hardware types (D11);
2. Computer architecture (D12);
3. Resources provided (D13);
4. Virtualization genre and technology (D14).

Category D12 groups works that relate to observations on the impact of architectural features on power consumption [49,50,67,71,76]. While these observations are useful, they are generally too coarse to be directly applicable to real-time power control. Their use emerges from guidance which they provide in the development of power models. For example, it was observed that when the number of threads that fully occupy a core's time (active 100%) exceeds the number of logical cores in the system, the energy efficiency (measured, in this case, in hash/J, a computational metric of energy efficiency) decreases [50]. Evidently, this is good guidance; equally evidently, it is not a directly applicable development.

Categories D11, D13 and D14 relate closely to (various aspects of) implementation, and as such are of particular interest to the ***system integrator***. Data of high quality from these developments inform and guide the tasks of gathering components into systems that meet the non-functional requirements obtained from concern with energy and power efficiency.

Category D11 gathers observations about the dependency of power consumption on specific processor hardware [49,69,76]. We have observed that these developments are gathered as a by-product of the process of research; they are rather incidental. Like any implementation, their usefulness is limited to the lifetime of the concerned device(s).

Category D13 [73] includes developments that regard the specific resource configuration of:

1. The instantiating host, i.e., the relationship between a VE's power consumption and the specific resources of its host hardware specifics such as the number of cores and amount of memory carried by the host instance;
2. The guest VE, i.e., the variation of a VE's power consumption with its resource assignment, on the same host.

Category D14 gathers observations about the dependency of power consumption on instances of virtualization genre and technology [55,56]. Developments in this category are less incidental than those in category D2 and are obtained with the focused intention of tackling challenges relating the power consumption of implementations. These developments relate to a less diversified group of implementations. For example, there are fewer virtualization platforms than processors to choose from. Direct use of these developments is mostly limited to the specific implementations concerned; however, some generalizable conclusions exist. For instance, it was observed that both hardware-assisted virtualization and paravirtualization are less efficient (in the specific empirical setup) than non-virtualized operation in the use of processor caches [56]. This empirical evidence favors the hypothesis that cache hit ratios suffer due to the greater thread rotation in virtualized and containerized environments.

Knowledge about dependency on power consumption while processing a networking workload (D15–D18): Development categories D15, D16, D17 and D18 are grouped here as their central characterization is knowledge about the behavior of power consumption by VEs while processing a networking workload. Categories D15, D16 and D17 reflect the challenges described in P11, P9 and P10, respectively.

Category D15 regards contributions to knowledge about the power consumption of software data planes [48,53]; D16 regards virtualization of network I/O [51,53], and D17 is about network functions [48,53,85]. Thus, in [51], the power efficiency of DPDK PMDs is demonstrated with respect to Netmap drivers, for packet transmission. This development is balanced by [48], where the power efficiency of transmission through a DPDK-enhanced Open vSwitch is shown to be worse (@500-byte Maximum Transmission Unit (MTU)) than that of the unenhanced Open vSwitch. In each of these categories, efforts are made to allocate burden through the isolation of power consumption and the attribution to the sub-system (data plane/virtualized IO/network function) under study.

Category D18's developments differ from those of D15–D17 because they cut across these categories' sub-system boundaries [56,65]. For example, in [65], the energy efficiency of web transactions executed on co-located VMs is found to be highest in the operating condition of processor-core over-subscription (more VMs than cores). In [56], it is shown that power consumed by packet delivery to a VM through a software packet switch is much higher than that required for delivery in a non-virtualized environment. In both these papers, the object of interest incorporates the virtualization of network I/O ***and*** the data plane. In [65], the object of interest encompasses the network function: a web service and accompanying application and database components.

All four categories are of keen interest to the system integrator. The emphasis is on the components in the integrator's set of building blocks, specifically on the behavior of power consumption of various implementations, and types thereof (in the scope of the categories).

2.4. P–A Dyads (Problem/Challenge-Approaches) Graphics

Over the following pages, we present a comprehensive set of graphics (Figures 6–17) that illustrate the approaches detected to tackle specific challenges (reference numbers for pertinent RUs appear in the shape tags). For example, consider Figure 7. The graphic shows the component approaches applied to discover "the impact of specific architectural attributes of the host system on power consumption of VEs" (see problem category P1).

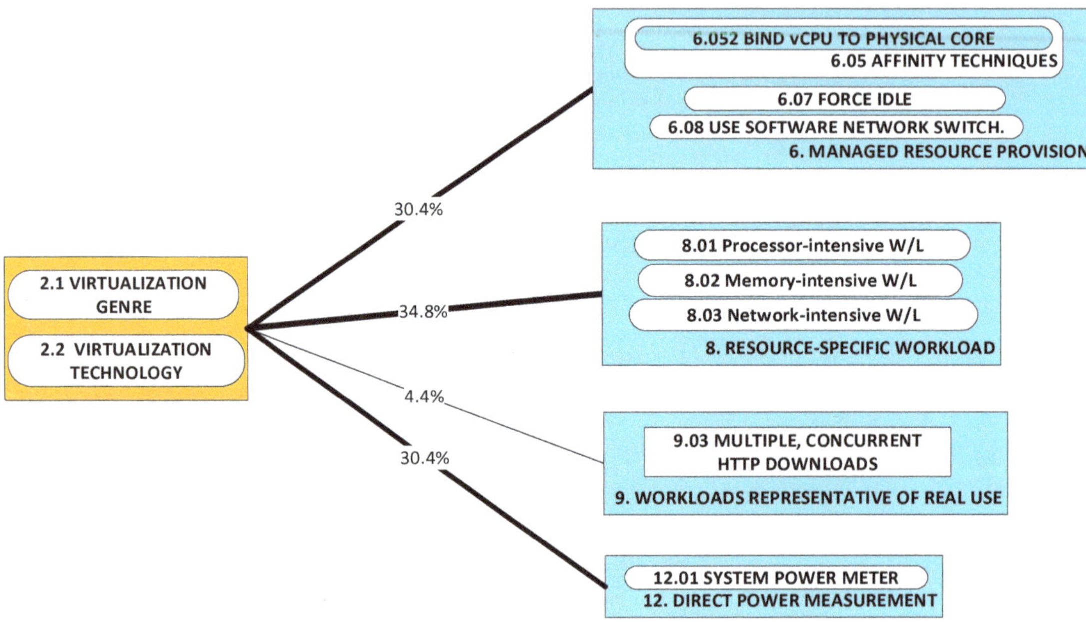

Figure 6. Approaches to solving challenges in category P2; utility metric U_{A_k} is shown in line thickness and as a percentage. RUs in the respective categories are the following: P2: [55,56]; A6: [55,56]; A8: [55,56]; A9: [55,56]; A12: [55,56].

The presence of approach categories A1, A5, A6, A8 and A10–A12 does not mean that ***every*** RU tackling this challenge uses a component from ***all*** of the approach categories. It does, however, mean that every RU tackling P1 uses a subset of the approaches shown. We proceed by giving examples (relevant to P1), with references.

1. ***Managed resource provision (A6):*** Prior to running experiments, researchers set the conditions for the experiment through this approach [48–51].
2. ***Resource-specific workload (A8):*** This may be used to stress the component implementing the architectural attribute under test [48–51].
3. ***Simple workload characterization and instrumentation (A11):*** The workload must be characterized by some parameter that serves to measure its demand for power [48,49].
4. ***Resource instrumentation (A10):*** This is an alternative to the use of workload profiling (A11) as a predictor of power consumption. Rather than use, say, the number of threads, or transmit bandwidth (for networking workloads) as predictors, this approach uses resource instrumentation [48,49].
5. ***Identification of metric of energy efficiency (A5):*** In certain cases [49,50], energy or power efficiency is investigated, rather than energy or power consumption. In these cases, the researchers identify and use a relevant metric of efficiency, rather than metrics of consumption (watts or joules).

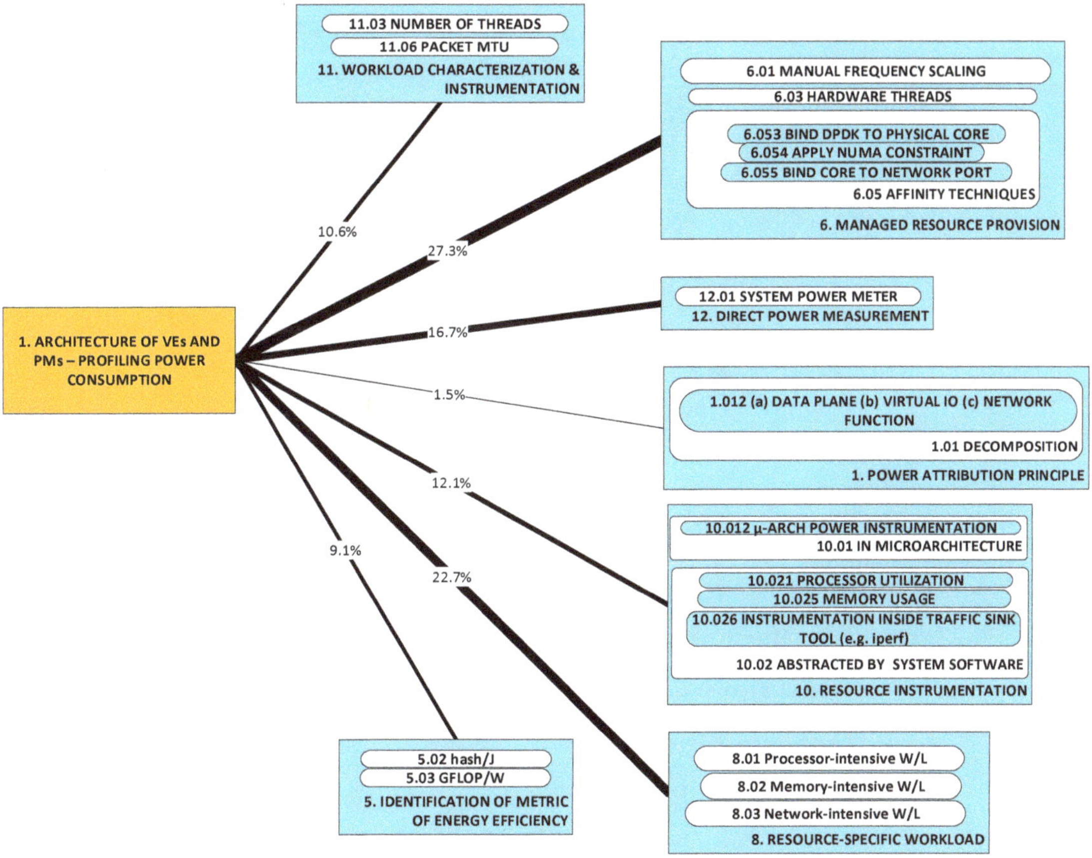

Figure 7. Approaches to solving challenges in category P1; utility metric U_{A_k} is shown in line thickness and as a percentage. RUs in the respective categories are the following: P1: [48–51]; A1: [48]; A5: [49,50]; A6: [48–51]; A8: [48–51]; A10: [48,49]; A11: [48,49]; A12: [48–50].

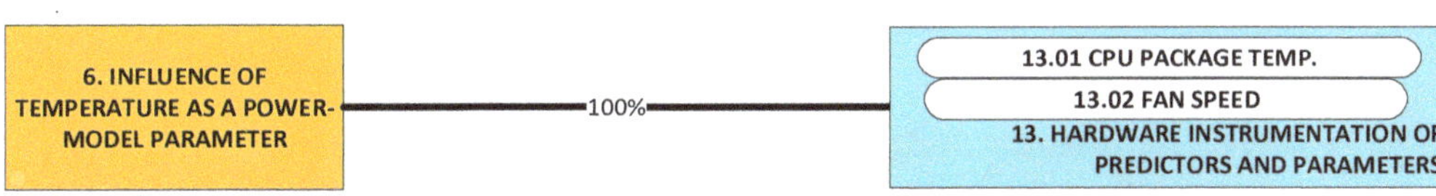

Figure 8. Approaches to solving challenges in category P6; utility metric U_{A_k} shown in line thickness and as a percentage. RUs in the respective categories are the following: P6: [60,81,82]; A13: [60,81,82].

On any of the dyad graphics, the approaches shown (inside the approach categories) include only those which are used by at least one RU that tackles the challenge category that is the root of the dyads. We delegate a repository with a full tabulation of all approaches in any approach category. Similarly, all individual developments within a category are delegated to the same repository (https://github.com/humaira-salam/PowerMeasurementAndModelingRawData, accessed: 30 October 2022).

2.5. Causality DAG

The Causality Directed Acyclic Graph (DAG) in Figure 18 shows a bird's eye-view of the proceedings of research in scope.

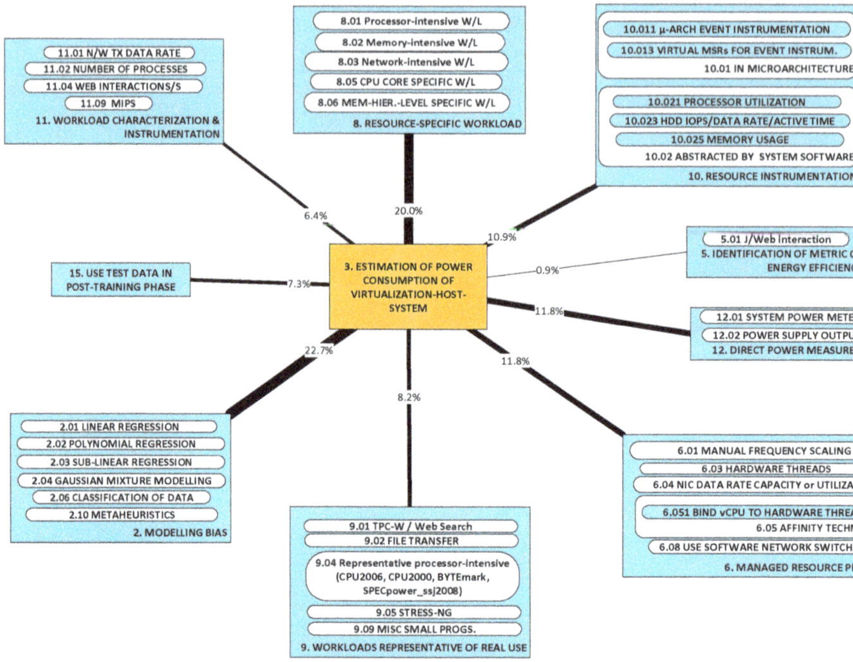

Figure 9. Approaches to solving challenges in category P3; utility metric U_{A_k} is shown in line thickness and as a percentage. RUs in the respective categories are the following: P3 [57–69,71]; A2 [57–64]; A5 [65]; A6 [60,66–68]; A8 [59–61,63,64,66–68,71]; A9 [57,58,60,62,65,68]; A10 [57–59,61–64,69]; A11 [64,65,67,68,71]; A12 [57–59,61–64,67,68,71]; A15 [57,62–64].

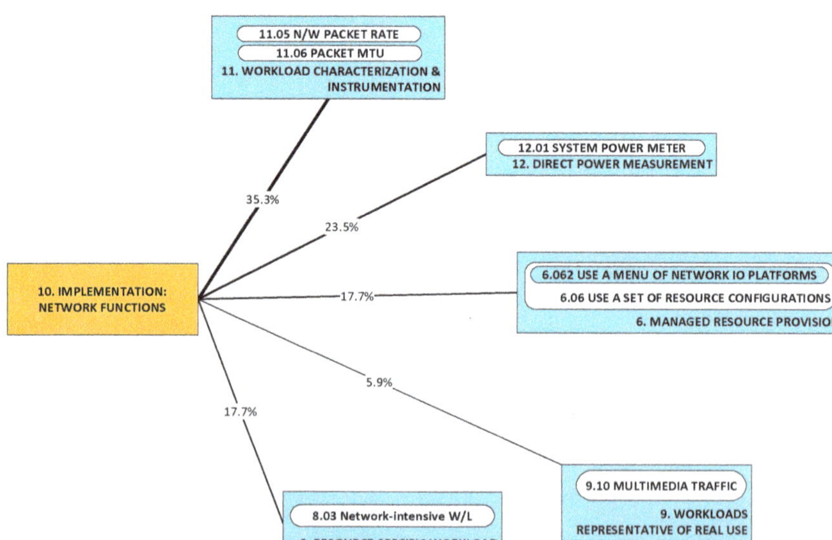

Figure 10. Approaches to solving challenges in category P10; utility metric U_{A_k} is shown in line thickness and as a percentage. RUs in the respective categories are the following: P10: [48,53]; A6 [53]; A8 [53]; A9 [48]; A11 [53]; A12 [48,53].

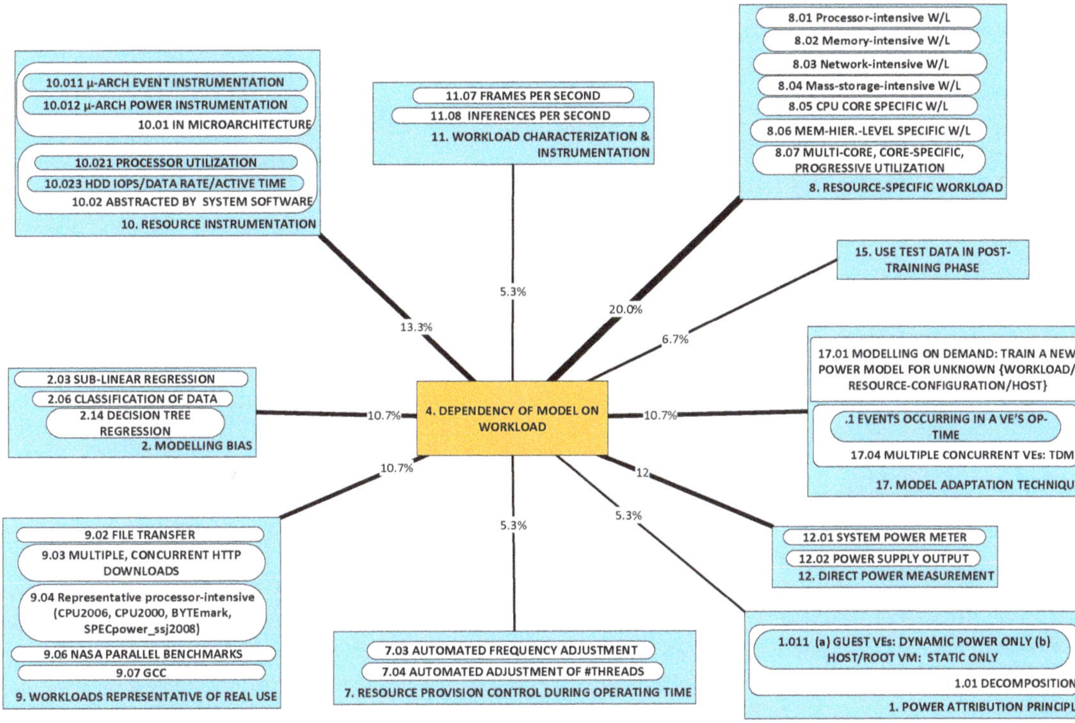

Figure 11. Approaches to solving challenges in category P4; utility metric U_{A_k} is shown in line thickness and as a percentage. RUs in the respective categories are the following: P4 [56,62,67–69,71,75–79]; A1 [62,77]; A2 [62,77]; A7 [78]; A8 [56,67,69,71,75,76]; A9 [56,62,68,75,79]; A10 [62,78,79]; A11 [78]; A12 [56,62,76,79]; A15 [62,77]; A17 [62,67,68,71,75,76,78].

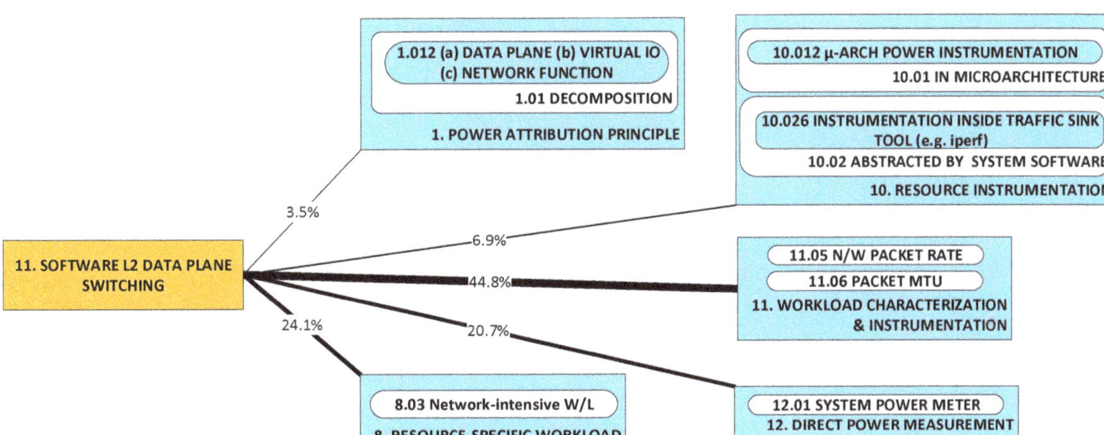

Figure 12. Approaches to solving challenges in category P11; utility metric U_{A_k} is shown in line thickness and as a percentage. RUs in the respective categories are the following: P11 [48,53]; A1 [48]; A8 [48,53]; A10 [48]; A11 [48,53]; A12 [48,53].

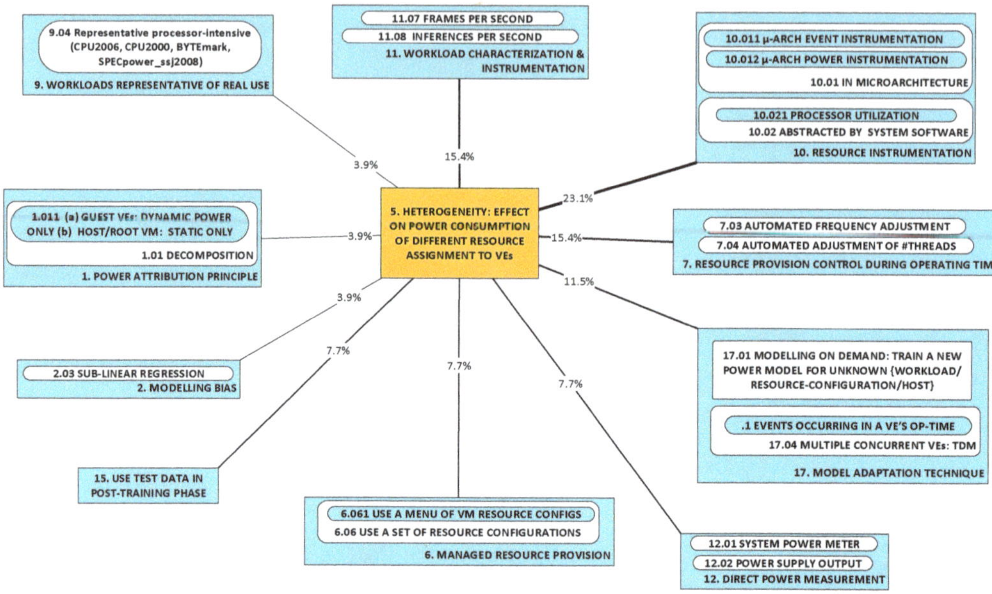

Figure 13. Approaches to solving challenges in category P5; utility metric U_{A_k} is shown in line thickness and as a percentage. RUs in the respective categories are the following: P5 [62,72,78]; A1 [62]; A2 [62]; A6 [72]; A7 [78]; A9 [62]; A10 [62,78]; A11 [78]; A12 [62,72]; A15 [62]; A17 [62,78].

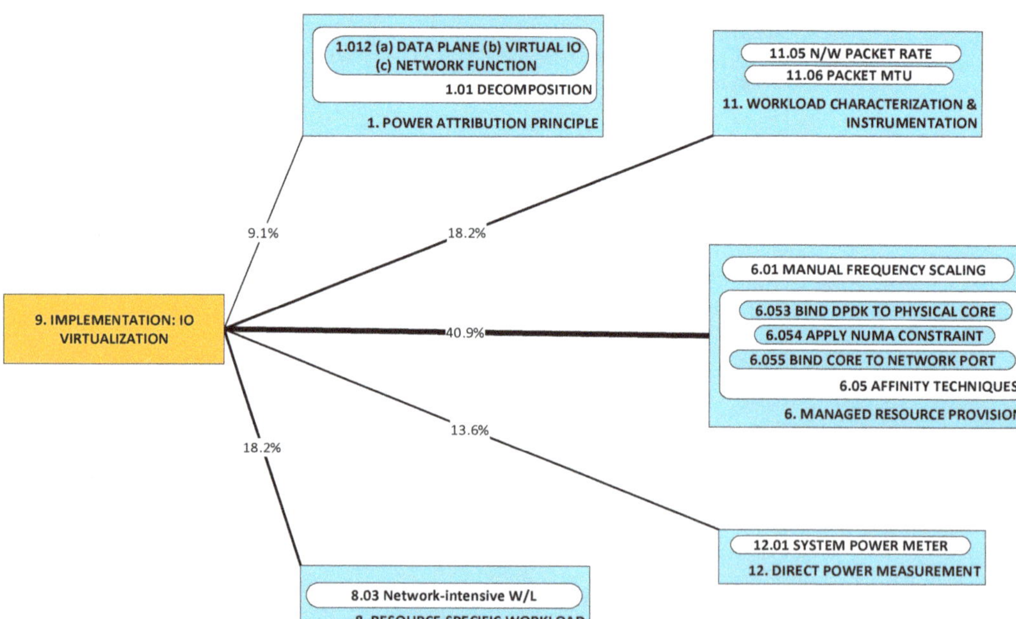

Figure 14. Approaches to solving challenges in category P9; utility metric U_{A_k} is shown in line thickness and as a percentage. RUs in the respective categories are the following: P9 [51,53]; A1 [53]; A6 [51]; A8 [51,53]; A11 [53]; A12 [53].

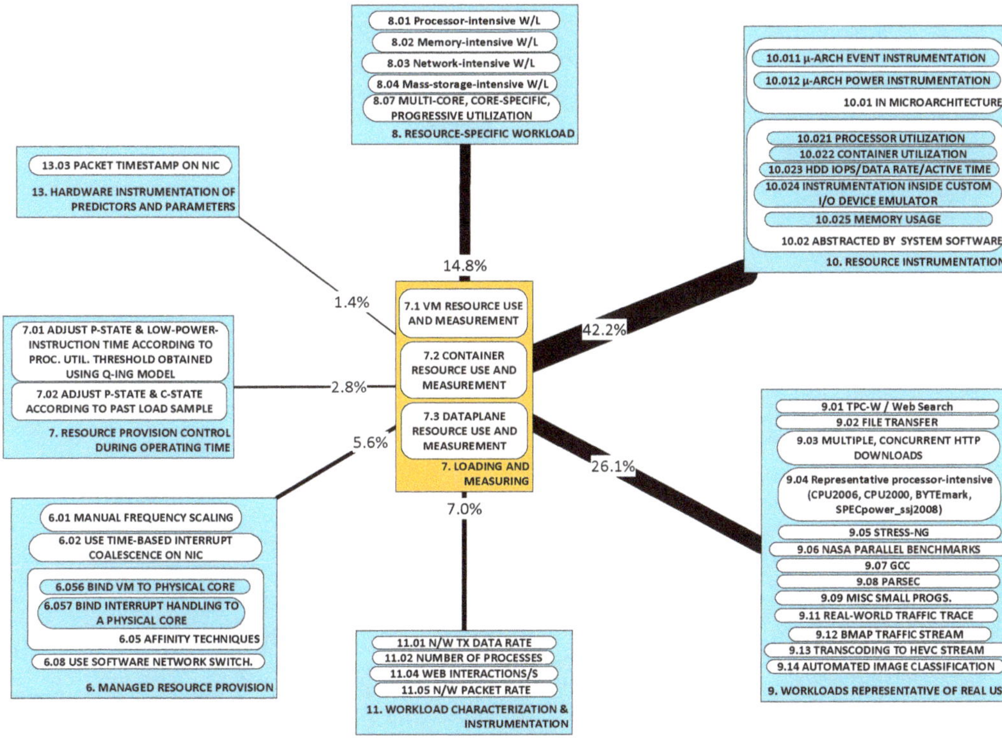

Figure 15. Approaches to solving challenges in category P7; utility metric U_{A_k} is shown in line thickness and as a percentage. RUs in the respective categories are the following: P7 [56–59,61,62,65,68,69,71,72,75–79,81–85,87]; A6 [56,68,81–83]; A8 [56,59,61,71,76,77,84,85]; A9 [56–58,62,65,68,69,72,78,79,81–84,87]; A10 [56–59,61,62,69,72,75–79,81–85,87]; A11 [65,68,71,83,84,87]; A13 [87].

2.6. Triads (Problem/Challenge-Approaches-Developments) Graphic

Figure 19 shows the triads graphic. To improve readability, we limit illustration to the triads that are in the top twenty percentile of a list ordered according to thickness. These triads comprise 49.2% of the total number of compiled triads.

2.7. Statistics

Bar charts that illustrate the category metrics described in [1, N. see section on "Statistics"] are presented below:

1. Challenges (Figure 20)
    a. Frequency of occurrence in the set of all RUs;
    b. Frequency of occurrence in the set of all challenges in all RUs;
    b. Frequency of occurrence, weighted by approach diversity, in the set of all challenges in all RUs.
2. Approaches: (Figure 21)
    a. Frequency of occurrence in the set of all approaches in all RUs;
    b. Frequency of occurrence in the set of all triads in all RUs.
3. Developments: frequency of occurrence in the set of all developments in all RUs: (Figure 22)

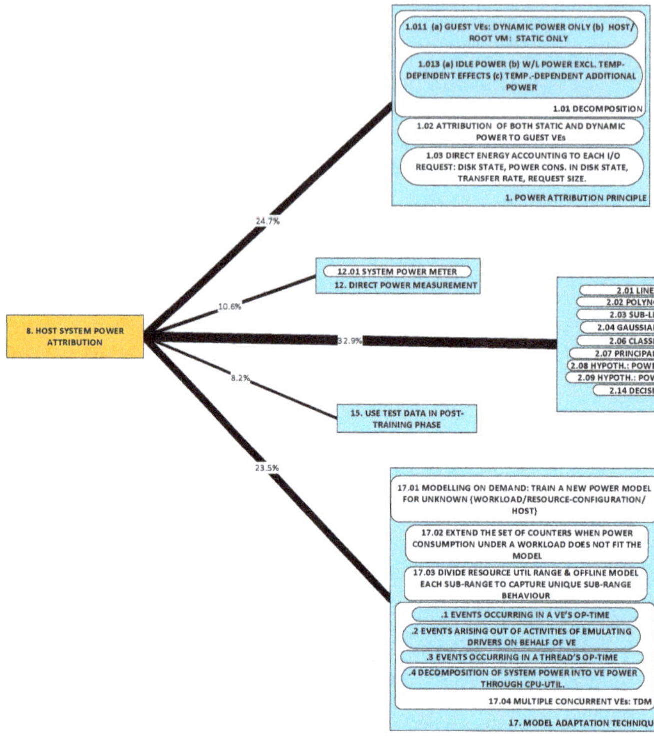

Figure 16. Approaches to solving challenges in category P8; utility metric U_{A_k} is shown in line thickness and as a percentage. RUs in the respective categories are the following: P8 [57–59,61,62,69,72,75–77,79,81,82,84,86]; A1 [57–59,61,62,69,72,75–77,79,81,82,84,86]; A2 [57–59,61,62,72,75–77,79,81,84]; A12 [57–59,61,72,76,84]; A15 [57,62,77]; A17 [57–59,61,62,69,72,75,76,79,82,84,86].

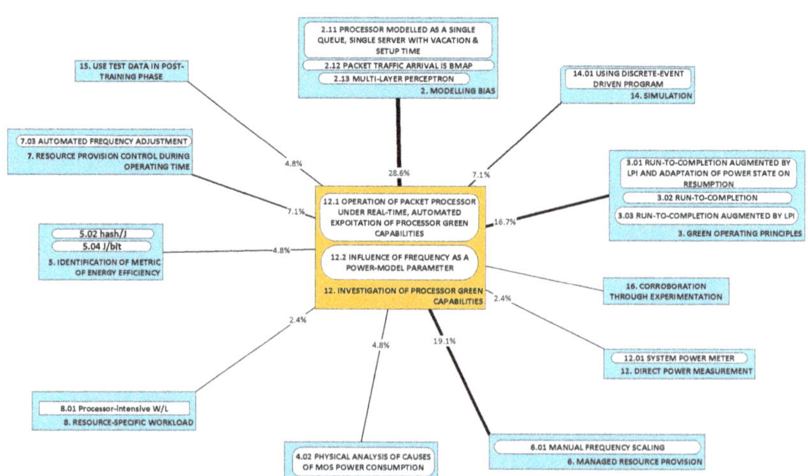

Figure 17. Approaches to solving challenges in category P12; utility metric U_{A_k} is shown in line thickness and as a percentage. RUs in the respective categories are the following: P12: [50,60,67,81–83,85,87,94]; A2 [83,85,87,94]; A3 [83,87,94]; A4 [60]; A5 [50,85]; A6 [50,60,67,81,82]; A7 [85]; A8 [50]; A12 [50]; A14 [87]; A15 [85]; A16 [83].

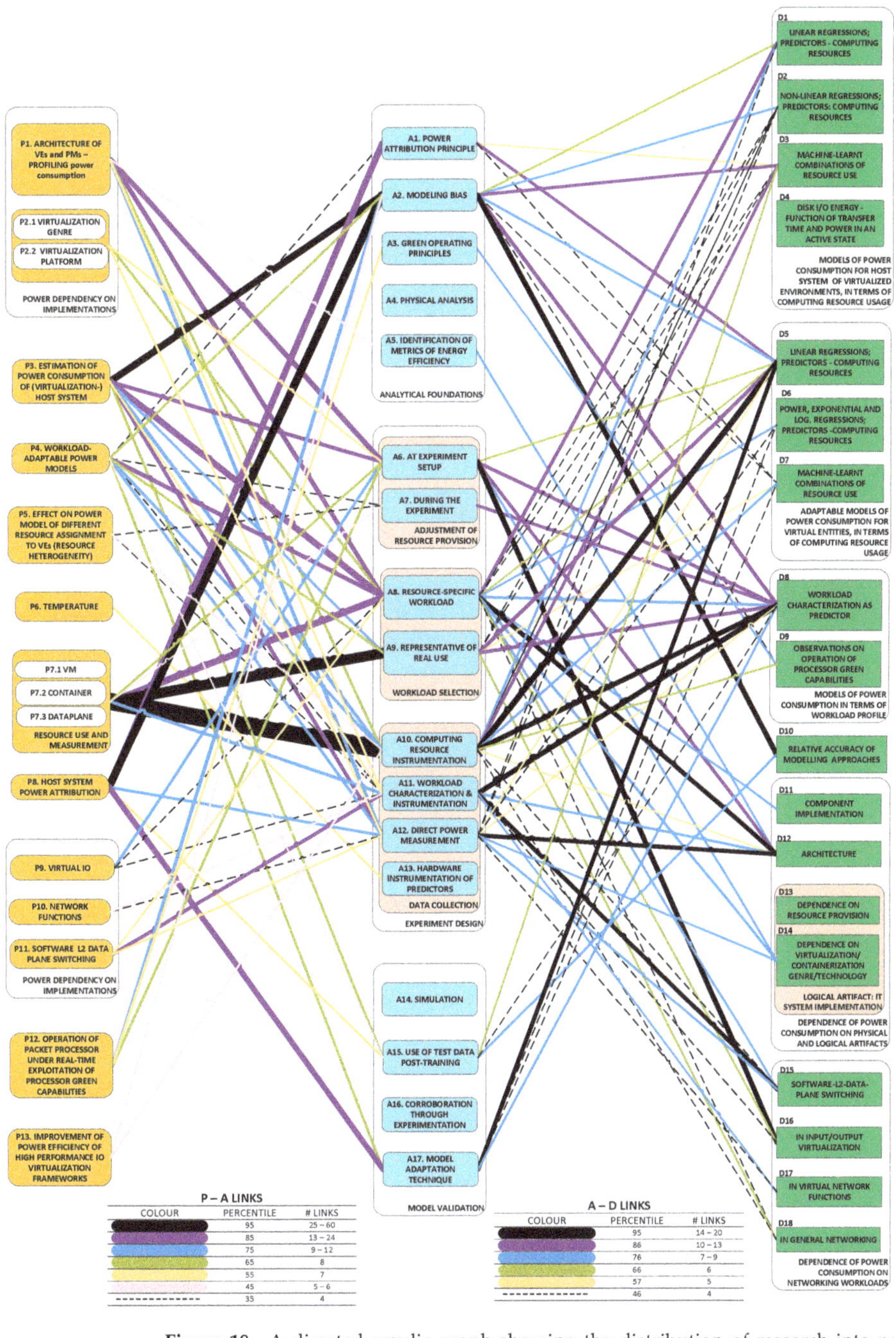

Figure 18. A directed acyclic graph showing the distribution of research into power measurement and power consumption models in virtualized networking and computing environments. D1 [57–59,69,75]; D2 [57,60,61,76,81]; D3 [62,77,79]; D5 [58,59,69,72,75,81,82,84,86]; D6 [61,66,76]; D7 [62,79]; D8 [63–68,71,78,83,85,87,94]; D9 [87]; D10 [60,62,63,77,79]; D11 [49,69,76]; D12 [49,50,67,68,76]; D13 [73]; D14 [55,56]; D16 [51,53]; D17 [48,53,85]; D18 [56,65].

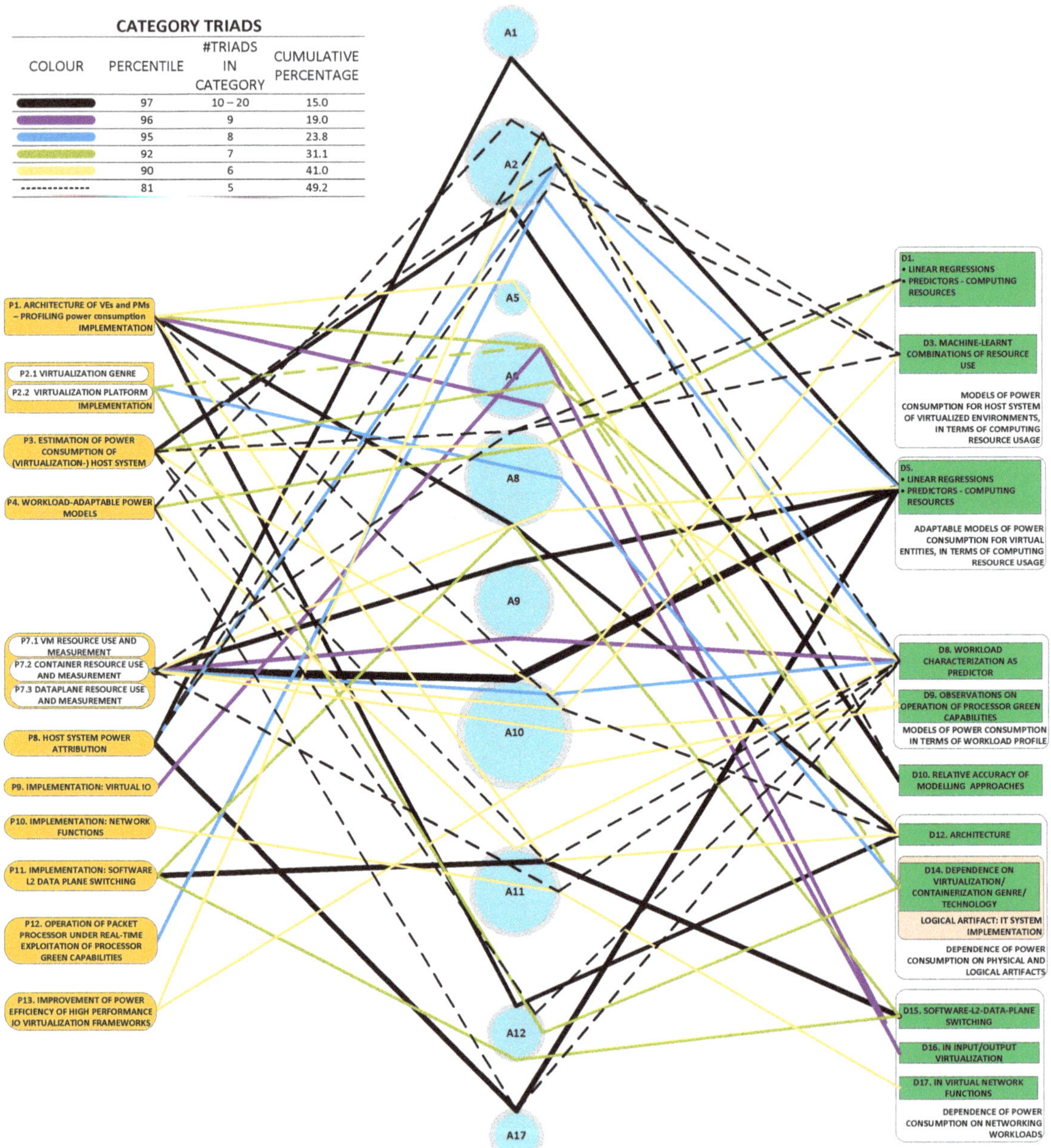

Figure 19. Triads. D1 [57–59,69,75]; D3 [62,77,79]; D5 [58,59,69,72,75,81,82,84,86]; D8 [63–68,71,78,83,85,87,94]; D9 [87]; D10 [60,62,63,77,79]; D11 [49,69,76]; D12 [49,50,67,68,76]; D14 [55,56]; D15 [48,53]; D16 [51,53]; D17 [48,53,85].

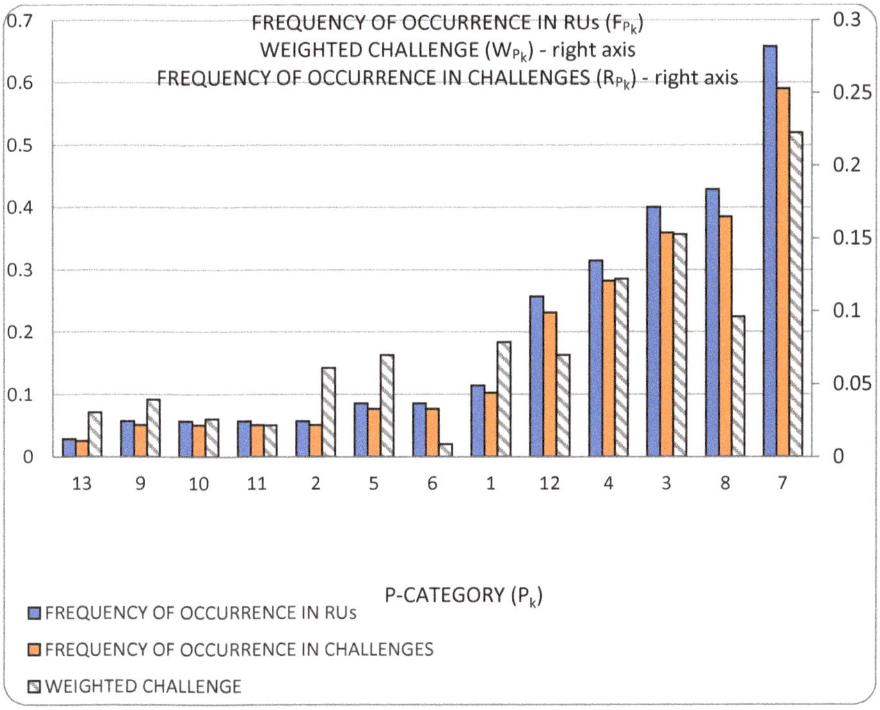

Figure 20. Frequency, Research Interest and weighted Challenge bar chart.

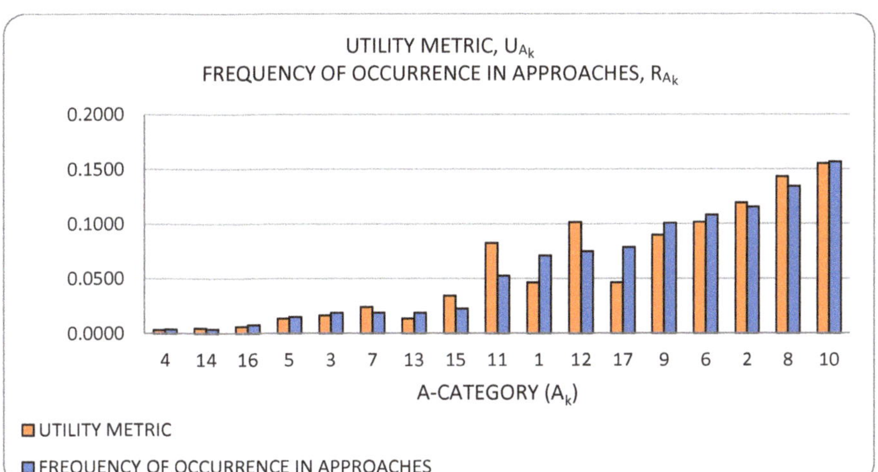

Figure 21. Approach metrics.

To improve legibility, the x-axis labels show category numbers only. The codes ("terse, dense representations of a verbose articulation of a concept", see [1, N. see sub-section "what are codes?"]) linked to the numbers are shown in Tables 2–4. In Table 2, we also show questions that help to clarify articulation of the challenge posed.

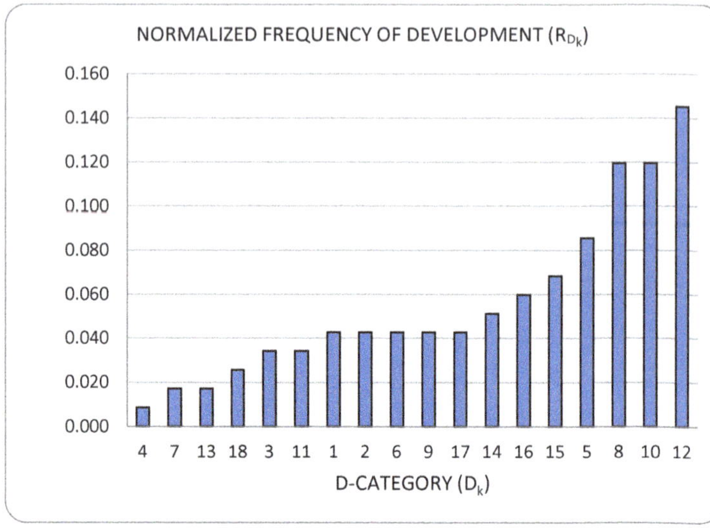

Figure 22. Normalized frequency of Developments.

Table 2. Challenge category codes and representative challenge questions.

CAT #	CODE	PROBLEM
P1	Architecture of VEs and PMs—profiling power consumption	What is the impact of specific major architectural features of computer hardware on the power consumption of VEs?
P2	Virtualization genre, platform	Is it possible to meaningfully rationalize the behavior of VE power consumption across different implementations of systems for virtualization?
P3	Estimation of power consumption of virtualization-host systems	Can host system power consumption be predicted on the basis of VE activity?
P4	Workload-adaptable power models	Can workload-adaptable power models be developed?
P5	Resource heterogeneity	How do a VE's power consumption and power model vary with resource configuration (heterogeneous VEs)?
P6	Influence of temperature on power model	How does temperature of operation affect VEs' power models?
P7	Resource use and measurement of VEs	How can load be targeted at specific VE resources? How can the actual resource consumption be measured?
P8	Host system power attribution	How can the (measured) power consumption of a host be attributed to the hosted VEs?
P9	IO Virtualization	What is the impact of specific major implementations of IO virtualization on the power consumption of VEs?
P10	Network functions	Which particular implementation of a network function is most power or energy efficient?
P11	Software L2 data plane switching	What is the impact of specific major implementations of software layer 2 data plane switching on the power consumption of VEs?
P12	Investigation of processor green capabilities	How can we model operation under real-time exploitation of processor green capabilities?
P13	Improvement of power efficiency of high-performance IO virtualization	How can we improve the power efficiency of high-performance packet IO frameworks?

Table 3. Approach category codes.

CAT #	CODE
A1	Power attribution principle
A2	Modeling bias
A3	Green operating principles
A4	Physical analysis
A5	Identification and use of metrics of energy efficiency
A6	Managed resource provision (setup time)
A7	Controlled resource provision (operation time)
A8	Resource-specific workloads
A9	Workloads representative of real use
A10	Computing resource instrumentation
A11	Workload characterization and instrumentation
A12	Direct power measurement
A13	Hardware instrumentation of predictors
A14	Simulation
A15	Use of test data in post-training phase
A16	Corroboration through experimentation
A17	Model adaptation technique

Table 4. Development category codes.

CAT #	CODE
D1	Host models -> linear regressions: predictors = computing resources
D2	Host models -> non-linear regressions: predictors = computing resources
D3	Host models -> machine-learnt: inputs = computing resources
D4	Host models -> mass storage energy consumption
D5	Adaptable VE models -> linear regressions: predictors = computing resources
D6	Adaptable VE models -> power, exponential and log regressions: predictors = computing resources
D7	Adaptable VE models -> machine-learnt: inputs = computing resources
D8	Host/VE models of power consumption -> predictors = workload characteristics
D9	Host/VE models of power consumption- > observations on operation of processor green capabilities
D10	Relative accuracy of modeling approaches
D11	Power's dependencies- > physical and logical artifacts -> component implementation
D12	Power's dependencies -> physical and logical artifacts -> architecture
D13	Power's dependencies -> physical and logical artifacts -> IT system implementation -> dependence on resource provision
D14	Power's dependencies -> physical and logical artifacts -> IT system implementation -> dependence on virtualization/containerization genre and technology
D15	Power's dependencies -> networking workloads -> software L2 data plane switching
D16	Power's dependencies- > networking workloads -> in I/O virtualization
D17	Power's dependencies- > networking workloads -> in VNFs
D18	Power's dependencies -> networking workloads -> in general networking

3. Thematic Analysis

The themes presented in this section are the product of a thematic analysis undertaken according to the method described in [1]. We first present an overview through a graphic (Figure 23) that groups the themes and then proceed to an exposition of the themes within the sub-sections.

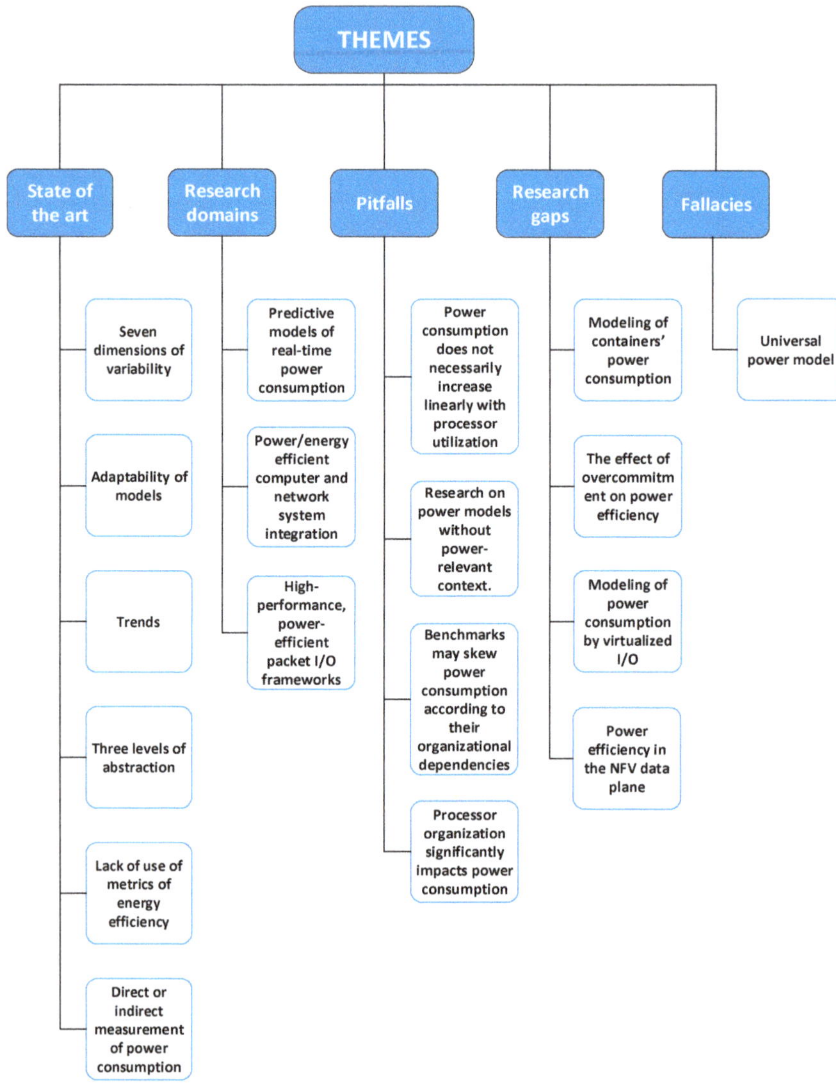

Figure 23. A graphical overview of the thematic analysis of the research space.

3.1. *State of the Art*

3.1.1. Seven Dimensions of Variability

Through our collation and resolution of the derivatives of the core challenge in this research space, we observe seven dimensions of variability in modeling power consumption by a virtual entity. Comparison with the power consumption of physical machines throws the core challenge into sharper relief. With physical machines:

- Power consumption can be measured directly;
- There is no virtualizing agent to consider;
- The activity of other physical machines (that do not send or receive workload) is irrelevant.

The seven dimensions of variability are shown in Table 5. It is not surprising that the problems we have seen researchers tackle are closely aligned with these dimensions. The research space is precisely about the need to obtain an understanding of the impact that these variables have on power consumption. The scope of models we found in the RUs covers only a subspace of the seven-dimensional space but the extent is not usually stated. We touch upon this issue briefly in our treatment of the pitfall of research without context and the fallacy of the universal power model.

Table 5. Seven dimensions of variability.

Dimension	Problem Category
1. Workload type	P4, P7
2. Virtualization agent	P2, P9, P11, P13
3. Host (resources and architecture)	P1, P3
4. Temperature	P6
5. Power attribution	P8
6. Co-hosted, concurrent VEs	P8, P5
7. Frequency	P12, P13

3.1.2. Adaptable Models

We observe ***three conditions*** which must be met for an automated modeling system to obtain a model of VE power consumption.

1. The first fundamental condition is common to all successful research into modeling of VE power consumption. ***Resources utilized by the VE*** (whether measured through architectural or microarchitectural instrumentation) ***must be strongly correlated to power consumption by the VE as well as power consumption in host system overheads***. If the overheads are uncorrelated to the VE's activity, or weakly so, then any significant power overhead must be modelled through a separate consideration of its driving causes, e.g., temperature [81].

Furthermore, in order that modeling may apply across a diversity of operating contexts, two other conditions must hold.

1. Any change in the parameters of correlation must be dynamically learnt and the model adjusted.
2. Any change in operating context that invalidates the parameters of correlation must be of finite duration. An indefinite transient precludes the formation of a model.

To this observation on the three conditions, we add another observation. Two of the seven dimensions of variability are commonly investigated in terms of the validation of the accuracy of modeling systems: ***workload type and co-hosted, concurrent VEs***. We suggest that a modeling system may be labelled adaptable if, minimally, ***it meets the above three conditions under these two dimensions, i.e., (a) operations of variable workload type per VE, and (b) a variable number of concurrent, co-hosted VEs.*** Before proceeding to refer to validations observed in the RUs, it is useful to draw attention here to the need for a ***modeling system***, rather than simply a model, in the estimation of power consumption. Dynamic adjustments are affected through the intervention of such a system, which adjusts model parameters to the operating context. We now proceed to describe how the variables in the two dimensions were varied in some studies.

Workload type: Different types of workload correspond to different use of resources; hence, the behavior of power consumption also differs. Several researchers [48,75,76] studied the effect of changing workload type either by categorizing the workload itself or by categorizing the resources consumed by a specific kind of workload. Both [75] and [76] characterized the use of CPU utilization for workloads, which is known to have a workload-type dependent relationship with power consumption [62]. Thus, in both these studies, the need arises to re-train on the change of workload type (second condition). Further, both carry an observation about the duration of re-training for refined models (third condition).

On the other hand, each one of [62,69,79,81,82] is capable of adapting to different workload types without re-training. All predictors here are event counters. However, events are not necessarily linear predictors of power consumption across all workload types. In [84], it was found that the model parameters of a linear regression of event counters onto power consumption are workload specific. Given the contrast with [73,74], it seems that the root cause is the selection of events for prediction.

Concurrent operation: In [61], a non-linear model of the dynamic power of a multi-(virtual)-core VM is obtained. The dynamic power p_{vm} is expressed in terms of the average utilization u_v which n virtual cores impose on a total of N physical cores:

$$p_{vm}(u_v, n) = \alpha \left(\frac{n}{N}\right)^{\beta} u_v^{\beta}$$

Parameters α and β are determined through linear regression of the logarithmic form. The relationship was successfully tested under the operating context of one VM and three VMs. The model is limited to processor-intensive workloads; indeed, this is not surprising, since the predictor is (virtual) processor utilization. However, a conclusion can be drawn: the accuracy of the model in predicting the dynamic power of an individual VM is independent of the concurrent operation of other VMs (albeit for a limited range of workload types).

In [75], it is shown that estimation based on processor utilization can accurately predict the power consumption of multiple concurrent VEs. The prerequisite is that a model for that VE's workload type has been learnt. The prediction accuracy for individual VEs is independent of concurrent operation. The same conclusion can be drawn for [61]; furthermore, here the range of workload types has expanded to a broader range. With regard to the results obtained in [84], the conclusion is yet again the same: individual prediction accuracy (albeit workload-type-dependent) is independent of concurrent operation.

Variability in both workload type and number of concurrent, co-hosted VEs is manifestly claimed in [62,69,81]. It is noteworthy, therefore, that not only is the modeling system adaptable, but it also produces a model that is itself adaptable without the need for real-time adjustment.

We conclude this sub-sub-section with the observation that limited adaptability (defined as independence of workload type and number of concurrent, co-hosted VEs) has been achieved. From the surveyed RUs, we conclude that, presently, the limits of adaptability are the following:

1. Workload type: the power consumption of processor-intensive, memory-intensive, disk-intensive workloads and mixes thereof has been modeled by a single modeling system in an automated manner;
2. Co-hosted, concurrent VEs have been modeled up to but not exceeding over-commitment of processor physical cores.

Caution has been exercised in claiming these limits. For example, while the commitment of logical cores (e.g., Intel Hyper-Threading logical cores) has been investigated, we have found no evidence that power consumption has been modeled in a manner that automatically adapts to a transition of consumption from physical to logical cores.

3.1.3. Lack of Use of Metrics of Energy Efficiency and Standards to Address This Deficiency

We note several experiments, e.g., [55,56,68,71,103–105], that target power consumption but less than 12% of the RUs in our corpus approach the problem in terms of some energy efficiency metric [49,50,65,85]. It is necessary to move beyond measurements of how much power was consumed, to measurements of how much power was consumed to carry out a specific task. This change in approach facilitates comparison between research works. More importantly, it directly addresses the question about the cost of operation of infrastructure.

This approach requires identification of a unit of comparison that transcends the boundaries of disparate systems that deliver this unit. This unit of comparison is referred to in the Life Cycle Assessment (LCA) framework (ISO 14040) [106] as the ***functional unit***. A definition specific to telecommunications equipment is given in [107]: the functional unit is defined as "a performance representation of the system under analysis". This definition is too broad; therefore, units specific to a variety of classes of equipment are defined too [107]. In our corpus, two approaches we have seen are hash/J [50] and J/Web Interaction [65]. The functional units in these cases are the performance of one crypto-hash and one web interaction, respectively. Another is to define a functional unit specific to a digital service delivered over a telecommunications network, e.g., ten minutes' time of browsing [108]. Note that L.1310 [107] recognizes both metrics where energy is in the numerator [65] as well as those where it is in the denominator [50]. Guidance is available: energy efficiency measurement for several NFV components has been standardized, as well as measurement standards for servers, switches and virtualization systems [109].

3.1.4. Trends

Here, we present trends which we have identified through our collation of problems, approaches and developments. This sub-sub-section is divided into four parts, regarding trends in: challenges, complexity of tackled problems, approaches and developments. Within each part, trends are numbered using Arabic numerals, to differentiate cleanly between them.

Challenges

1. The causality DAG (Figure 18) shows that the research space can be characterized succinctly, note the thickness of the links originating at P7 (VE resource use and measurement) and P8 (how to attribute host system power to VEs).
    a. Figures 1 and 20 show that the challenge-category tackled most frequently ($R_{P_7} = 25.3\%$) is how to load the VE's resources and measure the loading (P7). P7 also has the most frequent presence in RUs: $F_{P_7} = 65.7\%$) (Figure 20).
    b. The accurate specification and measurement of load is essential to model formation. These measurements provide the aggregated (i.e., indiscriminate of which VE is consuming) predictors (the predictors are the input side of the model). Disaggregating the predictors and attributing measured power consumption to the individual VEs (the output side of the model) constitutes the second most frequently tackled challenge ($R_{P_8} = 16.5\%$). P8 also has the second most frequent presence in RUs: $F_{P_8} = 42.9\%$.
    c. In the following, the vector (P, A, D, weight) will be used to indicate a path (P, A, D) and the number of triads (weight) through the path:
        i. A total of 34.2% of triads collected regard P7 and P8, respectively. The triads graphic (Figure 19) shows that efforts rooted in these two challenges converge on a common objective: building adaptable models of VEs' power consumption, notably using regressions to linear combinations of computing resource predictors;
        ii. The set of triads leading to linear regressions consists of {(7,10,5,20), (8,17,5,13), (7,9,5,12), (8,1,5,10), (8,2,5,8), (7,8,5,6), (8,12,5,4) and (7,6,5,3)} and these account for 11.3%.
2. The third most frequently tackled challenge is that of the estimation of the virtualization-host-system power consumption ($R_{P_8} = 15.4\%$). This category might be overlooked in a first inspection of the research space, as it might seem futile to attempt to estimate a power consumption which can be measured. However, in practice, the logistical

challenge of the measurement of horizontally scaled system deployments seems to be well known and several works have been undertaken to develop software meters.

3. Several significant links originate on P1 (profiling power consumption's dependency on architecture). The DAG (Figure 18) indicates that approaches to tackling P1 are spread across a mixture of managing resource provisions, the use of synthetic (resource-specific) workloads and predictions using system software's instrumentation.
4. Another large group of links originates on P3 (estimation of host system power consumption). The DAG (Figure 18) indicates that a primary concern in tackling P3 is the type of model to select. The most common choices are linear and non-linear; machine-learning techniques are the least common of the three. The thickness of the triad (3,2,10,11) (D10: relative accuracy of formal approaches) indicates that there is already significant interest in whether the advanced models are worth the effort to develop them and the computational resources required to operate them.
5. While $F_{P_7} = 65.7\%$ (loading VEs and measuring their use of resources), only 8.6% ($F_{P_{7.2}} = 8.6\%$) of all RUs investigate loading containers and measuring their use of resources. With virtual data plane devices, the figure is even lower: $F_{P_{7.3}} = 8.6\%$. This imbalance suggests that there is much room left for research into modeling power consumption by containers and data plane devices.

Challenge complexity

In [1], we suggest that the diversity of approaches is a metric of the complexity of the challenge. We note (Figure 20) that challenge complexity (W_{P_k}) generally follows the frequency with which a particular challenge-category is tackled. That is: the more frequently the challenge is addressed, the more diverse are the approaches applied to it. This can be verified by noting that the heights of both sets of bars in the chart (superimposed on the same graphic) follow roughly the same pattern. However, some categories do stand out. For example:

1. P6 regards the influence of temperature on the power model. Approach diversity is poor here because only the use of additional instrumentation can be attributed to this challenge. Model bias might be attributed to this challenge too but, largely, model bias is determined by other challenges within the scope of the RU.
2. While host system power attribution (P8) has the second highest research interest (and frequency of occurrence, F_{P_k}), the number of approaches taken to solve this challenge is relatively small.
3. On the other hand, power consumption's dependencies are tackled by a disproportionately large number of approaches. This is not surprising, as the objects of study (architecture, virtualization platform, virtualization genre) are multi-faceted and dependencies can be investigated through a variety of approaches.

Approaches

Our approach utility metric, U_{A_k} [1] (Figure 21), seems to be a useful one. It communicates clearly what we have observed in our surveying. Below, we draw attention to saliencies perceived during surveying and confirmed by the metrics.

1. Instrumentation of the consumption of computing resources (A10—which includes microarchitectural instrumentation and that abstracted by system software) is repeatedly adopted ($U_{A_{10}} = 15.5\%$, $R_{A_{10}} = 15.7\%$) in empirical work in this field. It is also the most utilized of all approaches. In comparison, workload instrumentation accounts for 8.2% utilization ($R_{A_{11}} = 5.2\%$).
2. Resource-specific workloads (A8) are the more utilized approach to loading VEs (14.3% of all triads). This approach category is the second most utilized. Workloads representative of real use (A9) account for 9.0% of all triads. The corresponding frequency of occurrence figures (R_{A_k}) are 13.4% and 10.0%, respectively.
3. Two other high-utility approaches are (a) modeling bias (A2, with $U_{A_2} = 11.9\%$, $R_{A_2} = 11.6\%$) and (b) the managed provision of resources (A6, with $U_{A_6} = 10.2\%$,

$R_{A_6} = 10.8\%$). The use of representative workloads (A9) follows at $U_{A_9} = 9\%$, $R_{A_9} = 10.1\%$.

4. A comparison of the patterns of bar height distribution for U_{A_k} and R_{A_k} reveals that some categories stand out.
   a. While A12 (direct power measurement) is not as frequent as the other software-based forms of instrumentation, this approach has a utility that sticks out of the pattern (Figure 21). The reason is that most, if not all, developments obtained in an RU that include this approach depend on the directly measured power.
   b. Similarly, while workload characterization and instrumentation (A11) are employed with a frequency that is about one-third that of its alternative (i.e., computing resource instrumentation, A10), it has a far better utility-to-frequency ratio than A10. The triads graphic (Figure 19) indicates that one important cause of this high utility is that software-L2-data plane switches and (virtual) network functions are investigated primarily using measurement of workload (and not measurement of computing resources consumed to process a workload).

Developments

1. A total of 55.6% of all identified developments are obtained in modeling power consumption. The remaining 44.4% regard how power consumption depends on implementations. The implementations investigated (for their impact on power consumption) range from entire virtualization platforms (e.g., KVM) to components (e.g., processors).
2. The most frequent ($R_{D_{12}} = 14.5\%$) category of developments is that which regards observations on the dependency of power consumption on architecture. The cause of this high frequency is that developments in the study of architecture establish directional (negative, positive, neutral) correlations rather than predictive forms. For example, in a single RU [67], all of the following developments emerge:
   a. D12.08: VM power consumption increases linearly with a vCPU frequency of operation when the vCPU is operating at 100% utilization;
   b. D12.09: Virtualization-host-system power consumption increases linearly with the number of physical cores operating at 100% utilization;
   c. D12.1: Virtualization-host-system power consumption increases with the number of VMs active on the same core.
3. Amongst developments of models, the most common category (D8) is of the type where power consumption (host or VE) is predicted in terms of workload characteristics ($R_{D_8} = 12.0\%$). The next most common category regards the prediction of power consumption as a linear function of computing resources ($R_{D_5} = 8.5\%$)). Machine-learnt models of VE power consumption comprise the second least frequent category, with $R_{D_7} = 1.7\%$.
4. Despite the frequency of developments in category D10 (Figure 22), few works [60,62,63,77,79] compare the accuracy of model types. Models have largely been treated as a means to an end, with little investigation of their relative accuracy and range of validity (in the seven dimensions of variability). This may be a reflection of researchers' interests. As the popularity of methods from the body of knowledge of data science increases, works (e.g., [63]) that span a broader range of model types may be expected to increase in concert.
5. Adaptable, non-linear VE models (D2) occupy a modest 4.3%. Such a distribution adds weight to the observation that this research space is ripe for exploration using advanced modeling techniques. Indeed, one attraction for data scientists is the relative ease with which data can be collected. However, polynomial and other types of regression to closed-form are problematic as suspicion of over-fitting increases with the order of the polynomial. For example, in [76], a sixth-order polynomial is suggested

to model the relationship between processor utilization and power consumption by a host.

3.1.5. Three Levels of Abstraction

We note that existing power models may be classified into one of three levels of abstraction. In ascending order of abstraction, these are:

- Microarchitecture and architecture;
- Simple characterization of workload;
- Complex characterization of workload.

The meaning of "abstraction" specific to our use here is perhaps most easily grasped by referring to the variables used as model inputs. In all cases, the variables are some measure of load. At the lowest level, inputs that quantify the operation of processor sub-units are used (event counters and event timers). The highest level uses inputs that quantify the demand for a telecommunications system or service. Clearly, the latter inputs are far more detached from the underlying, concrete implementation than the former.

With regard to the lowest level of abstraction, many power models use hardware resource consumption to estimate the power consumed by virtual components. One survey [105] (while comparing the available power models for processors, VMs and servers) observes that most of the power models for virtual machines use physical machine counters to estimate the corresponding resource utilization by the virtual components. We offer further insight on this matter. We concur in the observation that much current research is concerned with modeling power consumption of virtual machines. The approach may be succinctly described as estimates obtained from models trained out of either architectural or microarchitectural instrumentation data. Note that we distinguish between architecture and microarchitecture using the classical interpretation [110]. Now, the term "architecture" is severely overloaded and its interpretation can easily differ from that which we wish to exploit. In the following description of the levels of abstraction, we use the alternative "system software's instrumentation" to convey the same meaning as "architectural instrumentation" with less ambiguity.

Microarchitectural instrumentation is the lowest level of instrumentation. Power consumption is expressed in terms of variables that are defined at sub-CPU and sub-subsystem levels. The granularity of this level holds the greatest potential for accuracy, but the rate of change of the observed variables has deterred several researchers from pursuing this approach to instrumentation, citing concerns about communicational and computational overhead. This concern has been dismissed by three groups of researchers [70,86,98] who have indicated that acceptable accuracy can be achieved with negligible overheads. Given the proliferation of works based on this approach and the availability of high-level (programming) language code (HLL) that facilitates the use and the potential for the capture of physical behaviors, then ***our general recommendation is a preference to investigate the use of microarchitecture instrumentation***.

System software's instrumentation regards a class of instrumentation that has meaning across the spectrum of computer systems. The input variables, such as the CPU or network interface utilization, are produced by some digest (function) of intermediary system software. For example, an operating system's (OS) measurement of core utilization can roughly be described as the core's duty cycle on behalf of the OS. This is significantly removed from knowledge of activity within the core. While models at this level are more abstract, they are still low-level, especially when compared with the other levels.

Simple characterization of workload is a less frequently encountered abstraction, used by Enokido, Takizawa and various others with whom they have co-published [68,71,111]. These models describe power consumption in terms of fundamental descriptors of workload, e.g., the number of processes and transmit/receive data rate.

Complex characterization of workload is the least granular of the models in the survey [34,112]. The objective here is to quickly proceed to a good estimate of the power or energy required to produce the workload. This kind of model has no use in real-time control

but it is useful for macroscopic comparisons, i.e., comparisons between two disparate systems for the provision of a service. The comparison might regard two different paradigms of provision of the service, e.g., classical vs virtualized implementations. Thus in [34], the implied (system) metric is the amount of power required to deliver 1 million packets per second of throughput through an evolved packet-core's (EPC) serving gateway (SGW). In [112–114], the objective is to minimize the amount of consumed power by virtualizing baseband processing functions, evolved packet core, customer premises equipment and radio access network functions.

3.1.6. Service Determinism: A Criterion Particular to the Telco Cloud

The NFV data plane demands determinism [37,96,115]. Strictly, ***service determinism*** is sought since the packet arrival process is generally unconstrained. The root cause of this need is to correctly size equipment resources to meet load demands, whether throughput, latency or jitter. ***This need is intense***, as it impinges on a PTNO's obligations, specified in legally binding service-level agreements (SLAs). Service determinism has been approached through the augmentation of GP hardware, with domain-specific architectures (DSAs, which we first referred to while describing problem category P1). Hardware-centric DSAs for the NFV data plane are constructed out of elements (or systems thereof) that can be divided into three groups.

1. **IO system architecture:** This group comprises the abstraction, ***directly at a peripheral interface***, of functionality that facilitates the virtualization of hardware, e.g., SR-IOV (used in PCI—Peripheral Component Interconnect) and N-port ID virtualization (Fibre Channel).
2. **Processor architecture:** This comprises architectural change that facilitates the partitioning of processor resources, such as multi-core processors and NUMA.
3. **Co-processing:** Compression and decompression, encryption and decryption and packet header processing are examples of high-volume tasks that can be offloaded to co-processing subsystems. Examples include Intel QuickAssist Technology (QAT) and TCP Offload Engine (TOE).

We further observe a set of approaches that are complementary to DSAs in the quest for determinism. These include:

1. **Large memory pages:** exploitation of the facility to organize virtual memory into larger pages than the general-purpose 4 KiB;
2. **User space programming:** diversion of the control of hardware resources away from the multi-service kernel, to single-common-use user space programs, e.g., DPDK and AF_XDP.

Notably, resource partitioning (see processor architecture, above, i.e., the second hardware-centric approach), combined with user space programming (the second complementary approach), realize the run-to-completion scheduling model [96].

We conclude with an observation on ***the cost of the current realizations of service determinism***. The (aforementioned) combination of core partitioning and user space programming has been widely adopted through Intel's popularization of DPDK, via Intel's open-source liaison efforts. The multi-core processor enables service determinism through an approach that is anathema to the principles of multiprogramming: the dedication of hardware to a specific task.

3.1.7. Direct or Indirect Measurement of Power in Virtualized Environments?

We observe that most research in modeling power consumption seeks to obviate the need for direct measurement through indirect measurement. This indirection consists of a measurement of resource use which has a discoverable relationship with power consumption by the entity hosting the resources. Modeling, here, has the objective of indirect measurement of a variable that is not directly accessible (power consumption by VEs), through others which have convenient and reliable instrumentation. The accessible variables are referred to as ***power proxies***. The RAPL interface provides a unique approach to measurement as it directly addresses power consumption. However, notwithstanding appearances of direct measurement, RAPL is actually based on a software model that uses performance-monitoring counters (PMCs) as predictor variables to measure power consumption. It is available in processors starting from the Sandy Bridge microarchitecture. RAPL measures the power consumption of different physical domains, where each domain consists of either cores, sockets, caches or GPU. We briefly comment on its accuracy through references to research that has investigated them.

1. In [98] the advantages and drawbacks of using RAPL were investigated. Different Intel architectures such as Sandy Bridge, Haswell and Skylake were used in the experiments to analyze RAPL's accuracy and overhead. Data collected were modeled using a linear model and a Generalized Additive Model (GAM). The accuracy of the predicted results was compared with the measured power consumption from a precise external hardware power meter where RAPL based models show 1.8–4.3% of error for the various architectures. The prediction accuracy of RAPL-based power models was also compared with those based on OS counters, where OS-based models show a higher error of 5–16%. Furthermore, the performance overhead (in terms of time) of using RAPL was studied at different sampling frequencies and for different application runs. Results show that even with a high sampling frequency of 1100 Hz, RAPL incurs overhead of not more than 2%. Some limitations of using RAPL include poor driver support to read energy counters, the overflow of registers due to their 32-bit size and the measurement of energy consumed by individual cores.
2. Another study to analyze the precision of RAPL is presented in [116], where only the dynamic change in power consumption is observed. An external power measurement unit (WattsUp Pro) is used as a reference for power measurement values. Intel Haswell and Skylake servers were used in the experiments to run different applications and to find the reliability of RAPL with the help of external power meters. However, in this research work, only two power domain packages (power consumption of whole socket and DRAM domain of RAPL) were observed. Applications such as dense matrix multiplication and 2D Fast Fourier Transform were used for server power profiling. Results show that the power measurement error varies with changing applications and its workload size. For different applications the average measurement error using RAPL was in the range of 13–73% considering WattsUp power meter as the ground truth. It was concluded that with the modern multi-core parallel processing and resource contention for shared resources, there is a complex non-linear relation between performance, workload size and energy consumption. Hence it is difficult to attain low error percentage for power measurement using on-chip sensors.

3.2. Research Gaps

Three significant challenges remain unaddressed, while a fourth requires further attention:

1. The modeling of containers' power consumption;
2. The effect of overcommitment on power efficiency;
3. The investigation and classification of DPDK applications;
4. The fourth challenge, which is starting to receive some attention [83,85,87,94], regards the modeling of power consumption by virtualized I/O.

Note that our treatment of research gaps does not address improvements in approaches. As we have already indicated, it is in our treatment of developments that more adaptable modeling methods are required to tackle the dimensionality of the field. Similarly, we do not include the lack of use of metrics of energy efficiency with research gaps, as it is a deficiency in the approaches, not a challenge in itself. Rather, here our attention focuses on where the more pressing challenges lie for the development of power and energy control of VEs.

Gap #1: Modeling of containers' power consumption. Few works [78,81,82] tackle containers from the perspective of their power consumption. However, at least for the telco cloud, VMs are no longer the destination (see, for example, [40,117]. Containers have replaced virtual machines as the base for deployment of virtual network functions. In [78], the approach(-set) taken is to:

1. Use representative workloads, e.g., HEVC (High Efficiency Video Coding) transcoding and machine learning image classification;
2. Customize the set of low-level instruments used to correlate power and energy consumption with workload characteristic.

This work develops a profiling tool. It provides guidance that is specific to the application and both the hardware and software aspects of the containing platform. "[P]olicy" for "tradeoff between energy, power and application performance" is the cited objective. Given the high-dimensionality of the core challenge, this set of approaches to the modeling and measurement of power consumption, may well be more coherent with the European Telecommunications Standards Institute's (ETSI) Management and Orchestration (MANO) standard. Such information would then be included in the infrastructure-resource-requirement's meta-data descriptors in the VNF package [118].

Gap #2: The effect of overcommitment on power efficiency. Overcommitment consists of the allocation of more capacity of some compute resource to VEs, than is physically installed. The concept is very similar to the oversubscription of telecommunications capacity to subscribers, such as when the arithmetic sum of capacities of access links exceeds the aggregating device's backhauling capacity to a central office/local exchange. In this context, overcommitment principally concerns processing cycles and memory space. As with oversubscription, there is an optimization problem to solve. One problem of interest to this survey's scope is understanding the relationship (say, ratio) of committed virtual resources to installed physical resources that optimizes the total cost of ownership (TCO) of cloud infrastructure:

- On the one hand, the facility to overcommit has a direct impact on the density of the packing of VEs (number of concurrently active VEs) on a virtualization-host-system, thereby reducing the TCO;
- On the other hand, overcommitment may reduce the power efficiency of a workload.

Had this challenge been tackled in any depth, or at least in any breadth, it would have merited a category of its own. Currently, however, we are only aware of a single study [65] that tackled this challenge. The results obtained strongly justify overcommitment of processor cores to vCPUs, for the case of transaction web service workloads, with the increase in throughput (measured in web interactions per second, or WIPS) increasing at a faster rate than power consumption. This behavior was observed well into overcommitment ratios of processor cores to vCPUs equal to 3 (three). The overcommitment of physical to VM memory was not investigated.

Gap #3: Investigation and classification of DPDK applications. In Section 3.3.2, we address the relationship between DPDK and power consumption. In the course of a public discussion in the forum offered by the North American Network Operators' Group (NANOG) [119], there emerged a need for clarity on DPDK's association with inefficiency in power consumption. Interest was particularly expressed in the knowledge of a classification of extant DPDK applications according to their power consumption, and contribution to the code base to improve power-hungry applications. In summary, among high-performance packet I/O frameworks, a classification of applications that use DPDK APIs in order to assess their power consumption and correct usage would be of particular interest, given their diffusion and potential.

Gap #4: Modeling of power consumption by virtualized I/O. Power consumption of network I/O has been investigated to some extent as this is central to the feasibility of network functions decoupled from hardware. While software and hardware solutions are already available (see Section 3.3.2), they require frequency and idling control targeted to their specific operating conditions. Notably, naive DPDK runs the processor core at its maximum power consumption, regardless of load. The exploitation of adaptive-rate

(AR) processing and low-power idle (LPI) should provide a means to save power while processing high networking loads. However, effective control of these means is still elusive, despite both using Xeon Haswell microarchitectures, refs. [83,87] reach opposite conclusions about the feasibility of processor core C-states. The former [83] finds LPI an effective means of reducing power consumption of packet forwarding (with limitation on latency) while the latter [87] finds it ineffective, preferring the use of the **pause** instruction. Furthermore, in [120], performance state transitions (P-state) are found to impose a high transition latency, while in [87], P-state regulation is the preferred approach. There is scope for research in the dynamic adaptation of the processor's operating state to save power.

3.3. Pitfalls

3.3.1. Power Consumption Does Not in General Increase Linearly with Processor Utilization

Notwithstanding advances made in identifying operating contexts that manifest a sub-linear power-utilization relationship [62,69], more recent publications [57,121–128] persist in using the linear model without acknowledging its limitations. The model is simple to use and has some foundations in research [129]. It has three premises, described here with regard to the operation of Microsoft Windows:

1. When Windows has no threads to run on a logical core, it schedules the idle thread [130].
2. The idle thread keeps the processor in a low-power state [131]. The specific state depends on the processor's green capabilities.
3. In the complement (non-idle time), the processor issues instructions at a constant rate.

This simple model has limitations [62,132]. It fails to take into account diverse processor operating contexts, some of which are coming to bear on current use cases. Specifically, the third premise is true only to the extent to which instructions are being fetched and data are being loaded from/stored to an instruction and data cache, respectively. Consider the context of 90% and greater hit ratios. At such cache hit ratios, the rate of instruction issue is expected to be narrowly distributed about its mean. By contrast, the lower the hit ratio at the cache level before main memory, the lower the fraction of non-idle time at which power consumption saturates. This saturation is strikingly illustrated in [62]. The variation of power consumption due to the execution of tests from the SPEC CPU2000 benchmark suite is shown. The power consumption diverges at 25% CPU utilization and the consumption of the processor-bound test (mesa) is greater than that of the memory-bound test (mcf) ***by a factor of about 2.6***.

Another good (albeit broad) illustration of this pitfall is given in [133]. Data on power consumption and CPU utilization under a standardized benchmark are plotted for four different physical server models. None of the relationships are linear. Neither is there a single, common behavior.

Researchers align themselves into two groups with regard to CPU utilization. One group favors (operating-)system metrics (of which CPU utilization is one metric) and the other favors event counters (microarchitectural instrumentation). The arguments posed by each group against the other's approach can be summarized as follows. The "system metrics" group claims that the "event counters" group's work is (a) not portable (at least across microarchitecture families) and (b) cannot be exercised without low-level access to the host (therefore, this approach cannot be exploited by user-level privileges) (see, e.g., [134] and [75]). The "event counters" group claims that CPU utilization is a workload-dependent predictor (see, e.g., [59]) and therefore cannot be used without re-training the model. Indeed, this modification to the "system-counters" approach is employed in [76], where it is stated that "[b]ecause of changes of VM's internal applications ... parameters must [be] recalculated automatically". ***Given these arguments, it seems that the system metrics group argument is weak: both system metrics and event counters require re-training if hardware is changed but system metrics lack the granularity to discriminate between workloads*** (cf. [62]). This means that CPU utilization can only be used as the sole predictor if it is re-trained with a change in workload. We deal with this problem, which we have termed the fallacy of the universal model, in Section 3.4.1.

With regards to the use of hardware threads (Intel®Hyper-Threading), we have observed that various works concur on the operating context under which a linear relationship is subject to the lowest error. This includes at least the following two conditions:

1. The processor cores are increasing their instruction issue rate in proportion to the fraction of time they spend busy. This implies that instruction and data cache hit ratios are high. This is simply the third premise;
2. Only one logical core is active per physical core at any given time [62,63,69]. Expressed alternatively, actual utilization must lie below half maximum utilization. The underlying cause is that activation of the second logical core employs fewer organizational units of the processor than activation of the first logical core.

The first condition is particularly problematic, as cache miss ratios are likely to be much higher in the context of virtualized environments. In such environments, the number of runnable threads is the sum of runnable threads controlled by independent operating systems. Evidently, this is higher than the expected number of runnable threads on a single server instance.

Other evidence of this "utilization trap" is not hard to find. In [57], the compute resource is stressed using cpulimit and stress-ng. The "cpulimit" utility runs a specified process image, then pauses and resumes it until a certain percentage utilization is reached [127]. The repetitive execution of a single process is highly likely to create conditions for very high instruction- and data-cache hit ratios. Such favorable hit ratios skew results towards the linear relationship between CPU utilization and power consumption.

3.3.2. DPDK Is Not Intrinsically Inefficient in Power Consumption

Research on power efficiency in DPDK applications [48,51,53,87] has portrayed DPDK as power inefficient. Before proceeding to our exposition of this pitfall, we distinguish between data, control and management planes. "Data plane" is a term used to refer to the infrastructural means that provide the capacity for exchange of customer (or subscriber, or end-user) data. It is complemented by a control plane, which refers to those means that facilitate the dynamic setup, maintenance and tear-down of a functional data plane. Another complementary part is the management plane. This includes the infrastructural means for a network operator to configure and monitor the control plane and the data plane, as well as intervene to correct faults arising in either plane. Simpler networks may have no control plane.

Now, we proceed to the exposition of the pitfall. In one particular case [48], it is claimed that "we found that a poll mode driver (PMD) thread accounted for approximately 99.7 percent CPU occupancy (a full core utilization)." The implication that seems to emerge here is that the PMD itself is driving this power consumption.

This portrayal is problematic at best and incorrect at worst. The referenced investigations of DPDK have indicated a very low power efficiency, **but they do not clearly distinguish between the responsibility of the DPDK API and the application using it** (the API). A recent, public thread [119] has emphasized the responsibility of the application developer in the avoidance of the naïve, "default approach" of busy polling. Such an approach would, indeed, poll network IO hardware continuously [48], truly fitting the epithet "spinning-hot" [37]. However, a broader (in the sense of including industrial correspondents) investigation [119] suggests that:

- Contrary to claims in [48], it is the driving behavior of OvS that is inefficient in power consumption;
- There are simpler, technical means of throttling a polling loop, including, say, the use of program code to interleave ACPI C1 states with polls according to traffic demands.

These latter observations cast doubt on the claim that automated frequency control is outside the scope of current frequency governors, since "the OS won't be able to distinguish whether it's under a heavy load" [48]. On the other hand, savings through NUMA awareness [51] (where transmit/receive port, memory and processor core are kept within the same NUMA node) is affirmed in [119].

3.3.3. Research on Power Models without Power-Relevant Context

This pitfall traps readers who attempt to draw conclusions from published research which lacks a clear specification of context relevant to power consumption. The pitfall is best illustrated through examples.

1. **Failure to emphasize context: idle power consumption vs frequency**. The dependence of idle power consumption on clock frequency is context sensitive. In [49], it is explicitly stated "idle power consumption remains constant, regardless of the CPU frequency ... across the whole frequency range" (1.6–2.6 GHz). The CPU is an Intel Xeon E5620. In [81,82], a quadratic relationship between idle power consumption and frequency is observed. Here, the CPU is an Intel Core i5 Haswell. In these two instances, emphasizing the restricted scope of findings would suffice to spare a reader from excessively broad inferences.
2. **Failure to emphasize context: idle power consumption vs hardware and software specification**. Enokido's and Takizawa's work [71] derives a power consumption model for a server while VMs run computation-bound processes. The servers used run on Intel Core i5-3230M processors. These processors are used in the mobile device market [135]. They are capable of low-power idle states [136]. CentOS 6.5 uses a tickless kernel [137]. Combined, these facts, relevant to the context of power consumption, provide a plausible explanation for the observed increment in power (denoted, in [71], by $minC_t$), when a core in a package is activated. Again, therefore, the scope of findings is likely to be restricted.
3. **Failure to fully define context: Configuration of power-relevant parameters.** We use [138] as an example. No reference is made to whether Hyper-Threading is enabled. This is essential to understanding how the ESXi vCPUs are created. Neither is any information given about how the vCPUs are related to physical (or logical) cores. Nor are we told how virtual network interfaces and switching are set up. ESXi version 5 offers both paravirtualization ("vmxnet") and emulation ("e1000") to implement virtual network interfaces. The impact on energy consumption of selecting a virtual network interface implemented by emulation can be expected to be high [56].

The examples cited illustrate the importance for a researcher of power models to qualify his/her results ***with a well-defined physical context***. Research into power models involves hard components and a diligent characterization thereof is essential to the acceptance of work as scientific research.

3.3.4. Benchmarks May Skew Power Consumption According to Their Organizational Dependencies

We have seen that both "cpulimit" and "stress-ng" do not produce generally representative measurements of power consumption. This observation is not limited to the measurement of power consumption. The use of kernels, toy programs and synthetic benchmarks to measure performance has been identified as unrepresentative [110] of general performance. Benchmarks are standardized workload generators that are used for the comparison of computer systems for a specific class of application. Unless this application class is a good representative of the application of the computer system in productive use, the power consumption measured under test is not a reliable predictor of that obtained during productive use. It is necessary to plan test workload generators in advance and state the limits of the validity of results. In [65], TPC-W is used, which is a transactional web benchmark that can simulate the business-oriented online web servers. The MySQL++ Java version of TPC-W benchmark, suitable for cloud applications, is used to generate the online traffic, where three different traffic profiles based on the browsing, purchasing and ordering of books are generated. The throughput measure for these servers is observed through the metric Web Interactions Per Second (WIPS).

3.3.5. Processor Organization Significantly Impacts Power Consumption

We illustrate this point with a wide-ranging example [139] which compares the Intel Xeon X5670 and AMD Opteron 2435.

1. Different idle loops (using no operation, pause, repetition, etc.) were tested to see their effect on power consumption of both systems. It was observed that the Intel Xeon has a loop stream detector, which disables the processor's features such as fetch and decode. On the other hand, the AMD processor has no hint to process these loops efficiently; hence, it consumed more power than the Intel processor.
2. A processor consumes a different amount of power depending upon the instruction (such as load, addition, multiplication, etc.) and the level in the memory hierarchy which is accessed by the instruction.
    a. For the Xeon, data throughput of all instructions from a particular memory hierarchy level is almost the same, but there is a difference in their power consumption. The 'load' operation consumes the lowest power compared with other instructions, and this holds true for all memory hierarchy levels. The reason is that the 'load' instruction just needs to load the content on the processor registers whereas 'add' and 'mul' operations are more computationally demanding.
    b. However, the AMD processor's behavior is the opposite. When the 'load' operation accesses the L1 cache, it achieves almost one-and-a-half times the data throughput of other operations and hence also consumes more power. This difference in resource utilization is due to the different microarchitecture of AMD processors, where the 'load' instruction is handled by many floating-point pipelines. Other instructions just use a single pipeline for their operations. Moreover, AMD processors have an exclusive cache level design, which requires write-back when evicting data among different cache levels. Since Intel's inclusive cache design does not require this function, it consumes less power. Within higher memory hierarchy levels (L2 or L3 or main memory), the AMD's computation ('add' and 'mul') and data transfer operations ('load') deliver roughly the same data throughput and consume roughly the same power.

3.3.6. Isolation of VE for Power Modeling and Measurement

Isolation of any VE from its hardware counterparts cannot be achieved completely [140]; thus, the assumption of measuring power consumption of an individual virtual entity irrespective of the hardware on which it is implemented is an illusion. The virtual infrastructure is composed of several components at both hardware and software level, where the effect of underlying hardware, OS and VNF technology can significantly impact the power consumption. Hence, isolation as well as the modeling of power consumption for an individual virtual component is difficult to obtain.

3.4. *Fallacies*

3.4.1. A Universal Power Model

We have suggested that the core challenge in modeling power consumption by VEs lies in the number of dimensions of variability. This has been demonstrated throughout this survey, where a number of generalizations have been addressed. In summary, the literature shows that:

1. Host power consumption does not generally have a linear relationship with processor utilization;
2. CPU-intensive workloads that repeatedly execute the same code skew power consumption results;

3. Network-intensive workloads are power- and time-consuming because they employ emulations of network switches, but the root cause (emulation in the hypervisor software switch) disappears with SR-IOV [141];
4. Host saturation must be taken into account in predicting VEs' power consumption;
5. Processor utilization (an architectural attribute) is insufficient to predict host power consumption and microarchitectural attributes, such as LLC misses, are necessary to predict host power consumption even for the same level of processor utilization.

This list, while not exhaustive, amply illustrates that the several dimensions of variability are significant in the determination of VE power consumption. A model claiming to determine power consumption as a function of fewer variables than the dimensions we have pointed out ***must be accompanied by a scoping region that limits its use***. While a precise scope may be an unrealistic demand, it is essential that guidance be given about the conditions of the use of the model. We now illustrate this point by using two examples from the corpus.

Example #1: Khan [50] compares energy efficiency (hash/J) obtained by scheduling process threads on additional cores, with that obtained by scheduling them on hardware threads on active cores (through Intel Hyper-Threading). He shows that the former is greater than the latter. In apparent contrast, Enokido and Takizawa [68] show that for a given data transmission rate through the uplink of a software virtual switch, greater energy efficiency (W/bps) is obtained by operating an additional hardware thread on an active core (through Intel Hyper-Threading), than operating an otherwise idle core. An important difference lies in the task's processing "intensity", i.e., the rate of the supply of instructions. While Khan's operations are tightly bound to the processor (cryptographic hashing), Enokido's and Takizawa's operations are distributed over the processor and network input/output. Without delving into detail, it is realistic to hypothesize that the average instructions per second demanded are far lower in the networking application, since the transmission of a large file (as is the case here) does not take place in one processing burst. The operating time is divided between the processor and the media channel. In such a scenario, the added capacity of the same-core hardware thread suffices.

Example #2: At the time of writing, the scope of validity (where the scope is a subspace of the seven-dimensional space) is typically only implicit. Notably, in [75], a "refined model" is used as a means of the accurate prediction of power consumption by virtual machines while running very specific benchmarks. It is also noteworthy that the authors contemplate a type of onboarding process wherein "new" VM entrants to a cloud are modelled as a prerequisite to their inclusion in the power-prediction system. Indeed, such a process is already intrinsic to the management and orchestration of virtual network functions. Just as the virtual deployment unit (VDU) nodes (in virtual network function descriptors (VNFDs)) store VM properties describing computer system resource demands, so can the descriptor template be extended to provide properties regarding power consumption demands. This "onboarding" is necessary since the selected predictors and modeling do not cover a sufficiently broad range of workload types, and a specific model must be learnt online, i.e., on the fly.

On the other hand, we propose that a comprehensive power model for existing implementations may be possible, under two conditions.

1. Every resource that consumes power must own a counter that registers its usage, or lack thereof, during a specific clock cycle.
2. Usage of a specific resource during a specific clock cycle must consume a constant amount of energy. This has the following corollaries:
    a. Energy consumption by the specific resource is a linear function of the number of clock cycles for which the resource is active;
    b. Power consumption of a system can be expressed as a linear combination of the total set of such resources;

c. The amount of energy consumption by a specific resource during a specific clock cycle must be independent of usage of other resources during any other clock cycle.

4. Conclusions

We conclude by summarizing our contributions (Section 4.1) and suggest a framework for future research into real-time, predictive models of power consumption by VEs (Section 4.2).

4.1. Contributions

We have identified seven dimensions of variability (workload type; virtualization agent; host resources and architecture; temperature; power attribution; co-hosted, concurrent VEs; and (clock) frequency of operation) and observed that the challenges tackled have aligned themselves with these dimensions. This breadth has prompted us to emphasize the fallacy of the universal power model: no single power model can cover all seven dimensions through the inclusion of variables and parameters. It is essential that prospective users of any such power model be aware of the limits of its scope. On the other hand, we have pointed out that the state-of-the-art includes adaptable modeling systems that handle variability in more than one dimension. Moreover, at least limited variability in two of the seven dimensions—workload type and concurrent operation of (multiple) VEs—is commonly validated, i.e., whether the model is truly capable of predicting power consumption under variability in workload type and the number of concurrent VEs.

PAD elicits trends in its proceedings through a sample of a corpus. In particular, the following examples are among the most noteworthy (but not the only) saliencies.

1. The challenge category tackled most frequently is that of how to load the VE's resources and how to quantify and measure the load; disaggregating the predictors and attributing measured power consumption to the individual VEs is the second most frequently tackled.
2. The variety of approaches that tackle a (category of) challenge is positively correlated to the frequency with which it is tackled.
3. Instrumentation of computing resources (e.g., instrumentation of microarchitectural artifacts) is the most commonly adopted approach (towards developments), surpassing instrumentation of the workload.
4. Resource-specific workloads (e.g., processor-specific) are the most commonly utilized, surpassing workloads representative of real use (e.g., web applications).
5. In developments, the most commonly developed model type is that where the power consumption (of the host or VE) is predicted in terms of workload characteristics; power consumption as a linear function of computing resources is second.
6. At the other end of the frequency range of developments, machine-learnt models comprise the second least frequent category of developed models, and adaptable, non-linear VE models are also very infrequent.

The process of parsing works and aggregating their codes is, however, only the principal ingredient in the overall progression towards the end goal: a set of themes that suitably profile the works in an area of research. Indeed, these codes and their interrelationships have elicited several research gaps, pitfalls and a fallacy, as well as evidence of the state-of-the-art and of research domains.

4.2. A Framework for Development of Real-Time, Predictive Power Models

Evolution of the research space on power consumption in virtualized environments now suggests the following framework for the further development of power models:

1. Division of the problem into:
    a. A modeling concern:
        i. What components to include;

   ii. What workload(s) to consider;
   iii. What state factors (temperature, frequency, performance and idle states) to account for.

  b. An attribution concern, i.e., how to attribute host power to VEs.

2. Division of the approach into:
   a. Microarchitectural instrumentation, based on intimate knowledge of the microarchitecture and the memory system;
   b. Granular attribution based on time-division multiplexing;
   c. Model selection.
3. Development of parameterized models, subject to continuing (if not continuous) optimization of the parameters under machine learning.

Author Contributions: Conceptualization, E.-V.D. and H.R.; methodology, E.-V.D.; validation, F.D; formal analysis, E.-V.D.; investigation, E.-V.D., F.D and H.R.; resources, F.D, E.-V.D. and H.R.; data curation, H.R.; writing—original draft preparation, E.-V.D. and H.R.; writing—review and editing, E.-V.D. and F.D; visualization, E.-V.D.; supervision, F.D. All authors have read and agreed to the published version of the manuscript.

Funding: This research has been partially supported by the Horizon 2020 5G-PPP Innovation Action 5G-INDUCE (Grant Agreement no. 101016941).

Institutional Review Board Statement: Not applicable.

Informed Consent Statement: Not applicable.

Data Availability Statement: The data presented in this study are available in https://github.com/humaira-salam/PowerMeasurementAndModelingRawData.

Conflicts of Interest: The authors declare no conflict of interest.

Abbreviations

5GPPP	5G Infrastructure Public Private Partnership
ACPI	Advanced Configuration and Power Interface
and	Automatic Distribution Network
AMD®	Advanced Micro Devices
ANN	Artificial Neural Network
API	Application Programming Interface
AR	Adaptive Rate
AF_XDP	Address Family eXpress Data Path
BMAP	Batch Markov Arrival Process
BS	Base Station
BSS	Business Support System
CAGR	Compound Annual Growth Rate
CAPEX	Capital Expenditure
CNF	Containerized Network Function
COE	Cost Of Energy
CPU	Central Processing Unit
CRM	Customer Relationship Management
CSP	Communications Service Provider
DAG	Directed Acyclic Graph
DG	Diesel Generator
DPDK	Data Plane Development Kit
DSA	Domain Specific Architecture
DVFS	Dynamic Voltage and Frequency Scaling
EPC	Evolved Packet Core
ES	Energy Storage
ETSI	European Telecommunications Standards Institute
GAM	Generalized Additive Model

GHG	Greenhouse Gas
HEVC	High Efficiency Video Coding
HW	Hardware
IDS	Intrusion Detection System
IO	Input/Output
IP	Internet Protocol
ISO	International Organization for Standardization
IT	Information Technology
ITU	International Telecommunications Union
KVM	Kernel Virtual Machine
LCA	Life Cycle Assessment
LLC	Last-level-cache
LTE	Long Term Evolution
LTE-A	Long Term Evolution-Advanced
LPI	Low-Power Idle
LXC	Linux Containers
MANO	Management and Orchestration (ETSI)
MB	Megabyte
MBS	Macro-cell Base Station
MILP	Mixed Integer Linear Programming
MIMO	Multiple-Input-Multiple-Output
MIPS	Millions of Instructions Per Second
MOS	Metal-Oxide-Semiconductor
MEC	Multi-Access Edge Computing
ML	Machine Learning
MNO	Mobile Network Operator
MTU	Maximum Transmission Unit
NANOG	North American Network Operators Group
NFV	Network Functions Virtualization
NIC	Network Interface Card
NPC	Net Present Cost
NUMA	Non-Uniform Memory Architecture
OPEX	Operational Expenditure
OS	Operating System
OSS	Operational Support System
PAD	Problem–Approach–Development
PCI	Peripheral Component Interconnect
PDU	Power Distribution Unit
PMD	Poll Mode Driver
PTN	Public Telecommunications Network
PV	Photovoltaic
QAT	QuickAssist ®Technology
QoS	Quality of Service
RAM	Random Access Memory
RAN	Radio Access Network
RAPL	Running Average Power Limit
RE	Renewable Energy
RES	Renewable Energy Source
RRH	Remote Radio Head
RU	Research Unit
SBA	Service-based architecture
SCBS	Small Cell Base Station
SDN	Software Defined Networking
SGW	Serving Gateway
SPEC	Standard Performance Evaluation Corporation
SR-IOV	Single Root Input Output Virtualization
SVM	Support Vector Machine
TCO	Total Cost of Ownership

TCP	Transmission Control Protocol
TOE	TCP Offload Engine
TOU	Time-Of-Use
TPC-W	Transaction Processing Performance Council—Web
UA	User Association
vCPU	Virtual CPU
VDU	Virtual Deployment Unit
VE	Virtual Entity
VM	Virtual Machine
VNF	Virtual Network Function
VNI	Visual Networking Index
WIPS	Web Interactions Per Second
WSC	Warehouse Scale Computing

Appendix A. Application of PAD to the Scope of Renewable Energy Use in Radio Access Networks

In this appendix, we digress to elaborate on the application of PAD to a topical scope: research into the use of renewable energy sources (RESs) in radio access networks (RANs). Our purpose here is to shed light on the mechanics of the method within a context that complements the principal context of this work.

Nine research units (RUs) [142–150] have been mined. The motivation driving this research is presented first, followed by a tabulation of the P-, A- and D-node codes and a categorization of the codes. We do not proceed to thematic analysis, as this sample is not large enough to support it. Patterns are detected and presented as ***clusters*** (part of the PAD method), but enunciation of a thematic analysis requires the confidence emanating from a broader search through the corpus.

Appendix A.1. Motivation

A clear sense of motivation emerges: OPEX (energy cost) must be controlled and greenhouse gas (GHG) emissions must be reduced. The Jevons paradox (alternatively expressed in the Khazzoom–Brookes postulate) emerges again: the deployment density of 5G base stations (BSs) and 64/32-channel massive multiple-input multiple-output (MIMO) antenna arrays drives "the power consumption of a 5G BS to be 2-3 times that of an ordinary 4G BS" [144]. Other works are motivated by the need to control capital expenditure (CAPEX) [142,147,148] and understand how RES harvesters compare with conventional hydrocarbon-fueled generators, therein seeking either to reduce CAPEX or to determine the true cost of wholly autonomous operation under the condition of power supply from RES.

Motivation, therefore, may be tersely summarized as OPEX control, GHG emission control, CAPEX control and autonomous operation. Within Figure 1, motivation may be located at the center of the circle, corresponding to the core challenge driving a research area.

Appendix A.2. Tabulation and Clustering of P-, A-, and D-Node Codes

The tabulation below (Tables A1–A3) summarizes two rounds about the sample: a first round of coding (code: "terse, dense representations of a verbose articulation of a concept", see [1]) and a second round wherein a first categorization (note the prefixes, for example "RAN architecture", "Radio path", "Comparison", etc.) took place. Figures A1–A3 illustrate how the nodes have been clustered. An aspect of PAD that bears emphasis here is that the coding and clustering process is an ***iterative*** one: the final categorization only emerges at the end of several cycles of refinement, comprising consolidation, observation, disaggregation and re-aggregation as further similarities between the codes emerge. Hence, the categories shown in Figures A1–A3 are by no means final, but they are an essential transitory stage in the progression towards the final categorization.

Appendix A.2.1. Table of P-Node Codes

Table A1. Challenges and representative challenge questions.

Problem #	Problem Code
1 [142]	How can power supply system parameters be harvested in a convenient manner?
2 [142]	How does CAPEX of diesel fuel generators compare with that of RES plant?
3 [143]	How can the limitations of individual RES be overcome?
4 [144]	Spatio-temporal distribution of light intensity and communication demand misaligns energy supply and load.
5 [144]	Disparate ownership of RES plant and automatic distribution network (ADN) infrastructure leads to concerns with security and privacy.
6 [144]	How can participation of communications service providers (CSPs) in renewable energy (RE) harvesting be encouraged?
7 [144]	How can real-time energy scheduling be improved to avoid loss of RE, over-discharge of energy storage (ES) and balance between supply and demand?
8 [145]	How well do machine learning (ML) prediction models based on past traffic and RE harvesting support RE use and Quality of Service (QoS)?
9 [145]	Which long-term strategy of operation should be used over prediction intervals: (a) turn off/on micro-BSs to match network demand OR (b) turn off/on micro-BSs to match RE capacity?
10 [146]	How can Time-Of-Use (TOU) electricity pricing, RE production and fairness of allocation of throughput to users be integrated into a modeling scheme?
11 [147]	Which system of autonomous power supply, whether renewable or not, optimizes one or more of the following objectives: net present cost (NPC), cost of energy (COE) and GHG emissions?
12 [148]	How can we exploit multiple RES to design a RAN that minimizes grid energy consumption while serving at least 95% of users?
13 [149,150]	How can user association (UA) be exploited over a short timescale in a hybrid-powered (RES + hydrocarbons) RAN to minimize grid power consumption?
14 [150]	How can traffic and weather forecasts be exploited over a long timescale to minimize grid power consumption

Appendix A.2.2. Table of A-Node Codes

Table A2. Approaches.

Approach #	Approach Code
1 [142]	Use a wireless sensor network to measure parameters relevant to power supply.
2 [142]	Systematic survey of vendors and consultation with base station operators.
3 [143]	Broad comparison of research and industrial (technical specifications) output concerning use of RES in powering BSs.
5 [144]	Organize multiple 5G BSs carrying photovoltaic (PV) panels into an energy-aggregation system architecture.
6 [144]	Regulate the mid-term exchange of energy using contract theory.
7 [144]	Regulate the short-term use of energy by a Lyapunov optimization.
8 [145]	RAN architecture: a single service area served by one LTE-A macro-BS (MBS) and six micro-BSs providing additional capacity in the service area's hot spots.
9 [145]	Energy system architecture: PV panels + energy storage (ES – batteries) + power grid.

Table A2. *Cont.*

Approach #	Approach Code
10 [145]	Traffic model: Traffic data provided by a large Italian mobile network operator (MNO).
11 [145]	RE production data estimated using PVWATT.
12 [145]	QoS: percentage of lost traffic.
13 [146]	RAN architecture: a single service area served by one LTE-A MBS and two micro-BSs providing additional capacity in the service area's hot spots.
14 [146]	Energy system architecture: MBS uses power grid, while one micro-BS is supplied by the power grid and solar energy, and the other micro-BS is supplied by the power grid and wind energy.
15 [146]	Energy system economics: power-grid energy priced according to Time-Of-Use.
16 [146]	Game theory applied to balance opposing objectives of (a) minimization of energy cost or energy consumption and (b) maximization of user-throughput fairness.
17 [146]	Alternative energy cost optimization approaches: (a) reduction by decreasing BS energy consumption; (b) reduction through direct optimization of energy cost.
18 [146]	Simulation through bespoke software.
19 [147]	Energy system architecture: Alternative combinations: (a) PV + diesel generator (DG) + ES (b) PV + ES (c) DG + ES (d) DG.
20 [147]	Design, simulation and optimization, using HOMER; optimization objective is minimization of NPC.
21 [147]	Energy consumption model: power demand data provided by a Kuwaiti mobile network operator (MNO).
22 [147]	Energy system component selection: where relevant, mono-crystalline silicon (Mono-Si) is preferred to polycrystalline silicon (Poly-Si) and Lithium-Ion batteries is preferred over other technologies.
23 [148]	MILP formulation.
24 [148]	Traffic model: Traffic data provided by a mobile network operator (MNO), both voice and data, for a cell of 0.3 km^2, showing number of simultaneous active users at any time during a day.
25 [148]	RAN architecture: eight service areas, each served by one LTE-A MBS and four micro-BSs.
26 [148]	BS power consumption model: linear power model (developed in earlier work), with hourly load factor reflecting days of the week and of the weekend.
27 [148]	RE production data (hourly) obtained from Italian RES operator's (Terna) website.
28 [148]	Radio path: 3D map of suburban area.
29 [148]	Genetic algorithm.
30 [149]	RAN architecture: a single service area served by one 5G MBS and twenty-four small cell base stations (SCBSs), with microwave backhaul from MBS to core and SCBS to MBS.
31 [149]	Energy system architecture: MBS uses power grid, while SCBSs are supplied by RESs.
32 [149]	Random-uniform spatial distribution of low-data-rate and high-data-rate users, respectively.
33 [149]	Green heuristic: prioritize green SCBSs in the UA process.
34 [150]	RAN architecture: a single service area served by one LTE MBS and thirty-six/sixteen/four small cell base stations (SCBSs).

Table A2. *Cont.*

Approach #	Approach Code
35 [150]	Energy system architecture: MBS uses power grid, while SCBSs are supplied by wind turbines without ES.
36 [150]	Energy system architecture: MBS uses power grid, while SCBSs are supplied by PV with ES.
37 [150]	MILP formulation of problem of UA with minimal grid power consumption.
38 [150]	Short timescale control of UA through three alternative policies: green-SCBS-preferred vs minimization of energy consumption vs best-signal-level + transmission rate (greedy).
39 [150]	Long timescale control through use of model predictive control (MPC) based on traffic flow and weather forecast.
40 [150]	RE generation model: Weibull probability distribution to characterize wind speed, with average wind speed obtained from a two-year study of mean wind speed profiles in the Moscow air basin.
41 [150]	RE generation model: Solar radiation data from the Colombian Institute of Hydrology, Meteorology and Environmental Studies (IDEAM).
42 [150]	BS power consumption model: from ICT-EARTH (linear with traffic load specified for time slot t).
43 [150]	Traffic model: Short timescale, inhomogeneous Poisson point process (without temporal variability of traffic) and random-uniform spatial distribution of users in coverage area.
44 [150]	Traffic model: Traffic data provided by a mobile network operator (MNO).
45 [149]	Radio path: path loss calculation using rural microcell and data rate obtained using 5G R16.
46 [149,150]	Simulation, using Matlab.
47 [146]	RE generation model: solar energy production/15 minutes expressed using Gaussian distribution; wind-energy production expressed as constant.

Appendix A.2.3. Table of D-Node Codes

Table A3. Developments.

Development #	Development Code
1 [142]	Enablement of historical analysis of sensor data.
2 [142]	Diesel consumption model for BSs powered by a DG.
3 [142]	Comparison: energy efficiency of BSs powered by a DG with those powered by a RES
4 [142]	Capacity planning of RES power systems for autonomous operation.
5 [142]	CAPEX: PV system < DG system < PV+wind turbine system.
6 [143]	OPEX and energy efficiency of BS in rural/remote areas is significantly improved by using RES.
7 [143]	Autonomy improved, possibly indefinitely, using several RESs.
8 [144]	Optimal, long timescale contract between energy aggregator and CSP, based on availability, during a given time interval of ES discharge capacity at the CSP's BS site
9 [144]	Lyapunov optimization achieving short timescale objective of stable energy levels in ESs and full absorption of energy harvested from PV panels.
10 [144]	Case study, showing near-complete (99.87%) absorption of PV energy.

Table A3. *Cont.*

Development #	Development Code
11 [145]	ML algorithms: correct prediction of traffic only important around utilization threshold that activates/deactivates micro-BS sleep mode.
12 [145]	Comparison: ML algorithms: simple Block Linear Regression performs trade-off between energy consumption and QoS as well as more complex ones such as 24- and 48-Artificial-Neural-Network (ANN).
13 [145]	Energy consumption control strategy: turn micro-BSs on or off according to availability of RE (the sum of the predicted harvest and the stored) has less dependence on ML algorithm than according to demand.
14 [146]	Energy-cost model, integrating TOU grid electricity cost, solar energy and wind energy.
15 [146]	Comparison: algorithms: algorithm minimizing energy cost performs better than one minimizing energy consumption in the overall objective of optimizing energy cost and energy price and energy consumption and fairness of allocation of throughput to users.
16 [147]	Comparison: extensive characterization of the relative merits of four electricity generation systems: (a) PV + ES (b) PV + DG + ES (c) DG + ES (d) DG, in five dimensions: (i) optimal dispatch strategy (ii) NPC (iii) COE (iv) GHG (v) days of autonomy.
17 [148]	Genetic algorithm: minimization of energy cost and wastage in a multiple-RES system.
18 [148]	RES mix recommendation for a specific service area.
19 [149]	Application of green algorithm in 5G HetNet: for reactive, short timescale control, consisting of UA with a green (SC)BS.
20 [149]	Comparison: Green algorithm vs best-signal strength algorithm: UA with grid-powered MBS is reduced almost by half.
21 [149]	Comparison: Green algorithm vs best-signal strength algorithm: Users denied service significantly reduced.
22 [149]	Comparison: Green algorithm vs best-signal strength algorithm: Reduction in MBS energy consumption is limited by the need to backhaul SCBS traffic through the MBS.
23 [149]	Comparison: Green algorithm vs best-signal strength algorithm: Cell site's energy efficiency is improved (MBS's energy efficiency is worsened but SCBSs' improvement over-compensates).
24 [150]	Green algorithm: for reactive, short timescale control of UA with a green (SC)BS.
25 [150]	Comparison: Green algorithm vs on-grid only operation: Energy consumption reduced, and reduction improves markedly when growing from four RE-powered SCBSs (11.1%) to sixteen (26.6%), then to thirty-six (33.7%).
26 [150]	Comparison: Green algorithm vs best-signal-level policy vs best-signal-level + transmission-rate (greedy) heuristic: Energy consumption: significantly better than best-signal-level policy, slightly better than greedy heuristic.
27 [150]	Comparison: Green algorithm vs discrete (MILP) optimizer vs best-signal-level + transmission-rate (greedy) heuristic: Energy efficiency: better than MILP optimizer but roughly same as greedy heuristic.
28 [150]	Infrastructure dimensioning: Green algorithm's reduction of energy consumption does not decrease in direct proportion to number of SCBSs.
29 [150]	Comparison: Green algorithm vs discrete (MILP) optimizer vs best-signal-level policy: Throughput: slightly worse than best-signal-level, slightly better than discrete (MILP) optimizer.

Table A3. *Cont.*

Development #	Development Code
30 [150]	MPC for predictive, long timescale control, using 0- (reactive), 1- and 5-h prediction horizons.
31 [150]	Comparison: MPC vs on-grid only operation: Energy consumption: reduced by 13.1%.
32 [150]	Infrastructure dimensioning: optimal storage capacity can be determined and enables savings up to 22% with respect to on-grid only operation.

Appendix A.3. Clustering the P-, A- and D-Nodes

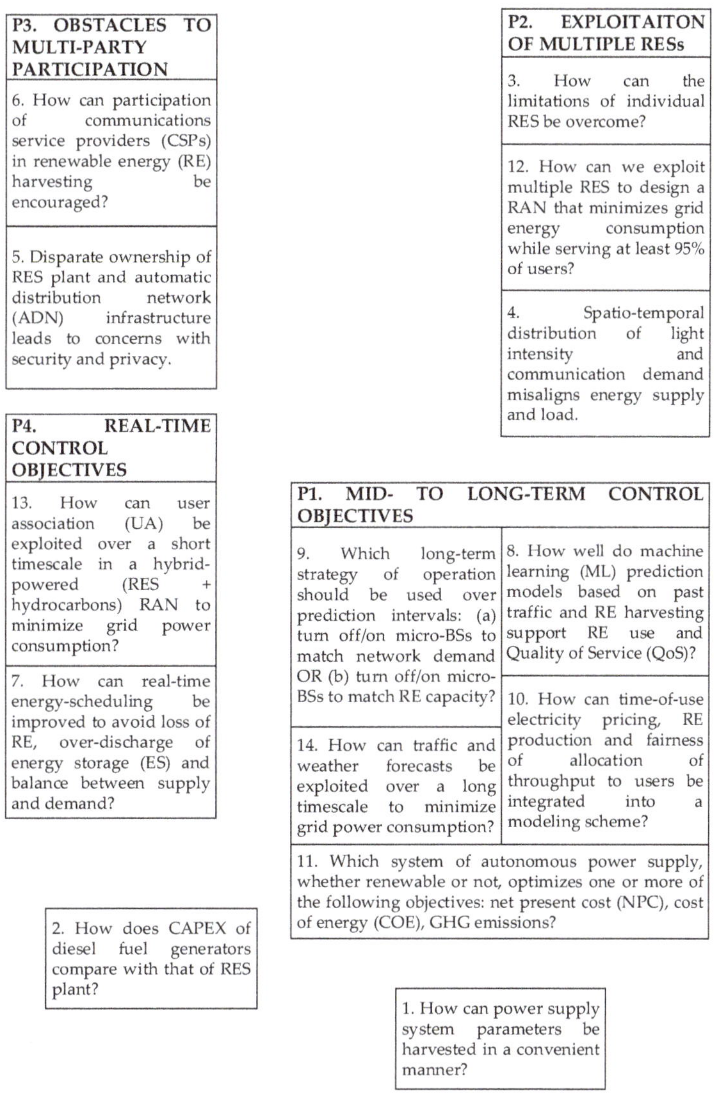

Figure A1. A first categorization of the problem nodes. P1: 8 [145], 9 [145], 10 [146], 11 [147], 14 [150]; P2 3 [143], 4 [144], 12 [148]; P3 5 [144], 6 [144]; P4 7 [144], 13 [149,150]. Unclustered: 1 [142], 2 [142].

A1. FORMAL METHODS

39. Long timescale control through use of model predictive control (MPC) based on traffic flow and weather forecast

38. Short timescale control of UA through three alternative policies: green-SCBS-preferred vs minimization of energy consumption vs best-signal-level + transmission rate (greedy)

37. MILP formulation of problem of UA with minimal grid power consumption

17. Alternative energy cost optimization approaches: (a) reduction by decreasing BS energy consumption; (b) reduction through direct optimization of energy cost

23. MILP formulation

29. Genetic algorithm

26. BS power consumption model: linear power model (developed in earlier work), with hourly load factor reflecting days of the week and of the weekend.

6. Regulate the mid-term exchange of energy using contract theory.

7. Regulate the short-term use of energy by a Lyapunov optimization.

16. Game theory applied to balance opposing objectives of (a) minimization of energy cost or energy consumption and (b) maximization of user-throughput fairness

33. Green heuristic: prioritize green SCBSs in the UA process

A2. TRAFFIC MODEL

43. Traffic model: Short timescale, inhomogeneous Poisson point process (without temporal variability of traffic) and random-uniform spatial distribution of users in coverage area

10. Traffic model: Traffic data provided by a large Italian mobile network operator (MNO)

24. Traffic model: Traffic data provided by a mobile network operator (MNO) – both voice and data, for a cell of 0.3km^2, showing number of simultaneous active users at any time during a day.

32. Random-uniform spatial distribution of low-data-rate and high-data-rate users respectively.

44. Traffic model: Traffic data provided by a mobile network operator (MNO)

A3. RAN ARCHITECTURE

8. RAN architecture: a single service area served by one LTE-A macro-BS (MBS) and six micro-BSs providing additional capacity in the service area's hot spots.

13. RAN architecture: a single service area served by one LTE-A MBS and two micro-BSs providing additional capacity in the service area's hot spots.

25. RAN architecture: eight service areas, each served by one LTE-A MBS and four micro-BSs

30. RAN architecture: a single service area served by one 5G MBS and 24 small cell base stations (SCBSs), with microwave backhaul from MBS to core and SCBS to MBS

34. RAN architecture: a single service area served by one LTE MBS and 36/16/4 small cell base stations (SCBSs)

A4. ENERGY SYSTEM ARCHITECTURE

5. Organize multiple 5G BSs carrying photovoltaic (PV) panels into an energy-aggregation system architecture.

14. Energy system architecture: MBS uses power grid, while one micro-BS is supplied by the power grid and solar energy, and the other micro-BS is supplied by the power grid and wind energy.

9. Energy system architecture: PV panels + energy storage (ES - batteries) + power grid

19. Energy system architecture: Alternative combinations: (a) PV + diesel generator (DG) + ES (b) PV + ES (c) DG + ES (d) DG

31. Energy system architecture: MBS uses power grid, while SCBSs are supplied by RESs.

35. Energy system architecture: MBS uses power grid, while SCBSs are supplied by wind turbines without ES

36. Energy system architecture: MBS uses power grid, while SCBSs are supplied by PV with ES

(a)

Figure A2. *Cont.*

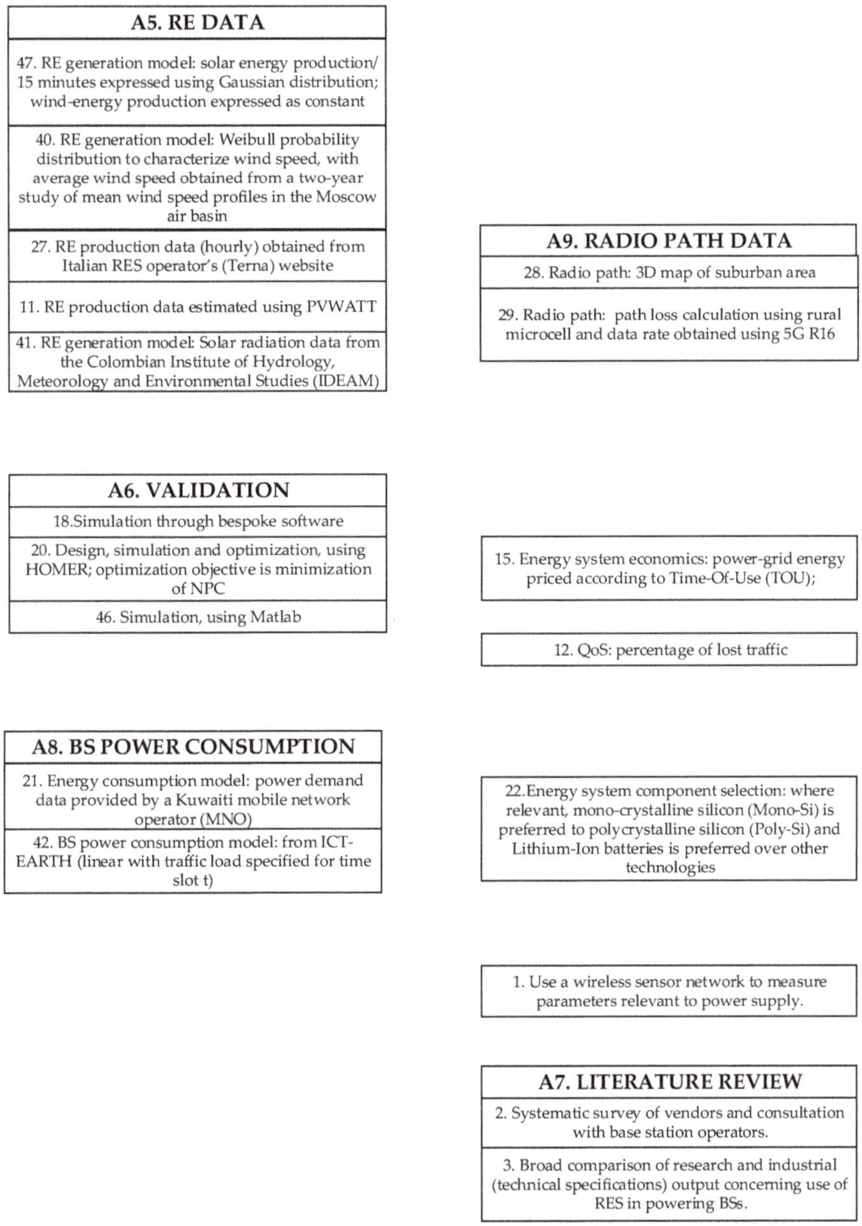

(**b**)

Figure A2. (**a**) A first categorization of the approach nodes. (**b**) A first categorization of the approach nodes. A1: 6 [144], 7 [144], 16 [146], 17 [146], 23 [148], 26 [148], 29 [148], 33 [149], 37 [150], 38 [150], 39 [150]; A2: 10 [145], 24 [148], 32 [149], 43 [150], 44 [150]; A3: 8 [145], 13 [146], 25 [148], 30 [149], 34 [150]; A4: 5 [144], 9 [145], 14 [146], 19 [147], 31 [149], 35 [150], 36 [150]. A5 11 [145], 27 [146], 40 [150], 41 [150], 47 [146]; A6 18 [146], 20 [147], 46 [149,150]; A7 2 [142], 3 [143]; A8 21 [147], 42 [150]; A9 28 [148], 29 [149]. Unclustered: 1 [142], 12 [145], 15 [146], 22 [147].

D1. RELATIVE MERITS OF ALGORITHMS IN ENERGY CONSUMPTION AND EFFICIENCY

20. Comparison: Green algorithm vs best-signal strength algorithm: UA with grid-powered MBS is reduced almost by half

12. Comparison: ML algorithms: simple Block Linear Regression performs trade-off between energy consumption and QoS as well as more complex ones like 24- and 48-Artificial-Neural-Network (ANN)

15. Comparison: algorithms: algorithm minimizing energy cost performs better than one minimizing energy consumption in the overall objective of optimizing energy cost & energy price & energy consumption & fairness of allocation of throughput to users

31. Comparison: MPC vs on-grid only operation: Energy consumption: reduced by 13.1%

29. Comparison: Green algorithm vs discrete (MILP) optimizer vs best-signal-level policy: Throughput: slightly worse than best-signal-level, slightly better than discrete (MILP) optimizer

21. Comparison: Green algorithm vs best-signal strength algorithm: Users denied service significantly reduced

22. Comparison: Green algorithm vs best-signal strength algorithm: Reduction in MBS energy consumption is limited by the need to backhaul SCBS traffic through the MBS

23. Comparison: Green algorithm vs best-signal strength algorithm: Cell site's energy efficiency is improved (MBS's energy efficiency is worsened but SCBSs' improvement over-compensates).

25. Comparison: Green algorithm vs on-grid only operation: Energy consumption reduced and reduction improves markedly when growing from 4 RE-powered SCBSs (11.1%) to 16 (26.6%), then to 36 (33.7%)

26 Comparison: Green algorithm vs best-signal-level policy vs best-signal-level + transmission-rate (greedy) heuristic: Energy consumption: significantly better than best-signal-level policy, slightly better than greedy heuristic

27. Comparison: Green algorithm vs discrete (MILP) optimizer vs best-signal-level + transmission-rate (greedy) heuristic: Energy efficiency: better than MILP optimizer but roughly same as greedy heuristic

D2. ALGORITHMS AND OTHER FORMAL SOLUTIONS

8. Optimal, long timescale contract between energy aggregator and CSP, based on availability, during a given time interval of ES discharge capacity at the CSP's BS site.

30. MPC for predictive, long timescale control, using 0-(reactive), 1- and 5-hour prediction horizons

17. Genetic algorithm: minimization of energy cost and wastage in a multiple-RES system

24. Green algorithm: for reactive, short timescale control of UA with a green (SC)BS

9. Lyapunov optimization achieving short timescale objective of stable energy levels in ESs and full absorption of energy harvested from PV panels

D3. AUTONOMOUS OPERATION ON RES(s)

7. Autonomy improved, possibly indefinitely, using several RESs

4. Capacity planning of RES power systems for autonomous operation

18. RES mix recommendation for a specific service area

10. Case study, showing near-complete (99.87%) absorption of PV energy.

(a)

Figure A3. *Cont.*

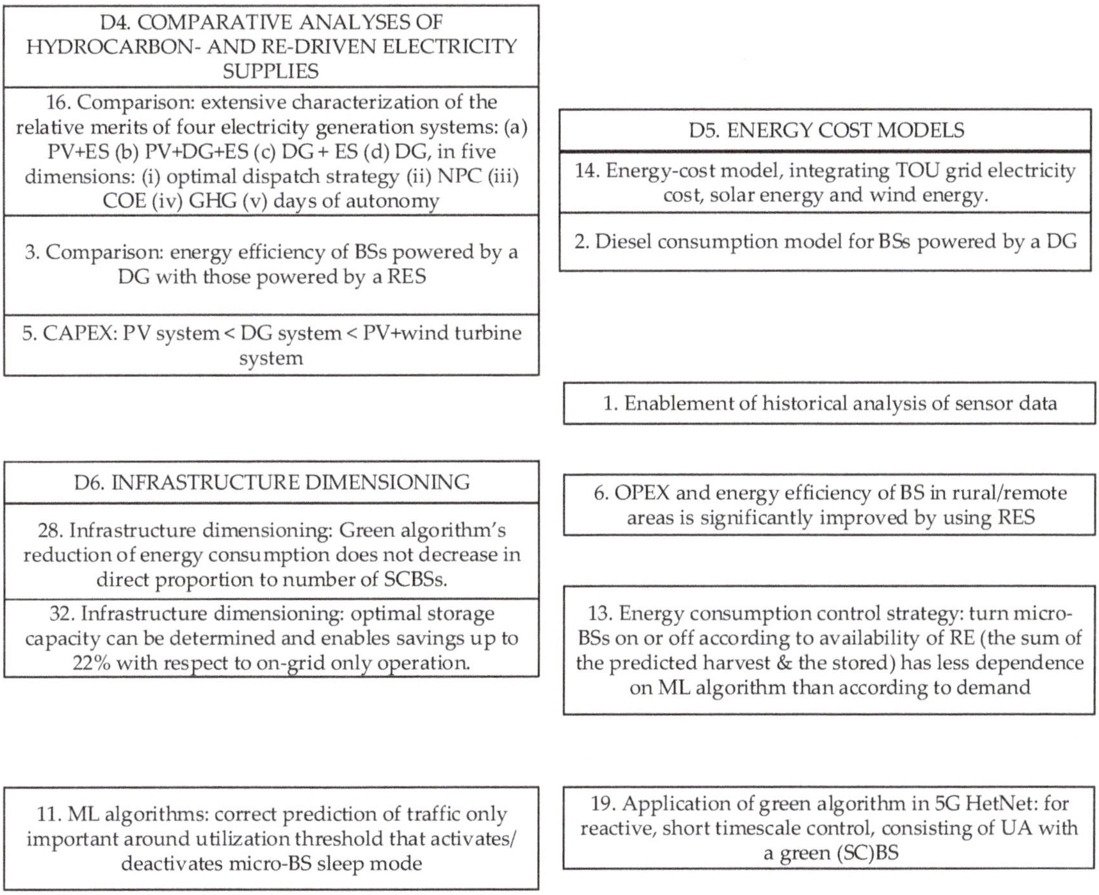

(**b**)

Figure A3. (**a**) A first categorization of the development nodes. (**b**) A first categorization of the development nodes. D1: 12 [145], 15 [146], 20 [149], 21 [149], 22 [149], 23 [149], 25 [150], 26 [150], 27 [150], 29 [150], 31 [150]; D2: 8 [144], 9 [144], 17 [148], 24 [150], 30 [150]. D3: 4 [142], 7 [143], 10 [144], 18 [148]. D4: 3 [142], 5 [142], 16 [147]; D5: 2 [142], 14 [146]; D6: 28 [150], 32 [150]. Unclustered: 1 [142], 6 [143], 11 [145], 13 [145], 19 [149].

References

1. Depasquale, E.-V.; Salam, H.A.; Davoli, F. PAD: A graphical and numerical enhancement of structural coding to facilitate thematic analysis of a literature corpus. *MethodsX* **2022**, *9*, 101633. [CrossRef]
2. Bolla, R.; Bruschi, R.; Davoli, F.; Cucchietti, F. Energy Efficiency in the Future Internet: A Survey of Existing Approaches and Trends in Energy-Aware Fixed Network Infrastructures. *IEEE Commun. Surv. Tutor.* **2011**, *13*, 223–244. [CrossRef]
3. Idzikowski, F.; Chiaraviglio, L.; Cianfrani, A.; Vizcaíno, J.L.; Polverini, M.; Ye, Y. A Survey on Energy-Aware Design and Operation of Core Networks. *IEEE Commun. Surv. Tutor.* **2016**, *18*, 1453–1499. [CrossRef]
4. Maaloul, R.; Chaari, L.; Cousin, B. Energy saving in carrier-grade networks: A survey. *Comput. Stand. Interfaces* **2018**, *55*, 8–26. [CrossRef]
5. Delforge, P.; Whitney, J. Data Center Efficiency Assessment—Scaling up Energy Efficiency Across the Data Center Industry: Evaluating Key Drivers and Barriers. Natural Resources Defense Council, IP:14.08-A, August 2014. Available online: https://www.nrdc.org/sites/default/files/data-center-efficiency-assessment-IP.pdf (accessed on 3 October 2020).
6. *Cisco Visual Networking Index: Forecast and Methodology, 2014–2019*; Cisco Systems: San Jose, CA, USA, 2015.
7. *Cisco Visual Networking Index: Forecast and Methodoolgy, 2015–2020*; Cisco Systems: San Jose, CA, USA, 2016.

8. *Cisco Visual Networking Index: Forecast and Methodology, 2016–2021*; Cisco Systems: San Jose, CA, USA, 2017.
9. *Cisco Visual Networking Index: Forecast and Trends, 2017–2022*; Cisco Systems: San Jose, CA, USA, 2019.
10. Ishii, K.; Kurumida, J.; Sato, K.; Kudoh, T.; Namiki, S. Unifying Top-Down and Bottom-Up Approaches to Evaluate Network Energy Consumption. *J. Light. Technol.* **2015**, *33*, 4395–4405. [CrossRef]
11. *Metro Network Traffic Growth: An Architecture Impact Study*; Alcatel-Lucent: Paris, France, 2013.
12. Norton, W.B. *The 2014 Internet Peering Playbook: Connecting to the Core of the Internet*; DrPeering Press: London, UK, 2014.
13. Global Internet Phenomena Report. Sandvine, 2013. Available online: https://www.sandvine.com/hubfs/downloads/archive/2013-2h-global-internet-phenomena-report.pdf (accessed on 4 July 2020).
14. Global Internet Phenomena—Africa, Middle East & North America. December 2015. Available online: https://www.sandvine.com/hubfs/downloads/archive/2015-global-internet-phenomena-report-africa-middle-east-and-north-america.pdf (accessed on 4 July 2020).
15. *Global Internet Phenomena Report*; Sandvine: Plano, TX, USA, 2019.
16. Lee, U.; Rimac, I.; Hilt, V. Greening the internet with content-centric networking. In Proceedings of the 1st International Conference on Energy-Efficient Computing and Networking, Passau, Germany, 13–15 April 2010; pp. 179–182. [CrossRef]
17. Guan, K.; Atkinson, G.; Kilper, D.C.; Gulsen, E. On the Energy Efficiency of Content Delivery Architectures. In Proceedings of the 2011 IEEE International Conference on Communications Workshops (ICC), Kyoto, Japan, 5–9 June 2011; pp. 1–6. [CrossRef]
18. Choi, N.; Guan, K.; Kilper, D.C.; Atkinson, G. In-network caching effect on optimal energy consumption in content-centric networking. In Proceedings of the 2012 IEEE International Conference on Communications (ICC), Ottawa, ON, Canada, 15 June 2012; pp. 2889–2894. [CrossRef]
19. Osman, N.I.; El-Gorashi, T.; Elmirghani, J.M.H. Reduction of energy consumption of Video-on-Demand services using cache size optimization. In Proceedings of the 2011 Eighth International Conference on Wireless and Optical Communications Networks, Toronto, ON, Canada, 1–4 May 2011; pp. 1–5. [CrossRef]
20. Llorca, J.; Tulino, A.M.; Guan, K.; Esteban, J.; Varvello, M.; Choi, N.; Kilper, D.C. Dynamic in-network caching for energy efficient content delivery. In Proceedings of the 2013 IEEE INFOCOM, Turin, Italy, 14–19 April 2013; pp. 245–249. [CrossRef]
21. Mandal, U.; Chowdhury, P.; Lange, C.; Gladisch, A.; Mukherjee, B. Energy-efficient networking for content distribution over telecom network infrastructure. *Opt. Switch. Netw.* **2013**, *10*, 393–405. [CrossRef]
22. Modrzejewski, R.; Chiaraviglio, L.; Tahiri, I.; Giroire, F.; Le Rouzic, E.; Bonetto, E.; Musumeci, F.; Gonzalez, R.; Guerrero, C. Energy efficient content distribution in an ISP network. In Proceedings of the 2013 IEEE Global Communications Conference (GLOBECOM), Atlanta, GA, USA, 9–13 December 2013; pp. 2859–2865. [CrossRef]
23. Abji, N.; Tizghadam, A.; Leon-Garcia, A. Energy efficient content delivery in service provider networks with content caching. In Proceedings of the 2015 IEEE Online Conference on Green Communications (OnlineGreenComm), Piscataway, NJ, USA, 10–12 November 2015; pp. 23–29. [CrossRef]
24. Savi, M.; Ayoub, O.; Musumeci, F.; Li, Z.; Verticale, G.; Tornatore, M. Energy-efficient caching for Video-on-Demand in Fixed-Mobile Convergent networks. In Proceedings of the 2015 IEEE Online Conference on Green Communications (OnlineGreenComm), Piscataway, NJ, USA, 10–12 November 2015; pp. 17–22. [CrossRef]
25. Jayasundara, C.; Nirmalathas, A.; Wong, E.; Chan, C. Improving Energy Efficiency of Video on Demand Services. *IEEEOSA J. Opt. Commun. Netw.* **2011**, *3*, 870–880. [CrossRef]
26. Fratini, R.; Savi, M.; Verticale, G.; Tornatore, M. Using replicated video servers for VoD traffic offloading in integrated metro/access networks. In Proceedings of the 2014 IEEE International Conference on Communications (ICC), Sydney, Australia, 10–14 June 2014; pp. 3438–3443. [CrossRef]
27. Pascale, E.D.; Payne, D.B.; Ruffini, M. Bandwidth and energy savings of locality-aware P2P Content Distribution in next-generation PONs. In Proceedings of the 2012 16th International Conference on Optical Network Design and Modelling (ONDM), Colchester, UK, 17–20 April 2012; pp. 1–6. [CrossRef]
28. Bolla, R.; Bruschi, R.; Davoli, F.; Depasquale, E.-V. Energy-efficient management and control in video distribution networks: 'legacy' hardware based solutions and perspectives of virtualized networking environments. In *Guide to Greening Video Distribution Networks—Energy-Efficient Internet Video Delivery*; Popescu, A., Ed.; Springer: Berlin/Heidelberg, Germany, 2018; pp. 25–57.
29. ITU-T, Y. *3001 Future Networks: Objectives and Design Goals*; International Telecommunication Union: Geneva, Switzerland, 2011.
30. Pervasive Mobile Virtual Services. Expert Advisory Group of the European Technology Platform Networld 2020, July 2016. Available online: https://www.networld2020.eu/wp-content/uploads/2014/02/SRIA_final.pdf (accessed on 23 December 2022).
31. Bouchat, C.; Paul, M.; Allan, D.; Dalle, G. 5G Wireless Wireline Convergence Atchitecture. Broadband Forum, TR-470. August 2020. Available online: https://www.broadband-forum.org/technical/download/TR-470.pdf (accessed on 3 October 2020).
32. *ETSI Standard ETSI ES 203 237 V1.1.1*; Environmental Engineering (EE); Green Abstraction Layer (GAL); Power management Capabilities of the Future Energy Telecommunication Fixed Network Nodes. ETSI: Sophia Antipolis, France, 2014.
33. Environmental Engineering (EE); Green Abstraction Layer (GAL); Power Management Capabilities of The Future Energy Telecommunication Fixed Network Nodes; Enhanced Interface for Power Management in Network Function Virtualisation (NFV) Environments. ETSI, Sophia Antipolis, ETSI ES 203 682 V1.1.0. December 2019. Available online: https://www.etsi.org/deliver/etsi_es/203600_203699/203682/01.01.00_50/es_203682v010100m.pdf (accessed on 31 October 2022).

34. Bolla, R.; Bruschi, R.; Davoli, F.; Lombardo, C.; Pajo, J.F.; Sanchez, O.R. The dark Side of Network Functions Virtualization: A Perspective on the Technological Sustainability. In Proceedings of the IEEE International Conference on Communications (ICC), Paris, France, 21–25 May 2017; pp. 1–7. [CrossRef]
35. *ETSI, RGR/NFV-001ed121*; Network Functions Virtualisation (NFV); Use Cases. ETSI: Sophia Antipolis, France, 2017.
36. Shi, W. Making Sense of the Telco Cloud. Telecoms.com, May 2020. Available online: https://telecoms.com/opinion/making-sense-of-the-telco-cloud/ (accessed on 27 June 2020).
37. CNF Friendly Networking for Telco/Edge Kubernetes Platforms. September 2020. Available online: https://www.brighttalk.com/webcast/12229/438035/cnf-friendly-networking-for-telco-edge-kubernetes-platforms (accessed on 28 September 2020).
38. Crawshaw, J. Deconstructing the Telco Cloud. Light Read. April 2019. Available online: https://www.lightreading.com/cloud/deconstructing-the-telco-cloud/a/d-id/753874 (accessed on 25 July 2020).
39. Ayoubi, S.E.E.; Jeux, S.; Marache, F.; Pujol, F.; Fallgren, M.; Spapis, P.; Yang, C.; Widaa, A.; Markendahl, J.; Ghanbari, A.; et al. Refined Scenarios and Requirements, Consolidated Use Cases, and Qualitative Techno-Economic Feasibility Assessment. D1.1. January 2016. Available online: https://metis-ii.5g-ppp.eu/wp-content/uploads/deliverables/METIS-II_D1.1_v1.0.pdf (accessed on 23 December 2022).
40. Muthurajan, J. Why Use Cloud-Native Containerized Network Functions (CNFs)? Intel, White paper. Available online: https://www.intel.com/content/www/uk/en/communications/why-containers-and-cloud-native-functions-paper.html (accessed on 25 July 2020).
41. Vettor, R.; Smith, S.A. Architecting Cloud Native .NET Applications for Azure. October 2020. Available online: https://docs.microsoft.com/en-us/dotnet/architecture/cloud-native/ (accessed on 30 October 2022).
42. Logan, M. CNCF F Technical Oversight Committee (TOOC). GitHub, March 2021. Available online: https://github.com/cncf/toc (accessed on 30 October 2022).
43. DPDK. DPDK, 2021. Available online: https://www.dpdk.org/ (accessed on 25 February 2021).
44. Dayarathna, M.; Wen, Y.; Fan, R. Data Center Energy Consumption Modeling: A Survey. *IEEE Commun. Surv. Tutor.* **2016**, *18*, 732–794. [CrossRef]
45. Costa, G.D.; Pierson, J.-M.; Fontoura-Cupertino, L. Effectiveness of Neural Networks for Power Modeling for Cloud and HPC: It's Worth It! *ACM Trans. Model. Perform. Eval. Comput. Syst.* **2020**, *5*, 1–36. [CrossRef]
46. Ismail, L.; Materwala, H. Computing Server Power Modeling in a Data Center: Survey, Taxonomy, and Performance Evaluation. *ACM Comput. Surv.* **2020**, *53*, 1–34. [CrossRef]
47. Lin, W.; Shi, F.; Wu, W.; Li, K.; Wu, G.; Mohammed, A.-A. A Taxonomy and Survey of Power Models and Power Modeling for Cloud Servers. *ACM Comput. Surv.* **2020**, *53*, 100:1–100:41. [CrossRef]
48. Fu, S.; Liu, J.; Zhu, W. Multimedia Content Delivery with Network Function Virtualization: The Energy Perspective. *IEEE Multimed.* **2017**, *24*, 38–47. [CrossRef]
49. Chen, Q.; Grosso, P.; van der Veldt, K.; de Laat, C.; Hofman, R.; Bal, H. Profiling Energy Consumption of VMs for Green Cloud Computing. In Proceedings of the 2011 IEEE Ninth International Conference on Dependable, Autonomic and Secure Computing, Sydney, Australia, 12–14 December 2011; pp. 768–775. [CrossRef]
50. Khan, N. Investigating Energy Efficiency of Physical and Virtual Machines in Cloud Computing. Univeristy of Oslo, 2017. Available online: https://www.duo.uio.no/handle/10852/61321 (accessed on 7 November 2019).
51. Li, G.; Zhang, D.; Li, Y.; Li, K. Toward energy-efficiency optimization of pktgen-DPDK for green network testbeds. *China Commun.* **2018**, *15*, 199–207. [CrossRef]
52. Intel®Ethernet Server Adapter I350: Product Brief. Intel, March 2021. Available online: https://www.intel.com/content/www/us/en/products/docs/network-io/ethernet/10-25-40-gigabit-adapters/ethernet-i350-server-adapter-brief.html (accessed on 9 February 2021).
53. Xu, Z.; Liu, F.; Wang, T.; Xu, H. Demystifying the energy efficiency of Network Function Virtualization. In Proceedings of the 2016 IEEE/ACM 24th International Symposium on Quality of Service (IWQoS), Beijing, China, 20–21 June 2016; pp. 1–10. [CrossRef]
54. Zhang, Z.; Fu, S. Macropower: A coarse-grain power profiling framework for energy-efficient cloud computing. In Proceedings of the 30th IEEE International Performance Computing and Communications Conference, Orlando, FL, USA, 17–19 November 2011; pp. 1–8. [CrossRef]
55. Morabito, R. Power Consumption of Virtualization Technologies: An Empirical Investigation. In Proceedings of the 2015 IEEE/ACM 8th International Conference on Utility and Cloud Computing (UCC), Limassol, Cyprus, 7–10 December 2015; pp. 522–527. [CrossRef]
56. Shea, R.; Wang, H.; Liu, J. Power consumption of virtual machines with network transactions: Measurement and improvements. In Proceedings of the IEEE INFOCOM 2014—IEEE Conference on Computer Communications, Toronto, ON, Canada, 27 April–2 May 2014; pp. 1051–1059. [CrossRef]
57. Aldossary, M.; Djemame, K.; Alzamil, I.; Kostopoulos, A.; Dimakis, A.; Agiatzidou, E. Energy-aware cost prediction and pricing of virtual machines in cloud computing environments. *Future Gener. Comput. Syst.* **2019**, *93*, 442–459. [CrossRef]
58. Li, Y.; Wang, Y.; Yin, B.; Guan, L. An Online Power Metering Model for Cloud Environment. In Proceedings of the 2012 IEEE 11th International Symposium on Network Computing and Applications, Cambridge, MA, USA, 23–25 August 2012; pp. 175–180. [CrossRef]

59. Wen, C.; Long, X.; Yang, Y.; Ni, F.; Mu, Y. System Power Model and Virtual Machine Power Metering for Cloud Computing Pricing. In Proceedings of the 2013 Third International Conference on Intelligent System Design and Engineering Applications, Hong Kong, 16–18 January 2013; pp. 1379–1382. [CrossRef]
60. Arroba, P.; Risco-Martín, J.L.; Zapater, M.; Moya, J.M.; Ayala, J.L.; Olcoz, K. Server Power Modeling for Run-time Energy Optimization of Cloud Computing Facilities. *Energy Procedia* **2014**, *62*, 401–410. [CrossRef]
61. Wu, W.; Lin, W.; Peng, Z. An intelligent power consumption model for virtual machines under CPU-intensive workload in cloud environment. *Soft Comput.* **2017**, *21*, 5755–5764. [CrossRef]
62. Dhiman, G.; Mihic, K.; Rosing, A. T system for online power prediction in virtualized environments using Gaussian mixture models. In Proceedings of the 47th Design Automation Conference, Anaheim, CA, USA, 13–18 June 2010; p. 807. [CrossRef]
63. Salam, H.A.; Davoli, F.; Timm-Giel, A. Improving Prediction Accuracy for Power Consumption in Virtual Environments. In Proceedings of the 2019 29th International Telecommunication Networks and Applications Conference (ITNAC), Auckland, New Zealand, 27–29 November 2019; pp. 1–6. [CrossRef]
64. Salam, H.A.; Davoli, F.; Carrega, A.; Timm-Giel, A. Towards Prediction of Power Consumption of Virtual Machines for Varying Loads. In Proceedings of the 2018 28th International Telecommunication Networks and Applications Conference (ITNAC), Sydney, Australia, 21–23 November 2018; pp. 1–6. [CrossRef]
65. MCallau-Zori; Samoila, L.; Orgerie, A.-C.; Pierre, G. An experiment-driven energy consumption model for virtual machine management systems. *Sustain. Comput. Inform. Syst.* **2018**, *18*, 163–174. [CrossRef]
66. Chinprasertsuk, S.; Gertphol, S. Power model for virtual machine in cloud computing. In Proceedings of the 2014 11th International Joint Conference on Computer Science and Software Engineering (JCSSE), Chonburi, Thailand, 14–16 May 2014; pp. 140–145. [CrossRef]
67. Enokido, T.; Takizawa, M. The Extended Power Consumption Model to Perform Computation Type Application Processes on Virtual Machines. In Proceedings of the 2016 10th International Conference on Complex, Intelligent, and Software Intensive Systems (CISIS), Fukuoka, Japan, 6–8 July 2016; pp. 15–22. [CrossRef]
68. Enokido, T.; Takizawa, M. Power Consumption Model of a Server to Perform Communication Type Application Processes on Virtual Machines. In Proceedings of the 2015 10th International Conference on Broadband and Wireless Computing, Communication and Applications (BWCCA), Krakow, Poland, 4–6 November 2015; pp. 275–282. [CrossRef]
69. Krishnan, B.; Amur, H.; Gavrilovska, A.; Schwan, K. VM power metering: Feasibility and challenges. *ACM SIGMETRICS Perform. Eval. Rev.* **2011**, *38*, 56. [CrossRef]
70. Isci, C.; Martonosi, M. Runtime power monitoring in high-end processors: Methodology and empirical data. In Proceedings of the 36th Annual IEEE/ACM International Symposium on Microarchitecture, 2003. MICRO-36, Wahington, DC, USA, 3–6 December 2003; pp. 93–104. [CrossRef]
71. Enokido, T.; Takizawa, M. Power Consumption and Computation Models of Virtual Machines to Perform Computation Type Application Processes. In Proceedings of the 2015 Ninth International Conference on Complex, Intelligent, and Software Intensive Systems, Santa Catarina, Brazil, 8–10 July 2015; pp. 126–133. [CrossRef]
72. Aldossary, M.; Djemame, K. Performance and Energy-based Cost Prediction of Virtual Machines Live Migration in Clouds. In Proceedings of the 8th International Conference on Cloud Computing and Services Science, Funchal, Madeira, Portugal, 19–21 March 2018; pp. 384–391. [CrossRef]
73. Aldossary, M.; Djemame, K. Energy-based Cost Model of Virtual Machines in a Cloud Environment. In Proceedings of the 2018 Fifth International Symposium on Innovation in Information and Communication Technology (ISIICT), Amman, Jordan, 31 October 2018; pp. 1–8. [CrossRef]
74. Alzamil, I.; Djemame, K. Energy Prediction for Cloud Workload Patterns. In *Economics of Grids, Clouds, Systems, and Services*; Springer: Cham, Switzerland, 2017; pp. 160–174. [CrossRef]
75. Kansal, A.; Zhao, F.; Liu, J.; Kothari, N.; Bhattacharya, A.A. Virtual machine power metering and provisioning. In Proceedings of the 1st ACM Symposium on Cloud Computing—SoCC '10, Indianapolis, IN, USA, 10–11 June 2010; p. 39. [CrossRef]
76. Waßmann, I.; Versick, D.; Tavangarian, D. Energy consumption estimation of virtual machines. In Proceedings of the 28th Annual ACM Symposium on Applied Computing, Coimbra, Portugal, 18–22 March 2013; pp. 1151–1156. [CrossRef]
77. Gu, C.; Shi, P.; Shi, S.; Huang, H.; Jia, X. A Tree Regression Based Approach for VM Power Metering. *IEEE Access* **2015**, *3*, 610–621. [CrossRef]
78. Silva-de-Souza, W.; Iranfar, A.; Bráulio, A.; Zapater, M.; Xavier-de-Souza, S.; Olcoz, K.; Atienza, D. Containergy—A Container-Based Energy and Performance Profiling Tool for Next Generation Workloads. *Energies* **2020**, *13*, 2162. [CrossRef]
79. Veni, T.; Bhanu, S.M.S. Prediction Model for Virtual Machine Power Consumption in Cloud Environments. *Procedia Comput. Sci.* **2016**, *87*, 122–127. [CrossRef]
80. SPEC Benchmarks. December 2020. Available online: https://www.spec.org/benchmarks.html (accessed on 30 October 2022).
81. Phung, J.; Young, C.L.; Zomaya, A.Y. Application-Agnostic Power Monitoring in Virtualized Environments. In Proceedings of the 2017 17th IEEE/ACM International Symposium on Cluster, Cloud and Grid Computing (CCGRID), Madrid, Spain, 14–17 May 2017; pp. 335–344. [CrossRef]
82. Phung, J.; Lee, Y.C.; Zomaya, A.Y. Lightweight Power Monitoring Framework for Virtualized Computing Environments. *IEEE Trans. Comput.* **2020**, *69*, 14–25. [CrossRef]

83. Bolla, R.; Bruschi, R.; Davoli, F.; Pajo, J.F. A Model-Based Approach Towards Real-Time Analytics in NFV Infrastructures. *IEEE Trans. Green Commun. Netw.* **2020**, *4*, 529–541. [CrossRef]
84. Bohra, A.E.H.; Chaudhary, V. VMeter: Power modelling for virtualized clouds. In Proceedings of the 2010 IEEE International Symposium on Parallel Distributed Processing, Workshops and Phd Forum (IPDPSW), Atlanta, GA, USA, 19–23 April 2010; pp. 1–8. [CrossRef]
85. Zorello, L.M.M.; Vieira, M.G.T.; Tejos, R.A.G.; Rojas, M.A.T.; Meirosu, C.; de Brito Carvalho, T.C.M. Improving Energy Efficiency in NFV Clouds with Machine Learning. In Proceedings of the 2018 IEEE 11th International Conference on Cloud Computing (CLOUD), San Francisco, CA, USA, 2–7 July 2018; pp. 710–717. [CrossRef]
86. Stoess, J.; Lang, C.; Bellosa, F. Energy Management for Hypervisor-Based Virtual Machines. In Proceedings of the USENIX Annual Technical Conference, Santa Clara, CA, USA, 17–22 June 2007.
87. Li, X.; Cheng, W.; Zhang, T.; Ren, F.; Yang, B. Towards Power Efficient High Performance Packet I/O. *IEEE Trans. Parallel Distrib. Syst.* **2020**, *31*, 981–996. [CrossRef]
88. Virtio—KVM. 2020. Available online: https://www.linux-kvm.org/page/Virtio (accessed on 27 August 2020).
89. Poll Mode Driver—Data Plane Development Kit 20.11.0-rc0 Documentation. December 2014. Available online: https://doc.dpdk.org/guides/prog_guide/poll_mode_drv.html (accessed on 24 December 2022).
90. The Zeek Network Security Monitor. Zeek, March 2020. Available online: https://zeek.org/ (accessed on 2 October 2020).
91. Snort—Network Intrusion Detection & Prevention System. May 2015. Available online: https://www.snort.org/ (accessed on 2 October 2020).
92. Open vSwitch. 2020. Available online: https://www.openvswitch.org/ (accessed on 27 August 2020).
93. Rizzo, L.; Lettieri, G. VALE, a switched ethernet for virtual machines. In Proceedings of the Proceedings of the 8th international conference on Emerging networking experiments and Technologies, New York, NY, USA, 6–9 December 2012; pp. 61–72. [CrossRef]
94. Bruschi, R.; Davoli, F.; Lago, P.; Pajo, J.F. Joint Power Scaling of Processing Resources and Consolidation of Virtual Network Functions. In Proceedings of the 5th IEEE International Conference on Cloud Networking (Cloudnet), Pisa, Italy, 3–6 October 2016; pp. 70–75. [CrossRef]
95. Wu, W.; Lin, W.; He, L.; Wu, G.; Hsu, C. A Power Consumption Model for Cloud Servers Based on Elman Neural Network. *IEEE Trans. Cloud Comput.* **2019**, *9*, 1268–1277. [CrossRef]
96. Łukasik, R. Run-to-completion model in Data Plane processing—Nokia Wrocław. December 2019. Available online: https://nokiawroclaw.pl/blog/post/run-to-completion-model-in-data-plane-processing/ (accessed on 30 September 2020).
97. Kai, C.; Blesson, V.; Peter, K.; Nikolopoulos, D.S. Power Modelling for Heterogeneous Cloud-Edge Data Centers. *Adv. Parallel Comput.* **2018**, *32*, 804–813. [CrossRef]
98. Khan, K.N.; Hirki, M.; Niemi, T.; Nurminen, J.K.; Ou, Z. RAPL in Action: Experiences in Using RAPL for Power Measurements. *ACM Trans. Model. Perform. Eval. Comput. Syst.* **2018**, *3*, 9:1–9:26. [CrossRef]
99. Molnar, I. Performance Counters for Linux, v8. 2009. Available online: https://lwn.net/Articles/336542/ (accessed on 23 July 2020).
100. Bolla, R.; Bruschi, R.; Carrega, A.; Davoli, F. Green Networking with Packet Processing Engines: Modeling and Optimization. *IEEE ACM Trans. Netw.* **2014**, *22*, 110–123, Correction in *IEEE ACM Trans. Netw.* **2017**, *22*, 110–123. [CrossRef]
101. TPC-W Homepage. 2021. Available online: http://www.tpc.org/tpcw/ (accessed on 11 February 2021).
102. Advanced Configuration and Power Interface (ACPI) Specification—ACPI Specification 6.4 Documentation. January 2021. Available online: https://uefi.org/specs/ACPI/6.4/ (accessed on 2 March 2021).
103. Keong, C.K.; Wei, K.T.; Ghani, A.A.A.; Sharif, K.Y. Toward using software metrics as indicator to measure power consumption of mobile application: A case study. In Proceedings of the 2015 9th Malaysian Software Engineering Conference (MySEC), Kuala Lumpur, Malaysia, 16–17 December 2015; pp. 172–177. [CrossRef]
104. Li, K. Optimal configuration of a multicore server processor for managing the power and performance tradeoff. *J. Supercomput.* **2012**, *61*, 189–214. [CrossRef]
105. Mobius, C.; Dargie, W.; Schill, A. Power Consumption Estimation Models for Processors, Virtual Machines, and Servers. *IEEE Trans. Parallel Distrib. Syst.* **2014**, *25*, 1600–1614. [CrossRef]
106. Technical Committee ISO/TC 207/SC 5 Life Cycle Assessment—Principles and Framework. Standard ISO 14040:2006. July 2006. Available online: https://www.iso.org/cms/render/live/en/sites/isoorg/contents/data/standard/03/74/37456.html (accessed on 27 July 2020).
107. L.1310: Energy Efficiency Metrics and Measurement Methods for Telecommunication Equipment. July 2017. Available online: https://www.itu.int/rec/T-REC-L.1310/en (accessed on 27 July 2020).
108. Schien, D.; Shabajee, P.; Yearworth, M.; Preist, C. Modeling and Assessing Variability in Energy Consumption During the Use Stage of Online Multimedia Services. *J. Ind. Ecol.* **2013**, *17*, 800–813. [CrossRef]
109. Measurement Method for Energy Efficiency of Network Functions Virtualization. November 2018. Available online: https://www.itu.int/rec/T-REC-L.1361/en (accessed on 23 July 2020).
110. Hennessy, J.L.; Patterson, D.A. *Computer Architecture: A Quantitative Approach*; Morgan Kaufmann: Burlington, MA, USA, 2017.

111. Inoue, T.; Aikebaier, A.; Enokido, T.; Takizawa, M. Algorithms for Selecting Energy-Efficient Storage Servers in Storage and Computation Oriented Applications. In Proceedings of the 2012 IEEE 26th International Conference on Advanced Information Networking and Applications, Fukuoka, Japan, 26–29 March 2012; pp. 920–927. [CrossRef]
112. Al-Zubaedi, W.; Al-Raweshidy, H.S. A parameterized and optimized BBU pool virtualization power model for C-RAN architecture. In Proceedings of the IEEE EUROCON 2017—17th International Conference on Smart Technologies, Ohrid, Macedonia, 6–8 July 2017; pp. 38–43. [CrossRef]
113. Mijumbi, R.; Serrat, J.; Gorricho, J.-L.; Rubio-Loyola, J. On the Energy Efficiency Prospects of Network Function Virtualization. ArXiv151200215 Cs. December 2015. Available online: http://arxiv.org/abs/1512.00215 (accessed on 15 July 2020).
114. Dlamini, T. Softwarization in Future Mobile Networks and Energy Efficient Networks. In *Mobile Computing*; Intech Open: London, UK, 2019; p. 51. [CrossRef]
115. Dumitrescu, C. Design Patterns for Packet Procesing Applications on Multi-core Intel Architecture. December 2008. Available online: https://www.intel.com/content/dam/www/public/us/en/documents/white-papers/ia-multicore-packet-processing-paper.pdf (accessed on 29 September 2020).
116. Hähnel, M.; Döbel, B.; Völp, M.; Härtig, H. Measuring energy consumption for short code paths using RAPL. *ACM SIGMETRICS Perform. Eval. Rev.* **2012**, *40*, 13–17. [CrossRef]
117. TelecomTV The 5G Core is Vital to Deliver the Promise of 5G. TelecomTV, July 2020. Available online: https://www.telecomtv.com/content/intel-vsummit-5g-ran-5g-core/the-5g-core-is-vital-to-deliver-the-promise-of-5g-39164/ (accessed on 17 February 2021).
118. Bhyrraju, R.; Druta, D.; Brenner, M.; Gautam, D.; Kitroser, I.; Wocalewski, T.; Wright, S.; Feng, A.; Shao, G.; Liu, J.; et al. 'VNF Descriptor and Packaging Specification', ETSI, Sophia Antipolis Cedex-FRANCE, ETSI GS NFV-IFA 011. September 2019. Available online: https://www.etsi.org/deliver/etsi_gs/NFV-IFA/001_099/011/02.07.01_60/gs_NFV-IFA011v020701p.pdf (accessed on 24 December 2022).
119. The NANOG February 2021 Archive by Thread. NANOG Mailing List Archive. 2021. Available online: https://mailman.nanog.org/pipermail/nanog/2021-February/thread.html (accessed on 5 March 2021).
120. Bolla, R.; Bruschi, R.; Carrega, A.; Davoli, F. Green network technologies and the art of trading-off. In Proceedings of the 2011 IEEE Conference on Computer Communications Workshops (INFOCOM WKSHPS), Shanghai, China, 10–15 April 2011; pp. 301–306. [CrossRef]
121. Dargie, W. A Stochastic Model for Estimating the Power Consumption of a Processor. *IEEE Trans. Comput.* **2015**, *64*, 1311–1322. [CrossRef]
122. Garraghan, P.; Al-Anii, Y.; Summers, J.; Thompson, H.; Kapur, N.; Djemame, K. A Unified Model for Holistic Power Usage in Cloud Datacenter Servers. In Proceedings of the 2016 IEEE/ACM 9th International Conference on Utility and Cloud Computing (UCC), Shanghai, China, 6–9 December 2016; pp. 11–19.
123. Monteiro, A.F.; Loques, O. Quantum Virtual Machine: A Scalable Model to Optimize Energy Savings and Resource Management. In Proceedings of the 2015 27th International Symposium on Computer Architecture and High Performance Computing (SBAC-PAD), Florianopolis, Brazil, 17–21 October 2015; pp. 194–201. [CrossRef]
124. Sharma, S.; Chang, V.; Tim, U.S.; Wong, J.L.H.; Gadia, S.K. Cloud-based emerging services systems. *Int. J. Inf. Manag.* **2016**, *640*, 71–91. [CrossRef]
125. Zakarya, M.; Gillam, L. An Energy Aware Cost Recovery Approach for Virtual Machine Migration. In *Economics of Grids, Clouds, Systems, and Services*; Springer: Cham, Switzerland, 2017; pp. 175–190. [CrossRef]
126. Zhao, T.; Wu, J.; Zhou, S.; Niu, Z. Energy-delay tradeoffs of virtual base stations with a computational-resource-aware energy consumption model. In Proceedings of the 2014 IEEE International Conference on Communication Systems, San Jose, CA, USA, 3–6 November 2014; pp. 26–30. [CrossRef]
127. Marletta, A. CPU Usage Limiter for Linux. 2012. Available online: http://cpulimit.sourceforge.net/ (accessed on 27 July 2020).
128. Son, J.; He, T.; Buyya, R. CloudSimSDN-NFV: Modeling and simulation of network function virtualization and service function chaining in edge computing environments. *Softw. Pract. Exp.* **2019**, *49*, 1748–1764. [CrossRef]
129. Vasan, A.; Sivasubramaniam, A.; Shimpi, V.; Sivabalan, T.; Subbiah, R. Worth their watts? An empirical study of datacenter servers. In Proceedings of the HPCA —16 2010 The Sixteenth International Symposium on High-Performance Computer Architecture, Bangalore, India, 9–14 January 2010; pp. 1–10. [CrossRef]
130. Slomon, D.A.; Yosifovich, P.; Russinovich, M.E.; Lonescu, A. System Architecture, Processes, Threads, Memory Management, and More, 7th ed. Pearson Education, 2017. Available online: https://www.oreilly.com/library/view/windows-internals-seventh/9780133986471/ (accessed on 27 July 2020).
131. PU Analysis. Micrsoft, May 2017. Available online: https://docs.microsoft.com/en-us/windows-hardware/test/wpt/cpu-analysis (accessed on 27 July 2020).
132. Goraczko, M. Joulemeter: Computational Energy Measurement and Optimization. Microsoft Research, February 2010. Available online: https://www.microsoft.com/en-us/research/project/joulemeter-computational-energy-measurement-and-optimization/ (accessed on 27 July 2020).
133. Lin, W.; Wu, W.; Wang, H.; Wang, J.Z.; Hsu, C.-H. Experimental and quantitative analysis of server power model for cloud data centers. *Future Gener. Comput. Syst.* **2018**, *86*, 940–950. [CrossRef]

134. Aldossary, M.; Alzamil, I.; Djemame, K. Towards Virtual Machine Energy-Aware Cost Prediction in Clouds. In *Economics of Grids, Clouds, Systems, and Services*; Springer: Cham, Switzerland, 2017; pp. 119–131. [CrossRef]
135. Intel®Core^{TM} i5-3230M Processor (3M Cache, up to 3.20 GHz) rPGA Product Specifications. May 2014. Available online: https://ark.intel.com/content/www/us/en/ark/products/72164/intel-core-i5-3230m-processor-3m-cache-up-to-3-20-ghz-rpga.html (accessed on 27 July 2020).
136. Mobile 3rd Generation Intel®CoreTM Processor Family, Mobile Intel® Pentium® Processor Family and Mobile Intel® Celeron® Processor Family. Datasheet 326768–006, June 2013. Available online: https://www.intel.com/content/dam/www/public/us/en/documents/datasheets/3rd-gen-core-family-mobile-vol-1-datasheet.pdf (accessed on 27 July 2020).
137. Doleželová, M.; Skarvada, J.; Heves, J.; Ruseva, Y.; Reed, J.; Landmann, R.; Domingo, D. Power Management Guide Red Hat Enterprise Linux 6. *Red Hat Customer Portal*. December 2010. Available online: https://access.redhat.com/documentation/en-us/red_hat_enterprise_linux/6/html/power_management_guide/index (accessed on 24 December 2022).
138. Chios, M. Network Functions Virtualisation. An Introduction, Benefits, Enablers, Challenges & Call for Action. In Proceedings of the SDN and OpenFlow World Congress, Darmstadt, Germany, 15–18 October 2012.
139. Molka, D.; Hackenberg, D.; Schöne, R.; Müller, M.S. Characterizing the energy consumption of data transfers and arithmetic operations on x86-64 processors. In Proceedings of the International Conference on Green Computing, Chicago, IL, USA, 15–18 August 2010; pp. 123–133. [CrossRef]
140. Rahman, A.; Lynch, P.; Aranda, P.; Bernardos, C.; Zuniga, J.; Contreras, L. Network Virtualization Research Challenges. Request For Comments (RFC)—Informational RFC 8568, ISSN: 2070–1721. April 2019. Available online: https://tools.ietf.org/html/rfc8568$\backslash$#page-18 (accessed on 10 July 2020).
141. Dahyun SR-IOV for NFV Solutions—Practical Considerations and Thoughts. October 2017. Available online: http://man-glove.blogspot.com/2017/10/sr-iov-for-nfv-solutions-practical.html (accessed on 27 July 2020).
142. Lorincz, J.; Bule, I.; Kapov, M. Performance Analyses of Renewable and Fuel Power Supply Systems for Different Base Station Sites. *Energies* **2014**, *7*, 7816. [CrossRef]
143. Lorincz, J.; Bule, I. Renewable Energy Sources for Power Supply of Base Station Sites. *Int. J. Bus. Data Commun. Netw. IJBDCN* **2013**, *9*, 53–74. [CrossRef]
144. Zhang, X.; Wang, Z.; Zhou, Z.; Liao, H.; Ma, X.; Yin, X.; Lv, G.; Wang, Z.; Lu, Z.; Liu, Y. Optimal Dispatch of Multiple Photovoltaic Integrated 5G Base Stations for Active Distribution Network Demand Response. *Front. Energy Res.* **2022**, *10*, 787. Available online: https://www.frontiersin.org/articles/10.3389/fenrg.2022.919197 (accessed on 17 December 2022). [CrossRef]
145. Vallero, G.; Renga, D.; Meo, M.; Marsan, M.A. Greener RAN Operation Through Machine Learning. *IEEE Trans. Netw. Serv. Manag.* **2019**, *16*, 896–908. [CrossRef]
146. Guo, W.; Shin, D.R.; Siddiqui, I.F.; Koo, J.; Qureshi, N.M.F. Cost-Effective Optimal Multi-Source Energy Management Technique in Heterogeneous Networks. In Proceedings of the 2022 IEEE 42nd International Conference on Distributed Computing Systems Workshops (ICDCSW), Bologna, Italy, 13 July 2022; pp. 181–186. [CrossRef]
147. Baidas, M.W.; Hasaneya, R.W.; Kamel, R.M.; Alanzi, S.S. Solar-Powered Cellular Base Stations in Kuwait: A Case Study. *Energies* **2021**, *14*, 7494. [CrossRef]
148. Deruyck, M.; Bova, S.; Vallero, G.; Meo, M.; Martens, L.; Joseph, W. Designing a hybrid renewable energy source system to feed the wireless access network. *Sustain. Energy Grids Netw.* **2022**, *31*, 100722. [CrossRef]
149. Samorzewski, A.; Kliks, A. User Allocation in Heterogeneous Network Supplied by Renewable Energy Sources. In Proceedings of the 2021 17th International Conference on Wireless and Mobile Computing, Networking and Communications (WiMob), Bologna, Italy, 11–13 October 2021; pp. 419–422. [CrossRef]
150. Fletscher, L.A.; Suárez, L.A.; Grace, D.; Peroni, C.V.; Maestre, J.M. Energy-Aware Resource Management in Heterogeneous Cellular Networks with Hybrid Energy Sources. *IEEE Trans. Netw. Serv. Manag.* **2019**, *16*, 279–293. [CrossRef]

sensors MDPI

Article

How Trend of Increasing Data Volume Affects the Energy Efficiency of 5G Networks

Josip Lorincz [1,*] and Zvonimir Klarin [2]

1 Faculty of Electrical Engineering, Mechanical Engineering and Naval Architecture (FESB), University of Split, R. Boskovica 32, 21000 Split, Croatia
2 Polytechnic of Sibenik, Trg Andrije Hebranga 11, 22000 Sibenik, Croatia; zklarin@vus.hr
* Correspondence: josip.lorincz@fesb.hr

Abstract: As the rapid growth of mobile users and Internet-of-Everything devices will continue in the upcoming decade, more and more network capacity will be needed to accommodate such a constant increase in data volumes (DVs). To satisfy such a vast DV increase, the implementation of the fifth-generation (5G) and future sixth-generation (6G) mobile networks will be based on heterogeneous networks (HetNets) composed of macro base stations (BSs) dedicated to ensuring basic signal coverage and capacity, and small BSs dedicated to satisfying capacity for increased DVs at locations of traffic hotspots. An approach that can accommodate constantly increasing DVs is based on adding additional capacity in the network through the deployment of new BSs as DV increases. Such an approach represents an implementation challenge to mobile network operators (MNOs), which is reflected in the increased power consumption of the radio access part of the mobile network and degradation of network energy efficiency (EE). In this study, the impact of the expected increase of DVs through the 2020s on the EE of the 5G radio access network (RAN) was analyzed by using standardized data and coverage EE metrics. An analysis was performed for five different macro and small 5G BS implementation and operation scenarios and for rural, urban, dense-urban and indoor-hotspot device density classes (areas). The results of analyses reveal a strong influence of increasing DV trends on standardized data and coverage EE metrics of 5G HetNets. For every device density class characterized with increased DVs, we here elaborate on the process of achieving the best and worse combination of data and coverage EE metrics for each of the analyzed 5G BSs deployment and operation approaches. This elaboration is further extended on the analyses of the impact of 5G RAN instant power consumption and 5G RAN yearly energy consumption on values of standardized EE metrics. The presented analyses can serve as a reference in the selection of the most appropriate 5G BS deployment and operation approach, which will simultaneously ensure the transfer of permanently increasing DVs in a specific device density class and the highest possible levels of data and coverage EE metrics.

Keywords: green communications; green networking; 5G; radio access network; base station; energy-efficiency; metric; data; coverage; power; mobile network operator; wireless

Citation: Lorincz, J.; Klarin, Z. How Trend of Increasing Data Volume Affects the Energy Efficiency of 5G Networks. *Sensors* **2022**, *22*, 255. https://doi.org/10.3390/s22010255

Academic Editors: Omprakash Kaiwartya and Peter Han Joo Chong

Received: 29 November 2021
Accepted: 29 December 2021
Published: 30 December 2021

1. Introduction

The trend of constantly increasing the number of mobile users consequently leads to an increase in traffic data volumes (DVs). It is projected that global mobile data traffic will exceed 300 EB per month in 2026 [1]. The overall number of Massive Machine-Type Communications (mMTC) and non-mMTC connected devices in the fifth-generation (5G) networks will increase from 165.6 million in 2020 to 3.256 billion in 2030, with a compound annual growth rate (CAGR) of 35% [2]. All 5G connected devices will, by 2030, account for 13% of the overall number of connected devices (25.4 billion) worldwide [2]. Although the fourth-generation (4G) mobile networks will remain the dominant mobile access technology by 2025, the market uptake of the 5G mobile network is faster than that of 4G, and 5G is expected to overtake 4G in the second half of the 2020s.

As the next step of wireless network evolution, the 5G network will enable new applications and services. Most of them will request greater network capacity and data rates. For some use cases, these requirements are accompanied by a demand for ultralow latencies, ultra-high reliability or exceptional availability [3]. Additionally, the rapid growth in the number of connected devices is expected to sustain in the future. This growth will mainly be due to emerging applications, such as ultrahigh-definition (UHD) video, augmented and virtual reality, fixed wireless access, intelligent transportation, remote healthcare, industrial automation, Internet of Everything (IoE) networks, smart agriculture, smart cities, etc. Such applications will be characterized by an enormous number of devices that will generate a huge amount of traffic DVs, which will contribute to the continued global growth of DVs during the 2020s.

As the existing microwave spectrum becomes severely congested, the 5G network, as the first among all previous mobile network generations, enables communication in a frequency spectrum above 6 GHz and in millimeter-wave (mmWave) spectrums (above 30 GHz). A larger amount of available bandwidth in above 6 GHz bands enables multi-gigabit data rates through wider wireless channels. Wider bands also enable higher channel reuse, which further allows denser implementation of macro and small base stations (BSs) having significantly higher capacities than those in previous generations of mobile networks.

However, the drawback of transmission in above 6 GHz bands is in high propagation losses and susceptibility to blockages of short-wavelength signals, which reduces the operational coverage of such communication systems. As a consequence, the 5G radio access network (RAN) deployment is envisioned as a heterogeneous network (HetNet). Such HetNets consist of a small number of 5G macro BSs and a huge number of 5G small BSs. Macro BSs support indoor and outdoor basic signal coverage and capacity demands for wider geographic areas, while accompanying small BSs (micro, pico or femto BSs) accommodate most of the capacity demands for both indoor and outdoor geographic areas characterized as hotspots with high traffic DVs.

Deployment of such HetNets composed of densely allocated macro and small BSs, accompanied with the need for satisfying transfer of constantly increasing traffic DVs, will have a non-negligible impact on energy consumption of radio access part of 5G and the future sixth-generation (6G) mobile networks. Therefore, in this study, the influence of the increasing traffic DVs on the energy efficiency of 5G networks was analyzed. The analyses were performed for energy-efficiency (EE) metrics standardized by the European Telecommunications Standards Institute (ETSI). Moreover, the DVs used for analyses are simulated DVs, which amount to specific device density classes (rural, urban, dense urban and indoor hotspot); they were selected according to the maximal expected number of 5G devices per square kilometer of each device density class defined in ETSI standards. Analyzed device density classes represent expected device densities which average uplink (UL) and downlink (DL) throughputs prescribed by ETSI standards and contribute to the total expected DV of one of four specific device density classes. These device density classes represent typical operating environments served by corresponding 5G HetNet deployments. Performed analyses in this study highlight the impact of such an increase in DVs on future trends of 5G network EE metrics, with respect to different deployment and operating HetNet scenarios and device density classes.

The rest of the paper is organized as follows: In Section 2, an overview of energy requirements of 5G networks is been presented. In Section 3, standardized metrics used for evaluating 5G network EE are elaborated. The types of different device density classes used in the analyses are explained in Section 4. Modeling of DVs' increase used for analyses of the impact of increasing DVs on EE of 5G networks is described in Section 5. The 5G radio access network (RAN) deployment and operation approaches and their performance in terms of EE are presented in Section 6. Section 7 analyses the results, which indicate how continuous increase in DVs impacts the values of EE metrics, instant power consumption and yearly energy consumption of the 5G network. In Section 8, a discussion on performed analyses and obtained results is performed. Finally, conclusions related to the analyses presented in the paper are given in Section 9.

2. Energy Requirements of 5G Networks

The deployment of a large number of new 5G macro and especially small BSs dedicated to satisfying trends of ascending DVs will have a significant impact in terms of increased energy requirements related to the power supply of BSs in RANs. Even the energy consumption (EC) reduction expected through the removal of equipment characteristic for some older network generations (e.g., second and especially third), due to obsolescence, will be diminished with the need for installing a vast number of 5G BSs. Therefore, the expected future increase of DVs in mobile access networks will undoubtedly affect the realization of 5G networks according to the goals of the green communications and networking paradigm. Green communications and networking is an emerging strategy dedicated to the development of novel energy-efficient solutions, which will minimize the usage of network resources for reducing network energy consumption whenever possible.

The design of previous cellular network generations was dominantly focused on improving network performance through maximizing throughput and spectral efficiency at the expense of EC. The EE improvements contributing to the realization of the green communications paradigm in pre-5G cellular or local wireless networks were mainly dedicated to solutions related to BSs transmit power and activity adjustments according to dynamic variations of DVs [4,5]. Moreover, green communication solutions related to the supplying of BSs site with renewable energy sources, which have been capacitated according to the daily and monthly traffic DV variations on BS site, have been envisioned as a possible solution to improve network energy efficiency [6].

Although hardware components of 5G BSs are more energy-efficient than those of previous generations of BSs (2G, 3G and 4G), the amount of EC of 5G radio access networks is expected to increase due to the need for fulfilling new 5G requirements, such as accommodation of vast DVs, ultrahigh data rates and/or ultralow latencies needed for ensuring some new services. This increase in the EC presents not just an economic issue reflected in an increase of operational costs for mobile network operators (MNOs), but also in an environmental issue due to increased carbon emissions and strategic issue due to alienation from the main goals of green communications paradigm. Therefore, ensuring the energy-efficient operation of RAN in 5G networks has become one of the major challenges analyzed in the literature [7,8].

This challenge is further contributed to by the fact that MNOs must adapt the number of BSs and corresponding capacities according to the expected number of users and their DVs. Classification of device density classes (areas) in 5G networks performed by ETSI in References [9,10] indicates that the number of users and corresponding DVs can significantly vary per specific device density class (Table 1).

Table 1. Parameter values for the various device density classes and analyzed BS types.

User Density Class	Average UL Throughput (Mbit/s)	Average DL Throughput (Mbit/s)	DVs for 10–100% of Max DV (Gb/km^2)	Maximal Density of User Devices (/km^2)	DV Percentage (%) Transmitted by Macro BSs for 10–100% of Max DV
Indoor hotspot	26.67	200	5667–56,668	250,000	0.420–0.042
Dense urban	50	300	875–8750	25,000	1.370–0.137
Urban	25	50	75–750	10,000	8.00–0.80
Rural	25	50	4.5–45	600	66.70–6.67

Moreover, due to the global population growth, the advent of new and more capacity-demanding applications, the proliferation of the Internet of Everything (IoE) concept and the increase in the number of devices per user, the DV increase in any device density class will have continuous progress throughout the 2020s. Following this progress in terms of satisfying the capacity demand of each device density class raises challenges on MNOs in terms of keeping 5G RAN deployment and operation as energy efficient as possible. These challenges can be precisely expressed in changes of EE metrics for specific device density classes of 5G networks. In our preliminary work, the impact of the increasing number of active users in 5G networks on network EE metrics was analyzed [11]. However, in mobile access networks, the increasing trend of changes in user densities is not the same as the increasing trend of changes in data volumes (DVs). Therefore, analyses dedicated to the explanation of how the expected increasing trend of DVs during the 2020s impacts EE of 5G RAN have not been analyzed yet.

Satisfying this increasing trend of DVs can be performed in 5G RANs through the exploitation of different small and macro BSs implementation and operation scenarios. They can differ in the number of used BSs and corresponding ECs, which ultimately influence on 5G network EE. The EE in the 5G RAN can be validated through recently standardized metrics which enable consistent evaluation of RAN EE in different device density classes. Hence, the impact of the increased DVs on EE metrics of 5G RANs in versatile device density classes was analyzed in this work for the first time. The analyses give an explanation of how the ascending trend of DVs impacts distinct EE metrics. Moreover, the analyses for the first time clarify the impact of increased DVs on instant power and yearly energy consumption of the 5G RAN.

3. Metrics for Evaluating 5G Network Energy Efficiency

Recently, the main standardization organizations, such as the 3rd Generation Partnership Project (3GPP), the ETSI and the International Telecommunication Union-Telecommunications Standardization Sector (ITU-T), accepted two types of the EE metrics as the key performance indicators (KPIs) for expressing the EE of mobile networks (MNs).

The first metric represents the data EE metric of MN expressed in bit/J, and it is defined as the amount of data transmitted per Joule of energy. For known mobile network area, A, data EE is expressed as follows:

$$EE_{DA} = \frac{DV_A}{EC_A} \text{ [bit/J]} \quad (1)$$

According to Equation (1), the data EE metric (EE_{DA}) is calculated as the ratio between the overall DV (DV_A) transferred by the analyzed MN area and energy (EC_A) needed by MN equipment installed in this area for transferring these DVs.

Another standardized EE metric expressed in m^2/J represents the coverage EE of the MN area, and it is defined as the unit area which can be covered with the 5G wireless signal, using Joule of consumed energy. For known mobile network area, A, the coverage EE is expressed as follows:

$$EE_{CA} = \frac{S_A}{EC_A} \left[\text{m}^2/\text{J} \right] \quad (2)$$

Based on Equation (2), the coverage EE metric (EE_{CA}) is calculated as the ratio between the overall size of the area (S_A) covered with the wireless signal of the analyzed MN and the total energy (EC_A) consumed by the equipment allocated in the analyzed MN area. The total EC (EC_A) of the analyzed MN area can include the sum of the energies consumed by the RAN equipment (BSs), transmission equipment (e.g., wired or wireless backhauling equipment and radio controllers) and all ancillary equipment (e.g., air-conditioning, power backup, etc.).

These two standardized metrics are used to effectively analyze the deployment and operation performance of MN in terms of EE. They can be used for the assessment of MN EE performance in the case of various deployment and operation strategies of different BS types and device density classes [1,2]. As different deployment and operation strategies for a new generation of MN can be implemented in practice, the EE assessment of mobile networks by using standardized metrics becomes crucial for MNOs when assessing approaches related to the deployment and operation of new generations of mobile networks.

4. Device Density Classes for Assessment of Energy Efficiency

The method for EE evaluation of 5G RAN was presented in ETSI standard (ES) 203 228 [9]. According to this method, when the measurement of parameters needed for EE assessment of complete MN on the large scale is not feasible, the total network can be divided into a set of smaller sub-networks that are classified according to device densities of a specific sub-network. For EE assessment of 5G RAN presented in this work, analysis was performed by using sub-networks categorized by device density classes for an area of one square kilometer (1 km^2). Figure 1 visualizes analyzed 5G HetNet deployments for different device density classes with the maximal number of installed small and macro BSs per square kilometer area.

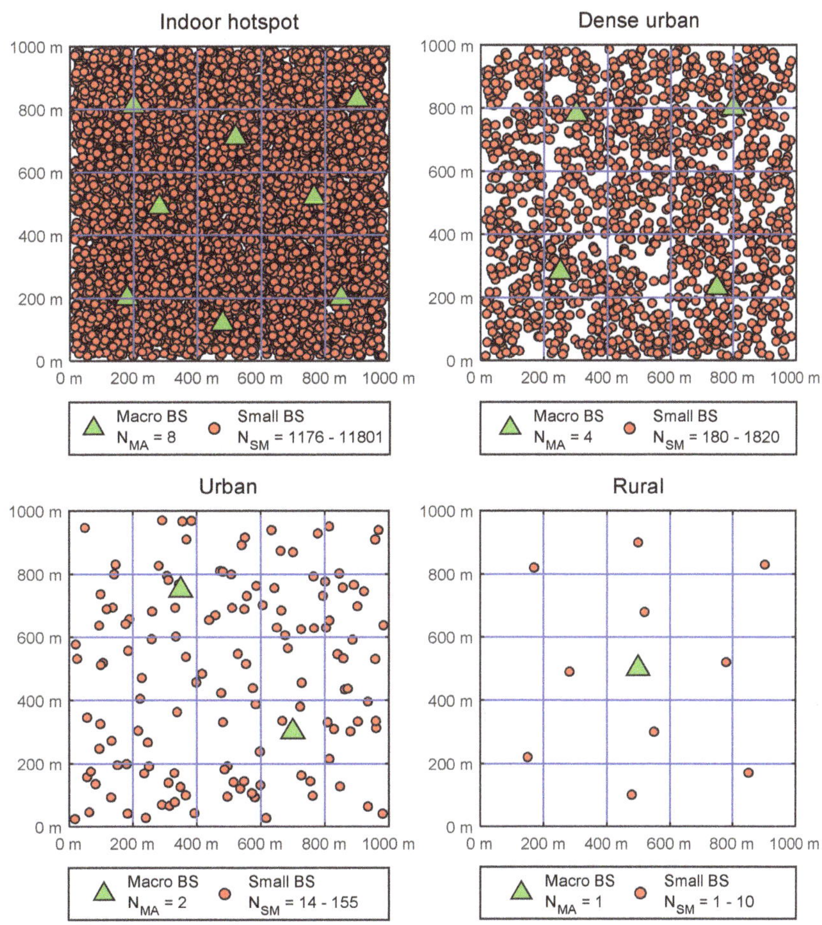

Figure 1. Visualization of analyzed 5G HetNet deployments for different device density classes with the maximal number of installed small and macro BSs per square kilometer area.

The impact of DVs on different EE metrics can be analyzed for different device density classes which are defined in ETSI standards [9,10]. They can range from the rarely populated rural device density class having up to hundreds of users per square kilometer (km^2) to the urban, dense-urban and densely populated indoor-hotspot device density classes, having up to a few hundreds of thousands of users per km^2. The maximal densities of user devices in specific device density classes for the analyzed area of one square kilometer are presented in Table 1. Hence, the device density classes presented in Table 1 are classified depending on the maximal user density of the square kilometer area (users/km^2).

As an example, Figure 1 visualizes analyzed device density classes of the 5G RAN deployment in terms of the number of allocated macro and small BSs. Analyzed HetNet 5G RAN deployment architectures are, for each of the device density classes, composed of a few macro BS(s) and a large number of small BSs (Figure 1). Figure 1 presents a maximal number of macro BSs allocated in each device density class. In the analyses, it is assumed that the number of macro BSs is constant during the 2020s and corresponds to the number of BSs allocated during the initial deployment of the 5G network. This approach to deployments of macro BSs is frequent in practice, since MNOs deploy macro 5G BSs to ensure initial coverage and capacity demands of a specific area. In areas expecting larger DVs, a larger number of macro BSs will be initially deployed, and this explains the differences in the maximal number of macro BSs deployed in different device density classes in Figure 1. Moreover, a number of small BSs will be deployed according to different deployment and operation approaches described and analyzed in further sections. Since the deployment of the number of small 5G BSs in some analyzed areas depends on different MNO deployment and operation approaches, in Figure 1, the span of small BSs that can be deployed is indicated in the range from minimal to the maximal number of small BSs that are considered for analyzes in different approaches. Therefore, Figure 1 illustrates basic MNO principles of RAN deployment analyzed in this work, according to which a few macro BSs are deployed for ensuring basic signal coverage and capacity of the analyzed area and a larger number of deployed small BSs are dedicated to satisfying capacity demands at locations of traffic hotspots.

5. Modeling of DVs Increase in 5G Networks

A continuous increase in DVs will ultimately affect the 5G RAN layout in terms of the needed number of BSs that must be deployed in a specific device density class (area). According to Reference [12], for each device density class, different requirements in terms of expected average uplink (UL) and downlink (DL) data rates per active user have been defined by the ETSI standard [1,2] (Table 1). Defined average UL/DL data rates assume that less populated areas (e.g., rural) require lower average throughputs and ultra-densely populated areas (e.g., indoor hotspot) require average throughputs that are significantly higher.

In this work, the DV increment was simulated for each year in the 2020s. The simulation is based on the total expected DV increase during the 2020s. This increase in DV is calculated for each year in the 2020s, based on the maximal expected number of 5G devices per square kilometer in each device density class prescribed by the ETSI standards (Table 1) [1].

According to values of UL and DL average throughputs and maximal user densities defined in References [9,10] (Table 1), the overall impact of trend in DV increase on EE metrics of specific device density classes was modeled. For analyses in this paper, this impact was modeled by scaling the DV in each device density class in the range from 10% to 100% of a total average UL and DL DV. The total UL and DL DV were obtained for the maximal number of user devices per square kilometer area of every device density class prescribed by the ETSI. More specifically, the 10% of maximal expected DV in 2030 was used for representing DV in 2021, and for each subsequent year in the 2020s, the increase in DVs is assumed to be incremented by 10% up to 2030. For the year 2030, the maximal average DV defined by total UL and DL traffic of the maximal number of user devices per square kilometer prescribed by ETSI in References [9,10] is used in the analyses (Table 1). Through modeled increases in DVs for each device density class, it is shown how an increasing trend of DVs affects the data and coverage EE of the specific device density class.

Based on the demands for the transfer of DVs in each device density class, a different number of installed macro (N_{MA}) and small (N_{SM}) BSs were allocated in the analyzed area of the square kilometer (Figure 1). An increase in DVs ranging from 10% to 100% of a total maximal DV in each device density class was accommodated through the deployment of an appropriate number of small BSs. For that reason, ranges of the number of small BSs in specific device density classes are also presented in Figure 1.

The overall number of installed macro (N_{MA}) and small (N_{SM}) BSs presented in Figure 1 depends on the increase in DVs of each device density class toward the maximal expected DV presented in Table 1. The installed number of macro and especially micro BSs in a specific device density class presented in Figure 1 is related to the overall capacity that active BSs in RAN must have for ensuring the transfer of expected DVs. The capacity of each BS type is defined based on operating parameters of macro and small BSs, as presented in Table 2 [13]. They are characteristic for the typical contemporary 5G BSs market models, and, based on them, the overall RAN capacity in terms of a minimal number of the macro and small BSs needed for the transfer of expected DVs in each device density class was allocated. Allocation of 5G BSs is performed in accordance with an increase of DV in the specific device density class. As the demand for transfer of higher DVs increases and the current capacity of installed 5G small BSs reaches maximal exploitation, the new small BSs are deployed in the network on positions of traffic hotspots (Figure 1). This concept of adding new BSs in the network simulates some of the analyzed approaches dedicated to the realization of future 5G networks, which will be realized through the gradual deployment of new small 5G BSs on positions of traffic hotspots.

Table 2. Operating parameters of different BS types used in the analyses.

BS Parameter	5G Macro BS	5G Small BS
Spectral efficiency (bit/s/Hz/cell)	10	6
Channel bandwidth (MHz)	100	800
Number of sectors (cells)/ BS capacity (Gbit/s)	3/3	1/4.8
Average power consumption in sleep mode (W)	N/A	5
Average power consumption in active mode (W)	2000	50
Average power consumption in Tx power scaling mode (W)	1600	40

According to Table 1, the percentage share of the transfer of DVs performed by macro BSs is lower for device density classes having higher DVs (user densities), and vice versa. Hence, in rarely populated rural device density classes, most of the DVs will be transferred over macro BSs. However, in the indoor-hotspot device density classes characterized by the necessity of transmitting huge DVs, macro BSs will transfer minor DVs, while the remaining DVs will be transferred by small BSs. Moreover, with the increase of DVs in specific device density classes, the involvement of macro BSs in the transfer of DVs will decrease on the account of small BS.

6. Simulation of 5G RAN Deployment and Operation Approaches

MNOs can exploit different approaches for the initial deployment of new BSs in 5G RANs. However, the selection of the best 5G network deployment approach in terms of keeping network EE at optimal levels is still an open question to MNOs. As in the case of previous generations of the cellular networks, implementation of the 5G network can last for years until full deployment in terms of needed capacity and coverage on the national level of complete countries will be achieved. According to the prediction in Reference [1], it is expected that 60% of the world population will be covered with 5G signal by 2026. By that period, installed 5G RAN resources will also need to accommodate increased DVs, which will be significantly higher than the ones at the first years of 2020s.

However, different deployment approaches of BSs in 5G RANs can be realized through versatile deployment and RRM techniques. They can have a different long-term impacts on the EC of the RAN. Selecting an appropriate deployment and RRM approach in terms of EE can contribute to the reduction of network EC and consequently to the improvement of the EE metrics. Hence, further analyses reveal how different 5G RAN deployment approaches and continuous increase of DVs in the 2020s influence the 5G network EE metrics. As the largest share of the 5G network EC is related to the EC of BSs in 5G RAN [6], further analyses are performed only for deployment and RRM approaches related to 5G BSs only.

Due to the known fact that many active BSs are lightly loaded for most of their operating time, during which active BSs consume energy [3], analyzed deployment and RRM approaches exploit this fact for potential energy savings. More specifically, an operating strategy based on the switching of small BSs in sleep operation mode during the idle traffic periods is considered in the analyses [14,15]. This operating strategy is considered, since the radio resource scaling (in terms of BS transmit (Tx) power levels, number of active transceivers, number of active subcarriers, etc.) according to dynamics of the user's activity has been proven as an approach that contributes to enhancement of the HetNets EE [4,5,16]. Such concepts dedicated to improving BSs EE are included in some contemporary types of the 5G small BSs [17], and even more advanced concepts will be implemented in the future releases of 5G BSs.

Three different 5G BSs modes of operation are analyzed in the paper. In Table 2, instant average power consumption of those operation modes, which can be active, sleep and Tx power scaling, are presented. The power consumption of 5G BSs in active, sleep and Tx power scaling modes were selected for analyses based on Reference [3]. The selected BSs power consumption and configuration parameters presented in Table 2 are inherent to the preliminary models of 5G small and macro BSs. The power consumption of BS working in active mode represents the average power consumption of BS for which operating capacities are fully exploited at the highest Tx powers [3]. The average EC of the small BSs in sleep mode is estimated to be 10% of the average EC characteristic for small BS operating in full active mode (Table 2). Due to the necessity of ensuring constant coverage of the area with 5G signal, the possibility that macro BSs can be in sleep mode is not considered in the analyses (Table 2). The average power consumption of BSs in Tx power scaling mode is assumed to be 80% of the average power consumption in active mode (Table 2). The Tx power scaling mode is analyzed, since it is confirmed in References [4,5] that scaling of the Tx power of BSs according to the space and time DV variations can additionally give a contribution to the enhancement of the EE of a mobile access networks. Hence, a very conservative assumption related to the reduction of average instant BSs power consumption for 20% on a daily basis in comparison with a power consumption of small BSs constantly operating in the active state at the highest Tx power is used in this analysis.

Another MN characteristic that is used in the presented analyses is the fact that the deployment of BSs in cellular RANs is based on ensuring the projected capacity needs for satisfying peak network traffic volumes during a BS exploitation period of approximately 10 years since initial BSs deployment. Besides the fact that such an implementation imposes large initial capital investments for MNO, it also imposes increased operational costs in terms of large monthly energy bills paid by MNO for operating RAN with significantly larger capacities than those needed during most of the 5G RAN equipment lifetime. These larger 5G RAN energy costs consequently result in reduced 5G network EE, which can be improved if an appropriate 5G BSs deployment and operation approach is exploited. Hence, finding an optimal deployment and RRM approach from the EE perspective represents a crucial challenge to MNOs.

6.1. Types of 5G RAN Deployment Approaches

As previously emphasized, a rapid increase in mobile users and the high-throughput requirements of future applications will contribute to the constant increase of the DVs in 5G HetNets during the upcoming decade. Therefore, the analyzed BS deployment approaches simulate this growth of DVs for the 10-year period starting with the year 2021 in every device density class (Table 1). In Table 3, the main characteristics of the analyzed network deployment and operation approaches are presented. The analysis takes into account broadly accepted MNOs practice based on the initial deployment of the fixed number of macro 5G BSs for each deployment approach in every device density class (Figure 1). The number of installed macro BSs corresponds to the number of BSs required to provide a minimal level of signal coverage and capacity requirements for the transfer of expected DVs in a specific device density class (Figure 1).

Table 3. Analyzed network deployment and operation approach.

	Approach 1—Variable Number of Small BSs in Active Mode	**Approach 2—Maximal Number of Small BSs in Sleep Mode**	**Approach 3—All BSs Constantly in Active Mode**	**Approach 4—Variable Number of Small BSs in Tx Power Scaling Mode**	**Approach 5—Variable Number of All BSs in Tx Power Scaling Mode**
Number of macro BSs in Tx power scaling mode	N/A	N/A	N/A	N/A	Based on DV changes
Number of small BSs in Tx power scaling mode	N/A	N/A	N/A	Based on DV changes	Based on DV changes
Number of small BSs in active mode	Based on DV changes	Based on DV changes	All	Based on DV changes	Based on DV changes
Number of small BSs in sleep mode	N/A	Maximal	N/A	N/A	N/A
Overall number of installed small BSs	Changes according to DV requirements	Maximal for satisfying full DV demand	Maximal for satisfying full DV demand	Changes according to DV requirements	Changes according to DV requirements

The number of installed small BSs and corresponding RRM principles differs among deployment approaches (Figure 1), and this constitutes the main difference among the analyzed deployment approaches (Table 3). In order to model various BSs deployment and RRM approaches and compare their impact on the standardized EE metrics, five different types of deployment and operation approaches were selected for analysis.

6.1.1. Approach 1—Variable Number of Small BSs in Active Mode

Deployment Approach 1 is characterized by the variable number of small BSs in active mode during their operational period (Table 3). This approach is based on 5G HetNet deployment strategy in which the number of installed small BSs is continuously increased over time. This increase is based on satisfying the demand for transfer of increasing DV traffic in specific locations of every device density class. This approach does not include any radio resource management (RRM) technique for improving BSs EE in periods when BSs are active (Table 3). Hence, those macro and small BSs that are installed (Figure 1) are constantly active with maximal EC during the entire working period. Their instant power consumption corresponds to the average instant power consumption of BSs working constantly in active mode (Table 2).

6.1.2. Approach 2—Maximal Number of Small BSs in Sleep Mode

This deployment approach is characterized by the possibility of having a maximal number of small BSs in sleep mode during RAN operation (Table 3). In this approach, all small BSs needed for accommodating expected DVs in the upcoming period of 10 years are initially installed in the RAN by MNO. This approach exploits the small BSs sleep mode strategy, which enables the preservation of energy through putting small BSs in a sleep operation mode during the periods lacking the data needed to be transferred by those BSs. When the demand for capacity exceeds the available capacity of the currently active macro and small BSs, the required number of small BS(s) that are in sleep mode are activated. Moreover, this approach does not assume the implementation of any RRM method for improving EE of macro BS(s).

6.1.3. Approach 3—All BSs Constantly in Active Mode

Deployment Approach 3 demonstrates the traditional approach to the deployment of BSs in which all (macro and small) BSs are constantly in active mode (Table 3). As in the case of deployment Approach 2, all small BSs needed for the accommodation of expected DVs in the upcoming 10-year exploitation period of 5G HetNet are initially installed in the RAN by MNO. The installed BSs operate without any adjustment of radio resources in accordance with time and space variations of DVs; thus, they consume constantly maximal energy. This approach characterizes the constant power consumption of small and macro BSs that corresponds to the average power consumption of BSs working in active mode (Table 2). This approach, which is traditionally exploited for the deployment of pre-5G BSs, is considered for comparison purposes with other more advanced BSs deployment and deployment approaches.

6.1.4. Approaches 4 and 5—Variable Number of Small/All BSs in Tx Power Scaling Mode

Deployment Approaches 4 and 5 correspond to deployment Approach 1 in terms of the deployment of small BSs according to a gradual increase of DV over time (Table 3). In these two approaches, a BSs Tx power scaling according to time and space variations of DV is exploited, since it is proven that such a technique can additionally preserve the EC of the RAN. The Tx power scaling technique used for the purpose of the EE assessment in this analysis follows a conservative approach, assuming that implementation of such technique reduces EC of the BS for 20% of the EC which BSs have in case of transmitting at the highest Tx power (Table 2). In the case of deployment Approach 4, the Tx power scaling is applied only to the small BSs, while, in the case of deployment Approach 5, the Tx power scaling mode is applied to both small and macro BSs (Table 3).

7. Results on the Impact of Increasing DVs on 5G Network EE

The impacts of the increase in DVs on data and coverage EE metrics of 5G RAN for different device density classes and five deployment approaches are shown in Figure 2a–d. Figure 2a–d indicated estimated DVs for each year during the 2020s, with 2030 as the last year for which DV estimation was performed. According to the presented simulation results, the increase in DVs has a significant impact on data and coverage EE metrics of 5G RANs. This impact is visible for every device density class and deployment approach. In the case of all deployment approaches and for every analyzed device density class, the data EE metric increases with the increase of DV that must be transferred in the 5G network (Figure 2a–d). Thus, an increase in DVs has a positive impact on data EE metrics of 5G networks. This implies that a higher amount of DVs can be transferred per Joule of energy consumed by the network BSs of the same area size. Therefore, the global trend of the constant increase of DVs in 5G networks will result in the improvement of the data EE metric. This improvement is a consequence of the fact that higher amounts of data will be transferred for the same unit of energy consumption of the BSs in the 5G RAN.

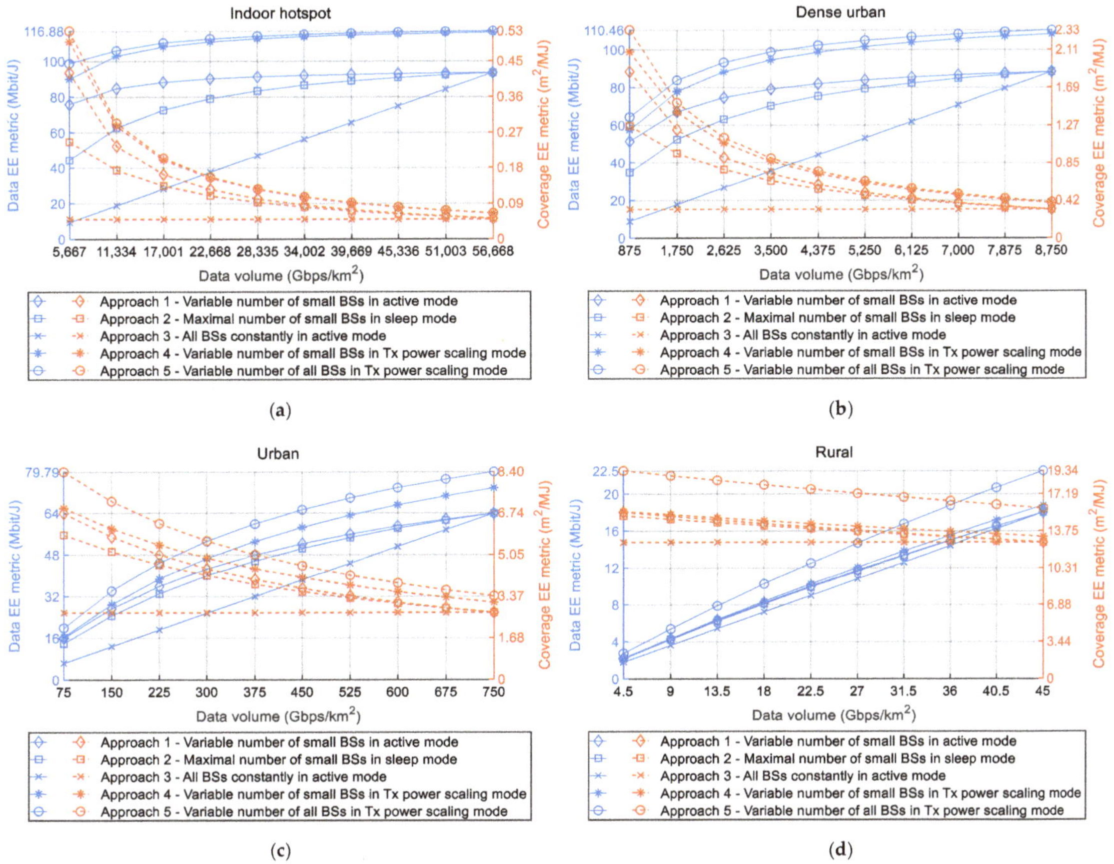

Figure 2. Impact of DV increase on data and coverage EE metrics for each year during the 2020s for (**a**) indoor-hotspot, (**b**) dense-urban, (**c**) urban and (**d**) rural device density class.

On the other hand, for every device density class and most of the deployment approaches, the coverage EE metric decreases as the DV of the analyzed square kilometer area increases (Figure 2a–d). This decrease is not perceived only for deployment Approach 3, which presents the traditional RAN implementation approach lacking any deployment or operational mechanisms dedicated to optimizing BSs energy consumption. In the case of other deployment approaches (Approaches 1, 2, 4 and 5), an increase in DVs has a negative influence on coverage EE metrics. This implies that ensuring the transfer of higher DVs over the same area requires more energy which will be consumed by the network elements (BSs) allocated in this area. Therefore, the global trend of the constant increase of DVs in 5G networks will result in the degradation of the coverage EE metric of the radio access part of the network. This decrease is a consequence of the fact that transferring higher amounts of data in a specific device density class demands an increase in the energy consumption of the BSs in the 5G RAN.

According to Figure 2a–d, an increase in DVs causes the opposite trends in changes of data and coverage EE metrics, where an increase in the data EE metric is followed by a decrease in the coverage EE metric. This is not the case only with deployment Approach 3, due to the above-explained reasons. Hence, there is no optimal 5G BSs deployment approach that can simultaneously contribute to the improvement of both standardized EE metrics. Favoring the data EE metric in terms of transferring higher amounts of DVs per unit of consumed energy will be on the cost of the degradation of coverage EE metric, and vice versa. However, to have as much energy-efficient 5G RAN as possible, both EE metrics must be simultaneously optimized in terms of obtaining network deployment and operation, which will have the highest possible data and coverage EE metric. This makes the realization of 5G HetNets in terms of satisfying both standardized EE metrics particularly challenging for MNOs. This challenge arises from the fact that there is no optimal DV around which MNOs should keep the traffic intensity in a specific device density class, for which both EE metrics will have the best possible values.

7.1. Impact of Increasing DVs on the Power Consumption of 5G Network

Since increasing DVs impact both EE metrics of 5G networks, an increase in DVs will also have an influence on instant 5G network power consumption. The instant network power consumption represents the total average instant power consumption of all (small and macro) BSs located in the analyzed device density class that are in an active and operating state (Table 2). Obtained results presenting the impact of DV increase on data EE metrics and instant 5G network power consumption for indoor hotspot, dense urban, urban and rural device density classes are shown in Figure 3a–d, respectively. Results presented in each figure have been obtained for all of the five different deployment and operation approaches and for DVs characteristic for every year in 2020s.

According to Figure 3a–d, for analyzed Approaches 1, 2, 4 and 5, an increase in DV of each device density class will impose an increase in the instant 5G network power consumption. This is due to the fact that transferring larger DVs requires more network resources in terms of activating an additional number of BSs, exploiting more capacity of active BSs (in terms of transceivers, channels, subcarriers and multiplexing slots) and transmitting at higher Tx power levels. This consequently results in higher instant power consumption of individual BSs, which jointly contributes to an increasing trend of total instant network power consumption in every device density class. The only exception from this power consumption trend is the instant power consumption of Approach 3 (Figure 3a–d). Since this approach lacks any adaptation of BSs deployment dynamics and operation activity according to DV variations, instant power consumption of the network will be constantly at maximal levels (Figure 3a–d). In comparison with other deployment and operation approaches, this results in the worse data EE metric of Approach 3.

Therefore, for any of the analyzed approaches (Approaches 1, 2, 4 and 5) which exploit adaptation of BSs deployment dynamics and operation activity according to increasing DV trend, this DV increase will have a negative impact in terms of increasing the instant power consumption of the 5G network. This negative impact is not translated to data EE metrics, which will increase with the increase of DV in all analyzed device density classes (Figure 3a–d). This is the consequence of the fact according to which instant power consumption and DV have different increasing rates. Although instant 5G network power consumption and, therefore, overall network energy consumption increase during the analyzed period of 10 years, the overall DV during this period also increases with higher rates, which in total contributes to the increase of data EE metrics calculated based on the Equation (1).

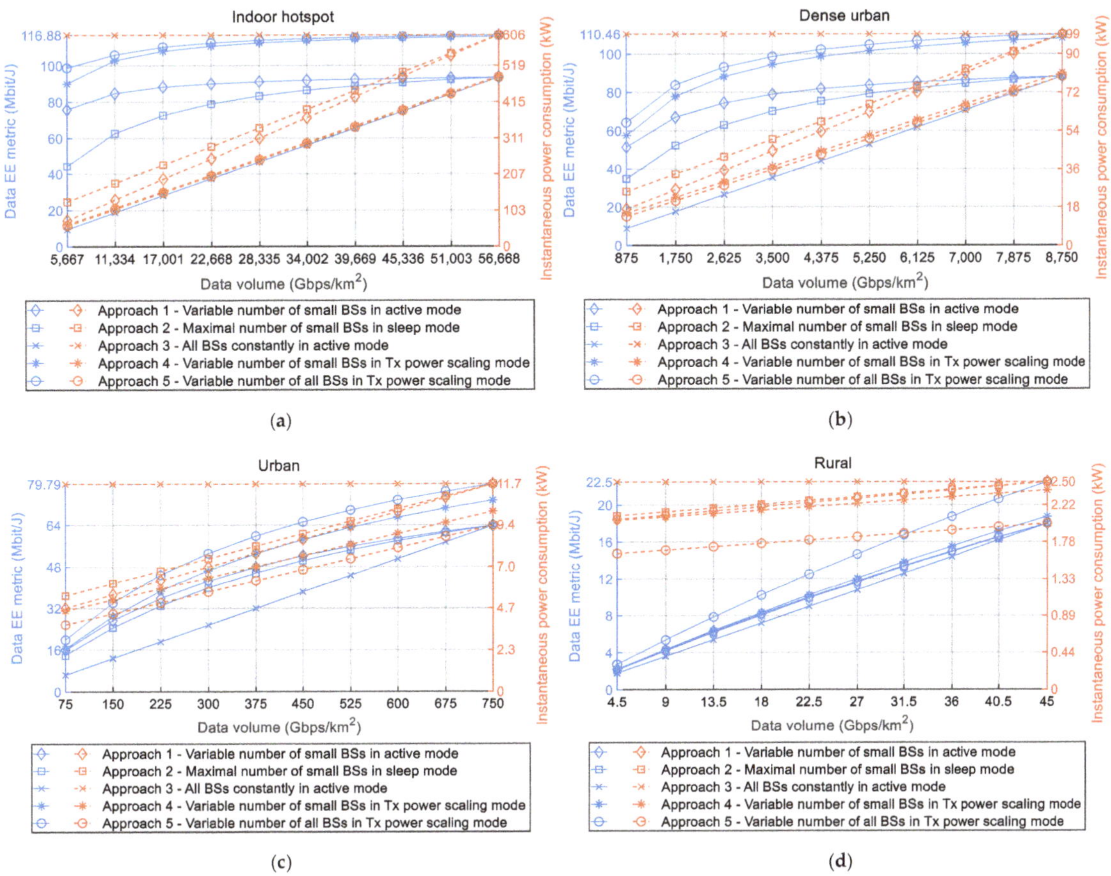

Figure 3. Impact of DV increase on data EE metrics and instant 5G network power consumption during the 2020s for (**a**) indoor-hotspot, (**b**) dense-urban, (**c**) urban and (**d**) rural device density class.

7.2. Impact of Increasing DVs on the Energy Consumption of 5G Network

Since trends in the necessity of transferring increasing DVs impact both EE metrics and power consumption of 5G networks, an increase in DVs will also have an influence on the total 5G network energy consumption. In the performed analyses, the total network energy consumption refers to the energy consumption of all (small and macro) BSs located in the area of the analyzed device density class during the time period of one year. The impacts of the DV increase through 2020s on coverage EE metrics and total yearly energy consumption of 5G network for indoor-hotspot, dense-urban, urban and rural device density classes are presented in Figure 4a–d, respectively.

According to Figure 4a–d, for analyzed Approaches 1, 2, 4 and 5, an increase in DV of each device density class will impose an increase in the yearly energy consumption of the 5G network. As explained in the previous section, this is due to the fact that transferring larger DVs requires the exploitation of more BSs and BSs resources, thus resulting in the higher total instant power consumption. This consequently results in an increase of the yearly energy consumption of the analyzed device density class area (Figure 4a–d). Due to explained reasons related to lack of any deployment and RRM approach dedicated to improving EE, only in the case of Approach 3 will the total yearly energy consumption be equal to maximal energy consumption for all years in every device density class.

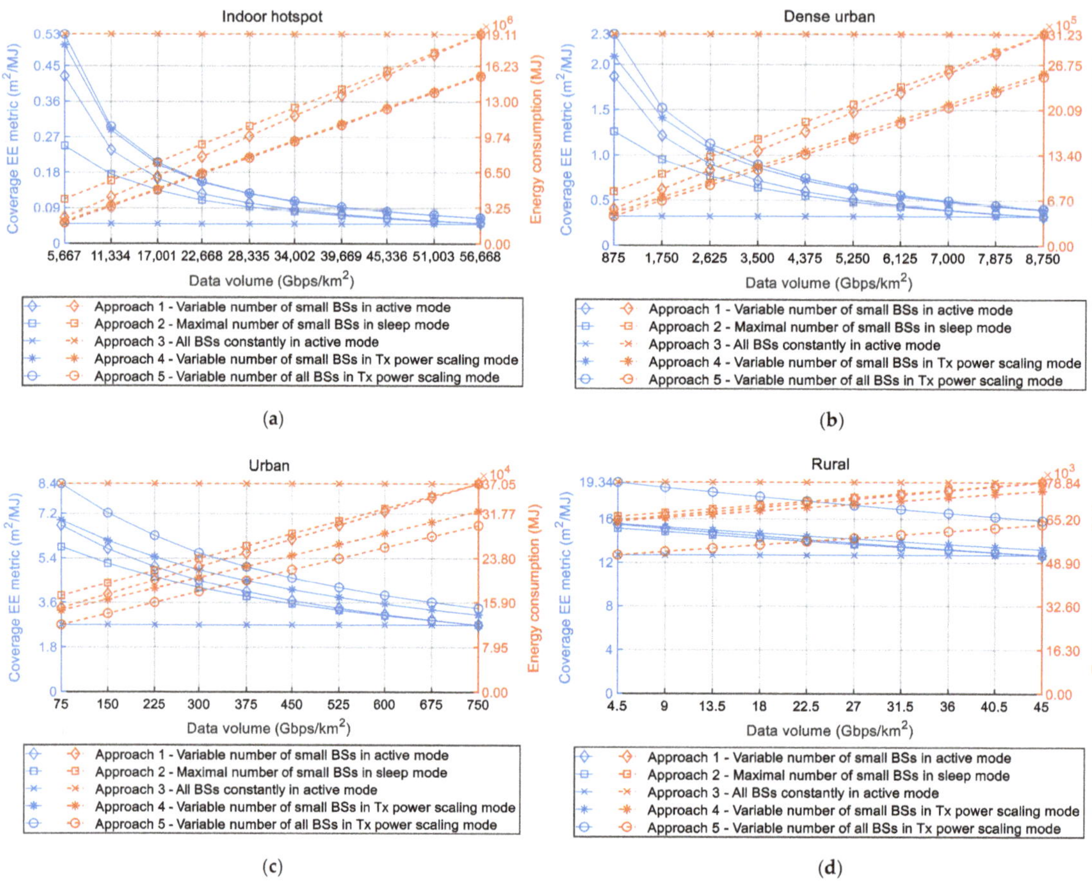

Figure 4. Impact of DV increase on coverage EE metrics and 5G network yearly energy consumption during the 2020s for (**a**) indoor-hotspot, (**b**) dense-urban, (**c**) urban and (**d**) rural device density class.

Although analyzed Approaches 1, 2, 4 and 5 are based on the adjustment of BSs deployment dynamics and/or operation activity with respect to the increasing DVs trend, this DV increase will have a negative impact in terms of increase of the yearly energy consumption of the 5G network. This negative impact also negatively impacts the coverage EE metrics, which will decrease with the increase of DV in all analyzed device density classes (Figure 4a–d). This is the consequence of the increase of network energy consumption when the DVs that must be transferred in the network increase. Consequently, larger energy consumption for the same size of the device density class area will, according to Equation (2), result in the degradation of the coverage EE metric.

8. Discussion on Performed Analyses and Obtained Results

The obtained results presented in Figures 2–4 show a significant impact on the coverage and data EE metrics of DV increase in the 5G HetNets. The results presented in Figure 2 reveal the reverse influence of increased DVs on the coverage and data EE metrics. These adverse changes of data and coverage EE metrics are noticed for all approaches which use deployment and operation policies that include an adaptation of BSs resources according to DV increase. While an increase of DV in the area of analyzed device density classes has an impact on the decrease of the coverage EE metrics (Figures 2 and 4), it also has an impact on the increase of the data EE metrics (Figures 2 and 3). Due to such contrary changes in trends of data and coverage EE metrics, there is no optimal DV that, transferred by the network, will ensure 5G network operation with the best possible combination of data and coverage EE metrics.

To simultaneously improve both data and coverage EE metrics, the best approach is to implement 5G BSs deployment and operation strategies that have the highest level of adaptation of installed BSs resources for accommodating the increasing DV trend in any device density class. Based on the results presented in Figure 3, for any device density class, approaches that will have better data EE metric are those which have the lower instant power consumption for the same DVs that must be transferred in the same network area. Those approaches are deployment and operation Approaches 4 and 5 (Table 3), which, in comparison with Approaches 1 and 2, have lower instant power consumption for transfer of the same DVs (Figure 3). Additionally, based on the results presented in Figure 4, it can be noticed that, for any device density class, approaches that have higher coverage EE metric are also approaches that have lower yearly energy consumption for the case when the same DV must be transferred in the same area of device density class. Obviously, approaches that have the lowest energy consumption (Figure 4) will also have the lowest instant power consumption (Figure 3).

Therefore, only optimization of network instant power consumption (and consequently energy consumption) will result in simultaneous improvement of both data and coverage EE metrics. Hence, those BSs' deployment and operation strategies that enable the best possible adaptation of BSs' resources according to the increase of network DV will result in the optimized power consumption, which contributes to the simultaneous increase of both EE metrics. This observation further raises the necessity of implementing 5G BSs deployment and operation concepts that will ensure the transfer of constantly increasing DVs, while keeping both data and coverage 5G HetNets EE metrics at higher values as much as possible.

The further discussion concerns the values of the coverage and data EE metrics shown in Figures 2–4. It is necessary to highlight that these values of EE metrics were obtained for the specific device density classes (Table 1), which consist of versatile numbers of small and macro BSs (Figure 1). Hence, every analyzed device density class has been defined in terms of the number of macro and small BSs, with the goal of satisfying the expected DV increase in the area of specific device density classes during the 2020s. However, different real-life deployments of 5G HetNets may vary in terms of the dynamics of DV increase through time and in the type and number of installed small and macro BSs. For that reason, versatile practical deployments of HetNets can differ in absolute values of coverage and data EE metrics when compared with those presented in this work. Nevertheless, the trends of the graphs presenting the impact of DV increase on data and coverage EE metrics will remain for every device density class, as with those in Figures 2–4. Therefore, conclusions presented in this paper about trends in changes of data EE metric, coverage EE metric, instant power and yearly energy consumption caused by an increase in DVs can be generalized for any practical implementation of 5G network segment which belongs to a specific device density class.

An additional discussion point is related to the assumption used in the analyses according to which the DV during a single year is constant in the area of every device density class (Table 1). In reality, the instant DVs that must be transferred in the area of every device density class will vary during a day and throughout the year. To approximate these DV variations, DVs used in the analyses represent the average yearly DVs expected to be transferred by the 5G BSs located in a square kilometer area of a specific device density class. They are calculated for every year in the 2020s (Figures 2–4) and for every device density class. The calculation takes into account an increase in the projected number of active user devices through the 2020s and the sum of their minimal uplink and downlink throughputs prescribed by ETSI standard for every device density class [1,2] Hence, it is reasonable to assume that the DVs used for modeling the effect of DV increase in square kilometer area of every analyzed device density class can be set to fixed DVs. For the purpose of analyses presented in this work, these fixed DVs express an average expected DVs in each year of the 2020s.

A final assumption is related to performing analyses with the excluded contribution to the total network energy consumption of other BS site elements, such as backhaul network equipment, cooling equipment and ancillary equipment. Moreover, the contribution to the overall network energy consumption of other types of BSs which can be collocated with 5G BSs at the same BSs site (e.g., 2G, 3G and 4G BSs) is not considered in the analyses. Although the energy consumption of all stated network elements contributes to the overall MN energy consumption and, therefore, impacts the network data and coverage EE metrics, this contribution was excluded from the analyses. The reason for this exclusion is in the main goal of this paper, which is dedicated to enlightening how the expected future increase in DVs will impact standardized data and coverage EE metrics of 5G networks. The analysis is, therefore, performed for 5G networks only, due to the fact that the cellular RAN has the greatest contribution to the overall MN power consumption and 5G networks will be dedicated to transferring the largest shares of DVs in the future transfer of DVs in mobile networks. For that reason, only 5G BSs and corresponding power consumption profiles were considered in the analyses. Therefore, the results of this analysis can be used for developing energy-efficient implementation and operation strategies of BSs in contemporary 5G networks.

9. Conclusions

The continuous increase of DVs on the global level caused by an increase in the number of mobile users and the introduction of novel and more bandwidth-demanding applications imposes challenges in deploying energy-efficient 5G networks. These challenges will additionally be contributed to by the fact that heterogeneous 5G networks must be deployed in areas of different device density classes, which will have different increases of DVs during the 2020s.

In this work, the influence of constant increase of DV during the 2020s on network EE was analyzed. The analysis was performed for different 5G network deployment and operation approaches implemented in rural, urban, urban-dense and indoor-hotspot device density classes. The two standardized EE metrics, known as data and coverage EE metrics were used for the assessment of the EE of the proposed 5G BSs deployment and operation approaches.

Obtained results show that an increase in DVs has an opposite effect on data and coverage EE metric of every device density class. This opposite trend in changes of data and coverage EE metric, caused by the increase in DV, indicates that there is no optimal amount of DVs in any of the analyzed device density classes for which a combination of both EE metrics will have the highest values. Obtained results reveal that only a reduction of 5G network instant power consumption and, consequently, energy consumption, will contribute to the simultaneous improvement of both data and coverage EE metrics. Therefore, the obtained results confirm that only those deployment and operation approaches that implement some of the RRM techniques (such as Tx power scaling and/or putting BSs in sleep mode) can bring the reduction of instant power and energy consumption, thus consequently contributing to the improvement of EE metrics of 5G RANs.

Additionally, the presented results indicate that the 5G BSs deployment and operation approaches that have the highest capabilities of adjusting BSs resources according to the increase of DV, will have the highest contribution to the improvement of both EE metrics in any device density class. It is further shown that, for any of the analyzed approaches which exploit adaptation of BSs deployment dynamics and operation activity according to increasing DV trend, the DV increase will cause an increase of the instant power consumption of the 5G network. Therefore, the analyses presented in the paper can serve as a basis in the future processes dedicated to the selection of deployment and operation strategies of 5G networks which will bring the highest network EE metrics, with respect to the permanently increasing DVs in a specific device density class.

Our future research activities will be dedicated to the investigation of the impact of DV increase on the 5G network EE of complete countries, which are composed of different in size and proportion device density classes having different DV patterns.

Author Contributions: Conceptualization, J.L.; methodology, J.L and Z.K.; validation, J.L. and Z.K.; formal analysis, J.L. and Z.K.; investigation, J.L. and Z.K.; resources, J.L. and Z.K., data curation, J.L. and Z.K.; writing—original draft preparation, J.L. and Z.K.; writing—review and editing, J.L.; visualization, J.L. and Z.K.; supervision, J.L. All authors have read and agreed to the published version of the manuscript.

Funding: This research received no external funding.

Institutional Review Board Statement: Not applicable.

Informed Consent Statement: Not applicable.

Conflicts of Interest: The authors declare no conflict of interest.

Abbreviations

The following abbreviations are used in this manuscript.

2G	second-generation mobile network
3G	third-generation mobile network
3GPP	3rd Generation Partnership Project
4G	fourth-generation mobile network
5G	fifth-generation mobile network
6G	sixth-generation mobile network
BS	base station
CAGR	compound annual growth rate
DL	uplink
DV	data volume
EB	Exabyte
EC	energy consumption
EE	energy efficiency
ES	ETSI standard
ETSI	European Telecommunications Standards Institute
GHz	Gigahertz
HetNet	heterogeneous network
IoE	Internet of Everything
ITU-T	International Telecommunication Union—Telecommunications Standardization Sector
KPI	key performance indicator
mmWave	millimeter-wave
MN	mobile network
MNO	mobile network operator
mMTC	Massive Machine-Type Communications
N/A	not applicable
RAN	radio access network
RRM	radio resource management
UHD	ultrahigh-definition
UL	uplink

References

1. Jejdling, F. *Ericsson Mobility Report*; Ericsson: Stockholm, Sweden, 2021.
2. Transforma Insights. Number of Internet of Things (IoT) Connected Devices Worldwide from 2019 to 2030, by Communications Technology (in Millions). 2020. Available online: https://www.statista.com/statistics/1194688/iot-connected-devices-communications-technology/ (accessed on 10 August 2021).

3. ITU-R. IMT Vision—Framework and Overall Objectives of the Future Development of IMT for 2020 and Beyond. 2015. Available online: https://www.itu.int/dms_pubrec/itu-r/rec/m/R-REC-M.2083-0-201509-I!!PDF-E.pdf (accessed on 4 August 2021).
4. Lorincz, J.; Matijevic, T.; Petrovic, G. On interdependence among transmit and consumed power of macro base station technologies. *Comput. Commun.* **2014**, *50*, 10–28. [CrossRef]
5. Lorincz, J.; Bogarelli, M.; Capone, A.; Begušić, D. Heuristic approach for optimized energy savings in wireless access networks. In Proceedings of the SoftCOM 2010, 18th International Conference on Software, Telecommunications and Computer Networks, Split, Croatia, 23–25 September 2010; IEEE: Manhattan, NY, USA, 2010; pp. 60–65.
6. Lorincz, J.; Bule, I.; Kapov, M. Performance Analyses of Renewable and Fuel Power Supply Systems for Different Base Station Sites. *Energies* **2014**, *7*, 7816–7846. [CrossRef]
7. Wu, Q.; Li, G.Y.; Chen, W.; Ng, D.W.K.; Schober, R. An Overview of Sustainable Green 5G Networks. *IEEE Wirel. Commun.* **2017**, *24*, 72–80. [CrossRef]
8. Chih-Lin, I.; Han, S.; Bian, S. Energy-efficient 5G for a greener future. *Nat. Electron.* **2020**, *3*, 182–184.
9. Environmental Engineering (EE). *Assessment of Mobile Network Energy Efficiency*; International Telecommunication Union: Geneva, Switzerland, 2020.
10. *5G Service Requirements for Next Generation New Services and Markets*; ETSI: Sophia Antipolis, France, 2018.
11. Lorincz, J.; Klarin, Z.; Begusic, D. Modeling and Analysis of Data and Coverage Energy Efficiency for Different Demographic Areas in 5G Networks. *IEEE Syst. J.* **2021**, 1–12. [CrossRef]
12. Alliance, N.G.M.N. *Recommendations for NGMN KPIs and Requirements for 5G*; Techreport: Berkshire, UK, 2016.
13. Holma, H.; Toskala, A.; Nakamura, T. *5G Technology: 3GPP New Radio*; Wiley: Hoboken, NJ, USA, 2020.
14. Wu, J.; Zhang, Y.; Zukerman, M.; Yung, E.K. Energy-Efficient BaseStations Sleep-Mode Techniques in Green Cellular Networks: A Survey. *IEEE Commun. Surv. Tutor.* **2015**, *17*, 803–826. [CrossRef]
15. Saker, L.; Elayoubi, S.E. Sleep mode implementation issues in green base stations. In Proceedings of the IEEE International Sympo-sium on Personal, Indoor and Mobile Radio Communications, Istanbul, Turkey, 26–30 September 2010; pp. 1683–1688.
16. Lorincz, J.; Matijevic, T. Energy-efficiency analyses of heterogeneous macro and micro base station sites. *Comput. Electr. Eng.* **2014**, *40*, 330–349.
17. Nokia. 5G Network Energy Efficiency. White Paper. 2016. Available online: https://global5g.5g-ppp.eu/sites/default/files/BookletA4_EnergyEfficiency.pdf (accessed on 4 August 2021).

sensors

Article

Analysis of the Impact of Detection Threshold Adjustments and Noise Uncertainty on Energy Detection Performance in MIMO-OFDM Cognitive Radio Systems

Josip Lorincz [1,*], Ivana Ramljak [2] and Dinko Begušić [1]

1 Faculty of Electrical Engineering, Mechanical Engineering and Naval Architecture (FESB), University of Split, R. Boskovica 32, 21000 Split, Croatia; dinko.begusic@fesb.hr
2 Elektroprenos—Elektroprijenos BiH" a.d. Banja Luka, 88000 Mostar, Bosnia and Herzegovina; ivana.ramljak@elprenos.ba
* Correspondence: josip.lorincz@fesb.hr

Abstract: Due to the capability of the effective usage of the radio frequency spectrum, a concept known as cognitive radio has undergone a broad exploitation in real implementations. Spectrum sensing as a core function of the cognitive radio enables secondary users to monitor the frequency band of primary users and its exploitation in periods of availability. In this work, the efficiency of spectrum sensing performed with the energy detection method realized through the square-law combining of the received signals at secondary users has been analyzed. Performance evaluation of the energy detection method was done for the wireless system in which signal transmission is based on Multiple-Input Multiple-Output—Orthogonal Frequency Division Multiplexing. Although such transmission brings different advantages to wireless communication systems, the impact of noise variations known as noise uncertainty and the inability of selecting an optimal signal level threshold for deciding upon the presence of the primary user signal can compromise the sensing precision of the energy detection method. Since the energy detection may be enhanced by dynamic detection threshold adjustments, this manuscript analyses the influence of detection threshold adjustments and noise uncertainty on the performance of the energy detection spectrum sensing method in single-cell cognitive radio systems. For the evaluation of an energy detection method based on the square-law combining technique, the mathematical expressions of the main performance parameters used for the assessment of spectrum sensing efficiency have been derived. The developed expressions were further assessed by executing the algorithm that enabled the simulation of the energy detection method based on the square-law combining technique in Multiple-Input Multiple-Output—Orthogonal Frequency Division Multiplexing cognitive radio systems. The obtained simulation results provide insights into how different levels of detection threshold adjustments and noise uncertainty affect the probability of detection of primary user signals. It is shown that higher signal-to-noise-ratios, the transmitting powers of primary user, the number of primary user transmitting and the secondary user receiving antennas, the number of sampling points and the false alarm probabilities improve detection probability. The presented analyses establish the basis for understanding the energy detection operation through the possibility of exploiting the different combinations of operating parameters which can contribute to the improvement of spectrum sensing efficiency of the energy detection method.

Keywords: spectrum sensing; energy detection; SLC; OFDM; noise uncertainty; dynamic threshold; MIMO; SISO; cognitive networks; SNR; probability; wireless; false alarm; transmit; receive; power

Citation: Lorincz, J.; Ramljak, I.; Begušić, D. Analysis of the Impact of Detection Threshold Adjustments and Noise Uncertainty on Energy Detection Performance in MIMO-OFDM Cognitive Radio Systems. *Sensors* **2022**, *22*, 631. https://doi.org/10.3390/s22020631

Academic Editors: Luís Castedo Ribas and Hsiao-Chun Wu

Received: 9 December 2021
Accepted: 12 January 2022
Published: 14 January 2022

1. Introduction

The increased popularity of wireless communication networks raises the need for an improvement of network capacity and the efficiency of spectrum usage. To address the problem of the efficient usage of the spectrum, the cognitive radio networks (CRN) concept was proposed as a promising solution that can be implemented in wireless communication systems. In CRNs, two types of users known as the primary users (PUs) and the secondary users (SUs) are known. The cognitive radio (CR) enables SU to perform dynamic spectrums access (DSA) in periods when PU does not use the spectrum. This means that the PU always has the priority when exploiting a dedicated licensed spectrum. SU may use licensed bands so long as it does not cause interference with PU. Therefore, the main goal of DSA in CRNs is to improve the efficiency of the spectrum usage [1,2].

Spectrum sensing (SS), as an essential function of CRN, enables users in cognitive networks to have information about its environment and spectrum availability. The most widely used SS method is energy detection (ED). The ED method is a local non-cooperative SS method that does not demand prior knowledge about the characteristics of the PU signal. Compared to other prominent local SS methods, ED has the least computational and implementation complexity [3,4]. However, the ED is very sensitive to fluctuations in noise power, low values of the signal-to-noise ratio (SNR), and fading [5]. Therefore, the performance of ED is confined by the noise power variations which is also defined as noise uncertainty (NU). In real wireless communication systems, the NU is caused by phenomena such as filtering effects, interference from surrounding sources, and thermal noise [6]. An additional disadvantage of the ED technique is in the lack of ability to distinguish between SU or PU and interference. Regardless of the presented disadvantages, ED is, due to its simple deployment and processing, the most applied SS method in practice [5,7,8].

In the ED process, the sensing of the energy of the signal transmitted by the PU in the licensed frequency spectrum is performed. The final goal of this sensing is to determine the test statistics that represent a measure of PU transmission activity. The test statistics are then compared to an in advance set detection threshold (DT). The DT is set from the energy of the noise. The value of the determined DT is the key to performing an accurate ED. The level of DT can be specified as a constant or dynamically adjusted value. The process of DT adjustment enables the dynamic selection of the value of DT according to the NU during the period of SS [3,9,10].

To accomplish a better level of detection performance in environments impacted with NU, employing dynamic DT adjustments is a promising solution. However, due to the influence of NU on the signal sensed at the location of SU, the practical realization of DT adjustments in the system exploiting Multiple-Input Multiple-Output (MIMO) Orthogonal Frequency Division Multiplexing (OFDM) transmission is very demanding. In this article, an assessment of the operational efficiency of the ED-based SS in the MIMO-OFDM CR system exploiting the DT adjustment according to NU at the position of SU is presented. The OFDM as technology is widely applied in many communication systems. Recent studies demonstrate that combining the OFDM with MIMO can improve the spectral efficiency in CRNs [11]. The transmission based on MIMO-OFDM technologies has the potential to achieve higher data rates and to alleviate the problem of Inter Symbol Interference in CRNs [12–14]. Therefore, combining MIMO and OFDM technologies in CRNs can contribute to the enhancement of spectral efficiency and transmission capacity and investigations dedicated to the performance efficiency of different SS methods in MIMO-OFDM CR systems have been done in [15–17].

While in Single-Input Single-Output (SISO) transmission systems a single transmission and reception chain or branch (antenna) is used for performing ED, in MIMO systems, multiple transmit (Tx) chains at the PU side and receive (Rx) chains at the SU side are employed. Therefore, the SS employing ED in the MIMO-OFDM CR system can be realized by exploiting various Tx-Rx antenna diversity techniques. Among the different diversity combining techniques, the Square-Law Combining (SLC) technique has the lowest complexity when implemented for the purpose of SS [18]. The ED employing the SLC technique belongs to the non-coherent SS method and its implementation in the MIMO-OFDM CR system does not require channel state information (CSI) for realization of ED.

Hence, ED employing the SLC technique represents a simple and efficient concept for the implementation in the MIMO-OFDM CR system, which motivates presenting the results of the performance analysis of such a concept in this paper. Further motivation is based on the massive exploitation of battery-powered and low-power devices in the emerging Internet of Things (IoT) concept. Massive practical implementation of the IoT concept will be supported with the implementation of the fifth-generation (5G) and upcoming sixth-generation (6G) mobile networks. For enabling the low-power IoT devices to exploit the concept of CR communications, the practical implementation of simple and low complexity SS techniques such as the ED method based on the SLC technique will be of particular interest. The implementation of such an SS technique does not require complex processing or significant device battery depletion, which makes ED based on the SLC technique a promising candidate for massive implementation in future IoT devices equipped with multiple antennas. Hence, the assessment of the ED performance based on the SLC technique in different operating environments affected with NU and performed with DT adjustments is of major significance for possible future realization of such an SS concept in IoT networks.

Therefore, in this paper, the impact of NU and DT adjustments on the efficiency of the ED employing SLC technique in the SISO and MIMO-OFDM CR systems was analyzed. This paper makes the following contributions:

- The development of the explicit analytic mathematical expressions for the performance assessment of ED process employing SLC method impacted by NU and DT adjustments in MIMO-OFDM CR systems.
- The introduction of the simulation algorithm for executing the ED process by exploiting the SLC method in MIMO-OFDM CR networks affected by different levels of DT adjustments and NUs.
- The comprehensive analyses of simulation results through investigation of the influence of various parameters including the OFDM modulations, the SNRs, the MIMO Tx-Rx chains number, the false alarm probabilities, the Tx powers of PU, the number of sampling points used in ED, and the different levels of NU and DT adjustments on the probability of detection of the PU transmission.

The remaining parts of the paper contain the following sections. A review of the topic associated with the exploitation of the ED method in the MIMO-OFDM CR systems is given in Section 2. Section 3 presents the mathematical expression of the ED principles which involve the effect of DT adjustments and NU on the performance of ED. Section 4 presents a simulation algorithm that enables the ED-based SS in MIMO-OFDM CR systems and the assessment of ED performance impacted with NU and DT adjustments. The comprehensive analysis of the extensive simulation results is given in Section 5. Section 6 concludes the paper.

2. Literature Overview

Table 1 presents the literature survey of related work. In the literature, the performance analysis of CR in MIMO-OFDM systems was performed in [18–23]. The authors in [22,23] show that the implementation of the MIMO-OFDM transmission contributes to the enhancement of SS efficiency performed using the ED method. In [22–26], the ED is performed employing the SLC method. In [22], the simulation results indicate that the ED employing SLC can enable precise signal detection for low to moderate SNRs. In [23], SS based on ED and cyclostationary feature detection with and without multiple Tx-Rx chains have been analyzed. The comparative performance results indicate that the Equal Gain Combining (EGC) method requires a precise CSI for performing SS, which consequently results in more efficient SS. On the contrary, the SLC method lacks the need for CSI, which results in lower SS efficiency. In addition, the ED method employing SLC requires detectors and combiners which additionally contribute to the increase in implementation cost. However, the SLC technique is still significantly less complex for implementation when it is compared with other diversity combining schemes which successful operation demands precise knowledge about CSI.

Table 1. Literature survey of related work.

Reference	Major Contribution
[18]	Improved SS at the SU side in a realistic environment by employing SLC and square-law selection (SLS) techniques.
[19]	Software radio implementation of MIMO-OFDM.
[20]	A comprehensive survey of OFDM transmission for wireless communications.
[21]	A detailed survey on the performance requirements of 5G wireless cellular communication systems in terms of capacity, data rate, spectral efficiency, latency, energy efficiency, and quality of service.
[22]	In comparison with single antenna CRs systems, significant improvement is observed in PU detection probability when ED based on the SLC technique is performed in MIMO CRs systems.
[23]	Multiple antenna techniques and cyclostationary feature detection-based systems are proposed for ED.
[24]	Analysis of cooperative spectrum sensing with ED over various fading channels using the SLC diversity scheme.
[25]	Analyses of the problem of ED of an unknown signal over a multipath channel by employing SLC and SLS techniques.
[26]	The tutorial presents a comprehensive overview of the ED-based SS and provides tools necessary for performing analyses of several SS algorithms.
[27]	A survey of the NU impact on ED in communication systems with different OFDM system designs has been presented.
[28]	A review of ED performance exploiting dynamic DT adaptations in the SISO-OFDM systems.
[29]	Presentation of a novel approach based on subchannel and transmission power allocation that adaptively assigns the radio resources considering the interference caused to the PUs in multi-cell wireless networks.
[30]	Analyses of the new communication approach based on the licensed shared access (LSA) spectrum sharing framework with in-band full-duplex multi-cell multi-user MIMO communication network as the licensee, which operates in the service region of a multi-user MIMO incumbent network.
[31]	Presentation of the simulation algorithm that enables the performance analysis of the ED method employing the SLC technique in MIMO-OFDM CR systems and analyses of simulation results.
[32]	Analyses of efficiency of ED SS - based on SLC technique in MIMO-OFDM Cognitive Radio Networks without the impact of NU and dynamic DT adjustments.
[33]	Presentation of novel transmission solution based on adaptive beamforming with the coding scheme based on STBCs in IEEE 802.11 n WLAN systems.
[34]	Presentation of the current state-of-the-art related to the research on SS by using ED with an extensive overview of basic theories in recent research, architectures for performing ED SS, the possible applications of ED and performance measurements of ED.
[35]	The analysis of optimal DT selection for SS in a CRN using the ED approach is performed for fixed detection and false alarm probabilities.
[36]	A survey of the fundamental concepts of CRN characteristics, functions, network architecture and applications is presented.
[37]	The introduction of the ED SS which reduces the SNR-wall problem caused by the NU effects through the cooperation of multiple receivers for adapting the DT at each sensing point to the noise power present at the moment of SS.
[38]	A new ED algorithm based on dynamic DT selection is presented and the relationship of detection sensitivity and ED performance with the impact of fluctuation of average noise power is investigated.
[39]	Analyses of the influence of DDT and NU factor in the case of ED SSs on the detection and false alarm probability with the significance of their ratio on the sensing technique is analyzed and the expression of the empirical relationship between the sampling number and SNR is also proposed.
[40]	Development of the analytical model for estimation of the statistical performance of the ED which can be used for setting the appropriate DT such that more spectrum sharing can be exploited, especially when combined with cooperative SS.

The challenges related to the hardware implementation of ED employing Square-Law Selection (SLS) and SLC methods are presented in [18]. The authors showed that the proposed solutions can facilitate hardware reliability and antenna diversity in a realistic implementation scenario.

The analyses in [6,27,28] show that the ED performance can be significantly impaired by NU. To improve the ED sensing efficiency degraded by the influence of NU, the authors in [6] proposed a kernelized ED concept based on a DT. For the PU signals affected by Gaussian noise, an assessment of ED performance in communication systems using MIMO transmission have been presented. The results indicate that by increasing the number of sampling points, the number of Rx chains, and SNR at the antennas of SU, the ED method achieves a good level of performance under the Gaussian mixture noise and exceeds the SS efficiency of the classical non-kernelized ED method.

In [29], the subchannel and Tx power allocation concept for downlink communication in multi-cell CR-OFDM access (OFDMA) networks with the adaptive fractional frequency reuse strategy have been proposed. The proposed concept enables maximization of the throughput of the CR SISO network by efficiently assigning OFDM subchannels to base station (BS) cells, while controlling the interference to the PUs through restricting the Tx power on the subchannels used by the PUs. It is shown that the proposed concept improves the system throughput by up to 50% for the same level of interference at locations of PUs. In [30], an even larger improvement of system throughput of 60% and interference reductions have been accomplished. Such improvement is obtained for the proposed licensed shared access (LSA) spectrum sharing framework with the in-band full-duplex transmission in multi-cell multi-user MIMO communication network as the licensee, which operates in the service region of a multi-user MIMO incumbent network. Such significant system throughput improvements have been obtained through the implementation of a hierarchical two-phases beamforming approach, where in the first phase, the LSA controller collects the quality of service demands from the licensee network and based on them, the beamformers are dedicated to reducing the cumulative interference towards the incumbents. In the second phase, the interference minimized in the first step is considered as a constraint for the new beamformer design problem that maximizes the cumulative downlink and uplink data rates of the licensee network. The analyses presented in this paper have been performed for single-cell MIMO-OFDM CR systems exploiting ED based on the SLC technique; however, it can be extended to multi-cell MIMO-OFDM systems.

In [27,28], we analyze the influence of NU on the ED performance in single-input single-output (SISO)-OFDM systems, whose operations are based on rate or margin-adaptive or combined transmission techniques. The obtained results indicate that the ED will predominantly be impacted by NU. To reduce the effect of NU, in [28] the analysis of ED performance exploiting DT adjustments in the SISO communications systems is performed. The obtained results indicate that the ED method can be improved by involving dynamic adjustments of the DT in the ED impacted with NU. Furthermore, in [31], a simulating algorithm that enabled the performance analysis of the ED method employing SLC in MIMO-OFDM CR systems is proposed. Based on the algorithm proposed, in [32] we present the preliminary results of the performance assessment of the ED method employing SLC in the MIMO communications systems. The results show that increasing the number of Rx and Tx chains in the MIMO system give a contribution to the enhancement of ED performance.

Although the earlier research works [18,22–26,31,32] show that using the ED method employing the SLC technique provides a positive contribution to the improvement of the efficiency of the ED, an investigation into how NU and DT adjustments impact the ED sensing efficiency in MIMO-OFDM CRs systems is missing. Therefore, in this paper for the first time, the mathematical equations which express how the NU and dynamic DT adjustments, the number of PU Tx and SU Rx chains, the probability of false alarms, the SNRs, and the number of sampling points impact the ED performance in the MIMO-OFDM CR system have been developed. In addition, an extensive analysis of the impact of NU and DT adjustments on the ED employing the SLC technique in MIMO-OFDM CR systems is presented in this paper. The presented analyses establish the basis for understanding the ED operation exploiting the capability of the dynamic DT adjustments in the MIMO-OFDM CR systems affected by NU.

3. System Design and Explanation of the Energy Detection Operation

The visualization of the block schema of the analyzed single-cell MIMO-OFDM CR system consisted of single PU and single SU is presented in Figure 1. The signal in the analyzed MIMO-OFDM CR system is transmitted using space-time block codes (STBCs) [33]. The PU is a licensed user and it has a higher priority when using the dedicated frequency spectrum. The SU has a lower priority and opportunistically accesses the spectrum in the absence of PU [3]. The SU is permitted to use the spectrum in a way that does not cause interference to the PUs. For that reason, the SU performs SS using the ED method employing the SLC technique (Figure 1). In the process of SLC, the SU equipped with an energy detector performs a signal squaring operation in the square-law device which is followed by combining the squared signals received at R Rx chains in the finite time integrator (Figure 1).

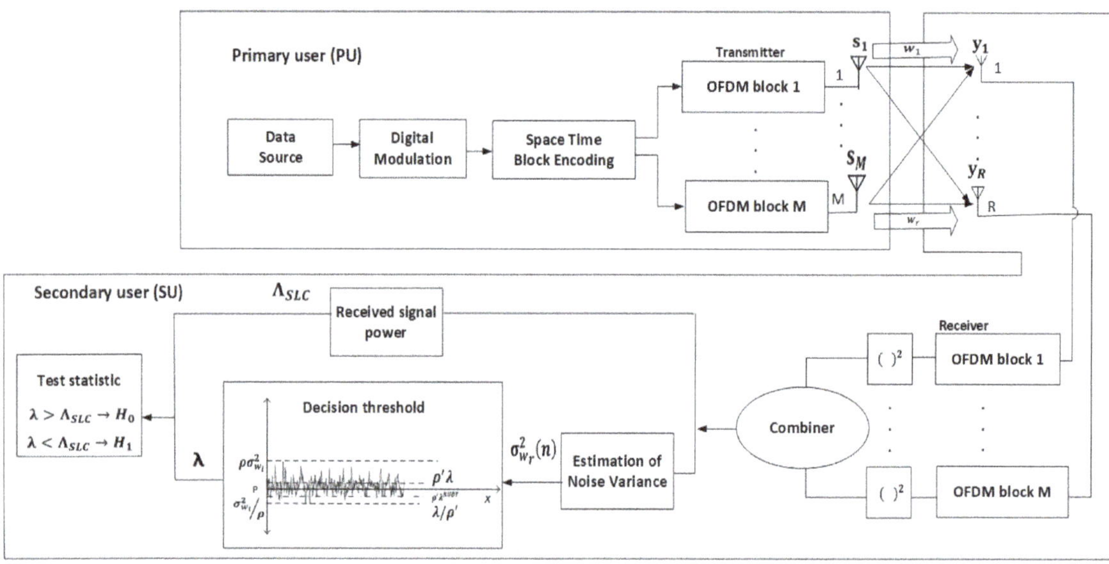

Figure 1. Main blocks of the MIMO-OFDM wireless communication system for SS based on ED employing SLC technique.

The PU Tx power in the MIMO system emitted over the m-th Tx chain (antenna) is formulated as P_m. The $P = \sum_{m=1}^{M} P_m$ defines the overall instant Tx power of the PU transmitted via M Tx chains (Figure 1). Table 2 lists the descriptions of all parameters used in the analysis. The complex signal defined as $s_m = s_{m,r} + js_{m,i}$ is assumed as the signal transmitted via m-th Tx chain of PU (Figure 1). Hence, the signals carried via the M Tx chains of PU are expressed as $s = \sum_{m=1}^{M} s_m$. The signal received by the SU at every R Rx chain (antenna) and sampled by n samples where $n = 1, \ldots, N$ can be formulated as:

$$\mathbf{y}_r(n) = \begin{cases} w_r(n) \\ \mathbf{h}_r(n)\, \mathbf{s}_r(n) + w_r(n) \end{cases} \tag{1}$$

The $\mathbf{h}_r(n)$ is of size $\mathbb{C}^{1XM}$ and it is a complex vector that represents the wireless channel gain among the M Tx chains of PU and r-th Rx chain of SU (Figure 1). Complex vector $\mathbf{s}_r(n)$ is of size $\mathbb{C}^{MX1}$ and represents the Tx signal of PU that is received at r-th Rx chain of the SU in n-th sample (sensing moment).

Table 2. Parameters used in the simulation analysis.

Index	Description
H_1	The hypothesis which defines the existence of the PU signal
H_0	The hypothesis which defines the non-existence of the PU signal
m	The number of Tx chains on the PU side
r	The number of Rx chains on the SU side
M	The total number of PU Tx chains
R	The total number of SU Rx chains
N	The overall number of sampling points utilized for ED without DT adjustment and influence of NU
N^{DT}	The overall number of sampling points utilized for ED with DT adjustment
N^{NU}	The overall number of sampling points utilized for ED influenced by NU
N^{NUDT}	The overall number of sampling points utilized for ED with DT adjustment and influence of NU
$\mathbf{s}_m$	The complex signal carried via the m-th Tx chain of the PU
$\mathbf{s}$	The complex signal of the PU transmitted over the M Tx chains
P	The total via M Tx chains transmitted instantaneous Tx power
P_m	Instantaneous Tx power transmitted on the PU m-th antenna chain
$\mathbf{y}_r(n)$	Vector of the signal detected at r-th Rx chain of the SU in the n-th SS period
$\mathbf{Y}(n)$	Vector of the signal received by all R Rx chains of the SU in the n-th SS period
$\boldsymbol{h}_r(n)$	Vector of channel gain among the M Tx chains and the r-th Rx chain in the n-th SS period
$\mathbf{s}_r(n)$	Vector of the signal detected within the n-th SS sample point at the SU r-th Tx chain
$\boldsymbol{w}_r(n)$	Vector of the noise impacting ED during the n-th SS period at the r-th Rx chain of the SU
$\sigma^2_{wr}(n)$	The variance of noise for the signal detected in n-th SS period at the SU r-th Rx chain
$\sigma^2_{s_r}(n)$	The variance of the received signal in the n-th SS period at the r-th Rx chain of the SU
$\sigma^2_{wNU}(n)$	AWGN variance used in the ED impacted with NU
$\sigma^2_{wNUDT}(n)$	AWGN variance used in the ED impacted with NU and DT adjustments
Λ_r	Test statistics for signals detected at the r-th Rx chain (antenna) of the SU
Λ_{SLC}	The overall test statistics of all signals detected via the R receive (Rx) chains of the SU
$\gamma_r(n)$	Signal-to-noise ratio at the r-th receive chain of the SU during the n-th SS period
$\gamma_{SLC}(n)$	The total signal-to-noise ratio associated with the R SU receive antennas (chains) in the n-th SS period
$\overline{\gamma_{SLC}}(n)$	The mean signal-to-noise ratio detected by the SU during the n-th SS period for all R receive chains
P_f	False alarm probability for ED performed without DT adjustments and impact of NU
P_d	Detection probability for ED performed without DT adjustments and impact of NU
P_{fa}^{NU}	False alarm probability for ED impacted with NU
P_d^{NU}	Detection probability for ED impacted with NU
P_{fa}^{DT}	False alarm probability for ED performed with DT adjustments
P_d^{DT}	Detection probability for performed with DT adjustments
P_{fa}^{NUDT}	False alarm probability for ED performed with DT adjustments and impact of NU
P_d^{NUDT}	Detection probability for ED performed with DT adjustments and impact of NU
$Q(x)$	Standard Gaussian Q function
λ	DT for ED performed without DT adjustments and impact of NU
λ_{fa}	False alarm threshold in the case of ED performed based on CFAR principles
λ_d	DT level for ED performed based on CDR principles
λ_d^{DT}	DT for SLC ED performed with DT adjustments
λ_{fa}^{DT}	False alarm threshold for ED performed with DT adjustments
λ_d^{NU}	DT for ED impacted with NU
λ_{fa}^{NU}	False alarm threshold for ED impacted with NU
λ_d^{NUDT}	DT for SLC ED performed with DT adjustments and impacted with NU
λ_{fa}^{NUDT}	False alarm threshold for ED performed with DT adjustments and impacted with NU
λ'^{DT}	DT for ED performed without NU
λ'^{NUDT}	DT for ED performed with DT adjustments and NU
ρ	NU factor
ρ'	DDT factor

Additionally, the complex noise signal $w_r(n)$ is the additive white Gaussian noise (AWGN) impacting the signal received at the r-th Rx chain of the SU. The impact of noise is assumed to be an identically distributed and independent random process having a mean value equal to zero and a variance of σ_w^2, which distribution is circularly symmetric and defined as $\mathcal{N}\ (0,\ 2\sigma_{wr}^2(n))$.

Since the PU signal is impacted by noise, the SNR at the SU r-th chain is formulated as:

$$SNR_r(n) \ = \ \gamma_r(n) \ = \ \frac{|\boldsymbol{h}_r(n)|^2 \frac{1}{N} \sum_{n=1}^{N} | \boldsymbol{s}_r(n)|^2}{2\sigma_{wr}^2(n)} \tag{2}$$

The total SNR at all M Rx chains (antennas) during the n-th SS period is expressed as: $\gamma_{SLC} \ = \ \sum_{r=1}^{R} \gamma_r(n)$. Furthermore, the mean SNR value at the antenna(s) of SU for all R Rx chains in the n-th sampling period is defined as: $\overline{\gamma_{SLC}} \ = \ \frac{1}{R} \sum_{r=1}^{R} \gamma_r(n) = \frac{1}{R} \gamma_{SLC}$.

Spectrum sensing is the process where the SUs continuously supervise the activity of the PUs in order to detect the spectrum holes. Detailed knowledge about spectrum availability obtained through the testing of a binary hypothesis H_0 and H_1 shown in Equation (3), represents the fundamental operation of the ED. This process aims to decide between two hypotheses, according to which the PU signal is assumed to be absent (denoted as H_0) or the PU signal is assumed to be present (denoted by H_1). Detailed knowledge about spectrum availability obtained through the testing of a binary hypothesis H_0 and H_1 shown in Equation (3), represents the fundamental operation of the ED. Therefore, the decision on SS occupancy is the result of the testing of the following hypothesis:

$$\boldsymbol{Y}(n) \ = \ \begin{cases} \sum\limits_{r=1}^{R} \boldsymbol{w}_r(n) : \ H_0 \\ \sum\limits_{r=1}^{R} \boldsymbol{h}_r\ (n)\ \boldsymbol{s}_r(n) + \sum\limits_{r=1}^{R} \boldsymbol{w}_r(n) : \ H_1 \end{cases} \tag{3}$$

where $\boldsymbol{Y}(n)$ is the total signal detected from all R Rx chains in the n-th SS period at a position of SU. The focus of the ED is making a decision on whether the detected signal $\boldsymbol{Y}(n)$ satisfies hypotheses H_0 or H_1. Therefore, in the process of deciding whether the PU is present or not, the threshold is compared with the sensed energy of the signal detected at the antennas of SU. The decision hypotheses H_1 is satisfied when the sensed energy of the signal detected at the antennas of SU is greater than the threshold. This results in the conclusion that the PU transmits in the dedicated band. The decision hypotheses H_0 is satisfied if the energy of the detected signal is lower than the DT. This results in the cognition that the signal of PU is absent. Thus, the result of this binary hypothesis test determines the SU activity in the terms of possible transmission in the PU frequency band.

3.1. Process of Energy Detection

The procedure of ED employing SLC technique in MIMO-OFDM CR systems aims to exploit the SS of the PU signal by means of all R Rx chains of SU. In accordance with the SLC method, the signals received on all R Rx chains are squared and combined to get the total received signal energy known as the test statistic. The overall test statistic can therefore be expressed as:

$$\Lambda_{SLC} \ = \ \sum_{r=1}^{R} \Lambda_r \ = \ \sum_{r=1}^{R} \sum_{n=1}^{N} |\boldsymbol{y}_r(n)|^2 \tag{4}$$

where the test statistics of the SU r-th Rx chain is expressed as Λ_r.

The decision regarding spectrum occupancy by the PU is performed through the comparison of the test statistic with a DT $(\lambda(n))$ that is dynamically selected for each sample n used in the SS process:

$$\Lambda_{SLC}(n) > \lambda(n) : \ H_0,\ \forall n \in \{1,\ \ldots,\ N\} \tag{5}$$

$$\Lambda_{SLC}(n) > \lambda(n) : \ H_1,\ \forall n \in \{1,\ \ldots,\ N\} \tag{6}$$

In accordance with [26,34], for an adequate number of sampling points N, the distribution of the total test statistic given in Equation (4) can be approximated using a normal distribution as:

$$\Lambda_{SLC} \sim \mathcal{N}\left(\begin{array}{c} \sum_{r=1}^{R} \sum_{n=1}^{N} \mathbb{E}\left[|\boldsymbol{y}_r(n)|^2 \right], \\ \sum_{r=1}^{R} \sum_{n=1}^{N} \mathbb{V}ar\left[|\boldsymbol{y}_r(n)|^2 \right] \end{array} \right), \tag{7}$$

where $\mathbb{E}\,[\cdot]$ expresses the expectation operator and $\mathbb{V}ar\,[\,\cdot\,]$ expresses the variance operator.

In each observed SS period n, the uniform gain of the wireless channel $\boldsymbol{h}_r(n)$ and the noise variance of the detected signal at the r-th Rx chain (antenna) $2\sigma^2_{wr}(n)$ can be formulated as:

$$\boldsymbol{h}_r(n) = \boldsymbol{h}\ \forall r = 1, \ldots, R;\ \forall n = 1, \ldots, N \tag{8}$$

$$2\sigma^2_{wr}(n) = 2\sigma^2_{w}\ \forall r = 1, \ldots, R;\ \forall n = 1, \ldots, N \tag{9}$$

where $\boldsymbol{h}$ represents the complex matrix of the channel gain for all R Rx chains.

Performing the SS using the ED method employing SLC demands knowledge about the average level of the power received at the antennas of SU Rx chains. Therefore, the overall instant PU Tx power of the signal transmitted using M Tx chains (antennas) within the n-th observation period, is equal to all signal variances at the r-th Rx chain of the SU. It can be formulated as $P = \sum_{r=1}^{R} |\boldsymbol{h}|^2 2\sigma^2_{s_r}(n)$. The interdependence among the mean values of SNR at the antennas of SU and the instant Tx power of the PU can be approximated with $\overline{\gamma_{SLC}} \approx \frac{P}{R2\sigma^2_w}$.

Considering the above assumptions, the total test statistics from Equation (7) can be approximated with:

$$\Lambda_{SLC} \sim \left\{ \begin{array}{r} \mathcal{N}\left(RN(2\sigma^2_w),\ RN(2\sigma^2_w)^2 \right):\ H_0 \\ \mathcal{N}\left(N(2\sigma^2_w)(R+\gamma_{SLC}),\ N(2\sigma^2_w)^2(R+2\gamma_{SLC}) \right):\ H_1 \end{array} \right. \tag{10}$$

According to the overall test statistic presented in Equation (10) for hypotheses H_1 and H_0, the false alarm and detection probability for ED SS employing SLC diversity technique in MIMO-OFDM CR systems was developed.

3.2. *Probabilities of False Alarm and Detection for MIMO-OFDM CR Systems*

The performance of ED SS techniques is exploited through two probabilities: detection probability (P_{di}) and false alarm probability (P_{fa}). The probability of sensing the transmitted PU signal at the position of SU when it is really transmitted is known as the probability of detection (P_d). It can be examined through the verification of hypothesis H_1 as $P_d[\Pr(\Lambda_{SLC} > \lambda)\,|H_1]$. The detection probability for the ED method employing SLC technique in MIMO communication systems can be defined as

$$\begin{aligned} P_d[\Pr(\Lambda_{SLC} > \lambda)\,|H_1] &\approx Q\left(\frac{\lambda_d - N(2\sigma^2_w)(R+\gamma_{SLC})}{\sqrt{N(R+2\gamma_{SLC})}\,(2\sigma^2_w)} \right) \approx Q\left(\frac{\lambda_d - RN(2\sigma^2_w)(1+\overline{\gamma_{SLC}})}{\sqrt{RN(1+2\overline{\gamma_{SLC}})}\,(2\sigma^2_w)} \right) \\ &\approx Q\left(\frac{\lambda_d - RN\left(2\sigma^2_{w_i}\right)\left(1+\frac{P}{2R\sigma^2_w}\right)}{\sqrt{RN\left(1+\frac{P}{R\sigma^2_w}\right)}\,(2\sigma^2_w)} \right) \end{aligned} \tag{11}$$

where Q (.) is the Gaussian-Q function and λ_d represents the DT level. For better PU signal detection, a higher detection probability (P_d) is required. A higher detection probability (P_d) improves the usage of spectrum and ED performance of the SU.

The false alarm probability (P_{fa}) is the probability of sensing a PU signal by the SU, when the signal of PU is not really transmitted. It can be examined through the verification of hypothesis H_0 as $P_{fa}[\Pr(\Lambda_{SLC} > \lambda)\,\big|H_0]$. The false alarm probability (P_{fa})

as a performance metric of ED employing the SLC method in MIMO communication systems can be defined as:

$$P_{fa}[\Pr(\Lambda_{SLC} > \lambda\ )\,|H_0] \approx Q\left(\frac{\lambda_{fa} - RN(2\sigma_w^2)}{\sqrt{RN}\ (2\sigma_w^2)}\right) \quad (12)$$

where λ_{fa} represents the false alarm threshold level. According to Equations (11) and (12), the detection probability and false alarm probability depend on the number of sampling points (N), the variance of noise (σ_w^2), the number of Rx chains (R) of SU, and the level of the defined detection or false alarm thresholds. Additionally, Equation (11) indicates that the detection probability will also depend on the level of PU Tx power P.

From the perspective of the SU, a lower P_{fa} means more chances that the channel can be reused when it is available. This consequently results in the possibility of achieving higher throughputs for the SU. Thus, a fundamental trade-off between the sensing capability and achievable throughput of the SU exists for SS based on ED. Therefore, a reliable energy detector should ensure a low false alarm probability (P_{fa}) and a high detection probability (P_d). This means that appropriate QoS for a SU when using a wireless network should be provided, while also establishing an appropriate level of PU protection during periods of transmitting should be guaranteed.

From Equations (11) and (12), it is possible to define the minimum number of sampling points (N_{min}) needed for achieving the precise detection of the PU signal, which is expressed as:

$$N_{min} = \frac{\left[\sqrt{R}Q^{-1}\left(P_{fa}\right) - \sqrt{(R+2\gamma_{SLC})}Q^{-1}(P_d)\right]^2}{\gamma_{SLC}{}^2} = \frac{\left[Q^{-1}\left(P_{fa}\right) - \sqrt{(1+2\overline{\gamma_{SLC}})}Q^{-1}(P_d)\right]^2}{R\overline{\gamma_{SLC}}^2} \quad (13)$$

Equation (13) indicates that the calculation of the minimum number of sampling points does not request knowledge of the DT level. For the number of sampling points higher than the minimal, the detection of the PU signal will be ensured for any DT level.

Furthermore, it is already known that for low SNR at the antennas of SU, a high number of sampling points is required for precise sensing of PU signal. The need for a larger number of sampling points results in a larger sensing duration, which negatively affects the battery discharge of battery-powered devices. By combining (11) and (12) and considering that in practice $\lambda_d = \lambda_{fa} = \lambda$, the correlation between a detection and false alarm probability can be formulated as

$$P_d = Q\left(\frac{Q^{-1}\left(P_{fa}\right) - \sqrt{\frac{N}{R}}\gamma_{SLC}}{\sqrt{(1+2\frac{\gamma_{SLC}}{R}\ )}}\right) = Q\left(\frac{Q^{-1}\left(P_{fa}\right) - \sqrt{RN}\ \overline{\gamma_{SLC}}}{\sqrt{(1+2\ \overline{\gamma_{SLC}})}}\right) = Q\left(\frac{Q^{-1}\left(P_{fa}\right) - \frac{\sqrt{NP}}{2\sqrt{R}\sigma_w^2}}{\sqrt{\left(1+\frac{P}{R\sigma_w^2}\right)}}\right) \quad (14)$$

Based on Equation (14), it can be seen that the detection probability can be expressed without a DT level, if the targeted probability of false alarm is known. An approach based on defining the operation of the CR network by setting a constant false alarm probability is known as the constant false alarm rate (CFAR) ED approach.

3.3. Detection Threshold Estimation

According to what has been presented in the previous section, the false alarm and detection probabilities are affected by the corresponding threshold values. Determining the level of DT is the main activity that influences the decision efficiency regarding the absence or presence of the PU signal. The fixed detection threshold (FDT) and the dynamic detection threshold (DDT) methods are mainly considered in the literature as two opposing approaches for the selection of DTs in the ED process. In the case of FDT, the set threshold has a constant value even when the signal fluctuations in the frequency channel are present. However, to take into account fluctuations in the state of the wireless channel, the method based on DDT adjustments has been introduced. In the DDT method, the level of DT is adjusted to the channel conditions to maximize the detection probability. The literature indicates that the DDT method provides better SS performance compared to that of the FDT method [35]. In addition, a well-chosen level of DT can minimize the SS error, ensuring that there is enough protection in the usage of the licensed band of the PU, which further contributes to enhancing the spectrum utilization [36].

According to (12), for a given noise variance σ_w^2, the number of Rx chains and the number of sampling points N used for SS in the ED process, the FDT level can be calculated for the fixed false alarm probability according to the following:

$$\lambda_{fa} = \left[Q^{-1}\left(P_{fa}\right) + \sqrt{RN}\right]\sqrt{RN}2\sigma_w^2 \quad (15)$$

Hence, such an approach known as the CFAR approach calculates the value of the false alarm threshold that needs to maximize the probability of detection. This CFAR approach is used for SS employing ED in systems that require the maximal utilization of the wireless channel.

Alternatively, to provide an appropriate priority and sufficient protection of the transmissions performed by PU, the selection of the DT based on the constant detection rate (CDR) can be defined from Equation (11):

$$\lambda_d = \left[Q^{-1}(P_d)\sqrt{\left(1 + \frac{P}{R\sigma_w^2}\right)} + \sqrt{RN}\left(1 + \frac{P}{2R\sigma_w^2}\right)\right]\sqrt{RN}2\sigma_w^2 \quad (16)$$

The CDR approach in SS based on ED is applied when it is important to eliminate the interference in the observed CR system.

By comparing Equations (15) and (16), it can be noticed that the CFAR approach does not request the information about the total instant PU Tx power (P) of the signal transmitted over M Tx chains. This makes the CFAR approach more applicable in practical implementations. Although the CFAR compared to the CDR approach improves the throughput of the SU in the CR systems, the CFAR approach is unable to ensure the adequate protection of the PU transmission in comparison to the CDR approach.

In addition, selecting a low value of false alarm probability (P_{fa}) results in need for a high level of the corresponding threshold (λ_{fa}). As a consequence, an interference among PU and SU can appear and the setting of the fixed false alarm threshold λ_{fa} based on the CFAR approach is not the most favorable. A more favorable approach is seen in the DDT adjustments based on dynamic changes in the level of DT according to the wireless channel conditions.

However, the DDT adjustments according to the conditions in the wireless channel between PU and SU are highly challenging when it comes to practical realization. In this work, to mathematically model the DDT adjustments, the DDT factor ρ' ($\rho' \geq 1$) is used. The DDT factor ρ' defines the level of DDT adjustments. The higher values of the DDT factor enable the modeling of the higher adjustment capabilities of the ED according to changes in the state of the wireless channel. Hence, instead of being permanently fixed as in the case of ED with FT, the dynamically selected DT values may be in the interval $[\lambda_d^{DT}/\rho', \rho'\lambda_d^{DT}]$.

When the ED employing SLC is performed with a DDT adaptation, the probability of detection (P_d^{DT}) and false alarm probability (P_{fa}^{DT}) is expressed as:

$$P_d^{DT} = \min_{\lambda'^{DT} \in [\frac{\lambda_d^{DT}}{\rho'}, \rho'\lambda_d^{DT}]} Q\left(\frac{\lambda'^{DT} - RN^{DT}(2\sigma_w^2)\left(1 + \frac{P}{2R\sigma_w^2}\right)}{\sqrt{RN^{DT}\left(1 + \frac{P}{R\sigma_w^2}\right)}\,(2\sigma_w^2)}\right) = Q\left(\frac{\frac{\lambda_d^{DT}}{\rho'} - RN^{DT}(2\sigma_w^2)\left(1 + \frac{P}{2R\sigma_w^2}\right)}{\sqrt{RN^{DT}\left(1 + \frac{P}{R\sigma_w^2}\right)}\,(2\sigma_w^2)}\right) \quad (17)$$

$$P_{fa}^{DT} = \max_{\lambda'^{DT} \in [\frac{\lambda_{fa}^{DT}}{\rho'} \quad \rho'\lambda_{fa}^{DT}]} Q\left(\frac{\lambda'^{DT} - RN^{DT}(2\sigma_w^2)}{\sqrt{RN^{DT}}\,(2\sigma_w^2)}\right) = Q\left(\frac{\rho'\lambda_{fa}^{DT} - RN^{DT}(2\sigma_w^2)}{\sqrt{RN^{DT}}\,(2\sigma_w^2)}\right) \quad (18)$$

where N^{DT} and λ_d^{DT} represents the number of sampling points and the level of DDT used in the ED exploiting DDT adjustments, respectively. Based on Equation (17), the level of DDT can be derived and expressed as:

$$\lambda_d^{DT} = \left[Q^{-1}\left(P_d^{DT}\right)\sqrt{\left(1+\frac{P}{R\sigma_w^2}\right)} + \sqrt{RN}\left(1+\frac{P}{2R\sigma_w^2}\right)\right]\sqrt{RN^{DT}}\left(2\sigma_w^2\right)\rho' \quad (19)$$

Similarly, the level of false alarm threshold can be derived from (18) and formulated as:

$$\lambda_{fa}^{DT} = \left[Q^{-1}\left(P_{fa}^{DT}\right) + \sqrt{RN^{DT}}\right]\sqrt{RN^{DT}}\left(\frac{2\sigma_w^2}{\rho'}\right) \quad (20)$$

A minimal number of sampling points (N^{DT}) for performing exact ED employing the SLC technique in MIMO systems with an implemented DDT, can be realized by adapting Equation (13) as follows:

$$N^{DT} = \frac{\left[Q^{-1}\left(P_{fa}^{DT}\right) - \sqrt{\rho'^2(1+2\overline{\gamma_{SLC}})}Q^{-1}\left(P_d^{DT}\right)\right]^2}{R\left[\rho'^2\overline{\gamma_{SLC}} + (\rho'^2-1)\right]^2} \quad (21)$$

The correlation between the probability of false alarm and detection in the case of systems performing ED with DDT adjustments can be formulated by combining Equations (17) and (18), and it can be expressed as:

$$\begin{aligned} P_d^{DT} &= Q\left(\frac{Q^{-1}\left(P_{fa}^{DT}\right) - \sqrt{RN^{DT}}\left(\rho'^2\overline{\gamma_{SLC}} + (\rho'^2-1)\right)}{\sqrt{\rho'^2(1+2\overline{\gamma_{SLC}})}}\right) \\ &= Q\left(\frac{Q^{-1}\left(P_{fa}^{DT}\right) - \sqrt{RN^{DT}}\left(\rho'^2\frac{P}{2R\sigma_w^2} + (\rho'^2-1)\right)}{\sqrt{\rho'^2\left(1+\frac{P}{R\sigma_w^2}\right)}}\right) \\ &= Q\left(\frac{Q^{-1}\left(P_{fa}^{DT}\right) - \frac{\rho'^2 P\sqrt{N^{DT}}}{2\sqrt{R}\sigma_w^2} - \sqrt{RN^{DT}}\left(\rho'^2-1\right)}{\sqrt{\rho'^2\left(1+\frac{P}{R\sigma_w^2}\right)}}\right) \end{aligned} \quad (22)$$

For the case where the DDT factor $\rho' = 1$, Equations (17), (18) and (22) are equal to Equations (11), (12) and (14), respectively. In the case where the DDT factor $\rho' > 1$, the process of signal detection is based on the DDT adjustments. For the larger values of the DDT factor, the selection of a larger range of DDT levels is possible.

3.4. Estimation of Noise Uncertainty

As shown in Equations (15) and (16), the calculation of false alarm and detection thresholds depends on the levels of noise variance σ_w^2. However, ED in a wireless MIMO-OFDM-based communication system is performed by exploiting an estimation of noise power. The lack of knowledge about noise fluctuations significantly contributes to the limited knowledge about the properties of the AWGN. This phenomenon of unknown random variations in the noise power σ_w^2 is known as NU. The NU negatively affects the ED performance in terms of reducing the precision of SS accuracy.

To have more realistic conditions for the performance analysis of ED sensing efficiency in the MIMO wireless communication system, the impact of the NU variations on the PU signal detection was included in the analysis presented in this paper. The impact of the fluctuations in noise power on ED performance is characterized by the NU factor (ρ). The

NU factor ρ ($\rho \geq 1$) is a positive parameter that defines the scope of the NU. To express the influence of NU on the sensing efficiency of the ED, the boundaries of the noise variance (σ^2_{wNU}) are defined by the finite interval $\sigma^2_{wNU} \epsilon \left[\frac{\sigma^2_{nw}}{\rho}, \rho\sigma^2_w\right]$. Based on Equations (11) and (12), the detection (P_d^{NU}) and false alarm (P_{fa}^{NU}) probabilities of the PU signal impacted by NU can be expressed for the ED method employing the SLC technique as:

$$\begin{aligned} P_d^{NU}(\Pr(\Lambda_{SLC} < \lambda)\,|H_1) &= \min_{\sigma^2_{wNU}\epsilon[\frac{\sigma^2_w}{\rho},\, \rho\sigma^2_w]} Q\left(\frac{\lambda - RN^{NU}\left(2\sigma^2_{wNU}\right)\left(1+\frac{P}{2R\sigma^2_{wNU}}\right)}{\sqrt{RN^{NU}\left(1+\frac{P}{R\sigma^2_{wNU}}\right)}\left(2\sigma^2_{wNU}\right)}\right) \\ &= Q\left(\frac{\lambda_d^{NU} - RN^{NU}\left(2\frac{\sigma^2_w}{\rho}\right)\left(1+\frac{P\rho}{2R\sigma^2_w}\right)}{\sqrt{RN^{NU}\left(1+\frac{P\rho}{R\sigma^2_w}\right)}\left(2\frac{\sigma^2_w}{\rho}\right)}\right) \end{aligned} \tag{23}$$

$$\begin{aligned} P_{fa}^{NU}(\Pr(\Lambda_{SLC} > \lambda)|H_0) &= \max_{\sigma^2_{wNU}\epsilon[\frac{\sigma^2_w}{\rho},\rho\sigma^2_w]} Q\left(\frac{\lambda - RN^{NU}\left(2\sigma^2_{wNU}\right)}{\sqrt{RN^{NU}}\left(2\sigma^2_{wNU}\right)}\right) \\ &= Q\left(\frac{\lambda_{fa}^{NU} - RN^{NU}\left(2\rho\sigma^2_w\right)}{\sqrt{RN^{NU}}\left(2\rho\sigma^2_w\right)}\right) \end{aligned} \tag{24}$$

where λ_d^{NU} represents the DT, λ_{fa}^{NU} represents the false alarm threshold and N^{NU} represents the number of sampling points used in the ED process impacted with NU. Equations (23) and (24) indicate that the total number of receiving chains (R) of SU, the number of sampling points (N^{NU}) and the NU variance (σ^2_{wNU}) with NU factor ρ are the parameters that influence both, the detection and false alarm probability in MIMO communication systems. Besides these parameters, the total instant Tx power of the PU signal (P) impacts the detection probability. Based on Equations (23) and (24), it can be noticed that setting appropriate DT or false alarm thresholds will significantly impact the detection and false alarm probabilities.

Based on Equations (23) and (24), the DT and false alarm threshold can be expressed as:

$$\lambda_d^{NU} = \left[Q^{-1}(P_d^{NU})\sqrt{\left(1+\frac{P\rho}{R\sigma^2_w}\right)} + \sqrt{RN^{NU}}\left(1+\frac{P\rho}{2R\sigma^2_w}\right)\right]\sqrt{RN^{NU}}\left(\frac{2\sigma^2_w}{\rho}\right) \tag{25}$$

$$\lambda_{fa}^{NU} = \left[Q^{-1}(P_{fa}^{NU}) + \sqrt{RN^{NU}}\right]\sqrt{RN^{NU}}\left(2\rho\sigma^2_w\right) \tag{26}$$

According to Equation (13), the minimal number of sampling points N^{NU} used in the ED which guarantees the accurate detection of the PU signal affected by NU can be formulated as

$$N^{NU} = \frac{\left[\rho Q^{-1}\left(P_{fa}^{NU}\right) - \sqrt{\left(\frac{1}{\rho} + 2\overline{\gamma_{SLC}}\right)}Q^{-1}\left(P_d^{NU}\right)\right]^2}{R\left(\overline{\gamma_{SLC}} - \frac{\rho-1}{\rho}\right)^2} \tag{27}$$

Equation (27) indicates that if the average SNR at antennas of SU is lower than the $\frac{\rho^2-1}{\rho}$, the energy detector cannot sense the signal. For that reason, the SNR level at the antennas of SU will have an important role in the accurate detection of PU signals.

According to Equations (22) and (23), when the ED is impacted by NU, the detection probability can be defined as a function of the false alarm probability:

$$P_d^{NU} = Q\left(\frac{\rho Q^{-1}\left(P_{fa}^{NU}\right)-\sqrt{RN^{NU}}\left(\overline{\gamma_{SLC}}-\frac{\rho-1}{\rho}\right)}{\sqrt{\frac{1}{\rho}+2\overline{\gamma_{SLC}}}}\right) = Q\left(\frac{\rho Q^{-1}\left(P_{fa}^{NU}\right)-\sqrt{RN^{NU}}\left(\frac{P}{2R\sigma_w^2}-\frac{\rho-1}{\rho}\right)}{\sqrt{\frac{1}{\rho}+\frac{P}{R\sigma_w^2}}}\right)$$
$$= Q\left(\frac{\rho Q^{-1}\left(P_{fa}^{NU}\right)-\frac{\sqrt{N^{NU}}P}{2\sqrt{R}\sigma_w^2}+\sqrt{RN^{NU}}\left(\frac{\rho-1}{\rho}\right)}{\sqrt{\frac{1}{\rho}+\frac{P}{R\sigma_w^2}}}\right) \tag{28}$$

The value of NU factor $\rho = 1.00$ indicates that fluctuations in noise power do not exist and that there is no impact of the NU on the ED process. In this case, Equations (22), (23) and (28) converge into (11), (12) and (14). The modelling of the influence of the NU on the ED is performed for the case where the NU factor $\rho > 1.00$. A higher value (than one) of the NU factor ρ means larger NU fluctuations, for which it is expected to have a more negative impact on the performance of the ED SS.

3.5. Energy Detection Process with NU and DT

An approach that can contribute to the minimization of the negative effects caused by NU on the efficiency of the ED is dedicated to involving the DT adjustments during the ED. This can contribute to the enhancement of SS performance of the ED method. The ED employing SLC which encompasses the DT adjustments for reducing the impact of NU, represents the most realistic approach to the analysis of the performance of ED in the MIMO-OFDM CR systems. However, this approach is the most difficult for either simulations or practical implementation due to the necessity of the continuous estimation of NU and the dynamic adjustments of DT during the ED according to the NU.

For the analysis of this ED approach, detection and false alarm probabilities are expressed as a function of NU variance and the DT adjustment factors. The boundaries of the NU are assumed to be in the interval $\sigma_{wNUDT}^2 \epsilon [\sigma_w^2/\rho, \rho\sigma_w^2]$, while the boundaries of the DTs are assumed to be in the interval $\lambda'^{NUDT} \epsilon [\lambda_d^{NUDT}/\rho', \rho'\lambda_d^{NUDT}]$. Considering these boundaries, the probabilities of false alarm (P_{fa}^{NUDT}) and detection (P_d^{NUDT}) for the ED approach which includes the impact of NU and DT adjustments can be formulated as:

$$P_d^{NUDT} = \min_{\lambda'^{NUDT} \epsilon [\frac{\lambda_d^{NUDT}}{\rho'} \rho'\lambda_d^{NUDT}]} \min_{\sigma_{wNUDT}^2 \epsilon [\frac{\sigma_w^2}{\rho}, \rho\sigma_w^2]} Q\left(\frac{\lambda_i'^{NUDT}-RN^{NUDT}\left(2\sigma_{wNUDT}^2\right)\left(1+\frac{P}{2R\sigma_{wNUDT}^2}\right)}{\sqrt{RN^{NUDT}\left(1+\frac{P}{R\sigma_{wNUDT}^2}\right)\left(2\sigma_{wNUDT}^2\right)}}\right)$$
$$= Q\left(\frac{\frac{\lambda_d^{NUDT}}{\rho'}-RN^{NUDT}\left(2\frac{\sigma_w^2}{\rho}\right)\left(1+\frac{P\rho}{2R\sigma_w^2}\right)}{\sqrt{RN^{NUDT}\left(1+\frac{P\rho}{R\sigma_w^2}\right)\left(2\frac{\sigma_w^2}{\rho}\right)}}\right) \tag{29}$$

$$P_{fa}^{NUDT} = \max_{\lambda'^{NUDT} \epsilon [\frac{\lambda_{fa_i}^{NUDT}}{\rho'}, \rho'\lambda_{fa}^{NUDT}]} \max_{\sigma_{wNUDT}^2 \epsilon [\frac{\sigma_{w_i}^2}{\rho}, \rho\sigma_w^2]} Q\left(\frac{\lambda'^{NUDT}-RN^{NUDT}\left(2\sigma_w^2\right)}{\sqrt{RN^{NUDT}}\left(2\sigma_w^2\right)}\right)$$
$$= Q\left(\frac{\lambda_{fa}^{NUDT}\rho'-RN^{NUDT}\left(2\sigma_w^2\right)\rho}{\sqrt{RN^{NUDT}}\left(2\sigma_w^2\right)\rho}\right) \tag{30}$$

The level of DT (λ_d^{NUDT}) can be derived from (29) as:

$$\lambda_d^{NUDT} = \left[Q^{-1}\left(P_d^{NUDT}\right)\sqrt{\left(1+\frac{P\rho}{R\sigma_w^2}\right)}+\sqrt{RN^{NUDT}}\left(1+\frac{P\rho}{2R\sigma_w^2}\right)\right]\sqrt{RN^{NUDT}}\left(\frac{2\sigma_w^2\rho'}{\rho}\right) \tag{31}$$

Similarly, for the defined value of false alarm probability, the corresponding threshold can be derived from (30) as:

$$\lambda_{fa}^{NUDT} = \left[Q^{-1}\left(P_{fa}^{NUDT}\right) + \sqrt{RN^{NUDT}} \right] \frac{2\rho\sigma_w^2}{\rho'} \sqrt{RN^{NUDT}} \tag{32}$$

In addition, the minimal number of sampling points for the successful detection of PU signal when ED is impacted by NU and performed based on DT adjustments can be formulated as:

$$N^{NUDT} = \frac{\left[\frac{\rho}{\rho'} Q^{-1}\left(P_{fa}^{NUDT}\right) - \sqrt{\rho'\left(\frac{1}{\rho} + 2\overline{\gamma_{SLC}}\right)} Q^{-1}\left(P_{di}^{NUDT}\right) \right]^2}{R\left[\rho' \overline{\gamma_{SLC}} + \frac{\rho'}{\rho} - \frac{\rho}{\rho'} \right]^2} \tag{33}$$

By combining Equations (28) and (29), the detection probability can be expressed as a function of false alarm probability as

$$\begin{aligned} P_d^{NUDT} &= Q\left(\frac{\frac{\rho}{\rho'} Q^{-1}\left(P_{fa}^{NUDT}\right) - \sqrt{RN^{NUDT}}\left(\rho' \, \overline{\gamma_{SLC}} + \frac{\rho'}{\rho} - \frac{\rho}{\rho'}\right)}{\sqrt{\rho'\left(\frac{1}{\rho} + 2\overline{\gamma_{SLC}}\right)}} \right) \\ &= Q\left(\frac{\frac{\rho}{\rho'} Q^{-1}\left(P_{fa}^{NUDT}\right) - \sqrt{RN^{NUDT}}\left(\rho' \, \frac{P}{2R\sigma_w^2} + \frac{\rho'}{\rho} - \frac{\rho}{\rho'}\right)}{\sqrt{\rho'\left(\frac{1}{\rho} + \frac{P}{R\sigma_w^2}\right)}} \right) \\ &= Q\left(\frac{\frac{\rho}{\rho'} Q^{-1}\left(P_{fa}^{NUDT}\right) - \frac{\rho'\sqrt{N^{NUDT}}P}{2\sqrt{R}\sigma_w^2} - \sqrt{RN^{NUDT}}\left(\frac{\rho'}{\rho} - \frac{\rho}{\rho'}\right)}{\sqrt{\rho'\left(\frac{1}{\rho} + \frac{P}{R\sigma_w^2}\right)}} \right) \end{aligned} \tag{34}$$

Equation (34) indicates that the NU and DDT factors have an affect on the detection probability when SS is performed in MIMO-OFDM cognitive radio systems based on ED. The joint impact of NU and DT adjustments on sensing performance can be modeled by selecting the corresponding values of DDT and NU factors ($\rho' > 1.00$ and $\rho > 1.00$). In the case of the DDT and NU factors equal to $\rho' = 1.00$ and $\rho > 1.00$, the channel is impacted by NU and there are no DDT adjustments during the ED process. For the same conditions in the channel and the same number of sampling points ($N^{NU} = N^{NUDT}$) used during the ED impacted by NU, Equations (29), (30) and (34) converge into Equations (23), (24) and (28), respectively. In case of $\rho' > 1.00$ and $\rho = 1.00$, the ED includes DT adjustments during the SS without an impact from NU. For the same conditions in the channel and the same number of sampling points ($N^{DT} = N^{NUDT}$) used in ED performed with DT adjustments, Equations (29), (30) and (34) converge into Equations (17), (18) and (22), respectively.

4. Simulation Algorithm for the ED Employing SLC

The ED performance employing the SLC technique in the MIMO-OFDM CRs systems has been tested through the executing the proposed simulation Algorithm 1. The algorithm is composed of two phases. In the first phase, the procedure of ED is executed according to the principles of the concept employing SLC of the signals received at the R RX chains of SU. In the second phase, to simulate the behavior of ED performance in different operating environments, the execution of the algorithm was continued through the selection of the impact of different combinations of NU and DDT factors on ED performance. Table 3 indicate parameters used in the process of simulation.

Algorithm 1 Simulation of the ED in distinct working environments of MIMO-OFDM CR systems.

INPUT: *MIMO_OFDM_M×r, noise variance* (σ_w^2), *number of sampling points* (N), *probability of false alarm* (P_{fa}), *number of Monte Carlo simulations(pp), SNR simulation range (SNR), length of the MIMO-OFDM data (mimo_ofdm_len), DDT factor* ($\rho\prime$), *and NU factor* (ρ),
OUTPUT: *Detection probability impacted by DT adjustment* (P_d^{DT}) *and Detection probability impacted by NU and DT adjustment* (P_d^{NUDT})
ON INITIALIZED: *MIMO-OFDM signal (MIMO_OFDM_M×r) do:*
Step 1: *Execution of simulation indicating detection probability impacted by DT adjustments* (P_d^{DT}) *and Detection probability impacted by DT adjustments and NU* (P_d^{NUDT}) vs. *SNR using* (14), (22), (28) *and* (34)
set pp = number of Monte Carlo simulations
set SNR = signal to noise ratio in interval [−25 dB, 25 dB]
FOR *b = 1:length* (SNR)
j1 = 0; j2 = 0;
FOR *pp = 1:10,000;*
Step 2: *Modeling AWGN noise with varince* $\sigma_w^2(n)$
Noise_DT (ρ = 1.00, ρ' > 1.00) = *sqrt*($\sigma_w^2(n)$ = 1.00).*randn (1, mimo_ofdm_len);
Noise_NUDT (ρ > 1.00, ρ' > 1.00) = *sqrt*($\sigma_w^2(n)$ > 1.00).*randn (1, mimo_ofdm_len);
Step 3: *Estimation of received signal y(t)*
finall_OFDM_M×r_DT = MIMO_OFDM_M×r + Noise_DT;
finall_OFDM_M×r_NUDT = MIMO_OFDM_M×r + Noise_NUDT;
Step 4: *Energy estimation of received signal using SLC concept*
REPEATE FOR r= 1:R
energy_calculation_DT = abs(finall_OFDM_M×r_DT).^2;
energy_calculation _NUDT= abs(finall_OFDM_M×r_NUDT).^2;
END
Step 5: *Estimation of test statistics based on mixing energies of R signals using* (4)
FOR r= 1:R
test_statistc_DT = sum(energy_calculation_DT);
test_statistic_NUDT = sum(energy_calculation_NUDT);
END
Step 6: *Threshold estimation using* (18), (20), *and* (30), (32))
threshold_DT (*b*) = ((*qfuncinv*(P_{fa}(*b*))./*sqrt*(N))+ 1)./ρ';
threshold_NUDT (*b*) = ((*qfuncinv*(P_{fa} (*b*)).* ρ./*sqrt*(N))+ ρ)./$\rho\prime$;
Step 7: *Making a final decision by using using* (5) *and* (6)
IF (*test_statistc_DT >= threshold_DT* (*b*));
j1 = j1 + 1;
END
IF (*test_statistic_NUDT >= threshold_NUDT* (*b*));
j2 = j2 + 1;
END
END
Step 8: *Evaluation* P_d^{DT} *and* P_d^{NUDT} *using Monte Carlo simulation (based on* (3))
P_d*_DT* (*b*) = *i1/pp;*
P_d*_NUDT* (*b*) = *i2/pp;*
END
UNTIL P_d^{DT}, P_d^{NUDT} = *[0, 1]*

Table 3. Parameters used in the process of simulation.

Parameters	Type/Quantity
PU signal modulation scheme	OFDM
Number of Tx chains (antennas) of the PU	1–4
Number of Rx chains (antennas) of the SU	1–6
OFDM modulation schemes	64 QAM, 16 QAM, QPSK
Model of the noise [28,34]	AWGN
Noise variance σ_w^2 for DT ($\rho = 1.00$, $\rho' > 1.00$) [37–40]	1.00
Noise variance σ_w^2 for NU and DT ($\rho > 1.00, \rho' > 1.00$) [37–40]	1.01
Number of sampling points for ED (FFT size) [28,34]	128, 512, 1024
SNRs range at SU position (dB) [28,34]	−25–25
DT factor $\rho\prime$ [37–40]	1.00, 1.03, 1.05
NU factor ρ [37–40]	1.00, 1.03, 1.05
Target false alarm probability [37–40]	0.01, 0.2
Overall number of Monte Carlo simulations	10,000

The pseudocode of the developed algorithm is shown in Algorithm 1. To obtain the statistical relevance of the simulation results for different ED operation scenarios, the Monte Carlo simulations were executed according to Algorithm 1.

Execution Steps of the Simulation Algorithm

The input parameters used in the performance evaluation of the ED method employing SLC technique are set in the first line of Algorithm 1. The parameters are the noise variance (σ_w^2), the received MIMO-OFDM signal (*MIMO_OFDM_M*×*r*), the overall number of sampling points (N), the length of the MIMO-OFDM data (*mimo_ofdm_len*), the false alarm probabilities (P_{fa}), the Monte Carlo simulation number (*pp*), the range of simulated SNRs (*SNR*), and the NU (ρ) and DDT factor (ρ').

The MIMO-OFDM signal (*MIMO_OFDM_M*×*r*) represents the signal sensed at SU antennas of the R Rx chains. This signal is generated for the different modulation types, PU Tx powers, the number of SU Rx and PU Tx chains (antennas), and the number of sampling points used in the ED. In simulated MIMO-OFDM CR systems, the received signal (*MIMO_OFDM_M*×*r*) is used as an input signal for the evaluation of the ED performance employing the SLC method. In Table 3, the exact values of the parameters used in the simulations are presented.

In Algorithm 1 lines 4–8, the overall number of Monte-Carlo simulations is set. The Monte Carlo simulations are executed for a span of different SNRs (Table 3). In lines 9–10 (Step 2 of Algorithm 1), the AWGN noise with a mean value equal to zero and a variance of σ_w^2 is generated. The chosen values of noise variance are typical of those in practical wireless environments (Table 3).

Lines 11–12 (Step 3) represent the ED of two types of sensed signal during the ED process employing the SLC technique. The first signal (*finall_OFDM_M*×*r_DT*) is the MIMO-OFDM signal detected with DT adjustments ($\rho' > 1.00$) and without the impact of noise variations ($\rho = 1.00$) on the ED process. The second received signal (*finall_OFDM_M*×*r_NUDT*) is the MIMO-OFDM signal detected with the implementation of DT adjustments ($\rho' > 1.00$) and the impact of NU ($\rho > 1.00$) on the ED process.

In lines 13–15 (Step 4), the calculation of signal energy after the SLC of the signals for each type of sensed signal (*energy_calculation_DT* and *energy_calculation_NUDT*) is performed. The tested statistic (energy) estimation for the signals sensed at the R Rx chains (antennas) of SU is shown in lines 17–20 (Step 5). In Algorithm 1, two cases of test statistic calculated according to Equation (4) are shown. The first one is the test statistic executed for the OFDM signals sensed without the impact of NU and with the DT adjustments (*test_statistc_DT*). In the second case, the signals are sensed for the simulation of the operating environment in which the impact of NU and with DT adjustments is considered during the ED (test_statistc_NUDT).

Algorithm 1 lines 21–22 (Step 6) represent two analyzed cases of the DT evaluation. In the first case (*threshold_DT* (*b*)), the DT evaluation is performed without the impact of NU and with DT adjustments. In the second case (*threshold_NUDT* (*b*)), the impact of both, the NU and ED exploiting the DT adjustments is considered. The mathematical expression for these two cases is presented in Equations (18), (20), (30) and (32), respectively.

Finally, the decision-making process which results in cognition about the exploitation of the frequency band by the SU is executed in lines 23–29 (Step 7). The decision-making process is performed based on testing the binary hypothesis presented by Equation (3). If the energy of the sensed signal is greater or the same as the set threshold, hypothesis H_1 is confirmed and the PU signal is present. Alternatively, if the energy is lower than a set DT, hypothesis H_0 points to an absence of PU and the presence of a spectrum hole.

In lines 30–33 (Step 8), a large number of repeated Monte Carlo simulations were executed for the different SNR ranges and for every simulation environment of the ED.

5. Results of Simulations

In Section 5, the simulation software used for performing comprehensive simulations for different ED operating environments is presented and the obtained simulation results have been thoroughly discussed. The process of SS was analyzed through the performance of the ED method based on the SLC technique in SISO and both asymmetric and symmetric MIMO-OFDM CR systems. The effect of NU on the transmitted MIMO-OFDM signal was modeled through simulations performed for the different NU factors. The process of signal detection at the location of SU was performed for various levels of the DT adjustment (DDT factors). The presented results analyze how different Tx-Rx antenna combinations at PU and SU sides, OFDM modulation constellations, the number of sampling points, Tx power levels, and SNRs influence the detection probability of the PU signal during the SS performed using the ED method employing the SLC technique.

5.1. Description of the Simulation Parameters and Software

For the simulation of the ED in different operating environments mathematically formulated in Section 3, the Matlab 2016 software was used. The values of all simulation parameters are presented in Table 3. The simulations were performed for the transmission of OFDM signals modulated using the most practically applied modulation schemes such as 16/64 quadrature amplitude modulation (16/64—QAM) and quadrature phase shift keying (QPSK). In addition, the ED performance was analyzed for the number of sampling points used in the ED equal to 128, 512, and 1024 (Table 3). Furthermore, the analysis examined the SNRs ranging between −25 dB and 25 dB in the position of SU. The values of SNR in this SNR range are being frequently experienced in real practical environments as SNRs of many communication systems based on OFDM. To improve the reliability and accuracy of the simulations, 10,000 Monte Carlo simulations per each simulation run were executed (Table 3). This number of Monte Carlo simulations were selected for analysis for the purpose of achieving balance among the simulation durations and simulation accuracies. This enables the execution of the simulation algorithm for all simulated ED operation environments in the time frame of a few tens of microseconds.

To analyze the impact of versatile levels of NU and DT adjustments on the performance of ED employing SLC principles, different combinations of values of DDT and NU factors have been selected for analysis (Table 3). More specifically, values of DDT and NU factors equal to 1.03 and 1.05 represent the lower and higher levels of DT adjustments and NU in the simulation analysis, respectively. To exclude the influence of DT adjustments and NU from the simulation analysis, DDT and NU factors equal to one ($\rho' = \rho = 1.00$) were used in the simulation analyses.

5.2. Effect of the Number of Transmit Chains on ED Efficiency

The simulation results shown in this section were used to analyze the impact of SISO, symmetric (Figure 2) and asymmetric (Figure 3) MIMO transmissions on the ED performance. Simulation results were obtained for the constant values of PU Tx power (P = 100 mW), the fixed number of sampling points used for the ED (N = 128), and the predefined false alarm (P_{fa}) probabilities equal to 0.1. The same results have been obtained independently of the used OFDM modulation schemes (64 QAM, 16 QAM and QPSK) and in Figures 2 and 3, this is denoted by the m-PSK/m- QAM notation. For each Tx-Rx diversity schema, the assessment of the ED performance for diverse levels of DDT ($\rho\prime$) and NU (ρ) factors was performed.

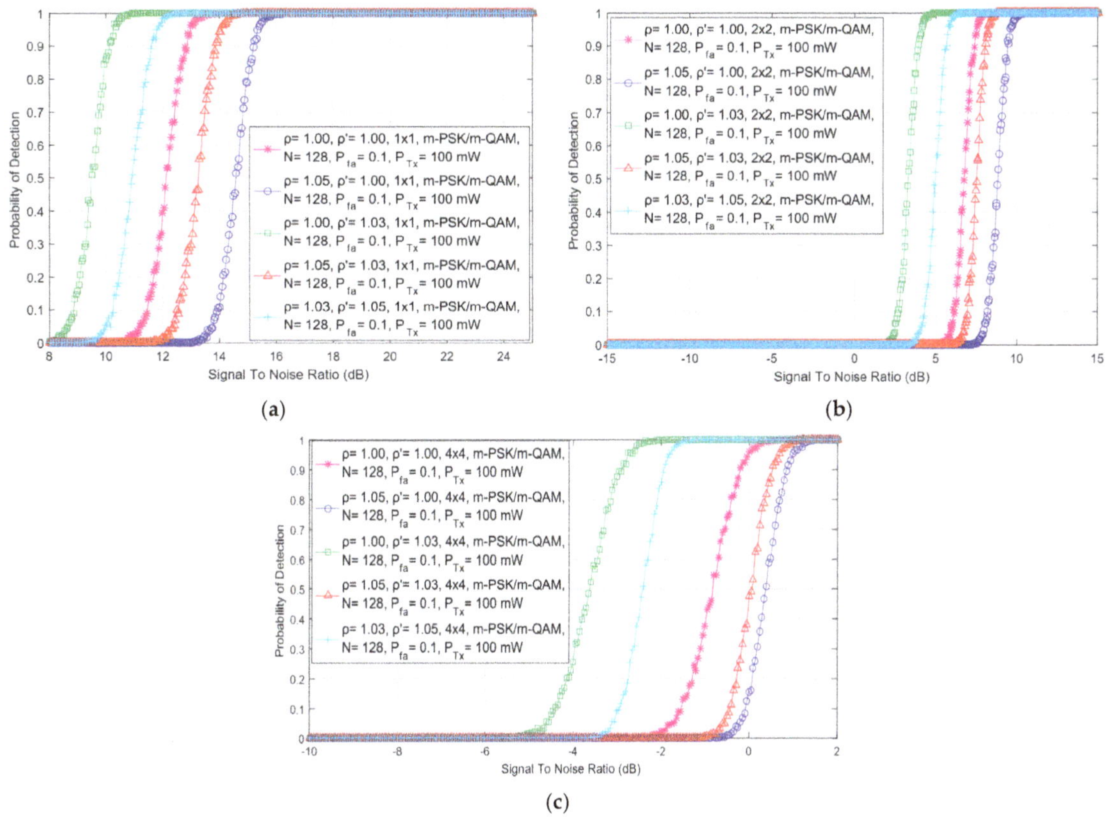

Figure 2. The dependency of detection probability on SNR for ED performed with different combinations of DDT and NU factors in (**a**) SISO, (**b**) symmetric 2 × 2 MIMO and (**c**) symmetric 4 × 4 MIMO communication systems.

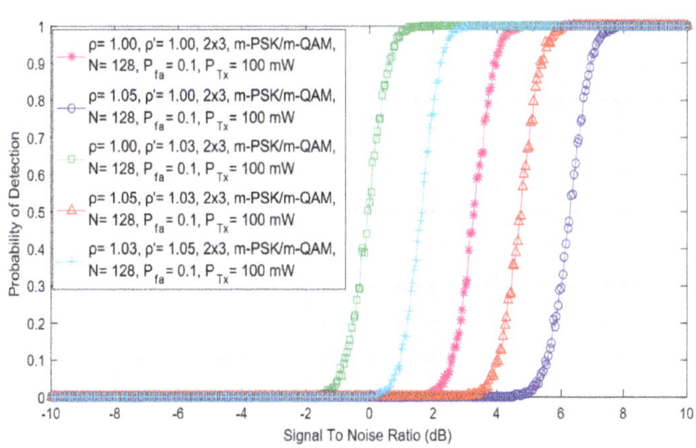

Figure 3. The dependency of detection probability on SNR for ED performed with different combinations of DDT and NU factors in asymmetric 2 × 3 MIMO communication system.

In Figure 2, the relationship among the SNR and detection probability for different combinations of DDT and NU factors and versatile symmetric MIMO communications systems (such as 1 × 1, 2 × 2 and 4 × 4) has been presented. Similar results have been presented in Figure 3 for the 2 × 3 asymmetric MIMO communications system. The results presented in Figures 2 and 3 indicate that for higher values of SNR, the detection probability will be better for any symmetric or asymmetric MIMO Tx-Rx diversity combination and vice versa. Since the performance of the ED method is poor for lower values of SNR, a better detection probability can be achieved in the case of communication systems with a higher number of Rx and Tx chains on the PU and SU side, respectively. In addition, results in Figure 2 show that for the equal levels of SNRs, higher detection probabilities of PU signal will be achieved for Tx-Rx diversity combinations with a higher number of SU Rx and PU Tx chains. This improvement in detection probability is a consequence of the transmission diversity which enhances the precision of PU signal detection in the case where the larger numbers of PU Tx and SU Rx chains (antennas) will be used in the MIMO communication systems.

5.3. Effect of NU and DT Adjustments on ED Efficiency

The results shown in Figure 3 further indicate the absence of the impact of the OFDM modulation scheme on the probability of detection (P_d). More specifically, for the same operating environment simulating the ED performance, the probability of detection is the same for any of the m-QAM or m-PSK constellations. These results indicate that the OFDM modulation type does not have an effect on the detection probability in the ED for any combination of the Tx-Rx chains at any SNR level (this can be also confirmed by Equations (11), (14), (17), (22), (23), (28), (29) and (34)). This is due to the dynamic adjustments of the OFDM modulation of the signal transmitted at a constant Tx power in rate-adaptive OFDM systems. In such transmission systems, at the location of SU, the signal is affected by the fluctuations in noise power (NU). According to this fact, the analysis further showed a strong influence of the NU and DT adjustments on the ED.

As shown in Figures 2 and 3, for the same channel conditions, the performance of the ED will be degraded when SS is performed in operating environments with a stronger NU and with a fixed DT lacking any DT adjustments (ρ = 1.05 and ρ' = 1.00). The results presented in Figures 2 and 3 indicate that the ED will achieve a better probability of detection for the same SNRs when a higher level of DT adjustments according to the NU (e.g., ρ = 1.03, ρ' = 1.05 compared to ρ = 1.05, ρ' = 1.03) will be performed during the ED (Figures 2 and 3). A better probability of detection will be obtained if ED is performed in operating environments characterized by a lower level of NU and a higher level of DT adjustment (ρ = 1.03, ρ' = 1.05) and vice versa (ρ = 1.05, ρ' = 1.03). The best detection performance was obtained by the operating environments characterized by the lack of any NU and by the implemented DT adjustments during the ED (ρ = 1.00, ρ' = 1.03). This operation scenario simulates the ED performed in the systems impacted by noise, which do not have fluctuations in time. This means that such fluctuations do not impact the ED. However, these operation scenarios are the least realistic since in real wireless communication systems, the noise from different sources and their power fluctuations (NU) occur as a frequent phenomenon.

Hence, for all modulations schemas, a better detection probability can be reached only if the ED is based on DT adjustment ($\rho' > 1.00$). However, setting the DT adjustment at a too high or too low level can result in exceptionally high or exceptionally low DTs. This can cause high misdetections of the PU signal or a high sensitivity of ED. In both cases, the ED performance will be degraded. Therefore, the DT adjustments must be performed in accordance with the level of NU. This means that higher values of NU variation need to be followed by a greater level of DT adjustment and vice versa.

5.4. Effect of the Transmit Power of PU on ED Sensing Efficiency

In this section, the presented results of the simulation explain the effect of the diverse PU Tx powers on the detection performance in the SISO and symmetric MIMO (2 × 2, 4 × 4) wireless communication systems. The results were obtained for the ED operating environment characterized by OFDM signals transmitted with QPSK modulation and sensed with a permanent number of sampling points (N = 128), the fixed false alarm probability (P_{fa}) of 0.1 and the different levels of DT adjustments and NU levels.

In Figure 4, the dependency of detection probability on SNR for ED performed with versatile combinations of DDT and NU factors in SISO and MIMO communication systems, has been analyzed for two Tx powers of PU equal to 100 mW and 10 W. The obtained results shown in Figure 4 indicate that in SS environments having equal SNR levels at the antennas of SU, a higher detection probability will be for the PU signals transmitted at higher Tx powers. The transmission at Tx powers having higher values consequently results in a higher amount of the PU signal energy that will be eventually sensed at the Rx Chains of SU through the process of SLC. Additionally in Figure 4, it can be seen that the results of simulation show that the transmission of OFDM signal, which combines a larger number of Tx-Rx chains (4 × 4) and larger PU Tx power (10 W), has a positive effect on ED performance. As the number of Tx-Rx chains and PU TX powers increase, the simulation results shown in Figure 4 present that better probabilities of detection can be achieved for lower SNR values. Hence, the transmission of PU at higher Tx powers and in communication systems with a higher number of Tx-Rx chains, will always have a positive effect on the performance of the ED employing the SLC method. In addition, in ED operating environments characterized with lower values of SNRs at the Rx antennas of SU, performing transmission of MIMO-OFDM signal with a higher number of Tx-Rx chains and a higher PU Tx power, yield to the enhancement of the probability of PU signal detection in the SS process based on the ED method.

5.5. Effect of Differences in the Number of MIMO Tx and Rx Chains on the ED Performance

The results presented in this section show the analysis of the impact of the differences in the number of SU Rx and PU Tx chains in relation to the ED performance in SISO and symmetric MIMO systems. The results were obtained for ED realized in the operating environment characterized by the transmission of QPSK-modulated MIMO-OFDM signal, a constant number of sampling points (N = 128), constant PU Tx power (P = 100 mW), and targeted false alarm probability equal to 1.01.

In Figure 5, the dependency of detection probability on SNR for ED performed in SISO and asymmetric MIMO 2 × 6 and 6 × 2 communication systems have been presented. The results shown in Figure 5 point that the differences among the number of MIMO Tx and Rx chains impact ED performance. More specifically, the MIMO communication system having a higher number of Rx chains, will achieve higher detection probabilities for the PU signal at the same level of SNRs in the position of SU.

These results are also confirmed in Figure 6, showing the dependency of the probability of detection on SNR for ED performed in asymmetric MIMO 4 × 6 and 6 × 4 communication systems. According to the results presented in Figure 6, for the same SNRs at SU antennas, the 4 × 6 MIMO communication system will achieve higher detection probabilities than the 6 × 4 communication system. Therefore, performing ED with SU having a larger number of Rx chains will contribute to the enhancement of ED performance. The main reason for this observation can be found in the fact that a larger number of Rx chains ensure the detection of more of the total energy of the PU signal in the process of SLC.

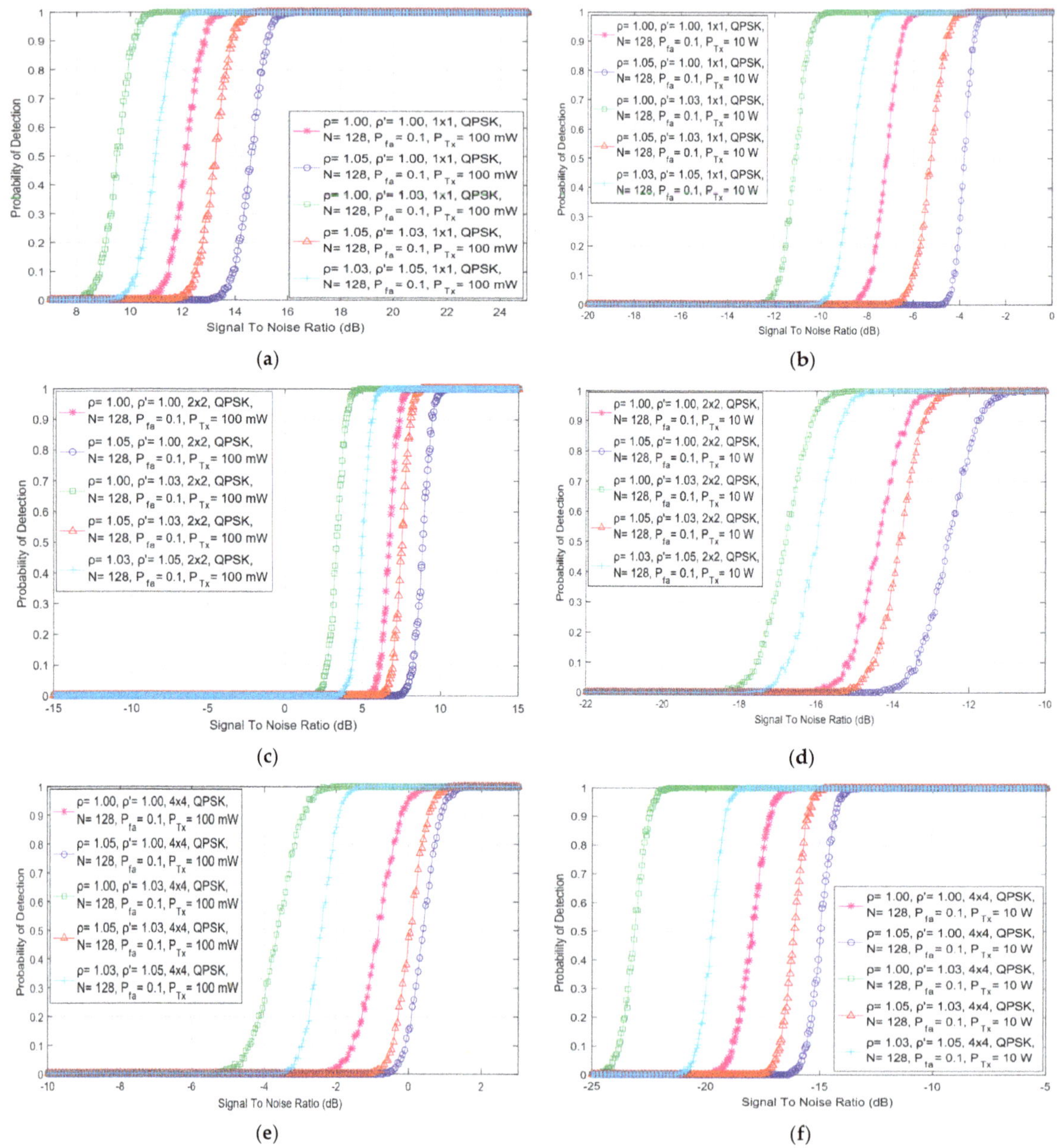

Figure 4. The dependency of detection probability on SNR of ED performed with different combinations of DDT and NU factors in communication: (**a**) SISO systems with 100 mW PU Tx power, (**b**) SISO systems with 10 W PU Tx power, (**c**) 2 × 2 MIMO systems with PU Tx power, (**d**) 2 × 2 MIMO systems with 10 W PU Tx power, (**e**) 4 × 4 MIMO systems with 100 mW PU Tx power and (**f**) 4 × 4 MIMO systems with 10 W PU Tx power.

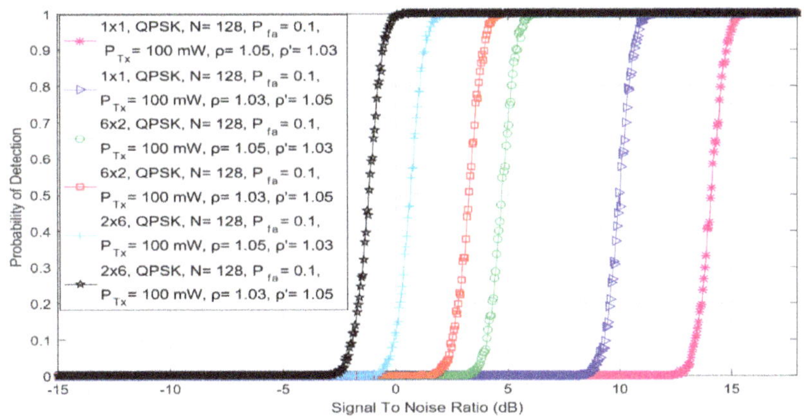

Figure 5. The dependency of detection probability on SNR for ED performed with versatile combinations of DDT and NU factors in SISO and asymmetric communication MIMO 2 × 6 and 6 × 2 systems.

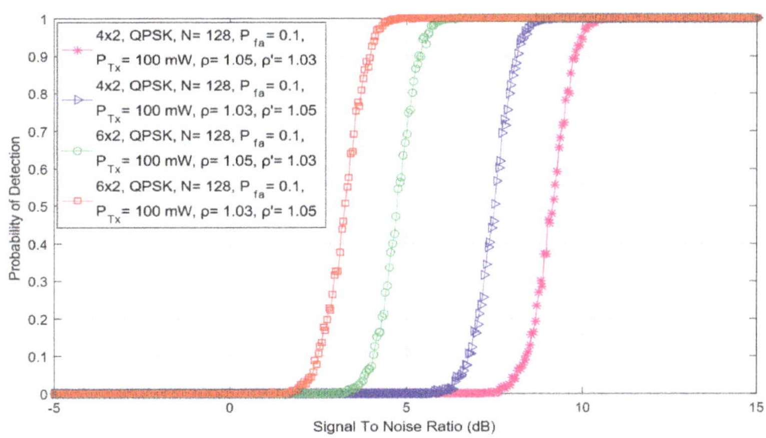

Figure 6. The dependency of detection probability on SNR for ED performed with different combinations of DDT and NU factors in asymmetric communication MIMO 4 × 6 and 6 × 4 systems.

Additionally, the results presented in Figures 5 and 6 indicate that the number of PU Tx chains also influences ED performance. The comparison of the results presented in Figures 5 and 6 show that for the fixed number of Rx chains (antennas), a larger number of PU Tx chains will improve the probability of detection in MIMO-OFDM CR systems (better detection probability is for 6 × 2 in comparison with 4 × 2 systems). Therefore, it is to be expected that the ED performance in terms of sensitivity and accuracy of PU detection will be improved for the new generation of user and network devices which will have more transmission chains and corresponding antennas.

5.6. Effect of the Number of Sampling Points on ed Efficiency

The impact of the various number of sampling points used in the ED based on the SLC technique in SISO and asymmetric MIMO systems (2 × 2 and 4 × 4) has been analyzed in this section. The analysis was performed for the ED operating environment characterized by the transmission of a QPSK-modulated PU signal, the constant PU Tx power (P = 100 mW) and the constant values of false alarm probability ($P_{fa} = 0.1$).

In Figure 7, the dependency of detection probability on SNR for ED performed with versatile numbers of samples and combinations of DDT and NU factors in SISO and symmetric MIMO 2 × 2 and 4 × 4 communication systems has been presented. The obtained results indicate that for any combination of MIMO Tx-Rx chains and NU/DDT factors, the detection probability will be improved with an enlargement of the number of sampling points used for performing ED employing the SLC method. This is also confirmed in Equations (13), (21), (27) and (33), showing that ED performed with a larger number of sampling points N results in higher detection probabilities for the PU signal.

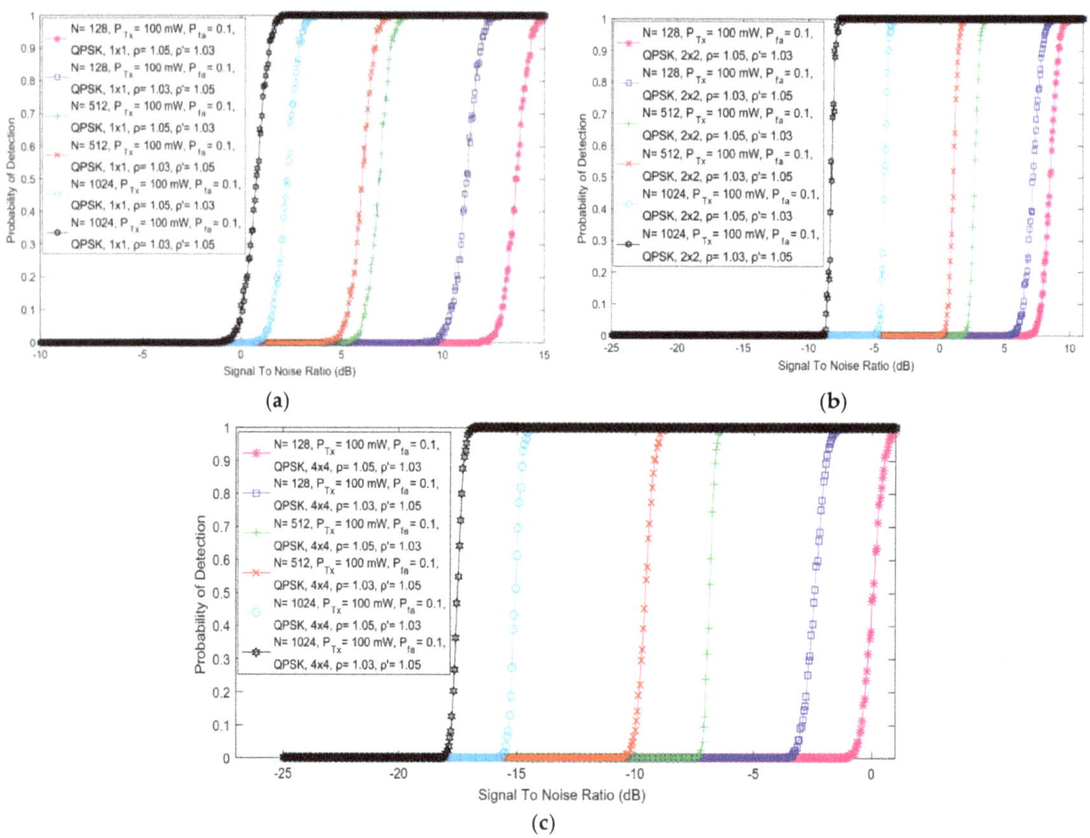

Figure 7. The dependency of the probability of detection on SNR for ED performed with the versatile number of sampling points and combinations of DDT and NU factors in (**a**) SISO, (**b**) 2 × 2 symmetric MIMO and (**c**) 4 × 4 symmetric MIMO communication systems.

This is due to the fact that ED performed with a larger number of sampling points in practice means performing the PU signal sensing with a higher number of sensing attempts during the ED, which increases the probability of PU signal detection.

Additionally, the results presented in Figures 2–6 including Figure 7 indicate that there exists an SNR-wall below which the detection probability cannot be ensured ($P_d = 0$). According to Figure 7, the levels of the SNR-walls shrink towards lower values of SNRs for ED performed in MIMO communication systems which have a larger number of Tx-Rx chains and for ED performed by the SU exploiting a higher number of sampling points. For example, the SNR-wall for a SISO system (1x1) performing ED with $N = 128$ samples (Figure 7a) will be significantly lower than the SNR-wall in symmetric MIMO 4 × 4 system performing ED with N = 1028 samples (Figure 7c). Consequently, the trade-off between the number of sampling points and the number of Tx-Rx chains on the PU and SU sides used in the ED, can have a significant influence on the probability of detection.

Furthermore, the results presented in Figure 7 were obtained for a different combination of NU and DDT factors ($\rho = 1.05$, $\rho' = 1.03$ and $\rho = 1.03$, $\rho' = 1.05$). The results indicate that for the same number of sampling points used in the ED, a better probability of detection will be obtained for the systems with a higher capability of DT adjustments (ρ' =1.05) during the ED impacted by moderate NU (ρ = 1.03), than the systems with the lower capability of DT adjustments (ρ' = 1.03) which are impacted by a high NU (ρ = 1.05). This is the consequence of the fact that lower levels of NU (factor) have a lower negative impact on ED performance. This can be compensated by the exploitation of the increased number of sampling points in the ED, which contributes to improving the detection probability of the PU signals. On the other hand, in ED operating environments impacted by large NU, an assertion of NU must be followed by an appropriate (dynamic) adjustment of the level of DT during the ED process. Therefore, the trade-off within the number of sampling points used for the ED, the appropriate DT adjustment during the ED and the number of SU Rx and PU Tx chains, strongly impact the effectiveness of the ED at the location of SU.

5.7. Effect of Probabilities of a False Alarm on the Efficiency of the ED Operation

In this section, the analysis of the results of simulations showing the effect of different levels of false alarm probabilities on detection probability in SISO and symmetric 4 × 4 MIMO-OFDM CR systems have been presented. The analysis was performed for the two different values of false alarm probabilities (0.01 and 0.2), the constant Tx power of PU (100 mW), the number of sampling points equal to 128, the QPSK modulation of PU signal, and the different levels of DT adjustments and NUs (ρ = 1.05, ρ' = 1.03 and ρ = 1.03, ρ' = 1.05).

In Figure 8, the dependency of detection probability on SNR for ED performed with distinct false alarm probabilities and combinations of DDT and NU factors in SISO and symmetric 2 × 2 and 4 × 4 MIMO communication systems have been presented. The results indicate that for the same SNR level, the detection probability becomes lower as the values of targeted false alarm probability (P_{fa}) decrease and vice versa. These trends are characteristic for both SISO and symmetric MIMO systems and for all analyzed combinations of NU and DDT factors. These results are a consequence of the fact that when there is no exploitation of spectrum by PU, the probability that the SUs incorrectly estimate that the transmission of PU exists increases.

As in the case of the simulation results presented in Figures 2–7, Figure 8 also shows that higher values of the DT factor (ρ' = 1.05) improve the detection probability for the equal false alarm probability (P_{fa}) at specific SNRs detected at the antennas of SU Rx chains. This cognition points to the importance of the implementation of appropriate DT adjustments in the ED process, for any targeted value of false alarm probability (P_{fa}) characteristic for ED based on CFAR principles. Furthermore, from Figure 8, it can be seen how the combination of the larger number of Tx-Rx chains and the higher values of requested false alarm probability (P_{fa}), lead to a better detection probability for the same SNR values at the position of SU. On the other hand, a higher false alarm probability can result in an incorrect decision being made about PU activity in the ED process performed by SU. For that reason, an appropriate selection of minimal targeted values of false alarm probability for a specific number of Tx-Rx chains should be considered in the case of any operating environments in which the ED occurs.

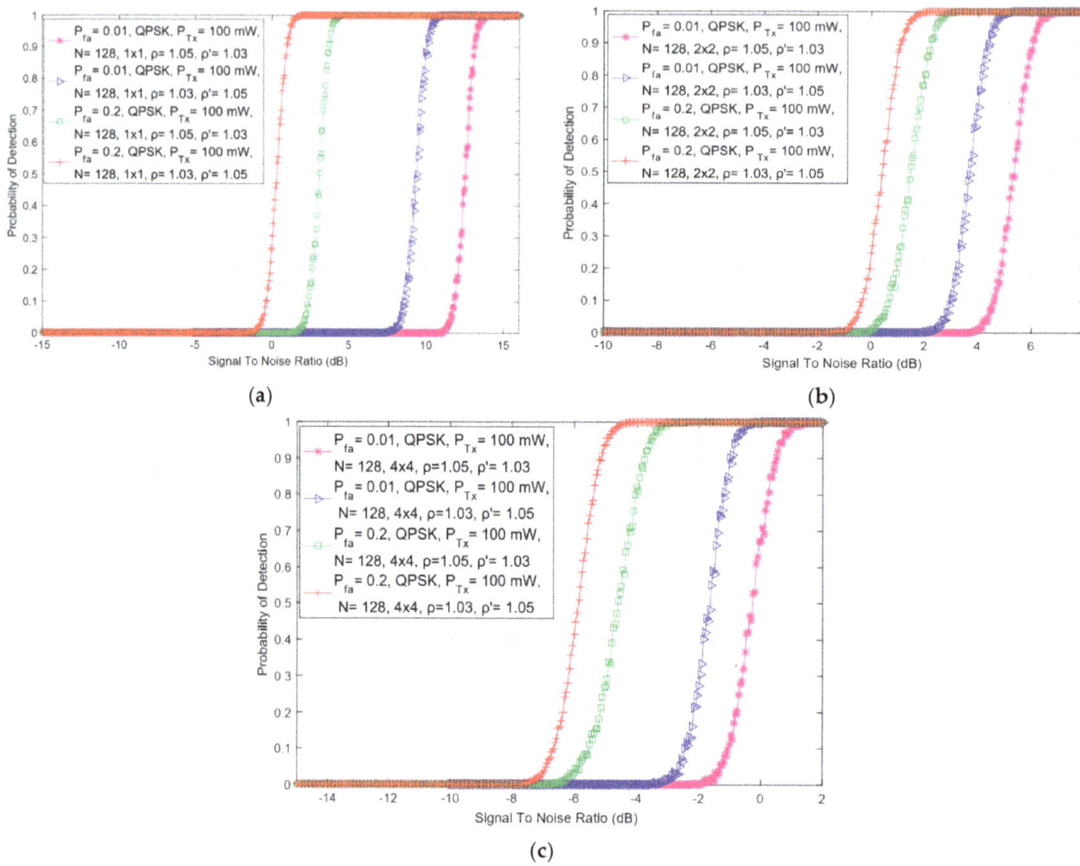

Figure 8. The dependency of detection probability on SNR for ED performed with different false alarm probabilities and combinations of DDT and NU factors in (**a**) SISO, (**b**) 2 × 2 symmetric MIMO and (**c**) 4 × 4 symmetric MIMO communication systems.

6. Conclusions

In this paper, the performance of the ED SS employing the SLC technique with DT adjustments according to NU variances in MIMO-OFDM CR systems has been analyzed. The mathematical expression of the main parameters used for the evaluation of the ED performance as a local SS technique employing SLC in MIMO-OFDM CR stems has been introduced. In addition, the algorithm for simulating the ED in versatile operating environments characterized by the influence of distinct levels of NU and performed with DT adjustments has been presented. The analysis of ED sensing efficiency has been performed through extensive simulations which indicates how different working parameters including the number of sampling points used in the ED, the Tx powers of PU, the DDT and NU factors, the probabilities of false alarm, and the SNRs impact the probability of the detection of PU signals in MIMO-OFDM CRs systems.

The results of the analysis reveal that a general improvement in the SS efficiency of the ED method employing the SLC technique can be achieved if the level of DT adjustments during the ED follows the intensity of the variances in NU. The analysis also shows that higher values for the Tx power of PU, the number of sampling points, the false alarm probability, and the number of SU Rx and PU Tx chains can positively impact ED performance. Exploiting the cognitions shown in this paper in terms of selecting appropriate operational parameters such as the Tx power levels of PU, the number of sampling points, the number of Tx-Rx chains at PU and SU side and the expected false alarm probabilities can enable the implementation of more efficient ED SS in MIMO-OFDM CR systems.

The obtained results further show that ED employing SLC in MIMO-OFDM CR systems containing multiple PU Tx and multiple SU Rx chains, outperforms in terms of ED efficiency the SISO communication systems. Additionally, in an asymmetric MIMO-OFDM CR system, configurations with a number of SU Rx chains larger than the number of PU TX chains yield better results with respect to ED performance of communication systems having an opposite ratio of Tx and Rx chains. Therefore, the proliferation of new user and network devices containing a higher number of build-in transmission chains will enable the improvement of ED performance in its practical implementation. Our future research work will be pointed towards the analysis of how massive-MIMO communications in terms of a large number of SU Rx and PU Tx chains in multi-cell CR networks, impact the performance of ED SS employing the SLC technique.

Author Contributions: Conceptualization, J.L.; methodology, J.L.; software, I.R.; validation, J.L. and D.B.; formal analysis, J.L. and I.R.; investigation, I.R.; writing—original draft preparation, J.L. and I.R.; writing—review and editing, J.L.; visualization, J.L. and I.R.; supervision, J.L. and D.B. All authors have read and agreed to the published version of the manuscript.

Funding: This research received no external funding.

Institutional Review Board Statement: Not applicable.

Informed Consent Statement: Not applicable.

Conflicts of Interest: The authors declare no conflict of interest.

Abbreviations

The following abbreviations are used in this manuscript:

AWGN	Additive white Gaussian noise
BS	Base station
CFAR	Constant false alarm rate
CR	Cognitive radio
CRN	Cognitive radio networks
CSI	Channel state information
DSA	Dynamic spectrum access
DT	Detection threshold
DDT	Dynamic detection threshold
ED	Energy detection
EGC	Equal Gain Combining
IoT	Internet of Things
ISI	Inter-symbol interference
LSA	Licensed shared access
MIMO	Multiple-input multiple-output
MISO	Multiple-input single-output
NU	Noise uncertainty
OFDM	Orthogonal frequency-division multiplexing
PU	Primary user
RF	Radiofrequency
SISO	Single-input-single-output
SIMO	Single-input multiple-output
SL	Square-law
SLC	Square-law combining

SLS	Square-Law Selection
SNR	Signal-to-noise ratio
SS	Spectrum sensing
STBC	Space-time block codes
SU	Secondary users
5G	Fifth-generation mobile network
6G	Sixth-generation mobile network

References

1. Pandit, S.; Singh, G. Spectrum sensing in cognitive radio networks. In *Spectrum Sensing in Cognitive Radio Networks: Potential Challenges and Future Perspective in Spectrum Sharing in Cognitive Radio Networks*, 1st ed.; Springer International Publishing AG: Cham, Switzerland, 2017; pp. 35–73.
2. Mahmoud, H.A.; Yucek, T.; Arslan, H. OFDM for cognitive radio: Merits and challenges. *IEEE Wireless Commun.* **2009**, *16*, 6–15. [CrossRef]
3. Rwodzi, J. Energy-Detection Based Spectrum Sensing for Cognitive Radio on a Real-Time SDR Platform. Master's Thesis, Department of Electrical Engineering, University of Cape Town, Cape Town, South Africa, 2016.
4. Xiao, Y.; Hu, F. *Cognitive Radio Networks*, 1st ed.; Auerbach Publications: Boca Raton, FL, USA, 2008; pp. 3–37.
5. Shbat, M.S.; Tuzlukov, V. SNR wall for generalized energy detector in the presence of noise uncertainty and fading. *Sensors* **2015**, *15*, 16105–16135. [CrossRef] [PubMed]
6. Zhang, J.; Liu, L.; Liu, M.; Yi, Y.; Yang, Q.; Gong, F. MIMO spectrum sensing for cognitive radio-based Internet of things. *IEEE Internet Things J.* **2020**, *7*, 8874–8885. [CrossRef]
7. Ji, G.; Zhu, H. Noise Uncertainty Study of the Low SNR Energy Detector in Cognitive Radio. In Proceedings of the International Conference on Artificial Intelligence and Computational Intelligence (AICI 2010), Sanya, China, 23–24 October 2010. [CrossRef]
8. Sun, H.; Nallanathan, A.; Wang, C.X.; Chen, Y. Wideband spectrum sensing for cognitive radio networks: A survey. *IEEE Wireless Commun.* **2013**, *20*, 74–81.
9. Arar, A.M.; Masri, A.M.; Ghannam, H.O.; Tumar, I.K. A proposed scheme for dynamic threshold versus noise uncertainty in cognitive radio networks (DTNU). *Wireless Pers. Commun.* **2017**, *96*, 4543–4555. [CrossRef]
10. Thuo, N. An Adaptive Threshold Energy Detection Technique with Noise Variance Estimation for Cognitive Radio Sensor Networks. Master's Thesis, Department of Electrical Engineering, University of Cape Town, Cape Town, South Africa, 2015.
11. Guerra, D.W.M.; Fukuda, R.M.; Kobayashi, R.T.; Abrao, T. Efficient detectors for MIMO-OFDM systems under spatial corequation antenna arrays. *Etri J.* **2018**, *40*, 570–581. [CrossRef]
12. Dang, W.; Tao, M.; Mu, H.; Huang, J. Subcarrier-pair based resource allocation for cooperative multi-relay OFDM systems. *IEEE Trans. Wirel. Commun.* **2010**, *9*, 1640–1649. [CrossRef]
13. Nandhakumar, P.; Kumar, A. Analysis of OFDM system with energy detection spectrum sensing. *Indian J. Sci. Technol.* **2016**, *9*, 1–6. [CrossRef]
14. Chen, Z.; Luan, T.; Zhang, X.D. Sensing orthogonal frequency division multiplexing systems for cognitive radio with cyclic prefix and pilot tones. *IET Commun.* **2012**, *6*, 97–106. [CrossRef]
15. Salam, A.O.A.; Sheriff, R.E.; Al-Araji, S.R.; Mezher, K.; Nasir, Q. Spectrum sensing in cognitive radio using multitaper method based on MIMO-OFDM techniques. *Ann. Telecommun.* **2019**, *74*, 727–736. [CrossRef]
16. Gesbert, D.; Shafi, M.; Shiu, D.-S.; Smith, P.J.; Naguib, A. From theory to practice: An overview of MIMO space-time coded wireless systems. *IEEE J. Sel. Areas Comm.* **2003**, *21*, 281–302. [CrossRef]
17. Kumar, R.; Saxena, R. Performance analysis of MIMO-STBC systems with higher coding rate using adaptive semiblind channel estimation scheme. *Res. Artic.* **2014**, *2014*, 304901. [CrossRef] [PubMed]
18. Rodes, L.; Kaushik, A.; Sharma, S.K.; Chatzinotas, S.; Jondral, F. Square-law Selector and Square-law Combiner for Cognitive Radio Systems: An Experimental Study. In Proceedings of the 2016 IEEE 84th Vehicular Technology Conference (VTC-Fall), Montreal, QC, Canada, 18–21 September 2016; pp. 1–5. [CrossRef]
19. Stuber, G.L.; Barry, J.R.; McLaughlin, S.W.; LI, Y.; Ingram, M.A.; Pratt, T.G. Broadband MIMO-OFDM wireless communications. *Proc. IEEE* **2004**, *92*, 271–294. [CrossRef]
20. Yang, C.; Wu, G.; Li, S.; Li, G.Y. OFDM and its wireless applications: A survey. *IEEE Trans. Veh. Tech.* **2009**, *58*, 1673–1694.
21. Gupta, A.; Jha, R.K. A survey of 5G network: Architecture and emerging technologies. *IEEE Access* **2015**, *3*, 1206–1232. [CrossRef]
22. Kuppusamy, V.; Mahapatra, R. Primary User Detection in OFDM Based MIMO Cognitive Radio. In Proceedings of the 2008 3rd International Conference on Cognitive Radio Oriented Wireless Networks and Communications (CrownCom 2008), Singapore, 15–17 May 2008; pp. 1–5. [CrossRef]
23. Ustok, R.F. Spectrum Sensing Techniques for Cognitive Radio Systems with Multiple Antennas. Master's Thesis, Department of Electrical Engineering, Izmir Institute of Technology, Izmir, Turkey, 2009.

24. Ranjeeth, M. Cooperative Spectrum Sensing with Square Law Combining Diversity Reception. In Proceedings of the 2015 3rd International Conference on Signal Processing, Communication and Networking (ICSCN), Chennai, India, 26–28 March 2015; pp. 1–6. [CrossRef]
25. Digham, F.F.; Alouini, M.S.; Simon, M.K. On the energy detection of unknown signals over fading channels. *IEEE Trans. Commun.* **2007**, *55*, 3575–3579. [CrossRef]
26. Tellambura, C. Spectrum sensing methods and their performance. In *Handbook of Cognitive Radio*, 1st ed.; Springer Nature: Singapore, 2018; pp. 163–185.
27. Lorincz, J.; Ramljak, I.; Begušić, D. A review of the noise uncertainty impact on energy detection with different OFDM system designs. *Comput. Commun.* **2019**, *148*, 185–207. [CrossRef]
28. Lorincz, J.; Ramljak, I.; Begušić, D. A survey on the energy detection of OFDM signals with dynamic threshold adaptation: Open Issues and Future Challenges. *Sensors* **2021**, *21*, 3080. [CrossRef]
29. Choi Won, K.; Hossain, E.; Kim In, D. Downlink Subchannel and Power Allocation in Multi-Cell OFDMA Cognitive Radio Networks. *IEEE Trans. Wireless Commun.* **2011**, *10*, 2259–2271. [CrossRef]
30. Singh, U.; Biswas, S.; Singh, K.; Kanaujia Kumar, B.; Li, C.-P. Beamforming Design for In-Band Full-Duplex Multi-Cell Multi-User MIMO LSA Cellular Networks. *IEEE Access* **2020**, *8*, 222355–222370. [CrossRef]
31. Lorincz, J.; Ramljak, I.; Begušić, D. Algorithm for Evaluating Energy Detection Spectrum Sensing Performance of Cognitive Radio MIMO-OFDM Systems. *Sensors* **2021**, *21*, 6881. [CrossRef]
32. Lorincz, J.; Ramljak, I.; Begušić, D. Performance Analyses of Energy Detection Based on Square-Law Combining in MIMO-OFDM Cognitive Radio Networks. *Sensors* **2021**, *21*, 7678. [CrossRef] [PubMed]
33. Lorincz, J.; Begusic, D. Adaptive Beamforming Structure with STBC for IEEE 802.11 n WLAN Systems. In Proceedings of the 16th International Conference on Software, Telecommunications and Computer Networks (SoftCOM 2008), Split, Croatia, 25–27 September 2008; pp. 258–263. [CrossRef]
34. Atapattu, S.; Tellambura, C.; Jiang, H. *Energy Detection for Spectrum Sensing in Cognitive Radio*; Springer International Publishing AG: New York, NY, USA; Heidelberg, Germany; Dordrecht, The Nertherland; London, UK, 2014; pp. 1–94.
35. Kumar, A.; Thakur, P.; Pandit, S.; Singh, G. Analysis of optimal threshold selection for spectrum sensing in a cognitive radio networks: An energy detection approach. *Wirel. Netw.* **2019**, *25*, 3917–3931. [CrossRef]
36. Wang, B.; Ray Liu, K.J. Advances in Cognitive Radio Networks: A Survey. *IEEE J. Sel. Top. Signal Process.* **2010**, *5*, 5–23. [CrossRef]
37. Martínez, D.M.; Andrade, Á.G. On the Reduction of the Noise Uncertainty Effects in Energy Detection for Spectrum Sensing in Cognitive Radios. In Proceedings of the 2014 IEEE 25th Annual International Symposium on Personal, Indoor, and Mobile Radio Communication (PIMRC), Washington, DC, USA, 2–5 September 2014; pp. 1975–1979.
38. Yu, G.; Long, C.; Xiang, M.; Xi, W. A novel energy detection scheme based on dynamic threshold in cognitive radio systems. *J. Comput. Inf. Syst.* **2012**, *8*, 2245–2252.
39. Chabbra, K.; Banerjee, P. Effect of Dynamic Threshold & Noise Uncertainty in Energy Detection Spectrum Sensing Technique for Cognitive Radio Systems. In Proceedings of the 2014 International Conference on Signal Processing and Integrated Networks (SPIN), Delhi, India, 20–21 February 2014; pp. 361–377.
40. Ye, Z.; Memik, G.; Grosspietsch, J. Energy Detection Using Estimated Noise Variance for Spectrum Sensing in Cognitive Radio Networks. In Proceedings of the 2008 IEEE Wireless Communications and Networking Conference, Las Vegas, NV, USA, 31 March–3 April 2008; pp. 711–716.

sensors

MDPI

Article

Forwarding in Energy-Constrained Wireless Information Centric Networks

Daniel Marques [1], Carlos Senna [1] and Miguel Luís [1,2,*]

[1] Instituto de Telecomunicações, 3810-193 Aveiro, Portugal; danielfmarques@av.it.pt (D.M.); cr.senna@av.it.pt (C.S.)
[2] ISEL—Instituto Superior de Engenharia de Lisboa, Instituto Politécnico de Lisboa, 1959-007 Lisboa, Portugal
* Correspondence: nmal@av.it.pt

Abstract: Information Centric Networks (ICNs) have been considered one of the most promising candidates to overcome the disadvantages of host-centric architectures when applied to IoT networks, having the potential to address the challenges of a smart city. One of the foundations of a smart city is its sensory capacity, which is obtained through devices associated with the IoT concept. The more sensors spread out, the greater the ability to sense the city. However, such a scale demands high energy requirements and an effective improvement in the energy management is unavoidable. To improve the energy management, we are proposing an efficient forwarding scheme in energy-constrained wireless ICNs. To achieve this goal, we consider the type of devices, their internal energy and the network context, among other parameters. The proposed forwarding strategy extends and adapts concepts of ICNs, by means of packet domain analysis, neighbourhood evaluation and node sleeping and waking strategies. The proposed solution takes advantage of the neighbourhood to be aware of the moments to listen and forward packets in order to consistently address mobility, improving the quality of content delivery. The evaluation is performed by simulation with real datasets of urban mobility, one from the lagoon of "Ria de Aveiro" and the other from a vehicular network in the city of Porto. The results show that the proposed forwarding scheme resulted in significant improvements in network content availability, in the overall energy saving and, consequently, in the network lifetime.

Keywords: energy consumption management; content forwarding; information-centric networks; internet of things; smart city

Citation: Marques, D.; Senna, C.; Luís, M. Forwarding in Energy-Constrained Wireless Information Centric Networks. *Sensors* **2022**, *22*, 1438. https://doi.org/10.3390/s22041438

Academic Editor: Josip Lorincz

Received: 28 January 2022
Accepted: 10 February 2022
Published: 13 February 2022

1. Introduction

In recent years we have seen a radical change in the way devices connect to the Internet. The best example is the IoT, a recent communication paradigm in which objects of everyday life will be able to, among others, communicate with one another, becoming an integral part of the Internet. In this context, wireless networking is expected to sustain the direct interaction between personal users' devices and also provide connectivity on a large scale for resource-constrained devices. However, conventional networking protocols, such as the traditional Transmission Control Protocol/Internet Protocol (TCP/IP) host-centric network, fail in large scale mobile wireless distributed environments, such as IoT scenarios, due to node mobility, dynamic topologies and intermittent connectivity, to name a few [1].

Information Centric Networks (ICNs) [2] have been considered one of the most promising candidates to overcome the drawbacks of host-centric architectures when applied to IoT networks. Conceptually, in ICNs, each piece of data has a unique, persistent and location-independent name that is directly used by the applications for content search and retrieval. Therefore, ICNs enable the deployment of in-network caching and content replication, thus facilitating the efficient and timely delivery of information. These features increase the challenge in building efficient energy management solutions for ICNs, as traditional solutions for IP networks are of little value in ICN-based environments. In this way, some questions arise: In mobile energy-constrained and multi-technology environments, which interface should be preferred for content resolution and content forwarding? Should we replicate all the Interest packets back from the same interface or should some contention be adopted to avoid unnecessary transmissions? Should energy-constrained devices adopt a conservative approach?

To answer these questions, we are proposing a forwarding strategy for energy-constrained wireless ICNs. Our solution considers the type of device, network context and ICN caching, among other parameters, to implement new content resolution for mobile wireless networks. The proposed forwarding strategy extends and adapts concepts of ICNs, by means of packet domain analysis, neighbourhood evaluation and node sleeping and waking strategies to increase the energy saving and reduce the use of resources in unnecessary situations. This strategy is particularly important in this scenario because IoT nodes may have small and limited batteries. Furthermore, the proposed solution takes advantage of each node's neighbourhood to be aware of the optimal moments to listen and forward packets in order to consistently address mobility, improving the quality of content delivery.

To bring our evaluations one step closer to a real scenario, we use two real traces of urban mobility, one with boat mobility, gathered from the Aveiro lagoon "Ria de Aveiro", in the city of Aveiro, Portugal, and the other from a vehicular network running in the city of Porto, Portugal, gathered from public buses from the Porto Collective Transport Society (STCP). STCP's buses form a sensing and data distribution platform composed of 600+ public transport vehicles equipped with On-Board Units (OBUs) for V2X (Vehicle to Everything) communications and in-vehicle internet connectivity for passengers. The mobility traces of STCP buses were collected through the Porto Living lab IoT platform [3] and the mobility of boats was collected by communication infrastructure of the Aveiro STEAM city project (https://uia-initiative.eu/en/uia-cities/aveiro (accessed on 28 January 2022)).

The results show that the energy-aware proposed forwarding scheme resulted in significant improvements in network content availability, in the overall energy saving of the nodes and, consequently, in the increase in the network lifetime. Moreover, it was possible to save energy previously spent in listening stages and forwarding packets in situations of no connectivity, resulting in an increase of 25 min of network lifetime in the Aveiro scenario and almost 5 h in the Porto scenario. The main contributions of our work are:

- Design of an efficient forwarding strategy for energy-constrained wireless ICNs that takes into account inputs from the network structures, as well as context, neighbouring environment and energy level in the decision process;
- Modification of the Named Data Networking (NDN) framework and the network simulator 3 (ns-3) energy framework, along with the redesign of the logging module, to include the proposed forwarding strategy;
- Assessment of the overall network performance and power consumption in IoT scenarios considering two mobility patterns.

The remainder of this paper is organised as follows: Section 2 discusses the related work; Section 3 shows the basic characteristics of the ICN paradigm; Section 4 details the proposed efficient forwarding in relation with energy management; Section 5 describes the scenarios and metrics and discusses the results obtained by simulation using real mobility traces; and, finally, Section 6 enumerates the conclusions and points out the future work.

2. Related Work

In the last years, several works have investigated a way to minimise the large broadcast storm problem and reverse path partitioning in mobile networks [4–6]. However, there are few works that take into consideration the energy-constrained wireless ICNs, more specifically, the energy use and management of each node in the routing decision process.

In the work of [7], a forwarding strategy was proposed to select a node from its neighbourhood as a cluster head responsible for forwarding the Interest packets, based on the information regarding the satisfied Interest rate between the neighbouring nodes. This required the addition of two new data structures, one to store these neighbour values and another to store its own values. In this case, the security barrier is broken by using the extraneous information of neighbouring nodes, a list of nodes that have a higher satisfied Interest rate. In the strategies of [8,9] , the Global Positioning System (GPS) positions are used by a geolocation mechanism that routes the Interest packets to the Producers. By using the Producer position, the content-centric paradigm ceases to exist, becoming host-centric. Moreover, in highly mobile environments, this Producer position is compromised, as there may be mobile Producers, not maintaining the same position over time. The works of [7,10] elect a node from their neighbourhood to forward their Interest packets according to their data delivery success rate. This requires adding new fields and structures for these values to be stored.

Although these solutions improve content accessibility, transmission reliability and robustness and minimise the broadcast storm problem, they still change the basics of the ICN architecture, bringing more complexity and overhead to the network. Therefore, it is essential to develop new and more efficient forwarding strategies that guarantee higher network performance while maintaining the basic ICN architecture.

On the other hand, the work in [11] proposes two mechanisms for packets transmission aiming to minimise the broadcast storm problem and the energy consumption. Two forwarding modes are combined and switched based on the lookup performed by the Forwarding Information Base (FIB) and efficient mechanisms are implemented to control flooding in order to save energy. However, this proposal does not consider mobility and very specific and static scenarios are evaluated. Similarly, [12] uses greedy packet forwarding, using Hello packets to send its neighbourhood information (ID, location and remaining power). In this proposal, the security barrier is broken through the sharing of confidential information from each node. The works [13,14] implement a sleep scheduler that is based on monitoring information from the environment and the operating state of the nodes, choosing one of two states (semi-sleep or sleep) at random times. However, this work is designed for Wireless Sensor Networks and does not apply to ICNs. The work in [15] proposes a multihop cooperative caching scheme for green wireless sensor networks over receiver-triggered ICNs. Using regional caches, the proposal saves energy while reducing the delay in fetching data. In [16], a Power Saving Mode is proposed to manage the time spent in idle listening state, consisting of a state machine (transmit, receive, idle, sleep) that is based on the Pending Interest Table (PIT) state and of a beacon that is sent by the Access Points containing the Traffic Indication Map (a list of nodes that have to wake up to be ready to receive data). In this proposal, the network depends on the beacon that is sent by the APs to be able to decide if the node should wake up to receive or transmit the content. The solution cannot be applied in fully mobile environments, where there is no interaction with a static network element and where mobile nodes may experience a long period of disconnection from the network. Furthermore, it is necessary for such beacons to circulate through the network and constantly update themselves, which increases the network overhead.

The works of [17,18] use a Software-Defined Network (SDN) that supports the ICN approach by having the knowledge of the network topology, communicating any information to the nodes and update routing information. The SDN also helps in efficient neighbourhood discovery and in the communication of the new nodes' positions. However, this architecture brings more complexity to the network system using external resources. Moreover, the communication of the new positions of certain nodes brings security issues and ends up with spatial and temporal non-reference of the nodes.

An important restriction found in the aforementioned solutions is the addition of new fields in the packets (Interest and Data), addition of new tables and structures to the native ICN architecture. These additional fields, which are added to the packets that will be transmitted through the network, usually refer to parameters or properties of the nodes, such as their position and remaining energy, which should be private information of each node. Many of the works conducted that achieve improvements in network energy saving do not take into account the mobility of the nodes or overcome the security barrier and exploit the positions of the Producers to route the Interest packets in the right direction, which exposes sensitive node information that should be private. Furthermore, they lack practical evaluation using real-world scenarios with resource-constrained mobile elements.

Our proposal aims to solve the various problems mentioned before. It does not compromise security with the sharing of information among the nodes, it is intended for application in highly mobile ICN environments, manages the nodes' internal energy in the packet forwarding decision process and maintains the basics of the ICN architecture without adding new fields to packets, creating new tables or changing structures. Finally, we evaluated our mechanisms and strategies using real data from boat and bus movements gathered from the cities of Aveiro and Porto, Portugal.

3. ICN Basics

In an ICN architecture, the Consumer makes a request for content without knowing where its Producer is. The communication between the Consumer and the Producer follows a request–response exchange model with native support of multicast, which can be especially useful both in a Vehicular ad hoc Network (VANET) environment and Internet of Things (IoT) [19]. During this process, as each piece of data has a unique name, persistent and independent of their location, the system takes care of mapping the requested data to their originating location, which facilitates applications in the process of searching and collecting future data with the same name. When the request arrives at the Producer's side, the requested Data follow the reverse path through which the request arrived, following a receiver-oriented approach. ICN has solved the need for in-network caching by replicating content across network nodes to improve content delivery by being faster and more efficient, as can be seen in Figure 1, where the mode of operation of an ICN is explained.

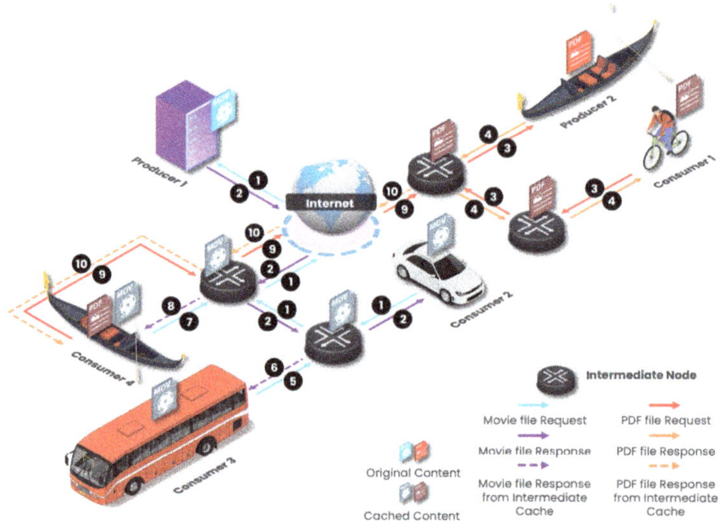

Figure 1. ICN operation mode. Consumers request specific content which is initially stored in the Producers. Then, as this content passes through the intermediary nodes, it is retained in the cache for faster response to future requests.

In the example of Figure 1, Consumers request content (be it a PDF document or a movie) initially stored at the respective Producers. The content request is forwarded to the Producers, and when the content comes back by the reverse path of the Interest packet, the content is cached in the intermediate nodes, enabling the content satisfaction of future requests.

ICN does not use source-destination host pair communication. Data are identified by unique names [20]. Naming schemes can be hierarchical, where content names are human readable and can be hierarchical and structured, such as Uniform Resource Locator (URL) [21]; flat, where names the contents use are small, unique and self-certified names (hashes, for example) [22]; attribute-based, where names are expressed through keywords that are extracted from the contents as attributes, such as type, version; and hybrid, where names the contents are using are a combination of all the schemes mentioned before to take advantage of the best parts of each one [23]. An example name for content that circulates a smart city network could be //SmartCityUrbanPlanning/TrafficManagement/Decision/ "ContentContext" /:::"ContentAttributes".

The ICN architecture uses two types of packets in all network communication: the Interest packet and the Data packet [2]. The Interest packet is used to define a request for a particular content and contains two mandatory fields: the Name, which identifies the intended content, and the Nonce, which uniquely identifies the packet circulating over the network. It may also contain Selectors that serve as input parameters to the routing and content resolution mechanisms, such as the indication that this Interest may receive stale Data, the lifetime of the Interest, the hop count limit before the Interest is discarded, among others. The creation of the Data packet is associated with the reception of an Interest packet, so the Name is the same in both packets (it is extracted from the Interest packet). It also has the data itself represented by content and the Signature that is divided into SignatureInfo, which is included in the calculation of the signature and describes itself, and SignatureValue, which is excluded from the signature calculation and represents the signature material.

ICN provides temporary storage of content on router nodes that can meet future requests [24]. Data packets are available closer to the Consumer, reducing the number of packets that travel through the network. This allows the reduction in the bottleneck of messages exchanged, greater availability of content spread over the network, reduction in response latency and, of course, energy savings.

One of the most promising architectures based on the ICN concept is the Named Data Network (NDN) [2]. The NDN architecture makes use of hierarchically structured content names, content source reliability (through authentication by the content provider through cryptographic mechanisms such as public keys) and a routing protocol that executes forwarding mechanisms using names instead of IP addresses. Due to the facilities provided by the NDN project [25], especially the NDN simulator (ndnSIM) [26], this architecture was the one we chose for the development of this work. The NDN allows users to make a request for data without knowing a priori which entity produces or possesses it. It also allows user mobility and security issues to be handled more efficiently than the current Internet Protocol (IP). The entities that exchange information in the NDN network can be separated into Producers, Consumers and router/intermediate nodes, forwarding the Interest and Data packets between Consumers and Producers. It is important to say that a node can assume all three roles concurrently. A node belonging to an NDN network maintains three important data structures: the Forwarding Information Base (FIB), the Pending Interest Table (PIT) and the Content Store (CS), illustrated in Figure 2.

The FIB is a table of name prefixes with their output interfaces that allows the forwarding of Interest packets by searching by their name. The PIT is a table that records the Interests that are requested, that are waiting for Data, and the interfaces through which they came. The CS is a cache for incoming Data packets, which can be used for faster responses to future Interests of that same Data. The Faces are the interfaces that receive and forward the packets.

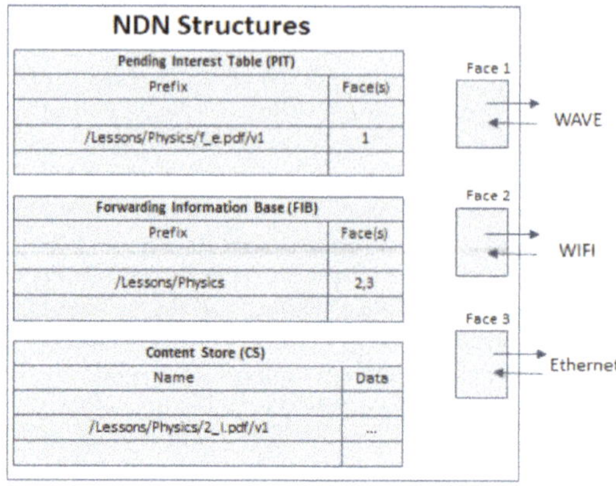

Figure 2. NDN data structures.

Let us now look on how these structures are used in the forwarding process. As shown in Figure 3, packet forwarding can follow an upstream or downstream direction according to the information present in the data structures and the type of packet concerned. When an Interest arrives at a node (upstream), the CS is checked for the intended Data. If the Data are not in the CS, the Interest is forwarded to the PIT where it is stored along with the interface through which it arrived (if this entry already exists, only the interface is stored and the Interest is discarded), and then uses the FIB to forward this packet if possible (otherwise it is discarded). If an Interest reaches a node that has the Data in the CS, that same node replies with the desired Data along with a signature from the Producer. This Data packet then follows the reverse path of the Interest packet, until it reaches the Consumer that requested that Data. The same procedure is performed with the Negative Acknowledgement (NACK) packets that will notify the Consumer of some congestion, duplication of Interest or in case no route was found.

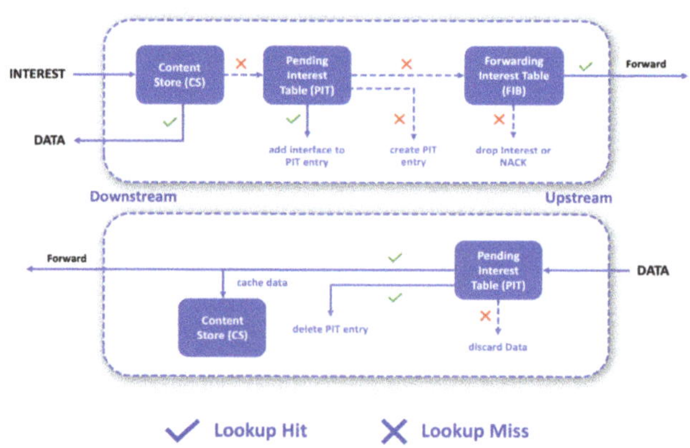

Figure 3. Forwarding process in NDN.

4. Energy-Efficient Wireless NDN Architecture

The main advantage of using the NDN design in a highly mobile environment is that it increases the availability of content on a network. Excessive power consumption in dense IoT scenarios brings problems to the forwarding process, causing unnecessary packet delays and decreasing network performance. To solve these problems, we propose strategies that aim to minimise excessive energy consumption, optimising the processes without harming the NDN functionalities. Figure 4 illustrates an NDN macro architecture with the main components of the base architecture (Storage, Forwarding and Interfaces and Mobility) [2] encompassing the proposed *Energy Consumption* component that supports the solution for efficient energy management.

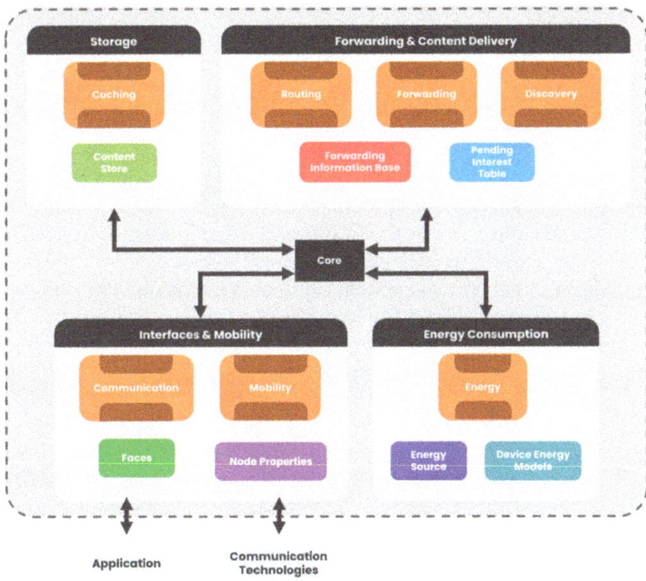

Figure 4. Energy-efficient wireless NDN-based architecture.

The central core module is responsible for the interaction between all the modules, allowing their logical and efficient operation. The Storage represents the Caching process of a node, which aims to store the contents in memory to be able to respond to future requests for the same contents. The Forwarding and Content Delivery is composed of three processes: Routing, Forwarding and Discovery. The Routing is responsible for keeping a register of the available paths to a given content. Forwarding allows the routing of Interest and Data packets in an efficient way through the available paths. Discovery complements Routing by determining the most logical path to forward Interest packets if no other path to the content is available. This is to make an effort to find a path to the content first, rather than discarding the packet outright. The main data structures used by these processes are FIB and PIT.

The module Interfaces and Mobility (I&M) consists of Communication and Mobility processes. The Communication is responsible for the connection and communication between devices and the application layer, and between other external devices. The Mobility is related to all node parameters that may have direct or indirect influence on its mobility. It allows the mobile characterisation of any mobile element, which is related to the Node Properties and that are used in the remaining forwarding and caching processes. Besides, the I&M processes use Faces that symbolise the communication interfaces of the NDN Stack and Node Properties that contain the node's properties, such as the mobility type, among others. Finally, our Energy Consumption module is composed of the Energy component responsible for measuring the energy consumption and how a node should react to certain energy situations, presenting two essential components: the Energy Source and the Device Energy Models. The Energy Source is responsible for the implementation of the power sources or batteries in a node that can be modelled in a linear or non-linear way. The Device Energy Models are associated with the way energy is consumed by different technologies (radio interfaces, processors, sensors, etc.). An example of the information collected and used by Device Energy Models for Wi-Fi technology is [27]: Sleep = 0.00132 A,

Transmit/Tx = 0.167 A, Receive/Rx = 0.310 A, Range = 50 m and Frequency = 2.4–5.0 GHz. Below, we highlight the implementations made for an efficient management of the energy resources of the network nodes.

4.1. Interfaces and Mobility Module

In the I&M module, we introduce the mechanism responsible for the discovery of the type and number of nodes in the neighbourhood as well as the mechanism responsible for processing the direction of the node to predict situations where energy management can be used.

4.1.1. Neighbourhood Status

The neighbourhood awareness is important for the decision of packet forwarding and for the sleep strategy. The state of a node's neighbourhood is based on four essential aspects: the available interfaces, the number and type of neighbours, the cost associated to the interfaces and the mobility metrics. A stationary or moving node can have different communication technologies. As a node is constantly moving, the wireless link conditions are constantly changing, and picking the best one to communicate is very important to the forwarding strategy and energy saving, bringing a great value to network efficiency.

The number and type of neighbours (mobile or static) per interface is an important factor, since it determines how many neighbours it has and what type of mobility they present, aspects that are considered when deciding which interface a packet should be forwarded through or whether it is a good time for the node to go in sleep mode.

The Mobility module is responsible for updating neighbourhood parameters, such as mobility type and Received Signal Strength Indicator (RSSI). However, there are other factors that influence the neighbourhood relationship such as the speed and direction of a node. To collect the aforementioned information needed to implement the neighbourhood awareness module, we present a mechanism whose flowchart is presented in Algorithm 1. The mechanism runs every 1 s for each mobile node and checks, for the NetDevice of the wireless interface (LINK_TYPE_ADD_HOC) (line 6), the RSSI value of a given neighbour (line 14) and the type of node (mobile or static) (line 16). In addition, we can also obtain the position of the neighbour (line 10), which can be obtained from the RSSI values [28,29], for example, or from periodic messages transmitted by the nodes through Cooperative Awareness Messages (CAM) of the Intelligent Transportation System (ITS) [30]. At the end, this information is communicated to the Forwarder module (line 16).

4.1.2. Node Heading and Direction

In order to determine situations where a node can save energy by entering in sleep mode, its direction has been considered. For each static node (Road Side Units—RSU) in the neighbourhood, its position is obtained, and the direction of the static node regarding the previous position of the node is calculated and then converted to an angle in degrees (α). The quadrant is also computed, as illustrated in Figure 5. The next step is the definition of the angles that define the possible directions named as "Towards an RSU" (TRSU) or "Outwards of an RSU" (ORSU). For this, the range of TRSU angles was considered to be all those between the RSU +90° and RSU −90° angle, giving a range of angles with a view of 180°. All angles that do not belong to this range are considered as ORSU. Whenever the angle of the node's direction is within the range TRSU, it means that it is heading towards that RSU. This proposal is very useful to predict the direction of the node and to estimate the time when it may be entering an area without connection/low RSSI.

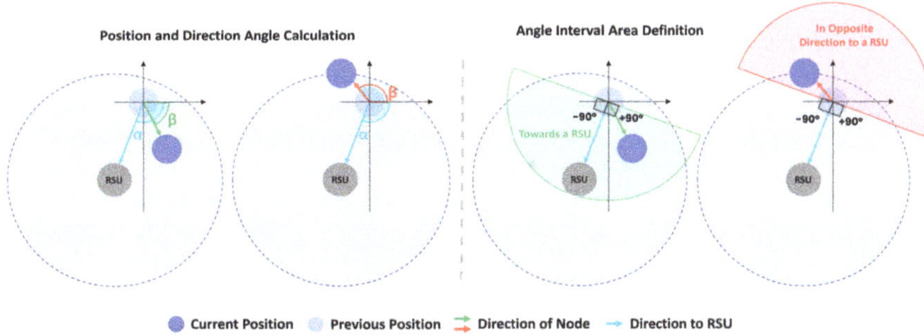

Figure 5. Node heading calculation: two distinct situations.

Algorithm 1 Neighbourhood status procedure.

```
procedure NEIGHFINDER
    number ← 0                                              ▷ neighbour count
    type ← []                                               ▷ list of node mobility types
    rssis ← []                                              ▷ list of RSSIs
    positions ← []                                          ▷ list of positions
    for each device do
        if device is TypeAdHoc then
            Obtain WifiNetDevice of device
            Obtain nodePosition to WifiNetDevice            ▷ Obtain node position
            positions(device).insert(nodePosition)          ▷ insert position on list
            Obtain phy of WifiNetDevice
            receiverMobility ← phy → GetMobility()
            Obtain RSSI to receiverMobility                 ▷ Obtain RSSI
            rssis(device).insert(RSSI)                      ▷ insert RSSI on list
            nodeType ← WifiNetDevice → GetMobilityType()    ▷ Obtain node type
(mobile or static)
            type(device).insert(nodeType)                   ▷ insert type on list
            number ← number + 1                             ▷ neighbour count increment
        end if
    end for
    Updates number, rssis, and type to Forwarder
end procedure
```

4.2. Energy Consumption Module

Energy management in communication networks is closely linked to the technology used. In Wi-Fi, a node can be in one of three general states: in the *Active Mode* state when a node transmits, receives and is listening for content, in the *Power-Save Mode* state when a node is asleep, with no content being transmitted or received, or in the *Off* state when the node is off [31]. During Active Mode, there are three ways in which power is consumed: in Transmitter/Tx mode, in Receiver/Rx mode or in Listening/Idle mode. On the other hand, during the Power-Save Mode state, the node will be sleeping, without any communication, and it will only consume the energy defined for that state (*Sleep Mode*). Finally, in the *Off Mode* state the node is disconnected, interrupting communication and not having any energy consumption. For each of the states defined above, it is important to define how much energy is consumed by each of them. For Wi-Fi, the following values have been considered: Sleep = 0.00132 A, Transmit/Tx = 0.167 A, Receive/Rx = 0.310 A, Range = 50 m and Frequency = 2.4–5.0 GHz.

The battery level is an important factor to consider in the decision to forward a packet, because in a device with limited portable battery, it is important to guarantee a minimum amount of battery. Whenever there is an update from the power state, the battery level (%) is updated in the Forwarder module of the respective node. In this way, the Forwarder uses this value in the forwarding decision and checks if it exceeds the minimum value.

An important part of any energy saving strategy is how to evaluate situations where the node can or cannot communicate. Situations where the mobile node is isolated or unable to communicate with its neighbours are energy saving opportunities by putting them to sleep. In this scheme, a node is put into sleep mode in two situations. When there are no neighbours, mobile or static, it sleeps for 1 s. After that, it wakes up and checks the channel for neighbourhood. If the same condition remains, it goes back to sleep another second, and so on. The other situation is when a node is within the range of at least one RSU, is heading in the opposite direction to it and exceeds a certain RSSI threshold.

Our proposal defines the RSSI threshold as a function of the distance between the mobile node and the static node [28,29]. The idea is to put the mobile node into sleep when it is expected to leave the wireless range of the static node in the next second. In addition, the sleep time will not be one second, but the amount of time needed by the mobile node to leave the range of the static node. Then, the sleep time T is determined by d_l/d_s, where d_l is the remaining distance until it reaches the limit of the wireless coverage of the static node and d_s is the expected distance travelled in the next second, and is related with the current average speed (Figure 6). Whenever a node finds itself in this condition (within range of at least one static node with low RSSI) but motion is stopped, the node cancels any sleep mode scheduled that it may have and schedules a new one for the new distance to the edge of the communication range. On the other hand, if the node changes direction, it just cancels any sleep mode scheduled. Regarding the transition from sleep mode to "wake up" it can happen in two situations: every second, periodically, and when it has urgent content to transmit.

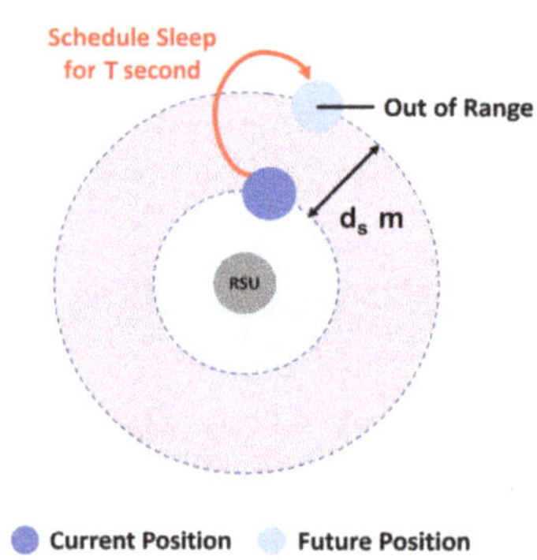

Figure 6. Illustration of the sleeping situation of a mobile node based.

4.3. Forwarding and Content Discovery Module

This module comprises the entire process regarding packet forwarding. Content is discovered through Interest packets, which contain the prefixes within the network domain of the content to be fetched, and this content is retrieved in the form of a Data packet. To make this module suitable for efficient node power management, we added content priority and refined the entire caching and forwarding strategy to make the module energy friendly.

4.3.1. Content Priority

In order to study the degree of importance/urgency of a requested content, they have been divided according to [32], specifying six domains with specific weights:

- **Emergency**. These are the most urgent applications that usually follow a pushed-based approach to be spread across all nodes in the network, being limited in time, such as an accident or dangerous weather condition. It is time constrained, and we give it the value 6.
- **Decision**. These are the applications where requests are important in decision making, being also time limited, such as traffic management. It is time constrained, and we give it the value 5.
- **Information (Stream)**. This includes applications that send and receive data that are limited in time, such as a video stream that should not have latency. It is time constrained, and we give it the value 4.
- **Feedback**. These are the applications that work as data notification, not being limited in time, as is the case of casual meteorological values. It is hybrid, and we give it the value 3.
- **Interaction**. These are the applications connected to the communication between nodes for the exchange of states. It has no time restriction, and we assign the value 2.
- **Information (Data)**. These are the applications that send and receive data which are not limited in time, for instance, the exchange of a file. It has no time restriction, and we assign the value 1.

The weight assigned to each domain distinguishes the degree of importance of a given content base. These values are used in the forwarding decision, along with the different neighbourhood scenarios and the internal energy state of a node. In the content forward decision, values equal to or greater than 4 are considered as priority.

4.3.2. Energy-Efficient Forwarding

All the aforementioned factors and metrics provide a set of information about the conditions around a node that serves as a heuristic to decide about the forwarding of a given packet. This process is driven by the Forwarding Strategy, through the assessment of the context, energy, neighbourhood, node properties and data structures. The workflow for this decision has been implemented in the same way for both Interest and Data packets and can be seen, in a general and hierarchical way, in Figure 7.

Every time a node has a packet to transmit, it starts by checking whether it is in *Sleep Mode*. If so, it cancels all schedulers that may have been created to wake up and wakes up. Otherwise, it moves to the next phase responsible for updating the neighbouring information. In the case of not having any neighbour, it enters in sleep mode and does not transmit the content. In the case of having a neighbour, and the connection is considered to be stable, the node checks the packet priority and it is transmitted only if the packet weight is higher than 4. If the connection is considered to be unstable, it checks if it is facing a mobile-only neighbourhood or with a single static node. If there are only mobile neighbours, the node forwards everything.

On the other hand, if the node has at least one RSU as a neighbour, the position of the RSU and its direction of movement regarding the mobile node is computer. If the node is stopped (direction 0), it cancels all schedulers that may have been created to sleep and sets a new scheduler to fall asleep after a time T, calculated as explained before. In addition, the packet weight is checked and if it is considered priority, the packet is forwarded.

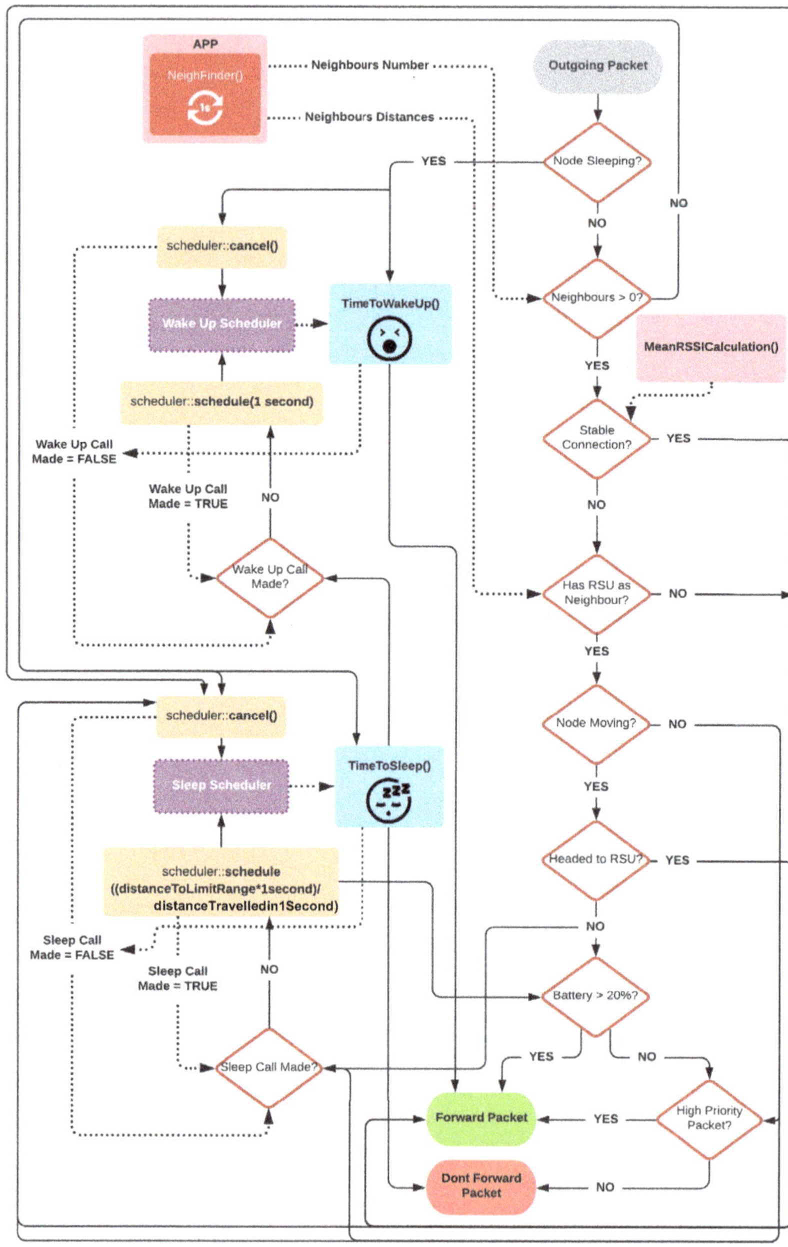

Figure 7. Packet forwarding decision process.

However, if the node is moving, it is checked if the direction of the node is in the direction to the RSU or in the opposite direction. If it is in the direction of the RSU, the node cancels all schedulers that may have been created to sleep and sends the packet. If the node is going in the opposite direction, it checks if any scheduler to fall asleep was made. If not, it sets a scheduler to fall asleep after a time T and it checks the remaining battery level. If it is greater than or equal to the predefined threshold of 20%, the packet is forwarded. Otherwise, the packet weight is checked and the packet is only forwarded if it is considered to be priority.

Finally, our solution mainly focuses on predicting transition times between static nodes (RSUs) and dead zones to put some nodes to sleep, thus reducing unnecessary packet forwarding, and listening times, resulting in energy savings. Not only that, but we also use the internal power state of each node to decide in which situations it is most feasible and efficient to send certain packets. This makes each node autonomous and the entire decision making on energy consumption happens in a decentralised way. It is up to the mobile node to decide what to do with the information at its disposal, without the need for any centralised consultation or consultation supported by the static part of the network. Moreover, it is important to highlight that, despite the breadth of innovations added to the NDN's base architecture, its paradigm has not been changed, i.e., the control structures (PIT, FIB and CS) have not been changed or new types of packets have been created.

5. Performance Evaluation

To bring our evaluations one step closer to a real scenario, we use two real traces of urban mobility, one with boat mobility gathered from a lagoon of the city of Aveiro and the other gathered from the vehicular network in operation in the city of Porto, both of them cities of Portugal. For both scenarios, the Data providers/producers are the static nodes (RSUs) and the Data consumers are the mobile nodes (OBUs). Each OBU has only one 802.11n (Wi-Fi) interface, while each RSU has two communication interfaces, an 802.11n (Wi-Fi) interface and a point-to-point connection (Ethernet) with the backend Routers. All IEEE 802.11n interfaces have been configured to have a range of approximately 50 metres, considering that they are in an urban environment, and use the same transmission rate for each packet sent, following a constant data transmission rate of 54 Mbps, while the point-to-point communication interfaces have a data transmission rate of 1 Gbps and a delay of 1 ms. All interfaces of all nodes have a cost of 1 that remains fixed until the end of the simulation.

The Aveiro scenario is composed by the mobility of nine tourist boats (named moliceiros, and simply denoted as OBUs) through the lagoon of the "Ria de Aveiro". On each route, the boats can head off in different directions, which means that the contact time between them is short and intermittent. These OBUs make contact with four RSUs installed along the boats route and two backend routers to create the connection between all RSUs. These mobility data were collected from 10:35 a.m. to 17:56 p.m. on 23 February 2018. The boats may request Urgent contents, such as information about accidents, adverse weather conditions, among others, or Non Urgent contents, such as Points of Interest (POI) information, traffic, among others.

The Porto scenario uses datasets collected by a vehicular network placed in the city [3]. The dataset with mobility traces is composed by 80 mobile nodes (buses—OBUs) that have circulated during four hours, from 9:00 a.m. to 13:00 p.m., on 23 January 2018. Besides the 80 mobile nodes, this scenario is also composed of a set of 26 RSUs spread throughout the city and 4 backend routers that connect the RSUs. The parameters of both scenarios are represented in Table 1.

The platform used for the evaluation of the proposed forwarding strategy was the ndnSIM [26] software, an Open-Source Simulator Platform extended from the well-established NS-3 research-oriented network simulator. To simulate the energy consumed by each of the PHY states, we defined 90.000 Joules for each node, which is equivalent to a battery of 5.000 mAh with a 5 V output. Different energy levels (% of battery) were defined for each OBU at the beginning of the simulation to evaluate the implemented forwarding strategy that depends on a threshold that delimits the low battery, and the battery levels were placed in the nine Consumers (mobile nodes) of the Aveiro scenario in the following order: 90%, 10%, 55%, 14%, 32%, 88%, 23%, 45% and 76%. For 80 Consumers (mobile nodes) of the Porto scenario, we have used the following battery pattern (first 10 nodes and repeats the cycle for the remaining ones): 90%, 56%, 35%, 55%, 19%, 32%, 88%, 20%, 45% and 76%.

Table 1. Parameters used in the Aveiro and Porto mobile scenarios.

Parameter	Aveiro	Porto
Number of nodes	15	100
OBUs	9	80
RSUs	4	26
Backend Routers (BR)	2	4
Content size	1024 bytes	1024 bytes
Cache size	20 OBUs, 60 RSUs, 100 BR	20 OBUs, 60 RSUs, 100 BR
Cache eviction policy	LRU	LRU
Forwarding strategy	Best-Route Modified	Best-Route Modified
Propagation delay model	ConstantSpeedPropagation	ConstantSpeedPropagation
Propagation loss model	RangePropagationLoss	RangePropagationLoss
Simulation time	14,400 s	14,400 s

In order to evaluate the proposed forwarding strategy, four key metrics were considered [33]: satisfied Interest rate, transmission delay and network overhead (through In/Out Interests and In/Out Data). In order to evaluate the energy saving of the network, we considered the total energy consumed in each PHY state. In order to have a better reference regarding the performance of our solution, called Efficient version, we compared our results with the results obtained by another NDN implementation (Integrated version [32]) that has similar characteristics but without any energy management.

The configuration of each network element is described in Table 2. It was assumed that mobile nodes have a limited cache size; thus, many cache replacements will occur throughout the simulation. As the OBUs are network elements that have limited resources, due to their mobility capability, they will obviously have a limited battery capacity and with that also a lower caching capacity compared to the static network elements. Finally, OBUs were separated into Consumers and intermediate nodes.

Table 2. Network element's configuration.

	OBUs		RSUs	Routers
Node Type	Consumer	Intermediate	Producer	Backend
Installed Technologies	Wi-Fi 802.11n		Wi-Fi 802.11n Ethernet	Ethernet
Mobility Type	Mobile		Static	
Cache Size	20	20	60	100
Aveiro's Devices	9	0	4	2
Porto's Devices	40	40	26	4

The Consumer nodes will request all the available content related to each domain, both Urgent (U) and Non Urgent (NU), with a frequency of one Interest per second, to ensure that the requests are being constant and equal for all metrics. The content requests popularity follows a Zipf–Mandelbrot Distribution [34]. The Producer nodes will provide 5 chunks for each content in the domains of Emergency, Decision, Feedback and Interaction, and 10 chunks for each content in the domain of Information, which is the content domain that comprises most of the communications within the Internet. All these contents have freshness values increasing in relation to the domains. About 900 s for content within the Emergency domain, 1800 s for Decision content, 3000 for Feedback and Interaction contents and 3600 for Information contents. Contents that have a higher update rate and that are volatile have their freshness reduced so that they are not kept as long in Content Stores. For example, longer freshness for contents within the Information domain (about 3600 s, equivalent to one hour which is a quarter of the simulation time), because this type of content is not so volatile and therefore it can stay longer in the Content Stores.

5.1. Satisfied Interests Rate

In the Aveiro scenario, illustrated in Figure 8a, it can be observed that our implementation brought an increase of about 13% in the satisfaction of Interests in comparison with the Integrated version, either in urgent and non urgent packets. The difference comes from the fact that our implementation does not forward non urgent packets in some specific situations; hence, there is an increase for urgent packets and a decrease in the non-urgent packets. As a node goes into sleep mode whenever the surrounding environment justifies it, it is expected that there is a lower number of expired Interests, due to reverse path partitioning or the broadcast storm problem that happens in this highly mobile environment where Data follows the reverse path.

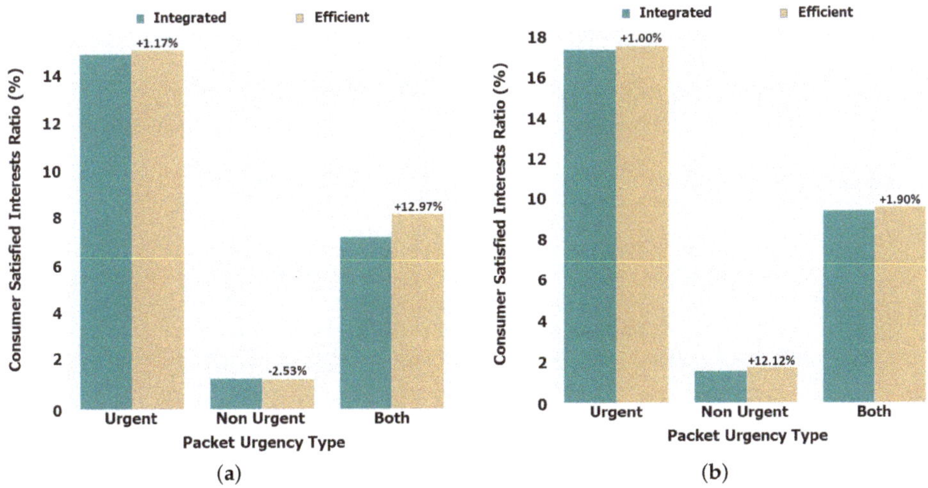

Figure 8. Satisfied Interest Rate (Urgent and Non Urgent). (**a**) Aveiro scenario. (**b**) Porto scenario.

In the Porto scenario, illustrated in Figure 8b, our proposal had an increase of about 2% in the satisfaction ratio of Interests in comparison with the Integrated version, either in urgent and non urgent packets. The difference in the number of satisfied Interests between this use case and the Aveiro's scenario (about 13%) is explained by the fact that, in the Porto scenario, as there is composed by buses moving at high speed compared to the boats, there is a higher mobility and consequently more intermittent disconnections between the nodes. That is, the contact between the nodes is faster and in a shorter time, while the slower boats have a longer connection between the nodes.

5.2. Transmission Delay

Figure 9 depicts the delay between the transmission of an Interest and the reception of the respective Data (boat scenario), during a period of 14,400 s divided into 34 min (2057 s) periods.

Our proposal shows, for most of the time, an increase in performance, showing slightly lower performance only in the first 85 min (2.5 interval) in relation to the Integrated version. The results obtained reveal that the content is closer to the Consumers since a Leave-Copy-Everywhere approach is followed in the cache placement. To complement this, the number of Interests circulating through the network is smaller, which justifies the fact that the content is closer to the Consumers. The fact that the nodes enter in Sleep mode when they are in adverse conditions contributes to this, thus reducing the number of packets transmissions that most likely would be expired and lost in the network.

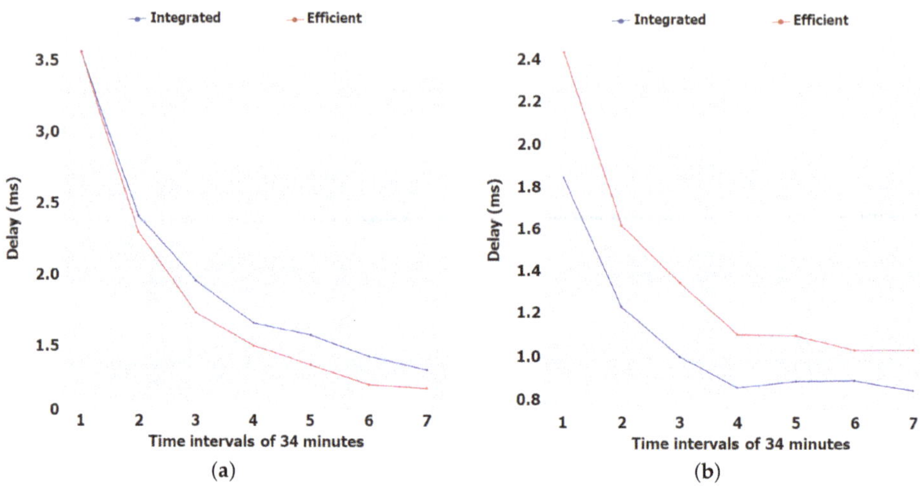

Figure 9. Transmission delay. (**a**) Aveiro scenario. (**b**) Porto scenario.

On the other hand, the delay observed in the Porto scenario (Figure 9b) presents a higher delay in our version because as the nodes are in a high mobility environment, most of them will be constantly faced with adverse situations that imply immediate falling asleep for a short period of time (1 s). During this period, a node that possibly in the Integrated version would be awake and could be part of the return route of data to a particular Consumer, would no longer receive packets, interrupting the of data, or Interests to their destinations. This factor leads that the full delay and last delay present higher values with our solution.

5.3. Network Overhead

Through Figure 10, we can evaluate the network overhead for the Porto scenario. It is possible to see that the overall number of Interests flowing through the network has increased by about 2% for In Interests (Integrated = 7,277,690, Efficient = 7,395,351) and reduced by about 50% for Out Interests (Integrated = 1,547,405, Efficient = 779,320) with the new implementation. The increase in the Interests received is due to the fact that there is more mobility in this scenario, being the contact between the nodes for the exchange of information reduced, so there will be packets that will be lost in the network or that will expire, originating retransmissions to obtain the content. On the other hand, the transmitted Interests are reduced by the reduction in the time in Idle Mode with the sleeping of the nodes and the fact that there are more nodes in the neighbourhood. Since sending only one packet through broadcast for the entire neighbourhood, all nodes will receive the same packet increasing the In Interest number, so there is an increase in In Interest and a reduction in Out Interests.

Regarding the Data packets, illustrated in Figure 11, it can be observed that although there are fewer Interest packets being transmitted due to the short contact and intermittent disconnections between nodes, there are more Data packets moving around the network. This attests that the Interests that were transmitted were successfully satisfied. In the Integrated version, the number of Interests satisfied is less, meaning that they could expire even before receiving Data, which resulted in discarding Data packets and consequently reducing their transmission by other nodes. Our version shows an increase in Data packets circulating in the network by 3.3% for In Data (Integrated = 34,073, Efficient = 35,188) and by 2% for Out Data (Integrated = 664,074, Efficient = 676,766).

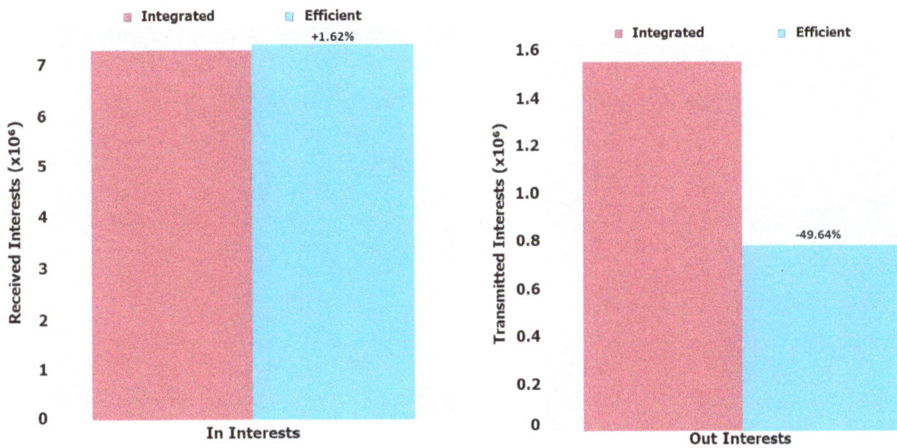

(a) (b)

Figure 10. Global received and transmitted Interests packets for Porto scenario. (**a**) In Interests. (**b**) Out Interests.

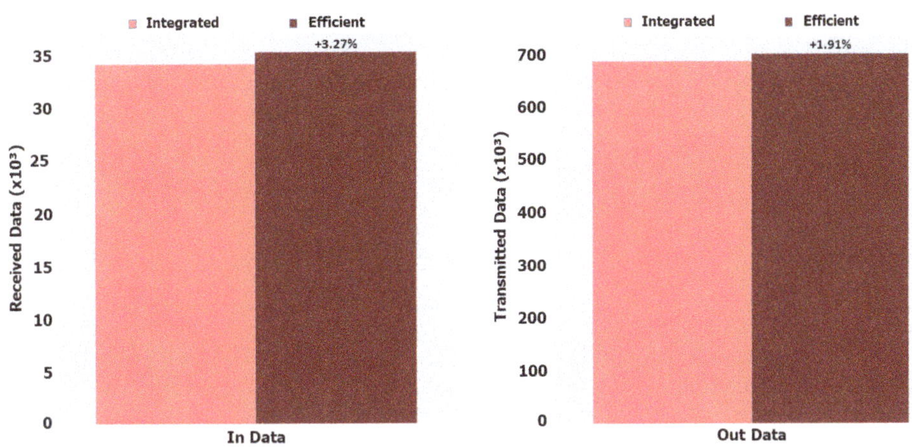

(a) (b)

Figure 11. Global received and transmitted Data packets for Porto scenario. (**a**) In Data. (**b**) Out Data.

In the Aveiro scenario, the total number of Interests flowing through the network has been reduced by about 0.02% for internal Interests and about 38% for external Interests with our implementation. This is mainly due to the fact that a node reduces its time in idle mode and goes to sleep when its internal state and the environment demands it, producing and receiving fewer Interest packets than in the Integrated version. In the Integrated version, the number of Interests satisfied is less, meaning that they could expire even before receiving Data, which resulted in discarding Data packets and consequently reducing their transmission by other nodes. The Efficient version shows an increase in Data packets circulating in the network by 0.90% for both In Data and Out Data.

Despite having a greater delay in response to the Data, the few Interests that are sent are satisfied by Data, attesting that our solution has better performance in terms of not forwarding unnecessary packets and reducing idle times, which will be reflected in the times in each state and energy saving.

5.4. Energy Consumption

The results presented in Figure 12a show a reduction of around 63% (Aveiro scenario) of the energy and time spent in the transmission of packets (TX Mode) in the Efficient version (Integrated TX = 587, Efficient TX = 215). Considering the increase in energy and time spent in the reception of packets (RX Mode), it can be concluded that the large reduction in energy and time spent in both idle Mode and TX Mode impacted only the packets that would be lost in the network because they could never reach the nodes, ensuring an almost accurate transmission and reception of the packets that matter and that actually have a chance of reaching their destination.

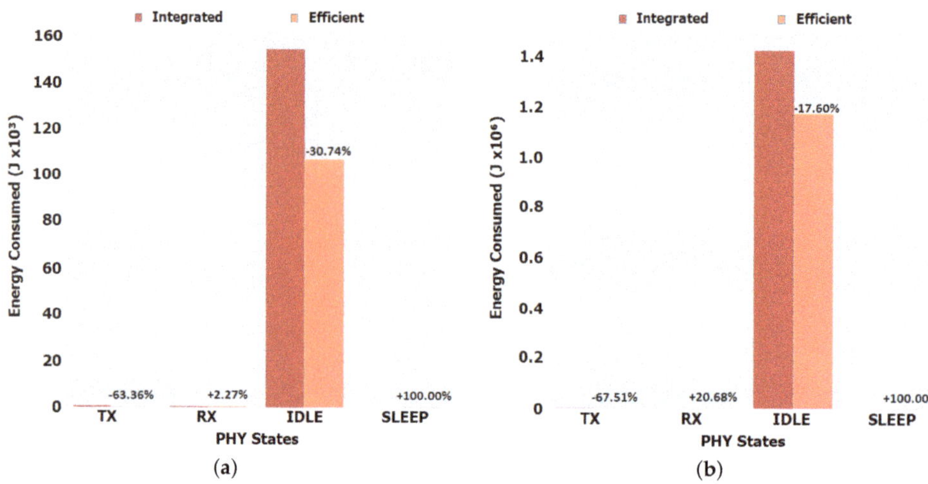

Figure 12. Total energy consumed in each PHY state. (**a**) Aveiro scenario. (**b**) Porto scenario.

The increase in the time in receiving mode (Rx) is due to the fact that there are fewer packets circulating in the network, meaning that the increase in the reception energy consumed is for the Data packets that successfully reach the Consumer. A reduction of between 17% (Porto scenario—Integrated IDLE = 1,420,309, Efficient IDLE = 1,170,393) and 30% (Aveiro scenario—Integrated IDLE = 154,149, Efficient IDLE = 106,768) in the energy consumption performed in idle mode is also visible, as the time spent in this state was converted into consumption in Sleep Mode.

With the Efficient implementation, a node goes to sleep mode depending on its internal energy state (battery level) and the environment it is in. This reflects in a reduction in the number of transmissions and receptions. However, a node wakes up whenever it produces an Interest request to be sent to the network, waking up the node for that purpose. The reduction in transmission energy between the Integrated and Efficient version is directly related to the reduction in the number of Interests circulating in the network, because the content is closer to the Consumers.

Regarding the network lifespan, which is due to the energy savings considered by the forwarding strategy and the sleep and wake mechanisms, there is more energy distributed to the mobile nodes of the network in our Efficient version. Our results show a 1.3% increase for the Aveiro scenario (boats) and a 1.5% increase for the Porto scenario (buses). These savings translate into an additional 25 min of network activity for the Aveiro scenario and 4 h and 40 min for the Porto scenario.

6. Conclusions

In this paper, we are proposing a forwarding strategy for energy-constrained wireless ICN networks. Our solution considers the mobility of the nodes, which directly impacts the forwarding decision of a packet and checks the environment around the node, which includes the number and type of neighbours, signal strength and direction of movement. Furthermore, it combines this information with the internal state of each node, both in its internal energy level and in state transitions between awake and sleep. The proposed forwarding strategy extends and adapts concepts of ICNs, by means of packet domain analysis, neighbourhood evaluation and node sleeping and waking strategies so that there is energy saving and non-use of resources in unnecessary situations, such as the moments when a node is listening when it is isolated. In addition, our proposal guarantees the privacy and security of information that ends up not being shared, such as their neighbourhood and their energetic state. We also ensure that packets are prioritised with a major impact on how the routing decision and sleeping strategy is made.

The evaluation of our solution was performed by simulation with real traces of urban mobility in two distinct scenarios: boats in the lagoon of Aveiro and public buses in the city of Porto, both in Portugal. These scenarios offer different conditions. Boats sailing in low speed represent less mobility and a longer contact time between the several nodes, while buses circulate at a higher speed, leading to a shorter time of contact between the several nodes due to intermittent and faster disconnections. Our tests considered the dissemination of Urgent and Non Urgent packets to evaluate the satisfied Interest rate, transmission delay, network overhead and energy consumption, important metrics used in ICN/NDN performance network evaluation.

Our results have shown that the proposed forwarding strategy resulted in significant improvements in network content availability, overall node power savings and increased network lifetime. In addition, we have shown that it was possible to save energy that was wasted listening and forwarding packets unnecessarily, generating a longer lifetime distributed by the network nodes, which was more than 25 min in the case of Aveiro and it was almost 5 h of network activity in the case of Porto.

As future work, we intend to expand our solution by adding mobile node speed assessment, handling for multiple communication interfaces and a dynamic low battery threshold among other improvements.

Author Contributions: All authors designed the solution, analysed the results and wrote the paper. D.M. implemented the solution, prepared the evaluation scenarios and the results. C.S. and M.L. supervised the entire research process. All authors have read and agreed to the published version of the manuscript.

Funding: This work is supported by the European Regional Development Fund (FEDER), through the Competitiveness and Internationalization Operational Programme (COMPETE 2020) of the Portugal 2020 Regional Operational Program of Lisbon (FEDER), and Foundation for Science and Technology, project InfoCent-IoT (POCI-01-0145-FEDER-030433).

Institutional Review Board Statement: Not applicable.

Informed Consent Statement: Not applicable.

Conflicts of Interest: The authors declare no conflict of interest.

References

1. Tariq, A.; Rehman, R.A.; Kim, B. Forwarding Strategies in NDN-Based Wireless Networks: A Survey. *IEEE Commun. Surv. Tutor.* **2020**, *22*, 68–95. [CrossRef]
2. Jacobson, V.; Smetters, D.K.; Thornton, J.D.; Plass, M.; Briggs, N.; Braynard, R. Networking Named Content. *Commun. ACM* **2012**, *55*, 117–124. [CrossRef]
3. Santos, P.M.; Rodrigues, J.G.P.; Cruz, S.B.; Lourenço, T.; d'Orey, P.M.; Luis, Y.; Rocha, C.; Sousa, S.; Crisóstomo, S.; Queirós, C.; et al. PortoLivingLab: an IoT-based Sensing Platform for Smart Cities. *IEEE Internet Things J.* **2018**, *5*, 523–532. [CrossRef]
4. Li, S.; Kim, J.G.; Han, D.H.; Lee, K.S. A Survey of Energy-Efficient Communication Protocols with QoS Guarantees in Wireless Multimedia Sensor Networks. *Sensors* **2019**, *19*, 199. [CrossRef] [PubMed]

5. Elhoseny, M.; Shankar, K., Energy Efficient Optimal Routing for Communication in VANETs via Clustering Model. In *Emerging Technologies for Connected Internet of Vehicles and Intelligent Transportation System Networks: Emerging Technologies for Connected and Smart Vehicles;* Elhoseny, M., Hassanien, A.E., Eds.; Springer International Publishing: Cham, Switzerland, 2020; pp. 1–14. [CrossRef]
6. Memon, I.; Hasan, M.K.; Shaikh, R.A.; Nebhen, J.; Bakar, K.A.A.; Hossain, E.; Tunio, M.H. Energy-Efficient Fuzzy Management System for Internet of Things Connected Vehicular Ad Hoc Networks. *Electronics* **2021**, *10*, 1068. [CrossRef]
7. Ahmed, S.H.; Bouk, S.H.; Kim, D. RUFS: RobUst Forwarder Selection in Vehicular Content-Centric Networks. *IEEE Commun. Lett.* **2015**, *19*, 1616–1619. [CrossRef]
8. Grassi, G.; Pesavento, D.; Pau, G.; Zhang, L.; Fdida, S. Navigo: Interest forwarding by geolocations in vehicular Named Data Networking. In Proceedings of the 2015 IEEE 16th International Symposium on A World of Wireless, Mobile and Multimedia Networks (WoWMoM), Boston, MA, USA, 14–17 June 2015; pp. 1–10. [CrossRef]
9. de Sousa, A.M.; Araújo, F.R.; Sampaio, L.N. A Link-Stability-Based Interest-Forwarding Strategy For Vehicular Named Data Networks. *IEEE Internet Comput.* **2018**, *22*, 16–26. [CrossRef]
10. Yu, Y.T.; Dilmaghani, R.B.; Calo, S.; Sanadidi, M.Y.; Gerla, M. Interest propagation in named data manets. In Proceedings of the 2013 International Conference on Computing, Networking and Communications (ICNC), San Diego, CA, USA, 28–31 January 2013; pp. 1118–1122. [CrossRef]
11. Jiang, X.; Gao, S.; Zhang, H.; Zhang, B. Energy Efficient Interest Forwarding in NDN-Based Wireless Sensor Networks. *Mob. Inf. Syst.* **2016** , *2016*, 3127029. [CrossRef]
12. Aboud, A.; Touati, H.; Hnich, B. Efficient forwarding strategy in a NDN-based internet of things. *Clust. Comput.* **2019**, *22* 805–818. [CrossRef]
13. Wu, F.; Yang, W.; Ren, J.; Lyu, F.; Yang, P.; Zhang, Y.; Shen, X. Cutting Down Idle Listening Time: A NDN-Enabled Power Saving Mode Design for WLAN. In Proceedings of the ICC 2019 IEEE International Conference on Communications (ICC), Shanghai, China, 22–24 May 2019; pp. 1–6. [CrossRef]
14. Wu, Y.; He, Y.; Shi, L. Energy-Saving Measurement in LoRaWAN-Based Wireless Sensor Networks by Using Compressed Sensing. *IEEE Access* **2020**, *8*, 49477–49486. [CrossRef]
15. Yang, Y.; Song, T. Energy-Efficient Cooperative Caching for Information-Centric Wireless Sensor Networking. *IEEE Internet Things J.* **2022**, *9*, 846–857. [CrossRef]
16. Wu, F.; Yang, W.; Ren, J.; Lyu, F.; Yang, P.; Zhang, Y.; Shen, X. Named Data Networking Enabled Power Saving Mode Design for WLAN. *IEEE Trans. Veh. Technol.* **2020**, *69*, 901–913. [CrossRef]
17. Benedetti, P.; Piro, G.; Grieco, L.A. An Energy Efficient and Software-Defined Information-Centric Networking Approach to Consumer Mobility. In Proceedings of the 2020 22nd International Conference on Transparent Optical Networks (ICTON), Bari, Italy, 19–23 July 2020; pp. 1–4. [CrossRef]
18. Tariq, A.; Rehman, R.A.; Kim, B.S. EPF—An Efficient Forwarding Mechanism in SDN Controller Enabled Named Data IoTs. *Appl. Sci.* **2020**, *10*, 7675. [CrossRef]
19. Gameiro, L.M.M. Context-Based Forwarding for Mobile ICNs. Master's Thesis, University of Aveiro, Aveiro, Portugal, 2019.
20. Mannes, E.; Maziero, C. Naming Content on the Network Layer: A Security Analysis of the Information-Centric Network Model. *ACM Comput. Surv.* **2019**, *52*, 1–28. [CrossRef]
21. Zhang, L.; Afanasyev, A.; Burke, J.; Jacobson, V.; Crowley, P.; Papadopoulos, C.; Wang, L.; Zhang, B. Named Data Networking. *SIGCOMM Comput. Commun. Rev.* **2014**, *52*, 66–73. [CrossRef]
22. Farrell, S.; Kutscher, D.; Dannewitz, C.; Ohlman, B.; Keranen, A.; Hallam-baker, P. *Naming Things with Hashes;* IETF: Wilmington, DE, USA, 2013; RFC 6920.
23. Bouk, S.; Ahmed, S.H.; Kim, D. Hierarchical and Hash based Naming with Compact Trie Name Management Scheme for Vehicular Content Centric Networks. *Comput. Commun.* **2015**, *71*, 73–83. [CrossRef]
24. Ud Din, I.; Hassan, S.; Khan, K.; Guizani, M.; Ghazali, O.; Habbal, A. Caching in Information-Centric Networking: Strategies, Challenges, and Future Research Directions. *IEEE Commun. Surv. Tutor.* **2017**, *20*, 1443–1474. [CrossRef]
25. Zhang, L.; Estrin, D.; Jacobson, J.B.V.; Thornton, J.D.; Smetters, D.K.; Zhang, B.; Tsudik, G.; Claffy, K.; Krioukov, D.; Massey, D.; et al. *Named Data Networking (NDN) Project, NDN-0001;* Technical Report; Named Data Networking (NDN): Los Angeles, CA, USA, 2010.
26. Mastorakis, S.; Afanasyev, A.; Zhang, L. On the Evolution of ndnSIM: An Open-Source Simulator for NDN Experimentation. *ACM Comput. Commun. Rev.* **2017**, *47*, 19–33. [CrossRef]
27. Gravogl, K.; Haase, J.; Grimm, C. Choosing the best wireless protocol for typical applications. In Proceedings of the 24th International Conference on Architecture of Computing System 2011 (ARCS 2011), Como, Italy, 22–23 February 2011.
28. Cardote, A.; Neves, F.; Sargento, S.; Steenkiste, P. A statistical channel model for realistic simulation in VANET. In Proceedings of the 2012 IEEE Vehicular Networking Conference (VNC), Seoul, Korea, 14–16 November 2012; pp. 48–55. [CrossRef]
29. Yokoyama, R.S.; Kimura, B.Y.L.; Villas, L.A.; Moreira, E.D.S. Measuring Distances with RSSI from Vehicular Short-Range Communications. In Proceedings of the 2015 IEEE International Conference on Computer and Information Technology; Ubiquitous Computing and Communications; Dependable, Autonomic and Secure Computing; Pervasive Intelligence and Computing, Liverpool, UK, 26–28 October 2015; pp. 100–107. [CrossRef]

30. ETSI. *ETSI EN 302 637-2 V1.4.1 (2019-04) Intelligent Transport Systems (ITS); Vehicular Communications; Basic Set of Applications; Part 2: Specification of Cooperative Awareness Basic Service*; white paper; European Telecommunications Standards Institute (ETSI): Sophia Antipolis, France, 2019.
31. Halperin, D.; Greenstein, B.; Sheth, A.; Wetherall, D. Demystifying 802.11n Power Consumption. In Proceedings of the 2010 International Conference on Power Aware Computing and Systems, Vancouver, BC, Canada, 6–8 October 2010.
32. Gameiro, L.; Senna, C.; Luís, M. Context-based Forwarding for Mobile ICNs, In Proceedings of the 2020 IEEE International Conference on Communications (ICC), Dublin, Ireland, 7–11 June 2020. [CrossRef]
33. Ahed, K.; Benamar, M.; Lahcen, A.A.; Ouazzani, R.E. Forwarding strategies in vehicular named data networks: A survey. *JKSU-Comput. Inf. Sci.* 2020, *in press*. [CrossRef]
34. Edward, G.; Coffman, J.; Denning, P.J. *Operating Systems Theory*; Prentice Hall: Hoboken, NJ, USA, 1973; p. 331.

sensors

MDPI

Article

Joint User Scheduling and Hybrid Beamforming Design for Massive MIMO LEO Satellite Multigroup Multicast Communication Systems

Yang Liu [1], Changqing Li [2,*], Jiong Li [2] and Lu Feng [1]

[1] Graduate School, Space Engineering University, Beijing 101416, China
[2] Space Information School, Space Engineering University, Beijing 101416, China
* Correspondence: lcqqcl5577@sohu.com

Abstract: In the satellite multigroup multicast communication systems based on the DVB-S2X standard, due to the limitation of the DVB-S2X frame structure, user scheduling and beamforming design have become the focus of academic research. In this work, we take the massive multi-input multi-output (MIMO) low earth orbit (LEO) satellite communication system adopting the DVB-S2X standard as the research scenario, and the LEO satellite adopts a uniform planar array (UPA) based on the fully connected hybrid structure. We focus on the coupling design of user scheduling and beamforming; meanwhile, the scheme design takes the influence of residual Doppler shift and phase disturbance on channel errors into account. Under the constraints of total transmission power and quality of service (QoS), we study the robust joint user scheduling and hybrid beamforming design aimed at maximizing the energy efficiency (EE). For this problem, we first adopt the hierarchical clustering algorithm to group users. Then, the semidefinite programming (SDP) algorithm and the concave convex process (CCCP) framework are applied to tackle the optimization of user scheduling and hybrid beamforming design. To handle the rank-one matrix constraint, the penalty iteration algorithm is proposed. To balance the performance and complexity of the algorithm, the user preselected step is added before joint design. Finally, to obtain the digital beamforming matrix and the analog beamforming matrix in a hybrid beamformer, the alternative optimization algorithm based on the majorization-minimization framework (MM-AltOpt) is proposed. Numerical simulation results show that the EE of the proposed joint user scheduling and beamforming design algorithm is higher than that of the traditional decoupling design algorithms.

Keywords: LEO satellite communications; massive MIMO; multigroup multicast; user scheduling; hybrid beamforming; robust; joint design; energy efficiency

Citation: Liu, Y.; Li, C.; Li, J.; Feng, L. Joint User Scheduling and Hybrid Beamforming Design for Massive MIMO LEO Satellite Multigroup Multicast Communication Systems. *Sensors* **2022**, *22*, 6858. https://doi.org/10.3390/s22186858

Academic Editor: Josip Lorincz

Received: 10 August 2022
Accepted: 7 September 2022
Published: 10 September 2022

1. Introduction

In recent years, LEO satellite communication systems have played an increasingly important role in wireless communication networks [1]. However, facing the high-performance requirements of future wireless communication systems, such as higher spectrum efficiency (SE) and increased EE, the performance of LEO satellite communication systems needs to be improved [2]. Applying the massive MIMO technology to LEO satellites is a good choice, and with the advantage of 5G technology [3] and the spatial multiplexing principle of MIMO technology [4], the performance of LEO satellite communication systems can be further improved. Meanwhile, by using the high-precision multiple beams generated by the massive MIMO technology and aggressive full frequency reuse scheme among beams, the performance of the communication system can be greatly improved. However, the full frequency reuse scheme will cause severe inter-beam interference [5], and adopting the beamforming design at the LEO transmitter side can efficiently manage it. In addition, the super-frame structure of multibeam satellite communications standards such as DVB-S2X [6] needs to apply the same beamformer to multiple users that share the same frame.

Therefore, the multigroup multicasting principle can be used in the beamformer design. In this paper, we concentrate on the massive MIMO LEO satellite communication system forward link multigroup multicasting beamforming scheme [7].

In the beamforming design of the massive MIMO LEO satellite multigroup multicast communication system, the following issues need to be considered:

- User scheduling: Due to that only a few users can be bound into a DVB-S2X frame, and there are a large number of active users in each multicast group, it is necessary to design the user scheduling algorithm. Meanwhile, it should be noted that the interference between scheduled users depends on the beamforming design, which in turn depends on the scheduled users in other beams. Therefore, user scheduling and beamforming design are coupled, and the joint design scheme of user scheduling and beamforming needs to be considered.
- Channel errors: The LEO satellite has high orbital speed, which will produce a large Doppler shift and result in the channel phase deviation [8]. Meanwhile, the factors such as distortion of high-frequency devices, expiration of the CSI and large propagation delay also can cause the channel phase disturbance. Therefore, it is difficult to obtain accurate channel state information (CSI) at the LEO satellite transmitter. Due to the existence of CSI errors, the designed beamforming vector does not match the actual CSI, resulting in the reduction in the receiving gain and signal to interference plus noise ratio (SINR) of the user terminal. Then, the QoS will not be guaranteed. Thus, it is of practical significance to study the robust user scheduling and beamforming design.
- EE optimization: Due to the limited energy load of LEO satellites, to prolong the service life of the LEO satellite and improve the stability of the LEO satellite communications system, under the consideration of green communications and economic benefits, we need to pay attention to the EE optimization [9].
- Beamforming scheme: In the beamforming architectures of the massive MIMO technology, although the digital beamforming design can significantly improve the SE, it would bring high hardware complexity and high power consumption. Although the hardware overhead of hybrid beamforming architecture based on full connection is slightly higher than that of the partial connection architecture, it can balance the hardware complexity and system performance, and has higher cost performance. Therefore, in this paper, we selected the hybrid beamforming technology based on the full connection structure [10].

2. Related Works and Main Contributions

2.1. Related Works

There is extensive literature regarding user scheduling and beamforming design in the wireless multigroup multicast communication system. In Ref. [11], based on the perfect CSI, taking the throughput maximization as the optimization objective the authors studied the precoding design of the multibeam satellite communication system, and proposed a decoupling scheme of the user scheduling and beamforming design. In Ref. [12], considering the influence of CSI errors and taking the minimizing transmission power as the optimization objective, the robust multigroup multicast transmission scheme of the multibeam satellite communication system was investigated, and the low complexity beamforming algorithm and the user grouping algorithm were proposed, but the length limit of the DVB-S2X frame was not considered. In Ref. [13], based on the perfect CSI and taking the maximizing SE as the optimization objective, the multigroup multicast transmission design scheme of the frame-based multibeam satellite communication system was investigated, and the authors proposed a joint design scheme of the user scheduling and beamforming. However, the influence of CSI errors was not considered, and the user grouping algorithm was simple. In Ref. [14], based on the perfect CSI, the authors studied the user scheduling problem of the multicast transmission in the high-throughput satellite communication system. The user scheduling was decoupled into intra-beam scheduling and inter-beam scheduling, and the correlation degree was calculated by using the equivalent CSI; therefore, the interaction of intra-beam and inter-beam scheduling cannot be fully considered. In Ref. [15], the authors studied the user scheduling problem of the multibeam satellite communication system but did not consider the inter-beam interference caused by scheduling. In Ref. [16], based on the perfect CSI, the authors studied the joint scheduling and beamforming design problem for multiuser MISO downlink, and the message-based user grouping and scheduling algorithm was mainly proposed, but the impact of user grouping on system performance was not fully considered. In Ref. [17], the user scheduling and hybrid beamforming design of the massive MIMO orthogonal frequency division multiple access (OFDMA) communication system was studied, in scheme design. First, a joint design algorithm of user scheduling and analog beamforming was proposed; then, the digital beamforming matrix was solved by the weighted minimum mean-square error (WMMSE) algorithm. In Ref. [18], the authors studied the design of user scheduling and subcarrier allocation in the downlink of a massive MIMO OFDMA communication system and proposed a hybrid beamforming scheme. First, based on the optimal solution of digital beamforming, the analog beamforming matrix was obtained by a singular value decomposition algorithm. Then, the authors proposed an algorithm to solve the digital beamforming matrix and its corresponding scheduling users.

Most of the above research took the geosynchronous earth orbit (GEO) satellite or the terrestrial cellular network as the research object, and less used the LEO satellite. In addition, the optimization objectives mainly focused on maximizing SE and minimizing transmission power, and less on the EE optimization. Meanwhile, the analysis of CSI errors was insufficient, as some research considered the CSI errors, but did not analyze the influence of the Doppler shift. In terms of the user scheduling, the common idea was to adopt the decoupling scheme of user scheduling and beamforming design, without fully considering the coupling relationship between the user scheduling and the beamforming design.

2.2. Main Contributions

Inspired by the above research, we focus on the downlink transmission design of the massive MIMO LEO satellite multigroup multicast communication system. In scheme design, comprehensively considering the influence of CSI errors caused by the residual Doppler shift and the phase disturbance, we mainly investigate the robust joint user scheduling and hybrid beamforming design to maximize the system EE. Meanwhile, we take the constraints of the transmission power and QoS into account. The main works are summarized as follows:

- We establish the downlink transmission system model and channel model of the massive MIMO LEO satellite multigroup multicast communication system and analyze the CSI errors.
- Based on the CSI, we adopt a low complexity hierarchical clustering algorithm based on the *Ward* connection method to group users, which can lay a foundation for the joint user scheduling and beamforming design.
- We establish the joint user scheduling and hybrid beamforming design problem model based on EE maximization, and binary variables are defined to represent whether the user is scheduled or not. Then, we transform the optimization problem into a Boolean fractional programming (BFP) problem, which is also a quadratic constraint quadratic programming (QCQP) form problem.
- For the BFP problem in QCQP form, we invoke the quadratic transformation algorithm to handle the fractional programming form problem in the objective function. Meanwhile, the SDP algorithm is invoked to convert the objective function in QCQP form into a concave function, and some nonconvex constraints can be converted into linear constraints. In addition, we adopt the relaxation and penalty algorithm to deal with the Boolean constraint. Then, the optimization problem is equivalently transformed into a difference of convex (DC) programming problem.
- For the DC programming problem, an iterative optimization algorithm based on the CCCP framework is proposed. For the rank-one matrix constraint introduced by the SDP algorithm, a penalty iterative algorithm is adopted.

- For the solution of the digital beamforming matrix and the analog beamforming matrix in the hybrid beamformer, the MM-AltOpt algorithm is proposed.

3. System Model and Problem Formulation

3.1. System Model

As shown in Figure 1, we focus on the downlink of the massive MIMO LEO satellite multigroup multicast communication system, and the LEO satellite uses a UPA, which is composed of $N = N_x \times N_y$ antennas and $L(L \leq N)$ RF links and covers L multicast groups and K active users, where $K \geq L, K \geq N$. Let the multicast group covered by the lth beam be U_l and the number of users in this multicast group be $|U_l|$, assuming that the number of users that can be accommodated in each DVB-S2X frame is U_s. It should be noted that each user terminal belongs to only one multicast group, i.e., $U_i \cap U_j = \emptyset, \forall i, j \in \{1, \ldots, L\}, i \neq j$. Meanwhile, we assume that the user terminal is equipped with a single antenna capable of data stream demodulation. According to the DVB-S2X standard, in a transmission slot, multiple users' data in a multicast group are multiplexed into a specific forward error correction (FEC) codeword to provide services for more users. The service process is shown in Figure 2.

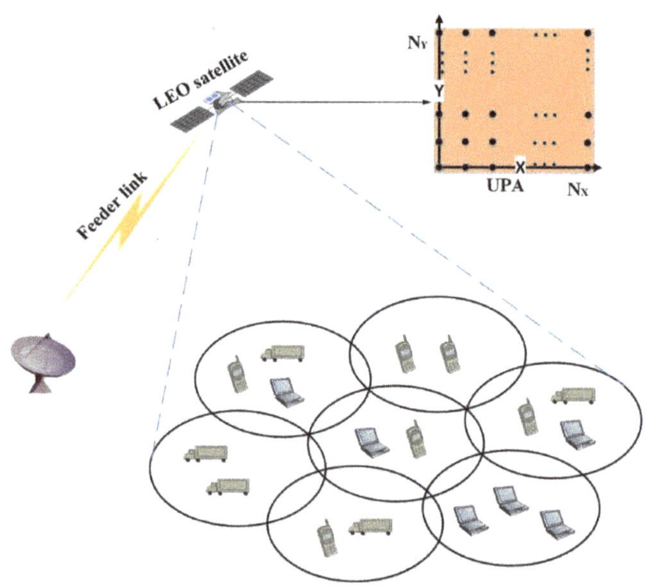

Figure 1. Transmission model of the massive MIMO LEO satellite multigroup multicast communication system.

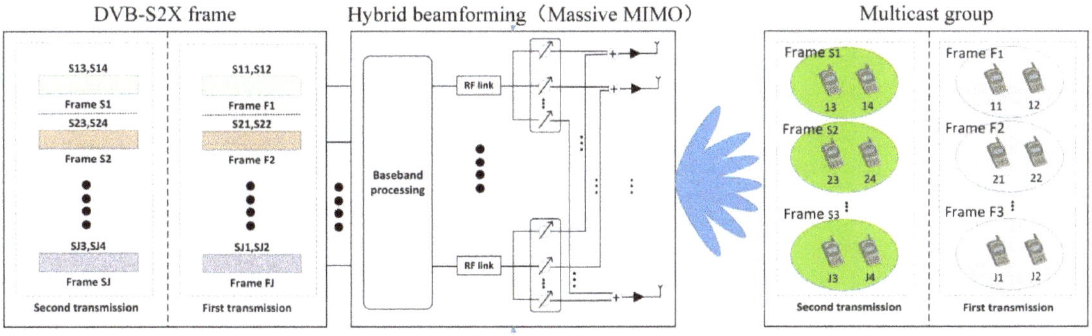

Figure 2. Service process of the LEO communication satellite multigroup multicast transmission based on the DVB-S2X.

The received signal $y_{k,l}$ of the kth user in the lth multicast group can be expressed as:

$$y_{k,l} = \boldsymbol{h}_{k,l}^H \boldsymbol{F}_{RF}\boldsymbol{F}_{BB}[:,l]s_l + \sum_{j=1,j\neq l}^{L} \boldsymbol{h}_{k,j}^H \boldsymbol{F}_{RF}\boldsymbol{F}_{BB}[:,j]s_l + n_{k,l}, k \in \{1,\ldots,|U_l|\}, l \in \{1,\ldots,L\}, \quad (1)$$

where the first term in (1) represents the expected received signal of the kth user in the lth multicast group, the second term represents the interference of other multicast groups and the third term represents the additive Gaussian white noise; $\boldsymbol{h}_{k,l} \in \mathbb{C}^{N\times 1}$ represents the channel vector of the kth user in the lth multicast group, $\boldsymbol{F}_{BB} \in \mathbb{C}^{L\times L}$ represents the digital beamforming matrix, $\boldsymbol{F}_{BB}[:,l] \in \mathbb{C}^{L\times 1}$ represents the digital beamforming vector of the lth multicast group, and $\boldsymbol{F}_{RF} \in \mathbb{C}^{N\times L}$ represents the analog beamforming matrix, where each element of $\boldsymbol{F}_{RF}$ should meet the unit modulus element [19], i.e., $\left|(\boldsymbol{F}_{RF})_{i,j}\right| = 1$. In addition, s_l represents the signal of the multicast group U_l, which meets the unit power constraint, i.e., $E\left\{|s_l|^2\right\} = 1$, $n_l \sim CN(0,\sigma^2)$ represents the additive Gaussian white noise, which is related to the Boltzmann constant κ, system bandwidth B and the noise temperature T.

For the convenience of analysis, we set $\boldsymbol{F} \in \mathbb{C}^{N\times L} = \boldsymbol{F}_{RF}\boldsymbol{F}_{BB} = [\boldsymbol{f}_1, \boldsymbol{f}_2, \ldots, \boldsymbol{f}_L]$ as the hybrid beamforming matrix, and $\boldsymbol{f}_l \in \mathbb{C}^{N\times 1} = \boldsymbol{F}_{RF}\boldsymbol{F}_{BB}[:,l]$ is the hybrid beamforming vector of the lth multicast group. Therefore, (1) can be rewritten as:

$$y_{k,l} = \boldsymbol{h}_{k,l}^H \boldsymbol{f}_l s_l + \sum_{j=1,j\neq l}^{L} \boldsymbol{h}_{k,j}^H \boldsymbol{f}_j s_l + n_{k,l}, k \in \{1,\ldots,|U_l|\}, l \in \{1,\ldots,L\} \quad (2)$$

Due to the high orbital speed of LEO satellites and the long transmission delay, it is difficult to obtain the precise instantaneous CSI. To cope with this problem, we adopt the statistical CSI, and the channel vector between the LEO satellite and the kth user in the lth multicast group at instant t and frequency f can be modeled as follows [20]:

$$\boldsymbol{h}_{k,l}(t,f) = \sum_{p=1}^{P_{k,l}} a_{k,l,p} e^{j2\pi(f_{d(k,l,p)}t - f\tau_{k,l,p})} \times \boldsymbol{V}_{k,l,p}, \quad (3)$$

where f denotes the carrier frequency, $a_{k,l,p}, f_{d(k,l,p)}, \tau_{k,l,p}$ are the complex channel gain, Doppler shift and propagation delay, respectively, $P_{k,l}$ denotes the number of propagation paths and $\boldsymbol{V}_{k,l,p} \in \mathbb{C}^{N\times 1}$ is the UPA array response vector, which can be given by

$$\boldsymbol{V}_{k,l,p} = \boldsymbol{V}(\varphi_{k,l,p}^x, \varphi_{k,l,p}^y) = \boldsymbol{v}_{N_x}\left(\sin\varphi_{k,l,p}^y \cos\varphi_{k,l,p}^x\right) \otimes \boldsymbol{v}_{N_y}\left(\cos\varphi_{k,l,p}^y\right), \quad (4)$$

$$\boldsymbol{v}_{N_x}(\sin\varphi^y_{k,l,p}\cos\varphi^x_{k,l,p}) = \frac{1}{\sqrt{N_x}}\left(1, e^{-j\frac{2\pi d}{\lambda}\sin\varphi^y_{k,l,p}\cos\varphi^x_{k,l,p}}, \ldots, e^{-j\frac{2\pi d}{\lambda}(N_x-1)\sin\varphi^y_{k,l,p}\cos\varphi^x_{k,l,p}}\right), \tag{5}$$

$$\boldsymbol{v}_{N_y}(\cos\varphi^y_{k,l,p}) = \frac{1}{\sqrt{N_y}}\left(1, e^{-j\frac{2\pi d}{\lambda}\cos\varphi^y_{k,l,p}}, \ldots, e^{-j\frac{2\pi d}{\lambda}(N_y-1)\cos\varphi^y_{k,l,p}}\right), \tag{6}$$

where $\varphi^x_{k,l,p}$, $\varphi^y_{k,l,p}$ represent azimuth angle and pitch angle associated with the propagation path p of the kth user in the lth multicast group, respectively, λ denotes the wavelength and d represents the spacing of antenna elements, the value of which is usually $\lambda/2$ [5].

Note that the LEO satellite communication system is usually operated under the line of sight (LOS) transmission, and the channel vector can be modeled using the widely accepted Rician distribution model as follows:

$$\boldsymbol{h}_{k,l} = \overline{\boldsymbol{h}}_{k,l} + \widetilde{\boldsymbol{h}}_{k,l}, \tag{7}$$

where $\overline{\boldsymbol{h}}_{k,l} = \sqrt{\frac{\kappa_{k,l}\gamma_{k,l}}{\kappa_{k,l}+1}} \times \boldsymbol{V}_{k,l}$ represents the LoS component, $\widetilde{\boldsymbol{h}}_{k,l} = \sqrt{\frac{\gamma_{k,l}}{\kappa_{k,l}+1}} \times \boldsymbol{V}_{k,l,c} \times \boldsymbol{V}^H_{k,l}$ represents the multipath component, $\kappa_{k,l}$ denotes the Rician factor, $\boldsymbol{V}_{k,l,c} \in \mathbb{C}^{N_{ut}\times 1} \sim \mathbb{CN}(0, \sum)$ represents Rician component, $Tr(\sum) = 1$ and $\gamma_{k,l}$ represents the average channel power, which mainly includes the transmit antenna gain G_{leo}, receiver antenna gain G_{ut} and link power loss. The link power loss is mainly caused by the free space path loss LP_{fs} and the atmospheric absorption loss LP_{at}. Therefore, the average channel power $\gamma_{k,l}$ can be written as

$$\gamma_{k,l} = G_{leo}[\text{dB}] + G_{ut}[\text{dB}] - LP_{at}[\text{dB}] - LP_{fs}[\text{dB}], \tag{8}$$

where LP_{fs} can be given by $LP_{fs} = 20(\log_{10}(D_{k,l})) + \log_{10}(f) + \log_{10}(4\pi/c)$, c is the speed of light, $D_{k,l}$ represents the transmission distance, LP_{at} is related to the carrier frequency, temperature $T(h)$, pressure $P(h)$ and humidity $\rho(h)$, which can be given by $LP_{at} = \int_{h_{ut}}^{h_{at}} LP_{at}(f, T(h), P(h), \rho(h))dh$, $LP_{at}(f, T(h), P(h), \rho(h))$ is the loss per meter, h_{ut} is the user's height and h_{at} is the atmosphere thickness. The specific calculation method of LP_{at} can be found in the literature [21].

For the convenience of analysis, it is assumed that the parameters in the channel vector $\boldsymbol{h}_{k,l}$ are constant within coherence time and change over time in a certain ergodic process. In (3), the Doppler shift $f_{d(k,l,p)}$ and the propagation delay $\tau_{k,l,p}$ usually cause CSI errors. Next, we focus on analyzing the influence of propagation delay and Doppler shift on CSI errors.

Doppler shift: In the LEO satellite communication systems, the Doppler shift is usually large, which is mainly composed of the Doppler shift $f^{leo}_{d(k,l,p)}$ generated by the LEO satellite motion and the Doppler shift $f^{ut}_{d(k,l,p)}$ generated by the users' motion [22]. Since the transmission between the LEO satellite and user terminals is mainly under LOS, $f^{leo}_{d(k,l,p)}$ of different transmission paths can be considered to be the same, and we omit the path index of $f^{leo}_{d(k,l,p)}$, i.e., $\left\{f^{leo}_{d(k,l,p)}\right\}_1^{P_k} = f^{leo}_{d(k,l)}$; $f^{leo}_{d(k,l)}$ can be calculated using the LEO satellite ephemeris information and the location information of user terminals, as shown in Figure 3.

$$f^{leo}_{d(k,l)} = -\frac{f}{c} \times \frac{\omega^{leo} r_e r \sin(\phi_t - \phi_{t_0})\mu(\theta_{\max})}{\sqrt{r_e^2 + r^2 - 2r_e r\cos(\phi_t - \phi_{t_0})\mu(\theta_{\max})}}, \tag{9}$$

where $\mu(\theta_{\max}) = \cos\left[\cos^{-1}\left(\frac{r_e}{r}\cos\theta_{\max}\right) - \theta_{\max}\right]$, r_e denotes the earth radius, r represents the distance between the LEO satellite track point and the earth center, $(\phi_t - \phi_{t_0})$ represents the angular distance of the earth's surface along the LEO satellite trajectory from instant t to instant t_0, ω represents the angular velocity of the LEO satellite and c represents the speed of light.

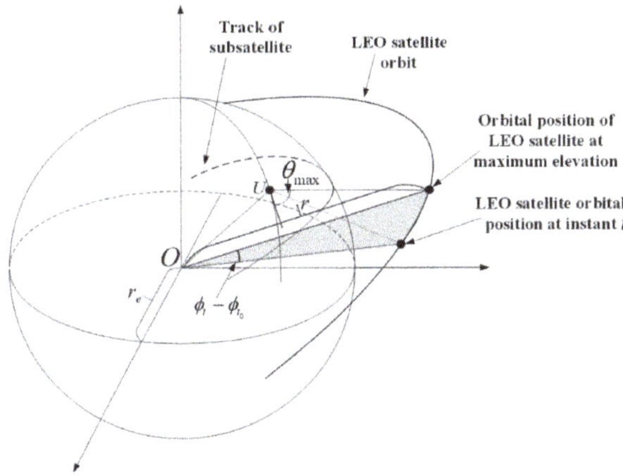

Figure 3. Geometric diagram of LEO satellite's orbital motion relative to the user terminal.

The value of $f^{ut}_{d(k,l,p)}$ in different transmission paths is different, which is mainly caused by the movement of user terminals and surrounding scatterers. The power spectrum of $f^{ut}_{d(k,l,p)}$ follows the *Jakes* power spectrum model, and the normalized power spectrum can be expressed as:

$$S(f^{ut}_{d(k,l,p)}) = \frac{1}{2\pi f^{ut}_{d(k,l,p),\max}\sqrt{1-(\frac{f^{ut}_{d(k,l,p)}}{f^{ut}_{d(k,l,p),\max}})^2}} \tag{10}$$

The large Doppler shift in LEO satellite communication systems can make it difficult to receive correctly and result in the degradation of communication performance. In application, to mitigate the impact of Doppler shift, the solution of estimation and compensation is usually adopted [23]. The Doppler shift estimation mainly includes two steps: coarse estimation and fine estimation, which can refer to the literature [24]. When we obtain the estimated value of Doppler shift, f^e_d, the Doppler shift compensation of size $f^{cps}_d = f^e_d$ can be implemented at the receivers. It should be noted that due to that the Doppler shift changes rapidly, in addition to the low SINR at the receivers and the limited pilot length in the DVB-S2X frame, the Doppler shift estimation is usually inaccurate, which can result in the incomplete compensation. Then, there would be the residual Doppler shift f^{rsd}_d, which can cause the sliding of channel phase. According to the Doppler shift estimation theory based the Cramer–Rao bound, the variance in the Doppler shift estimation can be expressed as the Cramer–Rao lower bound (CRLB) [25], i.e.,

$$\sigma^2_{f^e_d} = CRLB(f) = \frac{1}{\mathrm{SNR}}\frac{3}{2\pi^2T^2N(N^2-1)}, \tag{11}$$

where N is the pilot length, T is the sampling time and SNR is the signal-to-noise ratio. According to the properties of variance, after Doppler shift compensation of size f^{cps}_d, the variance of residual Doppler shift f^{rsd}_d is equal to the variance of f^e_d, i.e.,

$$\sigma^2_{f^{rsd}_d} = \sigma^2_{f^e_d} = \frac{1}{\mathrm{SNR}}\frac{3}{2\pi^2T^2N(N^2-1)}, \tag{12}$$

The influence of residual Doppler shift on channel phase errors can be expressed as $\phi_{f^{rsd}_d} = 2\pi f^{rsd}_d \Delta T$, where ΔT represents the sum of the downlink propagation delay and the duration of the DVB-S2X frame. Therefore, the variance of $\phi_{f^{rsd}_d}$ can be expressed as $\sigma^2_{\phi_{f^{rsd}_d}} = 4\pi^2\sigma^2_{f^{rsd}_d}\Delta T^2 = \frac{1}{SNR}\frac{6\Delta T^2}{T^2N(N^2-1)}$, and $\phi_{f^{rsd}_d}$ follows a real-valued Gaussian distribution with mean zero and variance $\sigma^2_{\phi_{f_d^{rsd}}}$, i.e., $\phi_{f_d^{rsd}} \sim N\left(0, \sigma^2_{\phi_{f_d^{rsd}}}\right)$. The channel error vector caused by $\phi_{f_d^{rsd}}$ can be expressed as $v_{\phi_{f_d^{rsd}}} = \left[e^{j\phi_{f_d^{rsd},1}}, e^{j\phi_{f_d^{rsd},2}}, \ldots, e^{j\phi_{f_d^{rsd},N}}\right]^T$.

Propagation Delay: The orbital height of LEO satellites is about 300 km to 2000 km, and the long transmission distance can cause the larger propagation delay. Note that the influence of atmosphere on propagation delay mainly includes ionospheric delay and tropospheric delay [26]. When the signal passes through the ionosphere, due to the refraction effect of the electromagnetic wave, the propagation path and speed of the signal will change. Meanwhile, the ionospheric delay is irregular, which is difficult to describe with a physical model. When the signal passes through the troposphere, the propagation speed, direction and path of the signal will change, which can result in propagation delay. The tropospheric delay is related to air pressure, air humidity and satellite elevation. The commonly used tropospheric delay correction model is given in the literature [27,28]. The round-trip delay of the LEO satellite with an orbit altitude of 1200 km is about 20 ms. The long propagation delay will lead to the expiration of CSI, which can result in CSI errors, phase disturbance and other problems. To handle this problem, the delay compensation of size $\tau^{cps} = \beta\tau_{\min} + (1-\beta)\tau_{\max}$, $(0 \leq \beta \leq 1)$ is usually implemented at the receivers, and $\tau_{k,l}^{\min} = \min\left\{\tau_{k,l,p}\right\}_1^{P_{k,l}}$ and $\tau_{k,l}^{\max} = \max\left\{\tau_{k,l,p}\right\}_1^{P_{k,l}}$ represent the minimum propagation delay and the maximum propagation delay of the kth user in the lth multicast group, respectively.

However, due to that the atmospheric propagation delay is irregular, the transmission delay cannot be fully compensated. Therefore, the incomplete delay compensation, expired CSI and distortion of high-frequency devices would cause the channel phase disturbance [29]. Let the phase disturbance be ϕ_τ, which follows a real-valued Gaussian distribution with mean zero and variance $\sigma^2_{\phi_\tau}$, i.e., $\phi_\tau \sim N\left(0, \sigma^2_{\phi_\tau}\right)$. The channel error vector caused by ϕ_τ can be expressed as $v_{\phi_\tau} = \left[e^{j\phi_{\tau,1}}, e^{j\phi_{\tau 2}}, \ldots, e^{j\phi_{\tau,N}}\right]^T$.

In conclusion, considering the influence of the residual Doppler shift and the phase disturbance, the relationship between the real channel vector $\boldsymbol{h}_{k,l}$ and the estimated channel vector $\hat{\boldsymbol{h}}_{k,l}$ can be expressed as:

$$\boldsymbol{h}_{k,l} = \hat{\boldsymbol{h}}_{k,l} \odot \boldsymbol{v}_{\phi_{f_{d,k,l}^{rsd}}} \odot \boldsymbol{v}_{\phi_{\tau,k,l}} = diag\left(diag(\hat{\boldsymbol{h}}_{k,l})\boldsymbol{v}_{\phi_{f_{d,k,l}^{rsd}}}\right)\boldsymbol{v}_{\phi_{\tau_{k,l}}}, \tag{13}$$

where $\odot$ represents the Hadamard product. Let the channel phase of the kth user in the lth multicast group be $\theta_{k,l} = [\theta_{k,l,1}, \theta_{k,l,2}, \ldots, \theta_{k,l,N}]^T$, which satisfies the uniform distribution between $0 \sim 2\pi$. Then, the real channel phase with phase errors at instant t_1 is as follows:

$$\theta_{k,l}(t_1) = \theta_{k,l}(t_0) + \phi_{f_{d,k,l}^{rsd}} + \phi_{\tau_{k,l}} \tag{14}$$

3.2. Problem Formulation

3.2.1. User Clustering

Before the joint user scheduling and hybrid beamforming design, it is necessary to group the active users within the coverage of the LEO satellite. Based on the CSI, the hierarchical clustering algorithm is adopted to group users [30]. As shown in Figures 4 and 5, the hierarchical clustering algorithm adopts the bottom-up method, where each user initially forms a group, and then according to the similarity measurement function, the user groups which meet the similarity threshold constraint are combined until the desired number of groups is formed.

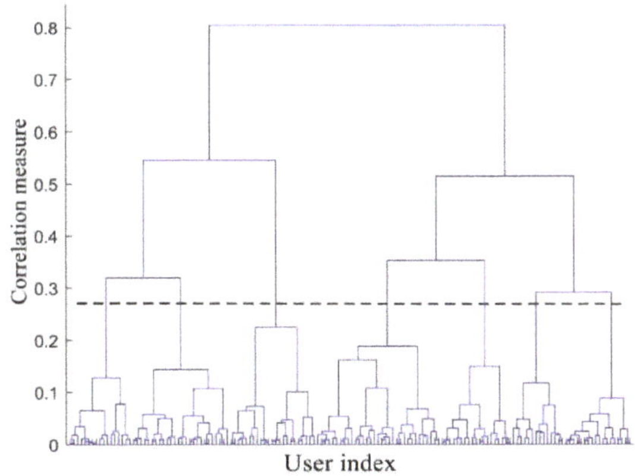

Figure 4. Schematic diagram of the hierarchical clustering algorithm.

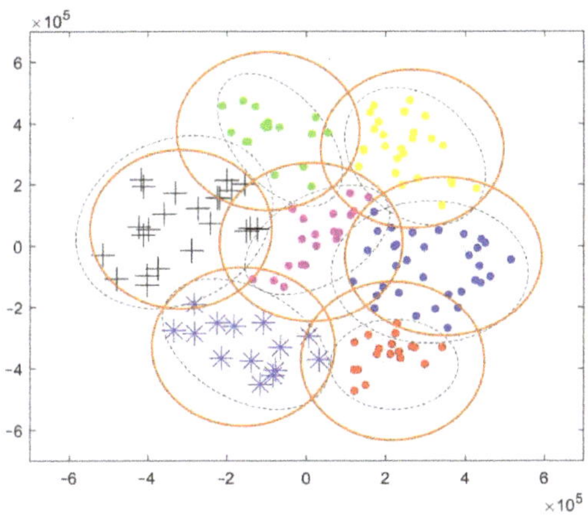

Figure 5. Schematic diagram of multibeam coverage with 7 multicast groups.

We adopt the similarity measurement function among multicast groups based on the *Ward* connection method, i.e.,

$$d(i,j) = \sqrt{\frac{2n_i n_j}{n_i + n_j}} dist(\boldsymbol{h}_i^{eq}, \boldsymbol{h}_j^{eq}), \tag{15}$$

where n_i, n_j represent the number of users of the group i and the group j, respectively, $\boldsymbol{h}_i^{eq}$, $\boldsymbol{h}_j^{eq}$ represent the equivalent CSI of the group i and the group j, respectively, and $dist(\boldsymbol{h}_i^{eq}, \boldsymbol{h}_j^{eq})$ represents the Euclidean distance between the vector $\boldsymbol{h}_i^{eq}$ and the vector $\boldsymbol{h}_i^{eq}$, i.e.,

$$dist(\boldsymbol{h}_i^{eq}, \boldsymbol{h}_j^{eq}) = \left\| \frac{\boldsymbol{h}_i^{eq}}{\left\| \boldsymbol{h}_i^{eq} \right\|} - \frac{\boldsymbol{h}_j^{eq}}{\left\| \boldsymbol{h}_j^{eq} \right\|} \right\| \tag{16}$$

3.2.2. System Rate

Affected by CSI errors, both the ergodic communication rate and the ergodic SINR do not admit explicit expressions. To handle this challenge, the statistical average method is adopted to model the SINR and the communication rate. Therefore, the SINR and the communication rate of the kth user in the lth multicast group can be expressed as:

$$\mathrm{E}\{\mathrm{SINR}_{k,l}\} \approx \frac{\mathrm{E}\left\{\left|\boldsymbol{h}_{k,l}^{H}\boldsymbol{f}_{l}\right|^{2}\right\}}{\mathrm{E}\left\{\sum\limits_{j=1,j\neq l}^{L}\left|\boldsymbol{h}_{k,l}^{H}\boldsymbol{f}_{j}\right|^{2}\right\}+\sigma^{2}}, \tag{17}$$

$$R_{k,l} \approx B\log_{2}\left(1+\frac{\mathrm{E}\left\{\left|\boldsymbol{h}_{k,l}^{H}\boldsymbol{f}_{l}\right|^{2}\right\}}{\mathrm{E}\left\{\sum\limits_{j=1,j\neq l}^{L}\left|\boldsymbol{h}_{k,l}^{H}\boldsymbol{f}_{j}\right|^{2}\right\}+\sigma^{2}}\right), \tag{18}$$

where B denotes system bandwidth. Equations (17) and (18) are approximations with closed form, the feasibility of which have been discussed in detail in Refs. [31,32].

3.2.3. Problem Description

We take the system EE as the optimization objective, and the EE is defined as the ratio of the system communication rate to the total power consumption, which can be modeled as:

$$\mathrm{EE} = \frac{\sum\limits_{l=1}^{L}\min\left(\{R_{k,l}\}_{k=1}^{|U_l|}\right)}{P_{total}} = \frac{B\sum\limits_{l=1}^{L}\log_{2}\left(1+\min\left\{\frac{\left|\boldsymbol{h}_{k,l}^{H}\boldsymbol{f}_{l}\right|^{2}}{\sum\limits_{j=1,j\neq l}^{L}\left|\boldsymbol{h}_{k,l}^{H}\boldsymbol{f}_{j}\right|^{2}+\sigma^{2}}\right\}_{k=1}^{U_s}\right)}{P_t+P_0}, \tag{19}$$

where P_{total} represents the total power consumption, $P_t = \sum\limits_{l}^{L}\left|f_l f_l^H\right|$ denotes the transmission power of the LEO satellite, and P_0 denotes the inherent power consumption of the communication system.

Let the Boolean variable $\eta_{k,l} \in \{0,1\}$ indicate whether the kth user in the lth multicast group is served, $\eta_{k,l} = 1$ and $\eta_{k,l} = 0$ indicate that the user can be served and not served, respectively, and $\boldsymbol{\eta} = [\boldsymbol{\eta}_1, \boldsymbol{\eta}_2, \ldots, \boldsymbol{\eta}_L]$, $\boldsymbol{\eta}_l = \left[\eta_{1,l}, \eta_{2,l}, \ldots, \eta_{|U_l|,L}\right]^T$. In conclusion, under the constraints of the transmission power and QoS, the problem of maximizing system EE can be modeled as:

$$Q_1: \max_{\eta, f_l, \mathrm{SINR}_l^{\min}} \mathrm{EE} = \frac{B\sum\limits_{l=1}^{L}\log_2\left(1+\mathrm{SINR}_l^{\min}\right)}{\sum\limits_{l}^{L}\left|f_l f_l^H\right|^2+P_0}, \tag{20}$$

$$s.t.\ C_1: \eta_{k,l} \in \{0,1\}, \forall k,l, \tag{21}$$

$$C_2: \mathrm{SINR}_{k,l} \geq \eta_{k,l}\mathrm{SINR}_l^{\min}, \forall k,l, \tag{22}$$

$$C_3: \mathrm{SINR}_l^{\min} \geq \mathrm{SINR}_0, \forall l, \tag{23}$$

$$C_4: \sum_{k=1}^{(U_l)} \eta_{k,l} = U_s, \forall l, \tag{24}$$

$$C_5: \sum_{l=1}^{L}\left|f_l f_l^H\right|^2 \leq P_T, \tag{25}$$

where $\mathrm{SINR}_l^{\min}$ represents the minimum SINR of the lth multicast group and constraint C_3 represents that $\mathrm{SINR}_l^{\min}$ should be greater than the minimum SINR constraint SINR_0. Constraint C_4 limits the number of scheduled users in each multicast group to U_s.

4. Joint User Scheduling and Hybrid Beamforming Design for Maximizing EE

In this section, we focus on the robust joint user scheduling and hybrid beamforming design strategy to maximize system EE. To handle the QCQP form problem and nonconvexity in optimization problem Q_1, the SDP method is applied to make the optimization problem more tractable. Then, we transform the optimization problem Q_1 into a DC programming problem. To address the DC programming problem, we adopt the CCCP algorithm. Finally, a penalty iterative algorithm is adopted to handle the rank-one matrix constraint.

4.1. SDP Algorithm

It is worth noting that the objective function and constraints C_2 and C_5 in the problem Q_1 involve the quadratic form of the variable $\boldsymbol{f}_l$, therefore, Q_1 is the QCQP form problem. To handle this problem, we invoke the SDP algorithm, a new variable $\boldsymbol{W}_l \triangleq \boldsymbol{f}_l\boldsymbol{f}_l^H$ is introduced, and the positive semidefinite matrix $\boldsymbol{W}_l$ needs to meet the constraints of $\boldsymbol{W}_l \succeq 0$ and $rank(\boldsymbol{W}_l) = 1$. Then, the problem Q_1 can be equivalent to:

$$Q_2: \max_{\eta, \boldsymbol{W}_l, \mathrm{SINR}_l^{\min}} \mathrm{EE} = \frac{B\sum\limits_{l=1}^{L}\log_2\left(1+\mathrm{SINR}_l^{\min}\right)}{\sum\limits_{l}^{L} Tr(\boldsymbol{W}_l) + P_0}, \tag{26}$$

$$s.t.\ C_1, C_2, C_3,\ C_4\ in\ Q_1, \tag{27}$$

$$C_5: \sum_{l}^{L} Tr(\boldsymbol{W}_l) \le P_T, \tag{28}$$

$$C_6: \boldsymbol{W}_l \succeq 0,\ \forall l, \tag{29}$$

$$C_7: rank(\boldsymbol{W}_l) = 1,\ \forall l, \tag{30}$$

Similarly, the SINR and the communication rate of the kth user in the lth multicast group can be equivalently converted to:

$$\mathrm{SINR}_{k,l} \approx \frac{\mathrm{E}\{Tr(\boldsymbol{H}_{k,l}\boldsymbol{W}_l)\}}{\mathrm{E}\left\{\sum\limits_{j=1,j\neq l}^{L} Tr(\boldsymbol{H}_{k,l}\boldsymbol{W}_j)\right\} + \sigma^2}, \tag{31}$$

$$\begin{aligned} R_{k,l} &= B\log_2\left(1+\frac{\mathrm{E}\{Tr(\boldsymbol{H}_{k,l}\boldsymbol{W}_l)\}}{\mathrm{E}\left\{\sum\limits_{j=1,j\neq l}^{L} Tr(\boldsymbol{H}_{k,l}\boldsymbol{W}_j)\right\}+\sigma^2}\right) = B\log_2\left(1+\frac{Tr(\mathrm{E}\{\boldsymbol{H}_{k,l}\boldsymbol{W}_l\})}{\sum\limits_{j=1,j\neq l}^{L} Tr(E\{\boldsymbol{H}_{k,l}\boldsymbol{W}_j\})+\sigma^2}\right) \\ &= B\log_2\left(1+\frac{Tr(\overline{\boldsymbol{H}}_{k,l}\boldsymbol{W}_l)}{\sum\limits_{j=1,j\neq l}^{L} Tr(\overline{\boldsymbol{H}}_{k,l}\boldsymbol{W}_j)+\sigma^2}\right) \end{aligned} \tag{32}$$

where $\boldsymbol{H}_{k,l} \in C^{M\times M}$ is the instantaneous channel autocorrelation matrix of the kth user in the lth multicast group and $\overline{\boldsymbol{H}}_{k,l} \in C^{M\times M}$ is the long-term channel autocorrelation matrix. The relationship between the two can be expressed as:

$$\overline{\boldsymbol{H}}_{k,l} = \mathrm{E}\{\boldsymbol{H}_{k,l}\} \triangleq \mathrm{E}\left\{\hat{\boldsymbol{h}}_{k,l}\hat{\boldsymbol{h}}_{k,l}^H\right\} = diag(\hat{\boldsymbol{h}}_{k,l})\boldsymbol{P}_{f_{d,k,l}^{rsd}}\boldsymbol{Q}_{\tau_{k,l}}diag\left(\hat{\boldsymbol{h}}_{k,l}^H\right), \tag{33}$$

where $\boldsymbol{P}_{f_{d,k,l}^{rsd}}$ and $\boldsymbol{Q}_{\tau_{k,l}}$ can be expressed as follows:

$$\begin{aligned}
P_{f_{d,k,l}^{rsd}} &= E\left\{ v_{\phi_{f_{d,k,l}^{rsd}}} v_{\phi_{f_{d,k,l}^{rsd}}}^{H} \right\} \\
&= \mathrm{E}\left\{ \left[e^{j\phi_{f_{d,k,l}^{rsd},1}}, e^{j\phi_{f_{d,k,l}^{rsd},2}}, \ldots, e^{j\phi_{f_{d,k,l}^{rsd},M}} \right]^{T} \left[e^{-j\phi_{f_{d,k,l}^{rsd},1}}, e^{-j\phi_{f_{d,k,l}^{rsd},2}}, \ldots, e^{-j\phi_{f_{d,k,l}^{rsd},M}} \right] \right\} \\
&= \mathrm{E}\left\{ \begin{array}{ccc} 1 & \cdots & e^{j\phi_{f_{d,k,l}^{rsd},1}} e^{-j\phi_{f_{d,k,l}^{rsd},M}} \\ \vdots & \ddots & \vdots \\ e^{j\phi_{f_{d,k,l}^{rsd},M}} e^{-j\phi_{f_{d,k,l}^{rsd},1}} & \cdots & 1 \end{array} \right\} \\
&= \left\{ \begin{array}{ccc} 1 & \cdots & \mathrm{E}\left\{ e^{j\phi_{f_{d,k,l}^{rsd},1}} e^{-j\phi_{f_{d,k,l}^{rsd},M}} \right\} \\ \vdots & \ddots & \vdots \\ \mathrm{E}\left\{ e^{j\phi_{f_{d,k,l}^{rsd},M}} e^{-j\phi_{f_{d,k,l}^{rsd},1}} \right\} & \cdots & 1 \end{array} \right\}
\end{aligned} \tag{34}$$

In (34), the diagonal elements of $\boldsymbol{P}_{f_{d,k,l}^{rsd}}$ are all 1, and the elements in row i and column j on the non-diagonal are $\mathrm{E}\left\{ e^{j\phi_{f_{d,k,l}^{rsd},i}} e^{-j\phi_{f_{d,k,l}^{rsd},j}} \right\} = \mathrm{E}\left\{ e^{j\phi_{f_{d,k,l}^{rsd},i}} \right\} \mathrm{E}\left\{ e^{-j\phi_{f_{d,k,l}^{rsd},j}} \right\}$, according to $\phi_{f_{d}^{rsd}} \sim N\left(0, \sigma_{\phi_{f_{d}^{rsd}}}^{2}\right)$,

$$\begin{aligned}
\mathrm{E}\left\{ e^{j\phi_{f_{d,k,l}^{rsd},i}} \right\} &= \int_{-\infty}^{\infty} e^{j\phi_{f_{d,k,l}^{rsd},i}} \frac{1}{\sqrt{2\pi}\sigma_{\phi_{f_{d,k,l}^{rsd}}}} e^{-\frac{\phi_{f_{d,k,l}^{rsd}}^{2}}{2\sigma_{\phi_{f_{d}^{rsd}}}^{2}}} d\phi_{f_{d,k,l}^{rsd},i} \\
&= e^{-\frac{\sigma_{f_{d,k,l}^{rsd}}^{2}}{2}} \int_{-\infty}^{\infty} \frac{1}{\sqrt{2\pi}\sigma_{\phi_{f_{d,k,l}^{rsd}}}} e^{-\frac{\left(\phi_{f_{d,k,l}^{rsd},i} - j\sigma_{\phi_{f_{d,k,l}^{rsd}}}\right)^{2}}{2\sigma_{\phi_{f_{d}^{rsd}}}^{2}}} d\phi_{f_{d,k,l}^{rsd},i}
\end{aligned} \tag{35}$$

Similarly, $\mathrm{E}\left\{ e^{-j\phi_{f_{d,k,l}^{rsd},j}} \right\} = e^{-\frac{\sigma_{\phi_{f_{d}^{rsd}}}^{2}}{2}}$. Therefore, $\mathrm{E}\left\{ e^{j\phi_{f_{d,k,l}^{rsd},i}} e^{-j\phi_{f_{d,k,l}^{rsd},j}} \right\} = \mathrm{E}\left\{ e^{j\phi_{f_{d,k,l}^{rsd},i}} \right\} \mathrm{E}\left\{ e^{-j\phi_{f_{d,k,l}^{rsd},j}} \right\} = e^{-\sigma_{\phi_{f_{d}^{rsd}}}^{2}}$.

$$\begin{aligned}
\boldsymbol{Q}_{\tau_{k,l}} &= E\left\{ \boldsymbol{v}_{\phi_{\tau_{k,l}}} \boldsymbol{v}_{\phi_{t_{k,l}}}^{H} \right\} \\
&= \mathrm{E}\left\{ \left[e^{j\phi_{\tau_{k,l},1}}, e^{j\phi_{\tau_{k,l}2}}, \ldots, e^{j\phi_{\tau_{k,l},M}} \right]^{T} \left[e^{-j\phi_{\tau_{k,l},1}}, e^{-j\phi_{\tau_{k,l}2}}, \ldots, e^{-j\phi_{\tau_{k,l},M}} \right] \right\} \\
&= \mathrm{E}\left\{ \begin{array}{ccc} 1 & \cdots & e^{j\phi_{\tau_{k,l},1}} e^{-j\phi_{\tau_{k,l},M}} \\ \vdots & \ddots & \vdots \\ e^{j\phi_{\tau_{k,l},M}} e^{-j\phi_{\tau_{k,l},1}} & \cdots & 1 \end{array} \right\} \\
&= \left\{ \begin{array}{ccc} 1 & \cdots & \mathrm{E}\left\{ e^{j\phi_{\tau_{k,l},1}} e^{-j\phi_{\tau_{k,l},M}} \right\} \\ \vdots & \ddots & \vdots \\ \mathrm{E}\left\{ e^{j\phi_{\tau_{k,l},M}} e^{-j\phi_{\tau_{k,l},1}} \right\} & \cdots & 1 \end{array} \right\}
\end{aligned} \tag{36}$$

In (36), the diagonal elements of $\boldsymbol{Q}_{\tau_{k,l}}$ are all 1, and the elements in row i and column j on the non-diagonal are $\mathrm{E}\left\{e^{j\phi_{\tau_{k,l},i}}e^{-j\phi_{\tau_{k,l},j}}\right\} = \mathrm{E}\left\{e^{j\phi_{\tau_{k,l},i}}\right\}\mathrm{E}\left\{e^{-j\phi_{\tau_{k,l},j}}\right\}$, according to $\phi_\tau \sim N\left(0, \sigma_{\phi_\tau}^2\right)$,

$$\begin{aligned}\mathrm{E}\left\{e^{j\phi_{\tau_{k,l},i}}\right\} &= \int_{-\infty}^{\infty} e^{j\phi_{\tau_{k,l},i}} \frac{1}{\sqrt{2\pi}\sigma_{\phi_{\tau_{k,l}}}} e^{-\frac{\phi_{\tau_{k,l},i}^2}{2\sigma_{\phi_{\tau_{k,l}}}^2}} d\phi_{\tau_{k,l},i} \\ &= e^{-\frac{\sigma_{\phi_{\tau_{k,l}}}^2}{2}} \int_{-\infty}^{\infty} \frac{1}{\sqrt{2\pi}\sigma_{\phi_{\tau_{k,l}}}} e^{-\frac{(\phi_{\tau_{k,l},i}-j\sigma_{\phi_{\tau_{k,l}}})^2}{2\sigma_{\phi_{\tau_{k,l}}}^2}} d\phi_{\tau_{k,l},i} \\ &= e^{-\frac{\sigma_{\phi_{\tau_{k,l}}}^2}{2}}\end{aligned} \tag{37}$$

Similarly, $\mathrm{E}\left\{e^{j\phi_{\tau_{k,l},i}}\right\} = e^{-\frac{\sigma_{\phi_{\tau_{k,l}}}^2}{2}}$. Therefore, $\mathrm{E}\left\{e^{j\phi_{\tau_{k,l},i}}e^{-j\phi_{\tau_{k,l},i}}\right\} = \mathrm{E}\left\{e^{j\phi_{\tau_{k,l},i}}\right\}\mathrm{E}\left\{e^{-j\phi_{\tau_{k,l},i}}\right\}$ $= e^{-\sigma_{\phi_{\tau_{k,l}}}^2}$.

4.2. DC Programming

Since the constraint C_1 is a Boolean constraint and the constraint C_2 is a nonconvex constraint, the problem Q_2 is a nonconvex and nonsmooth combinatorial optimization problem. To handle this challenge, we can transform the problem Q_2 into a DC programming problem [33]. Therefore, the relaxation variable $\zeta_{k,l}$ is introduced as the lower bound of the SINR of the kth user in the lth multicast group, $\zeta = [\zeta_1, \zeta_2, \ldots, \zeta_L]$, $\zeta_l = [\zeta_{l,1}, \zeta_{l,2}, \ldots, \zeta_{l,|U_l|}]^T$. The problem Q_2 can be equivalently converted to:

$$Q_3: \max_{\eta, \boldsymbol{W}, \mathrm{SINR}_l^{\min}, \zeta} \mathrm{EE} = \frac{B\sum_{l=1}^{L}\log_2(1+\mathrm{SINR}_l^{\min})}{\sum_{l=1}^{L} Tr(\boldsymbol{W}_l) + P_0}, \tag{38}$$

$$s.t.\ C_1,\ C_3,\ C_4,\ C_5, C_6, C_7\ in\ Q_1, \tag{39}$$

$$C_2: \mathrm{SINR}_{k,l} \geq \zeta_{k,l}, \forall k, l, \tag{40}$$

$$C_8: \zeta_{k,l} \geq \eta_{k,l}\mathrm{SINR}_l^{\min}, \forall k, l, \tag{41}$$

where the constraint C_2 can be equivalently converted to:

$$C_2 \Rightarrow 1 + \mathrm{SINR}_{k,l} \geq 1 + \zeta_{k,l}, \tag{42}$$

To further express (42) in the form of DC programming, we introduce new function variables $\Gamma_{k,l}(\boldsymbol{W})$ and $\mathrm{I}_{k,l}(\boldsymbol{W}, \zeta_{k,l})$:

$$\Gamma_{k,l}(\boldsymbol{W}) = \sigma^2 + \sum_{j=1.j\neq l}^{L} Tr(\boldsymbol{H}_{k,l}\boldsymbol{W}_j), \tag{43}$$

$$\mathrm{I}_{k,l}(\boldsymbol{W}, \zeta_{k,l}) = \frac{\sigma^2 + \sum_{j=1}^{L} Tr(\boldsymbol{H}_{k,l}\boldsymbol{W}_j)}{1+\zeta_{k,l}}, \tag{44}$$

Therefore, the constraint C_2 can be rewritten as:

$$C_2 \Rightarrow \Gamma(\boldsymbol{W}) - \mathrm{I}(\boldsymbol{W}, \zeta_{k,l}) \leq 0, \tag{45}$$

In (45), $\Gamma_{k,l}(\mathbf{W}_l)$ is the affine function of $\mathbf{W}$, $I_{k,l}(\mathbf{W},\zeta_{k,l})$ is the concave function of $\mathbf{W}$ and $\zeta_{k,l}$. The transformed constraint C_2 is a typical DC constraint.

Similarly, the constraint C_8 can be equivalently converted into the following DC form:

$$C_8 \Rightarrow 4\zeta_{k,l} + \left(\eta_{k,l} - \mathrm{SINR}_l^{\min}\right)^2 \geq \left(\eta_{k,l} + \mathrm{SINR}_l^{\min}\right)^2, \tag{46}$$

In (38), the objective function in the problem Q_3 is a fractional programming problem with the sum-of-ratios form. To handle this problem, we invoke the quadratic transformation algorithm [34] and convert the problem Q_3 into the following form:

$$Q_4: \max_{\eta,\mathbf{W},SINR_l^{\min},\zeta} \mathrm{EE} = 2q\left(B\sum_{l=1}^{L} Y_l(\mathrm{SINR}_l^{\min})\right)^{\frac{1}{2}} - q^2\left(\sum_{l=1}^{L} Tr(\mathbf{W}_l) + P_0\right), \tag{47}$$

$$s.t.\ C_1,\ C_2, C_3,\ C_4,\ C_5, C_6, C_7, C_8\ in\ Q_3, \tag{48}$$

where q is the introduced auxiliary variable, and $Y_l(\mathrm{SINR}_l^{\min})$ is the introduced auxiliary function, which can be expressed as:

$$Y_l(\mathrm{SINR}_l^{\min}) = \log_2\left(1 + \mathrm{SINR}_l^{\min}\right), \tag{49}$$

$$q = \frac{\sqrt{\sum_{l=1}^{L} Y_l(\mathrm{SINR}_l^{\min})}}{\sum_{l=1}^{L} Tr(\mathbf{W}_l) + P_0}, \tag{50}$$

In addition, for the nonsmooth combinatorial optimization problem caused by constraint C_1, we invoke a relaxation and penalty algorithm. Firstly, we relax constraint C_1 into $C_1 \Rightarrow 0 \leq \eta_{k,l} \leq 1,\ \forall k,l$. Meanwhile, to avoid the non-duality of the solution of $\eta_{k,l}$ caused by the relaxation, the penalty term $\mathrm{P}(\eta_{k,l}) = \eta_{k,l}\log\eta_{k,l} + (1-\eta_{k,l})\log(1-\eta_{k,l})$ is introduced into the objective function: let $\lambda_1 > 0$ be the penalty factor, and the problem Q_4 can be equivalently converted to:

$$Q_5: \max_{\eta,\mathbf{W},\mathrm{SINR}_l^{\min},\zeta} \mathrm{EE} = 2q\left(B\sum_{l=1}^{L} Y_l(\mathrm{SINR}_l^{\min})\right)^{\frac{1}{2}} - q^2\left(\sum_{l=1}^{L} Tr(\mathbf{W}_l) + P_0\right) + \lambda_1\sum_{l=1}^{L}\mathrm{P}(\eta_{k,l}), \tag{51}$$

$$s.t.\ C_2: \Gamma_{k,l}(\mathbf{W}) - \mathrm{I}_{k,l}(\mathbf{W},\zeta_{k,l}) \leq 0, \forall k,l, \tag{52}$$

$$C_3,\ C_4,\ C_5, C_6, C_7\ in\ Q_4, \tag{53}$$

$$C_8: 4\zeta_{k,l} + \left(\eta_{k,l} - \mathrm{SINR}_l^{\min}\right)^2 \geq \left(\eta_{k,l} + \mathrm{SINR}_l^{\min}\right)^2 \forall k,l, \tag{54}$$

4.3. CCCP Algorithm

From (51), (52) and (54), it can be seen that the problem Q_5 is a DC programming problem. To handle this challenge, the CCCP framework algorithm is a common method to solve the DC programming problem [35], which is an iterative framework including two operations: convexification and optimization. In the convexification step, by adopting the first-order Taylor expansion, the convex part of the objective function and the concave part of the constraint function can be linearized; then, the DC programming problem is transformed into a convex problem. It should be noted that the convex problem obtained from the convexification step provides a global lower bound for the original problem, and the optimization step is mainly to maximize the lower bound. Meanwhile, the performance of the CCCP algorithm is closely related to the initial point of the variables, but the equality constraint C_4 of the problem Q_5 limits the selection of the initial point. To find a feasible initial point, we substitute the constraint C_4 into the objective function and set the penalty factor $\lambda_2 > 0$. Then, the problem Q_5 can be equivalently converted to:

$$Q_6: \max_{\eta, W, \mathrm{SINR}_l^{\min}, \zeta} \mathrm{EE} = 2q\left(B\sum_{l=1}^{L} \Upsilon_l(\mathrm{SINR}_l^{\min})\right)^{\frac{1}{2}} - q^2\left(\sum_{l=1}^{L} Tr(\boldsymbol{W}_l) + P_0\right) + \lambda_1 \sum_{l=1}^{L} \mathrm{P}(\eta_{k,l}) - \sum_{l}^{L} \lambda_2\left(\sum_{k}^{|U_l|} \eta_{k,l} - U_s\right)^2 \tag{55}$$

$$s.t.\ C_2, C_3,\ C_5, C_6, C_7, C_8\ in\ Q_5, \tag{56}$$

Convexification: Let $\left(\eta_{k,l}, \mathrm{SINR}_l^{\min}, \boldsymbol{W}, \zeta\right)^{(t-1)}$ be the estimated value of variables $\left(\eta_{k,l}, \mathrm{SINR}_l^{\min}, \boldsymbol{W}, \zeta\right)$ in iteration $t-1$ of the problem Q_6. In iteration t, for the convex part $\lambda_1 \sum_{l=1}^{L} \mathrm{P}(\eta_{k,l})$, we adopt the first-order Taylor expansion to replace it, which is reflected in line three of Algorithm 1. The first-order Taylor expansion of $\mathrm{P}(\eta_{k,l})$ can be expressed as:

$$\mathrm{P}(\eta_{k,l})^{te} = \mathrm{P}\left(\eta_{k,l}^{(t-1)}\right) + \left(\eta_{k,l} - \eta_{k,l}^{(t-1)}\right)\nabla \mathrm{P}\left(\eta_{k,l}^{(t-1)}\right), \tag{57}$$

$$\nabla \mathrm{P}\left(\eta_{k,l}^{(t-1)}\right) = \log\left(\eta_{k,l}^{(t-1)}\right) - \log\left(1 - \eta_{k,l}^{(t-1)}\right), \tag{58}$$

Similarly, in the constraint $C_2: \Gamma_{k,l}(\boldsymbol{W}) - \mathrm{I}_{k,l}(\boldsymbol{W}, \zeta_{k,l}) \le 0$, we replace the concave function $\mathrm{I}_{k,l}(\boldsymbol{W}, \zeta_{k,l})$ with its first-order Taylor expansion, which is reflected in line three of Algorithm 1. The first-order Taylor expansion of $\mathrm{I}_{k,l}(\boldsymbol{W}, \zeta_{k,l})$ can be expressed as:

$$\mathrm{I}_{k,l}(\boldsymbol{W}, \zeta_{k,l})^{te} = \mathrm{I}_{k,l}\left(\boldsymbol{W}^{(t-1)}, \zeta_{k,l}^{(t-1)}\right) + \nabla^T \mathrm{I}_{k,l}\left(\boldsymbol{W}^{(t-1)}, \zeta_{k,l}^{(t-1)}\right)\begin{bmatrix}\left\{\boldsymbol{W}_l - \boldsymbol{W}_l^{(t-1)}\right\}_{l=1}^{L} \\ \zeta_{k,l} - \zeta_{k,l}^{(t-1)}\end{bmatrix}, \tag{59}$$

$$\nabla \mathrm{I}_{k,l}\left(\boldsymbol{W}^{(t-1)}, \zeta_{k,l}^{(t-1)}\right) = \left[\left\{\frac{\left(\overline{\boldsymbol{H}}_{k,l}\right)^T}{1+\zeta_{k,l}^{(t-1)}}\right\}_{l=1}^{L}, -\frac{\sigma^2 + \sum_{l=1}^{L} Tr\left(\overline{\boldsymbol{H}}_{k,l}\boldsymbol{W}_l^{(t-1)}\right)}{\left(1+\zeta_{k,l}^{(t-1)}\right)^2}\right]^T, \tag{60}$$

In the constraint $C_8: 4\zeta_{k,l} + \left(\eta_{k,l} - \mathrm{SINR}_l^{\min}\right)^2 \ge \left(\eta_{k,l} + \mathrm{SINR}_l^{\min}\right)^2 \forall k, l$, we replace the concave function $\left(\eta_{k,l} - \mathrm{SINR}_l^{\min}\right)^2$ with its first-order Taylor expansion, which is reflected in line 3 of Algorithm 1. The first-order Taylor expansion of $\left(\eta_{k,l} - \mathrm{SINR}_l^{\min}\right)^2$ can be expressed as:

$$\left(\eta_{k,l} - \mathrm{SINR}_l^{\min}\right)^{2,te} = \left(\eta_{k,l}^{(t-1)} - \mathrm{SINR}_l^{\min,(t-1)}\right)^2 + \begin{bmatrix}2\left(\eta_{k,l}^{(t-1)} - \mathrm{SINR}_l^{\min,(t-1)}\right) \\ -2\left(\eta_{k,l}^{(t-1)} - \mathrm{SINR}_l^{\min,(t-1)}\right)\end{bmatrix}^T \begin{bmatrix}\eta_{k,l} - \eta_{k,l}^{(t-1)} \\ \mathrm{SINR}_l^{\min} - \mathrm{SINR}_l^{\min,(t-1)}\end{bmatrix}, \tag{61}$$

Optimization: The optimization step is reflected in line nine of Algorithm 1. According to (57), (59) and (61), the problem Q_6 can be equivalently converted to

$$Q_7: \max_{\eta, W, \mathrm{SINR}_l^{\min}, \zeta} \mathrm{EE} = 2q\left(B\sum_{l=1}^{L} \Upsilon_l(\mathrm{SINR}_l^{\min})\right)^{\frac{1}{2}} - q^2\left(\sum_{l=1}^{L} Tr(\boldsymbol{W}_l) + P_0\right) + \lambda_1 \sum_{l=1}^{L} \mathrm{P}(\eta_{k,l})^{te} - \sum_{l}^{L} \lambda_2\left(\sum_{k}^{|U_l|} \eta_{k,l} - U_s\right)^2 \tag{62}$$

$$s.t.\ C_3,\ C_5\ ,C_6\ ,C_7\ in\ Q_6, \tag{63}$$

$$C_2 : \Gamma_{k,l}(\boldsymbol{W}) - \mathrm{I}_{k,l}(\boldsymbol{W}, \zeta_{k,l})^{te} \leq 0, \forall k, l, \tag{64}$$

$$C_8 : 4\zeta_{k,l} + \left(\eta_{k,l} - \mathrm{SINR}_l^{\min}\right)^{2,te} \geq \left(\eta_{k,l} + \mathrm{SINR}_l^{\min}\right)^2 \forall k, l, \tag{65}$$

For the problem Q_7, the variables $\left(\eta_{k,l}, \mathrm{SINR}_l^{\min}, \boldsymbol{W}, \zeta\right)^{(t+1)}$ can be updated by the iterative optimization.

Feasible Initial Point: It should be noted that the CCCP algorithm needs a feasible initial point to ensure that the algorithm converges to a stationary point, as the selection of the initial point can affect the performance of the CCCP algorithm. To find a better initial point, we adopt the following method, which is reflected in line one of Algorithm 1.

- Initialize $\eta_{k,l}^{(0)} \approx 0, \mathrm{SINR}_0 = 1$;
- Find $\boldsymbol{W}^{(0)}$, the following optimization problem are modeled:

$$P_{FES} : \left\{\boldsymbol{W}^{(0)}\right\} : \min_{\boldsymbol{W}} \sum_{l=1}^{L} Tr(\boldsymbol{W}_l), \tag{66}$$

$$s.t.\ C_1 : \left\|\sigma^2 \ldots Tr(\overline{\boldsymbol{H}}_{k,l}\boldsymbol{W}_j)_{j\neq l} \cdots\right\| \leq \frac{Tr(\overline{\boldsymbol{H}}_{k,l}\boldsymbol{W}_l)}{\eta_{k,l}\mathrm{SINR}_0}, \forall k, l, \tag{67}$$

$$C_2 : \sum_{l=1}^{L} Tr(\boldsymbol{W}_l) \leq P_T, \tag{68}$$

- If P_{FES} is feasible, proceed to the next step, otherwise, update $\eta_{k,l}^{(0)} = \delta\eta_{k,l}^{(0)}, 0 < \delta < 1$ and repeat step 2;
- Based on the $\boldsymbol{W}^{(0)}$ obtained in step 2, calculate the SINR of each user, i.e., $\mathrm{SINR}_{k,l}^{(0)}, \forall k, l$, and update $\eta_{k,l}^{(0)}$ according to $\eta_{k,l}^{(0)} = \min\left\{1, \frac{\mathrm{SINR}_{k,l}^{(0)}}{\mathrm{SINR}_0}\right\}$;
- Based on $\eta_{k,l}^{(0)}$ and $\boldsymbol{W}^{(0)}$, calculate $\zeta^{(0)}$ and $\left\{\mathrm{SINR}_l^{\min,(0)}\right\}_{l=1}^{L}$.

4.4. Penalty Iteration Algorithm

It should be noted that the SDP algorithm brings the nonconvex and nonsmooth constraint, i.e., $C_7 : rank(\boldsymbol{W}_l) = 1$. To solve the rank-one constraint, many existing research directly relaxes the rank-one constraint in the optimization step [36], and then judges whether the optimization solution $\{\boldsymbol{W}_l\}_{l=1}^{L}$ meets the rank-one constraint. If so, the eigenvalue decomposition (EVD) algorithm is directly adopted to obtain the hybrid beamforming vectors $\{\boldsymbol{f}_l\}_{l=1}^{L}$ according to $\boldsymbol{W}_l = \boldsymbol{f}_l\boldsymbol{f}_l^H$, and if the optimization solution $\{\boldsymbol{W}_l\}_{l=1}^{L}$ does not meet the rank-one constraint, the Gaussian randomization algorithm (GRA) is usually adopted. The basic idea of the GRA is as follows: Firstly, a set of candidate Gaussian vectors $\left\{\left\{\boldsymbol{w}_{g,l}\right\}_{g=1}^{G}\right\}_{l=1}^{L}$ are generated based on the optimization solution $\{\boldsymbol{W}_l\}_{l=1}^{L}$, where G represents the number of the Gaussian randomization. Secondly, from the generated G-group candidate Gaussian vector pool, combined with the power redistribution among the multicast groups, a group of Gaussian vectors is selected as the optimal hybrid beamforming matrix to maximize the objective function in the problem Q_7. It should be noted that in the case of the high-dimensional matrix, GRA has high complexity and large performance loss, resulting in poor availability. To this end, we adopt a feasible algorithm with the better performance: the penalty iteration algorithm.

According to the properties of the matrix, $rank(\boldsymbol{W}_l) = 1$ is equivalent to $Tr(\boldsymbol{W}_l) - \lambda_{\max}(\boldsymbol{W}_l) = 0$. Therefore, the nonsmooth method is adopted to transform the constraint $rank(\boldsymbol{W}_l) = 1$ in the problem Q_7 into the following form:

$$C_7 : Tr(\boldsymbol{W}_l) - \lambda_{\max}(\boldsymbol{W}_l) \leq 0, \tag{69}$$

where $\lambda_{\max}(\boldsymbol{W}_l)$ is the function of solving the maximum eigenvalue. It should be noted that for any positive semidefinite matrix $\boldsymbol{W}_l \succeq 0$, the inequality $Tr(\boldsymbol{W}_l) - \lambda_{\max}(\boldsymbol{W}_l) \geq 0$ is always true, which means that the transformed constraint C_7 and $Tr(\boldsymbol{W}_l) - \lambda_{\max}(\boldsymbol{W}_l) = 0$ are equivalent. Then, we can obtain that the matrix $\boldsymbol{W}_l$ has only one non-zero eigenvalue and can be given by

$$\boldsymbol{W}_l = \lambda_{\max}(\boldsymbol{W}_l)\boldsymbol{w}_{l,\max}\boldsymbol{w}_{l,\max}^H, \tag{70}$$

where $\boldsymbol{w}_{l,\max}$ is the corresponding unit eigenvector. Therefore, the problem Q_7 can be converted into

$$\begin{aligned} Q_8 : \max_{\boldsymbol{\eta},\boldsymbol{W},\mathrm{SINR}_l^{\min},\zeta} \mathrm{EE} = \; & 2q\left(B\sum_{l=1}^{L} \Upsilon_l(\mathrm{SINR}_l^{\min})\right)^{\frac{1}{2}} - q^2\left(\sum_{l=1}^{L} Tr(\boldsymbol{W}_l) + P_0\right) \\ & + \lambda_1 \sum_{l=1}^{L} \mathrm{P}(\eta_{k,l})^{te} - \sum_{l}^{L} \lambda_2 \left(\sum_{k}^{|U_l|} \eta_{k,l} - U_s\right)^2 \end{aligned} \tag{71}$$

$$s.t.\ C_2,\ C_3,\ C_5,C_6,C_8\ in\ Q_7, \tag{72}$$

$$C_7 : Tr(\boldsymbol{W}_l) - \lambda_{\max}(\boldsymbol{W}_l) \leq 0, \tag{73}$$

In the iterative calculation of the problem Q_8, based on the obtained $\boldsymbol{W}_l$, if the value of $Tr(\boldsymbol{W}_l) - \lambda_{\max}(\boldsymbol{W}_l)$ is small enough, the matrix $\boldsymbol{W}_l$ can be considered to meet the rank-one constraint, which is reflected in line seven of Algorithm 1. Therefore, to make the value of $Tr(\boldsymbol{W}_l) - \lambda_{\max}(\boldsymbol{W}_l)$ as small as possible, we adopt the penalty iteration algorithm and substitute the constraint C_7 into the objective function in the problem Q_8. Therefore, the problem Q_8 can be converted into

$$Q_9 : \max_{\boldsymbol{\eta},\boldsymbol{W},\mathrm{SINR}_l^{\min},\zeta} \mathrm{EE} = F\left(\boldsymbol{\eta}, \boldsymbol{W}, \mathrm{SINR}_l^{\min}, \zeta\right) - \lambda_3 \sum_{l}^{L} (Tr(\boldsymbol{W}_l) - \lambda_{\max}(\boldsymbol{W}_l)), \tag{74}$$

$$s.t.\ C_2,\ C_3,\ C_5,C_6,C_8\ in\ Q_8 \tag{75}$$

where $F\left(\boldsymbol{\eta}, \boldsymbol{W}, \mathrm{SINR}_l^{\min}, \zeta\right) = 2q\left(B\sum_{l=1}^{L} \Upsilon_l(\mathrm{SINR}_l^{\min})\right)^{\frac{1}{2}} - q^2\left(\sum_{l=1}^{L} Tr(\boldsymbol{W}_l) + P_0\right)$ $+ \lambda_1 \sum_{l=1}^{L} \mathrm{P}(\eta_{k,l})^{te} - \sum_{l}^{L} \lambda_2 \left(\sum_{k}^{|U_l|} \eta_{k,l} - U_s\right)^2$ and $\lambda_3 > 0$ is the penalty factor, which is generally larger enough to ensure that a smaller value of $Tr(\boldsymbol{W}_l) - \lambda_{\max}(\boldsymbol{W}_l)$ can be obtained.

According to (74), the iterative calculation of the problem Q_9 can maximize function $F\left(\boldsymbol{\eta}, \boldsymbol{W}, \mathrm{SINR}_l^{\min}, \zeta\right)$ and minimize function $Tr(\boldsymbol{W}_l) - \lambda_{\max}(\boldsymbol{W}_l)$. It should be noted that $Tr(\boldsymbol{W}_l)$ is an affine function and $\lambda_{\max}(\boldsymbol{W}_l)$ is nonsmooth, which can result in the nonsmoothness of the objective function in the problem Q_9. To handle the challenge, we replace $\lambda_{\max}(\boldsymbol{W}_l)$ with its first-order Taylor expansion. The subgradient of $\lambda_{\max}(\boldsymbol{W}_l)$ is $\frac{\partial \lambda_{\max}(\boldsymbol{W}_l)}{\partial \boldsymbol{W}_l} = \boldsymbol{w}_{l,\max}\boldsymbol{w}_{l,\max}^H$ and its first-order Taylor expansion can be expressed as follows, which is reflected in line eight of Algorithm 1.

$$\lambda_{\max}\left(\boldsymbol{W}_l^{(t)}\right) \geq \lambda_{\max}\left(\boldsymbol{W}_l^{(t-1)}\right) + \left\langle \boldsymbol{w}_{l,\max}\boldsymbol{w}_{l,\max}^H, \boldsymbol{W}_l^{(t)} - \boldsymbol{W}_l^{(t-1)} \right\rangle, \tag{76}$$

where $\left\langle \boldsymbol{w}_{l,\max}\boldsymbol{w}_{l,\max}^H, \boldsymbol{W}_l^{(t)} - \boldsymbol{W}_l^{(t-1)} \right\rangle = Tr\left(\left(\boldsymbol{w}_{l,\max}\boldsymbol{w}_{l,\max}^H\right)^H \left(\boldsymbol{W}_l^{(t)} - \boldsymbol{W}_l^{(t-1)}\right)\right)$.

We substitute (76) into the objective function in the problem Q_9 to replace $\lambda_{\max}(\boldsymbol{W}_l)$, and the problem Q_9 can be expressed as:

$$Q_{10}: \max_{\boldsymbol{\eta},\boldsymbol{W},\mathrm{SINR}_l^{\min},\zeta} \mathrm{EE} = F\left(\boldsymbol{\eta},\boldsymbol{W},\mathrm{SINR}_l^{\min},\zeta\right) - \lambda_3 \sum_l^L \left(Tr(\boldsymbol{W}_l) - \lambda_{\max}\left(\boldsymbol{W}_l^{(t-1)}\right) + \left\langle \boldsymbol{w}_{l,\max}\boldsymbol{w}_{l,\max}^H, \boldsymbol{W}_l^{(t)} - \boldsymbol{W}_l^{(t-1)} \right\rangle \right), \quad (77)$$

$$s.t.\ C_2,\ C_3,\ C_5\ ,C_6\ ,C_8\ in\ Q_9, \quad (78)$$

In conclusion, the robust joint user scheduling and hybrid beamforming design algorithm for the massive MIMO LEO satellite multigroup multicast communication system is shown in Algorithm 1.

Algorithm 1: Joint user scheduling and hybrid beamforming design algorithm.

Input: CCCP algorithm iteration index k, thresholds ε_1, penalty iteration algorithm iteration index m, thresholds ε_2, penalty factor λ_1, λ_2, λ_3.

1. Initial: $\left(\boldsymbol{\eta},\boldsymbol{W},\mathrm{SINR}_l^{\min},\zeta\right)^{(k=0)}, q^{(k=0)}$.
2. while $\left|\mathrm{EE}^{(k)} - \mathrm{EE}^{(k-1)}\right| \geq \varepsilon_1$
3. Convexification step by (57), (59), (61).
4. Calculation $q^{(k)}$, substitute $q^{(k)}$ into (77).
5. Optimization step.
6. Let $\left(\boldsymbol{\eta},\boldsymbol{W},\mathrm{SINR}_l^{\min},\zeta\right)^{(m=0)} = \left(\boldsymbol{\eta},\boldsymbol{W},\mathrm{SINR}_l^{\min},\zeta\right)^{(k=0)}$.
7. while $\left|\left\{Tr\left(\boldsymbol{W}_l^{(m)}\right) - \lambda_{\max}\left(\boldsymbol{W}_l^{(m)}\right)\right\}_{l=1}^L\right| \geq \varepsilon_2$
8. Calculate the maximum eigenvalue $\lambda_{\max}(\boldsymbol{W}_l)$ of $\boldsymbol{W}_l^{(m)}$ and the corresponding eigenvector $\boldsymbol{w}_{l,\max}^{(m)}$.
9. Using CVX toolbox, calculate the variables $\left(\boldsymbol{\eta},\boldsymbol{W},\mathrm{SINR}_l^{\min},\zeta\right)_{opt}^{(m)}$ at the mth iteration according to (77).
10. If $\left\{\boldsymbol{W}_l^{(m+1)}\right\}_l^L \approx \left\{\boldsymbol{W}_l^{(m)}\right\}_l^L$, then
11. Update $\lambda_3 = 2\lambda_3$.
12. else
13. Update $m = m + 1$.
14. end
15. end
16. Update $\left(\boldsymbol{\eta},\boldsymbol{W},\mathrm{SINR}_l^{\min},\zeta\right)^{(k+1)} = \left(\boldsymbol{\eta},\boldsymbol{W},\mathrm{SINR}_l^{\min},\zeta\right)^{(m)}, k = k+1, \lambda_1 = \lambda_1 + 1$, $\lambda_2 = \lambda_2 + 1$.
17. end

Output: $(\boldsymbol{\eta},\boldsymbol{W},SINR_l^{\min},\zeta)_{opt}$.

5. Convergence and Complexity Analysis

5.1. Convergence

The effectiveness of the algorithm depends on its convergence. For the convergence of the CCCP algorithm, the convergence has been proven by [37]. To prove the convergence of the penalty iteration algorithm, let the variable solution and objective function value of the optimization problem Q_{10} be $\left(\boldsymbol{\eta},\boldsymbol{W},\mathrm{SINR}_l^{\min},\zeta\right)^{(k+1)}$ and $F_{EE}\left(\left(\boldsymbol{\eta},\boldsymbol{W},\mathrm{SINR}_l^{\min},\zeta\right)^{(k+1)}\right)$ at the kth iteration. Therefore, the convergence can be proved as follows:

$$
\begin{aligned}
&F_{EE}\left(\left(\eta, W, \mathrm{SINR}_l^{\min}, \zeta\right)^{(k+1)}\right) = F\left(\left(\eta, W, \mathrm{SINR}_l^{\min}, \zeta\right)^{(k+1)}\right) - \lambda_3 \sum_l^L \left(Tr\left(W_l^{(k+1)}\right) - \lambda_{\max}\left(W_l^{(k+1)}\right)\right) \\
&\geq F\left(\left(\eta, W, \mathrm{SINR}_l^{\min}, \zeta\right)^{(k+1)}\right) - \lambda_3 \sum_l^L \left(Tr\left(W_l^{(k+1)}\right) - \lambda_{\max}\left(W_l^{(k)}\right) - \left\langle w_{l,\max} w_{l,\max}^H, W_l^{(k+1)} - W_l^{(k)} \right\rangle\right) \\
&\overset{by(75)}{\geq} F\left(\left(\eta, W, \mathrm{SINR}_l^{\min}, \zeta\right)^{(k)}\right) - \lambda_3 \sum_l^L \left(Tr\left(W_l^{(k)}\right) - \lambda_{\max}\left(W_l^{(k)}\right)\right) \\
&= F_{EE}\left(\left(\eta, W, \mathrm{SINR}_l^{\min}, \zeta\right)^{(k)}\right)
\end{aligned}
\tag{79}
$$

The convergence can be proved according to (79). Therefore, after initializing the values of $\left(\eta, W, \mathrm{SINR}_l^{\min}, \zeta\right)^{(k=0)}$, $\lambda_1^{(k=0)}$, $\lambda_2^{(k=0)}$ and $\lambda_3^{(k=0)}$, the proposed algorithm can iteratively converge to an optimal solution by setting a reasonable convergence threshold.

5.2. Complexity

The complexity of the algorithm directly affects its performance. In the algorithms adopted, the complexity of the hierarchical clustering algorithm can be calculated according to the connection algorithm, similarity measurement criteria and hierarchical grouping process, and the algorithm complexity can be expressed as $O(LK^2N)$. The complexity of the joint user scheduling and hybrid beamforming design algorithm is closely related to the number of multicast groups and scheduling users. In addition, the number of optimization variables and constraints in the CCCP algorithm and the penalty iteration algorithm can also affect the complexity [38]. In the problem Q_{10}, the number of optimization variables is $2\sum_{l=1}^{L}|U_l| + 2L$, the number of convex constraints is $\sum_{l=1}^{L}|U_l|$ and the number of linear constraints is $\sum_{l=1}^{L}|U_l| + 3L$. Let the number of iterations in the penalty iteration algorithm and the CCCP algorithm be I_p and I_c, respectively. In conclusion, the overall complexity of the proposed algorithm is $O\left(LK^2N + I_p I_c\left(2\sum_{l=1}^{L}|U_l| + 2L\right)\left(2\sum_{l=1}^{L}|U_l| + 3L\right)\right)$.

According to the algorithm complexity, the proposed joint user scheduling and hybrid beamforming design algorithm has a strong timeliness in small dimensional communication systems. However, for large dimensional communication systems, such as the satellite communication system, the number of active users is usually large. According to the complexity analysis, with the increase in the total number of active users, the convergence speed of the algorithm would gradually slow down, and the complexity would gradually increase. Considering the characteristics of the LEO satellite communication system, the delay caused by the high complexity is unacceptable, which would affect the overall performance of the communication system. To handle this problem, considering the balance of the algorithm performance and complexity, before the joint user scheduling and hybrid beamforming design, we can appropriately reduce the system dimension by adding the user preselection step in the algorithm process. The user preselection step can reduce the total number of active users in each transmission, and then we carry out the joint user scheduling and hybrid beamforming design for preselected users.

6. User Preselection Algorithm

In the user preselection step, $U_{l,p}$ users in the lth multicast group are preselected as the user representatives, where $U_s < U_{l,p} \leq |U_l|$. The user preselection process is shown in Figure 6. The symbols (circles, squares and triangles) represent the users in different the multicast group. The red circles represent preselected users and scheduling users.

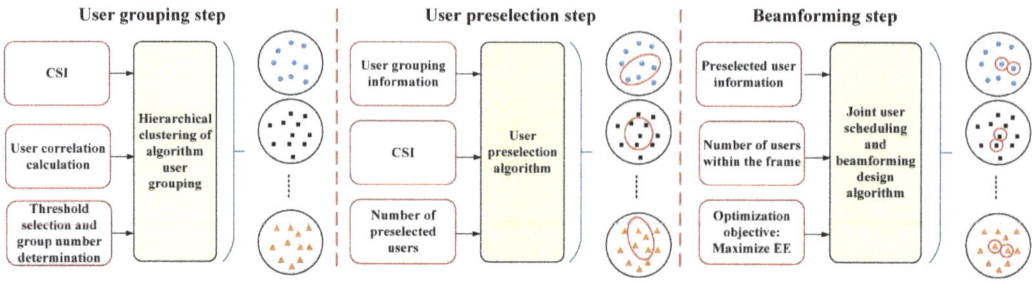

Figure 6. Design process of joint user scheduling and beamforming with the user preselection.

The selection of preselected users can affect the performance of the joint user scheduling and hybrid beamforming design algorithm, which depends on the preselected algorithm. In the beamforming design of the multigroup multicast communication system, the beamforming vector is oriented to multiple users in the multicast group. Therefore, in the process of user preselection, to maximize the receive gain of each user, i.e., $\left|\boldsymbol{h}_{k,l}^{H}\boldsymbol{f}_{l}\right|$, the beamforming vector $\boldsymbol{f}_l$ of the lth multicast group should be collinear with the users' channel vectors in the multicast group as far as possible. Therefore, the channel vectors of the preselected users in the same multicast group should also be strongly linearly correlated. Meanwhile, the interference among multicast groups should also be taken into account in the user preselection stage. To reduce the interference among multicast groups, the channel vectors of preselected users among different multicast groups should be orthogonal. Similarly, the beamforming vector of the multicast group should be orthogonal to the users' channel vectors in other multicast groups.

In conclusion, we adopt a low complexity user preselection algorithm, which can preselect orthogonal users among the different multicast groups and linearly correlated users in the same multicast group. The proposed algorithm is divided into two steps, as follows:

- The first step: according to the orthogonal criterion [11], a user is preselected for each multicast group in turn, which is reflected in line three and line four of **Step 1** in Algorithm 2;
- The second step: based on the users of each multicast group selected in the first step, linearly correlated users are selected for each multicast group, which is reflected in line two and line three of **Step 2** in Algorithm 2.

The specific preselection process of the two steps is as follows:

Algorithm 2: User preselection algorithm.

Step 1: Orthogonal user preselection algorithm among the different multicast groups.
Input: CSI.
1. Let $\mathrm{Id}_{(1)} = Index\{\max(\|\boldsymbol{h}_{k,l}\|), \forall k, l\}$, select the user with the largest channel gain, $\mathrm{Id}_{(1)}$ is the index of the user.
2. while $l \leq L, l \neq \mathrm{Id}_{(1)}$
3. For all users in the lth multicast group, calculate $Z_{k,l} = \boldsymbol{h}_{k,l}\left(I_N - \sum_{j=\mathrm{Id}_{(1)}}^{\mathrm{Id}_{(l)}} \frac{\boldsymbol{h}_{(j)}^H \boldsymbol{h}_{(j)}}{\|\boldsymbol{h}_{(j)}\|_2^2}\right)$ in turn.
4. $\mathrm{Id}_{(l)} = Index\{\max(\|Z_{k,l}\|), \forall k \in l\}$, the user with index $\mathrm{Id}_{(l)}$ is the preselected orthogonal user of the lth multicast group.
5. end
Output: Orthogonal users among the different multicast groups.
Step 2: User preselection algorithm in each multicast group.
Input: Orthogonal users among the different multicast groups, CSI.
1. For $l = 1 : L$
2. For other users in the lth multicast group except the orthogonal user preselected in step 1, calculate the linear correlation value between each user and the preselected orthogonal user of the multicast group in turn, i.e., $C_{k,l} = \boldsymbol{h}_{k,l}^H \frac{\boldsymbol{h}_{\mathrm{Id}_{(l)}} \boldsymbol{h}_{\mathrm{Id}_{(l)}}^H}{\left\|\boldsymbol{h}_{\mathrm{Id}_{(l)}}^H\right\|_2^2}$.
3. Based on the $C_{k,l}$ of users in each multicast group, select top $\left(U_{l,p} - 1\right)$ largest users, plus the orthogonal users in step 1 as the preselected users of each multicast group.
4. end
5. end
Output: Preselected users for each multicast group.

After the user preselection, the joint user scheduling and hybrid beamforming design is for the preselected users. Therefore, the dimension of the LEO satellite communication system will be reduced, and the algorithm complexity will be reduced. Although the algorithm performance has a slight loss, compared with the decoupling design of user scheduling and beamforming, the performance is greatly improved. In conclusion, the joint user scheduling and hybrid beamforming design with the user preselection step is a better choice after balancing performance and complexity.

7. Solution of The Digital Beamforming Matrix and The Analog Beamforming Matrix

In this section, we aim to investigate the design of digital beamforming matrix $\boldsymbol{F}_{BB}$ and analog beamforming matrix $\boldsymbol{F}_{RF}$ in a hybrid beamformer. After obtaining $\boldsymbol{W}$, we need to further solve $\boldsymbol{F}_{BB}$ and $\boldsymbol{F}_{RF}$. The solution method of $\boldsymbol{F}_{BB}$ and $\boldsymbol{F}_{RF}$ can be divided into two steps:

- The first step: we adopt the EVD algorithm to solve the hybrid beamforming matrix $\boldsymbol{F}$ from $\boldsymbol{W}$.
- The second step: we propose the MM-AltOpt algorithm to obtain $\boldsymbol{F}_{BB}$ and $\boldsymbol{F}_{RF}$.

7.1. Solution of The Hybrid Beamforming Matrix

Before calculating $\boldsymbol{F}_{BB}$ and $\boldsymbol{F}_{RF}$, it is necessary to obtain the hybrid beamforming matrix $\boldsymbol{F}$. For the solution of $\boldsymbol{F}$, we can adopt the EVD algorithm based on the previously obtained optimization variable $\boldsymbol{W}$. According to the relationship between $\boldsymbol{W}$ and $\boldsymbol{F}$, i.e., $\boldsymbol{W}_l \triangleq \boldsymbol{f}_l \boldsymbol{f}_l^H$, the solution of $\boldsymbol{F}$ can be modeled as follows:

$$\min_{\boldsymbol{f}_l} \left\|\boldsymbol{W}_l - \boldsymbol{f}_l \boldsymbol{f}_l^H\right\|_F^2, \tag{80}$$

where the hybrid beamforming vector $\boldsymbol{f}_l$ can be given by

$$\boldsymbol{f}_l = \sqrt{v_l}\boldsymbol{u}_l, \forall l, \tag{81}$$

where v_l is the maximum eigenvalue of the matrix $\boldsymbol{W}_l$ and $\boldsymbol{u}_l$ is the maximum eigenvector of the matrix $\boldsymbol{W}_l$.

7.2. MM-AltOpt Algorithm: Solution of $\boldsymbol{F}_{BB}$ and $\boldsymbol{F}_{RF}$

According to $\boldsymbol{F} = \boldsymbol{F}_{RF}\boldsymbol{F}_{BB}$ and $\left|(\boldsymbol{F}_{RF})_{i,j}\right| = 1$, the solution of $\boldsymbol{F}_{BB}$ and $\boldsymbol{F}_{BB}$ can be modeled as a joint optimization problem with the power and constant modulus constraints, as follows:

$$P_1 : \min_{\boldsymbol{F}_{BB},\boldsymbol{F}_{RF}} \|\boldsymbol{F}^{opt} - \boldsymbol{F}_{RF}\boldsymbol{F}_{BB}\|_F^2, \tag{82}$$

$$s.t.\ C_1 : \boldsymbol{F}_{RF} \in \mathbb{F}, \tag{83}$$

$$C_2 : \|\boldsymbol{F}_{RF}\boldsymbol{F}_{BB}\|_F^2 = P_T, \tag{84}$$

where $\mathbb{F} = \left\{\boldsymbol{F}_{RF} \in C^{N\times L} \middle| \left|\boldsymbol{F}_{RF(i,j)}\right|^2 = 1, 1 \le i \le N, 1 \le j \le L\right\}$ represents the unit modulus constraint, which is determined by the phase shifter in UPA, and the constraint C_2 represents the power constraint.

It is worth noting that the problem P_1 is a matrix decomposition problem with the constant modulus constraint and the equality constraint. The objective function is a nonconvex function of variables $\boldsymbol{F}_{BB}$ and $\boldsymbol{F}_{RF}$, and the constraints C_1 and C_2 are also nonconvex. Meanwhile, it can be seen that when one of the two variables is given, the objective function is the convex function of the other variable. To solve the problem P_1, we invoke the alternating optimization algorithm. The alternating optimization algorithm can decompose the multivariable joint optimization problem into multiple subproblems according to the partial convexity of the problem P_1, and one of the variables can be iteratively solved by fixing the residual variables.

It should be noted that the nonconvexity of constraints is still a challenge. To this end, we first relax the constraint C_2, and then use the scale factor to adjust the digital beamforming matrix $\boldsymbol{F}_{BB}$ to meet the power constraint. Then, for the solution of the analog beamforming matrix $\boldsymbol{F}_{RF}$ with the unit modulus constraint, the MM algorithm is adopted [33].

7.3. Solution of The Analog Beamforming Matrix Based on The MM Algorithm

According to the solution process of the alternating optimization algorithm, we first solve the analog beamforming matrix $\boldsymbol{F}_{RF}$ based on the digital beamforming matrix F_{BB}. Thus, the problem P_1 can be expressed as:

$$P_2 : \min_{\boldsymbol{F}_{RF}} \left\|\boldsymbol{F} - \boldsymbol{F}_{RF}\boldsymbol{F}_{BB}^{(n)}\right\|_F^2, \tag{85}$$

$$s.t.\ C_1 : \boldsymbol{F}_{RF} \in \mathbb{F}, \tag{86}$$

where $\boldsymbol{F}_{BB}^{(n)}$ represents the estimated value of the digital beamforming matrix $\boldsymbol{F}_{BB}$ at the nth iteration. Due to the unit module constraint of elements in $\boldsymbol{F}_{RF}$, the problem P_2 is a nonconvex optimization problem.

According to the MM framework theory, the key step is constructing a surrogate function of the objective function in the optimization problem [39]. To construct the surrogate function, we decompose the matrix $\boldsymbol{F}$ by rows. According to the equivalence of the F-norm and L_2-norm of the vector, the problem P_2 can be rewritten as:

$$P_3 : \min_{\boldsymbol{F}_{RF}} \sum_{i=1}^{N} \boldsymbol{F}_i^H \boldsymbol{F}_i - 2\Re\left(\boldsymbol{F}_i^H \boldsymbol{F}_{BB}^{(n)} \boldsymbol{F}_{RF,i}\right) + \boldsymbol{F}_{RF,i}^H \boldsymbol{F}_{BB}^{(n)} \boldsymbol{F}_{BB}^{(n)H} \boldsymbol{F}_{RF,i}, \tag{87}$$

$$s.t.\ C_1 : \boldsymbol{F}_{RF} \in \mathbb{F}, \tag{88}$$

where $\boldsymbol{F}_i^H$ represents the ith row vector of the matrix $\boldsymbol{F}$, and $\boldsymbol{F}_{RF,i}^H$ represents the ith row vector of the matrix $\boldsymbol{F}_{RF}$.

It should be noted that the third term $\boldsymbol{F}_{RF,i}^H \boldsymbol{F}_{BB}^{(n)} \boldsymbol{F}_{BB}^{(n)H} \boldsymbol{F}_{RF,i}$ in (86) is a convex function term, which needs further conversion. According to the first-order Taylor expansion, $\boldsymbol{F}_{RF,i}^H \boldsymbol{F}_{BB}^{(n)} \boldsymbol{F}_{BB}^{(n)H} \boldsymbol{F}_{RF,i}$ can be converted into:

$$\begin{aligned} \boldsymbol{F}_{RF,i}^H \boldsymbol{F}_{BB}^{(n)} \boldsymbol{F}_{BB}^{(n)H} \boldsymbol{F}_{RF,i} &= \boldsymbol{F}_{RF,i}^{(q)H} \boldsymbol{F}_{BB}^{(n)} \boldsymbol{F}_{BB}^{(n)H} \boldsymbol{F}_{RF,i}^{(q)} + 2\Re\left(\boldsymbol{F}_{RF,i}^{(q)H} \boldsymbol{F}_{BB}^{(n)} \boldsymbol{F}_{BB}^{(n)H} \left(\boldsymbol{F}_{RF,i} - \boldsymbol{F}_{RF,i}^{(q)} \right) \right) \\ &+ \left(\boldsymbol{F}_{RF,i} - \boldsymbol{F}_{RF,i}^{(q)} \right)^H \boldsymbol{F}_{BB}^{(n)} \boldsymbol{F}_{BB}^{(n)H} \left(\boldsymbol{F}_{RF,i} - \boldsymbol{F}_{RF,i}^{(q)} \right) \end{aligned} \tag{89}$$

where $\boldsymbol{F}_{RF,i}^{(q)}$ represents the estimated value of $\boldsymbol{F}_{RF,i}$ at the qth iteration. According to the MM algorithm, the surrogate of $\boldsymbol{F}_{RF,i}^H \boldsymbol{F}_{BB}^{(n)} \boldsymbol{F}_{BB}^{(n)H} \boldsymbol{F}_{RF,i}$ can be expressed as follows:

$$\begin{aligned} \boldsymbol{F}_{RF,i}^H \boldsymbol{F}_{BB}^{(n)} \boldsymbol{F}_{BB}^{(n)H} \boldsymbol{F}_{RF,i} &\le \boldsymbol{F}_{RF,i}^{(q)H} \boldsymbol{F}_{BB}^{(n)} \boldsymbol{F}_{BB}^{(n)H} \boldsymbol{F}_{RF,i}^{(q)} + 2\Re\left(\boldsymbol{F}_{RF,i}^{(q)H} \boldsymbol{F}_{BB}^{(n)} \boldsymbol{F}_{BB}^{(n)H} \left(\boldsymbol{F}_{RF,i} - \boldsymbol{F}_{RF,i}^{(q)} \right) \right) \\ &+ \left(\boldsymbol{F}_{RF,i} - \boldsymbol{F}_{RF,i}^{(q)} \right)^H \boldsymbol{X}^{(n)} \left(\boldsymbol{F}_{RF,i} - \boldsymbol{F}_{RF,i}^{(q)} \right) \end{aligned} \tag{90}$$

where $\boldsymbol{X}^{(n)}$ is a positive semidefinite matrix and satisfies the constraint $\boldsymbol{X}^{(n)} \succeq \boldsymbol{F}_{BB}^{(n)} \boldsymbol{F}_{BB}^{(n)H}$; here, we let $\boldsymbol{X}^{(n)} = \lambda_{\max}\left(\boldsymbol{F}_{BB}^{(n)} \boldsymbol{F}_{BB}^{(n)H} \right) \boldsymbol{I}$ and $\lambda_{\max}\left(\boldsymbol{F}_{BB}^{(n)} \boldsymbol{F}_{BB}^{(n)H} \right)$ represents the maximum eigenvalue of the matrix $\boldsymbol{F}_{BB}^{(n)} \boldsymbol{F}_{BB}^{(n)H}$. In conclusion, (89) can be further expressed as:

$$\begin{aligned} \boldsymbol{F}_{RF,i}^H \boldsymbol{F}_{BB}^{(n)} \boldsymbol{F}_{BB}^{(n)H} \boldsymbol{F}_{RF,i} &\le \lambda_{\max}\left(\boldsymbol{F}_{BB}^{(n)} \boldsymbol{F}_{BB}^{(n)H} \right) \boldsymbol{F}_{RF,i}^H \boldsymbol{F}_{RF,i} + 2\Re\left(\boldsymbol{F}_{RF,i}^H \left(\boldsymbol{F}_{BB}^{(n)} \boldsymbol{F}_{BB}^{(n)H} - \lambda_{\max}\left(\boldsymbol{F}_{BB}^{(n)} \boldsymbol{F}_{BB}^{(n)H} \right) \boldsymbol{I} \right) \boldsymbol{F}_{RF,i} \right) \\ &+ \boldsymbol{F}_{RF,i}^{(q)H} \left(\lambda_{\max}\left(\boldsymbol{F}_{BB}^{(n)} \boldsymbol{F}_{BB}^{(n)H} \right) \boldsymbol{I} - \boldsymbol{F}_{BB}^{(n)} \boldsymbol{F}_{BB}^{(n)H} \right) \boldsymbol{F}_{RF,i} \end{aligned} \tag{91}$$

According to (90), the surrogate function of the objective function in the problem P_3 can be expressed as:

$$\begin{aligned} &P_3 : \min_{F_{RF}} \sum_{i=1}^{N} F_i^H F_i - 2\Re\left(F_i^H F_{BB}^{(n)} F_{RF,i} \right) + F_{RF,i}^H F_{BB}^{(n)} F_{BB}^{(n)H} F_{RF,i} \Rightarrow \\ &P_4 : \min_{F_{RF}} \sum_{i=1}^{N} F_i^H F_i - 2\Re\left(F_i^H F_{BB}^{(n)} F_{RF,i} \right) + \lambda \left(F_{BB}^{(n)} F_{BB}^{(n)H} \right)_{RF,i}^{H}{}_{RF,i\,max} \\ &\qquad + 2\Re(F_{RF,i}^H (F_{BB}^{(n)} F_{BB}^{(n)H} - \\ &\lambda (F_{BB}^{(n)} F_{BB}^{(n)H})_{max} ()_{RF,i}) ()_{RF,i}^{(q)H} (\lambda (F_{BB}^{(n)} F_{BB}^{(n)H})_{BB\,BB}^{(n)(n)H}{}_{max} ()_{RF,i})), \end{aligned} \tag{92}$$

It is worth noting that the first and third terms of the objective function in the problem P_4 are constant terms, and the last term is independent of the variable $\boldsymbol{F}_{RF,i}^H$. After ignoring the above three items, the problem P_4 can be converted to the following projection problem:

$$P_5 : \min_{F_{RF}} \sum_{i=1}^{N} \left\| \boldsymbol{F}_{RF,i} - \boldsymbol{c}_i^{(q)} \right\|_2^2, \tag{93}$$

$$s.t.\ C_1 : \boldsymbol{F}_{RF} \in \mathbb{F}, \tag{94}$$

where $\boldsymbol{c}_i^{(q)} = \boldsymbol{F}_{BB}^{(n)} \boldsymbol{F}_i - \left(\boldsymbol{F}_{BB}^{(n)} \boldsymbol{F}_{BB}^{(n)H} - \lambda_{\max}\left(\boldsymbol{F}_{BB}^{(n)} \boldsymbol{F}_{BB}^{(n)H} \right) \boldsymbol{I} \right) \boldsymbol{F}_{RF,i}^{(q)H}$.

Therefore, the following closed form solution can be obtained for the problem P_5, which is reflected in line three of the **inner algorithm** in Algorithm 3:

$$\boldsymbol{F}_{RF,i} = e^{j\arg(c_i^{(q)})}, \forall i, \tag{95}$$

$$\boldsymbol{F}_{RF} = e^{-j\arg(C^{(q)T})}, \tag{96}$$

where $C^{(q)} = \boldsymbol{F}_{BB}^{(n)} \boldsymbol{F}^H - \left(\boldsymbol{F}_{BB}^{(n)} \boldsymbol{F}_{BB}^{(n)H} - \lambda_{\max}\left(\boldsymbol{F}_{BB}^{(n)} \boldsymbol{F}_{BB}^{(n)H} \right) \boldsymbol{I} \right) \boldsymbol{F}_{RF}^{(q)H}$, which is reflected in line two of the **inner algorithm** in Algorithm 3.

7.4. Solution of The Digital Beamforming Matrix

Based on the analog beamforming matrix $\boldsymbol{F}_{RF}$ obtained in the previous section, the solution problem of the digital beamforming matrix $\boldsymbol{F}_{BB}$ can be modeled as follows:

$$P_6 : \min_{\boldsymbol{F}_{BB}} \left\| \boldsymbol{F} - \boldsymbol{F}_{RF}^{(n)} \boldsymbol{F}_{BB} \right\|_F^2, \tag{97}$$

$$s.t. \left\| \boldsymbol{F}_{RF}^{(n)} \boldsymbol{F}_{BB} \right\|_F^2 = P_T, \tag{98}$$

where $\boldsymbol{F}_{RF}^{(n)}$ represents the estimated value of the analog beamforming matrix $\boldsymbol{F}_{RF}$ at the nth iteration. Due to the quadratic form and the convex equality constraint in the problem P_6, the problem P_6 is a nonconvex QCQP form problem.

One of the ways to solve the problem P_6 is to relax the equality constraint into the inequality constraint, and then convert the problem P_6 into a convex minimization problem, which can be solved with the CVX toolbox, but the complexity of this method is high. To this end, based on the fact that the hybrid beamforming matrix $\boldsymbol{F}$ satisfies the power constraint, i.e., $\|\boldsymbol{F}\|_F^2 = P_T$, the following closed form solution can be obtained for the problem P_6, which is reflected in line two of the **main algorithm** in Algorithm 3:

$$\boldsymbol{F}_{BB} = \left(\boldsymbol{F}_{RF}^{(n)H} \boldsymbol{F}_{RF}^{(n)} \right)^{-1} \boldsymbol{F}_{RF}^{(n)H} \boldsymbol{F}, \tag{99}$$

In conclusion, the MM-AltOpt algorithm for solving $\boldsymbol{F}_{BB}$ and $\boldsymbol{F}_{RF}$ can be described as the main algorithm and the inner algorithm, as follows:

Algorithm 3: Design algorithm of the digital beamforming matrix and the analog beamforming matrix.

Main algorithm: MM-AltOpt algorithm.
Input: Hybrid beamforming matrix $\boldsymbol{F}$, initial: $\boldsymbol{F}_{RF}^{(n=0)} \in \mathbb{F}$, iteration index $n = 0$, threshold $\varepsilon_3 = 10^{-3}$, the solution of the objective function of the problem P_1 in the nth iteration is $\delta^{(n)}$.
1. while $\left| \delta^{(n)} - \delta^{(n-1)} \right| \geq \varepsilon_3$
2. Based on $\boldsymbol{F}_{RF}^{(n)}$, calculate $\boldsymbol{F}_{BB}^{(n+1)}$ according to $\boldsymbol{F}_{BB} = \left(\boldsymbol{F}_{RF}^{(n)H} \boldsymbol{F}_{RF}^{((n))} \right)^{-1} \boldsymbol{F}_{RF}^{(n)H} \boldsymbol{F}$.
3. Based on $\boldsymbol{F}_{BB}^{(n+1)}$, calculate $\boldsymbol{F}_{RF}^{(n+1)}$ according to the **inner algorithm**.
4. Set $n = n + 1$.
5. end
Output: $\boldsymbol{F}_{RF}$,$\boldsymbol{F}_{BB}$, normalize $\boldsymbol{F}_{BB} = \frac{\sqrt{P_T}}{\left\| \boldsymbol{F}_{RF} \boldsymbol{F}_{BB} \right\|_F} \boldsymbol{F}_{BB}$.
Inner algorithm: Algorithm for solving the analog beamforming matrix.
Input: Hybrid beamforming matrix $\boldsymbol{F}$, $\boldsymbol{F}_{BB}^{(n)}$, $\boldsymbol{F}_{RF}^{(q=0)} \in \mathbb{F}$, iteration index $q = 0$, threshold $\varepsilon_4 = 10^{-3}$, the solution of the objective function of the problem P_5 in the qth iteration is $\delta_1^{(q)}$.
1. while $\left| \delta^{(n)} - \delta^{(n-1)} \right| \geq \varepsilon_4$
2. Calculate $C^{(q)} = \boldsymbol{F}_{BB}^{(n)} \boldsymbol{F}^H - \left(\boldsymbol{F}_{BB}^{(n)} \boldsymbol{F}_{BB}^{(n)H} - \lambda_{\max} \left(\boldsymbol{F}_{BB}^{(n)} \boldsymbol{F}_{BB}^{(n)H} \right) \boldsymbol{I} \right) \boldsymbol{F}_{RF}^{(q)H}$.
3. Calculate $\boldsymbol{F}_{RF} = e^{-j\arg(C^{(q)T})}$.
4. Set $q = q + 1$.
5. end
Output: $\boldsymbol{F}_{RF}^{(n)}$.

---

8. Results and Discussion

In this section, we evaluate the performance of the proposed joint user scheduling and hybrid beamforming design algorithm by numerical simulations. In the numerical simulations, we set the number of multicast groups to $L = 7$, which cover 150 active users, and set the SINR constraint threshold of each multicast group to $\mathrm{SINR}_0 = 1$. To facilitate analysis, we assume that the CSI errors of different multicast groups are the same, which are expressed as $\sigma^2_{f^{rsd}_{d,k,l}} = \sigma^2_{f^{rsd}_d}$ and $\sigma^2_{\phi_{\tau_{k,l}}} = \sigma^2_{\phi_\tau}$. The value of P_0 can be calculated by [31]. In addition, the system parameters used in the numerical simulations are shown in Table 1.

Table 1. Simulation parameters.

Parameters	Values	Parameters	Values
N	$8 \times 8 = 64$	κ	1.38×10^{-23} J·K^{-1}
L	7	P_0	21.5 W
$\kappa_{k,l}$	10	T	300 K
Bandwidth	50 MHz	G_{leo}	3 dB
Orbit altitude	1000 km	G_{ut}	3 dB
Beam radius	250 km	f	20 GHz
LP_{at}	0.017 dB	K	150

Figure 7 shows the convergence trajectory of the EE of the massive MIMO LEO satellite multigroup multicast communication system, versus the number of iterations for different CSI errors, different numbers of preselected users and different scheduling algorithms. In this simulation, two groups of channel errors are set according to Refs. [23,29], i.e., $\sigma^2_{f^{rsd}_d} = 25$, $\sigma^2_{\phi_\tau} = 10$ and $\sigma^2_{f^{rsd}_d} = 20$, $\sigma^2_{\phi_\tau} = 5$. In addition, we set $U_s = 2$ and $P_T = 50$ W. Meanwhile, we set two different numbers of the preselected users, i.e., $U_{l,p}/U_s = 2, U_{l,p} = 4$ and $U_{l,p}/U_s = 3, U_{l,p} = 6$. It can be seen that the proposed robust algorithm has higher performance gain than the traditional nonrobust algorithm, which shows the effectiveness of the robust algorithm. Meanwhile, it can be seen that when $\sigma^2_{f^{rsd}_d} = 25$, $\sigma^2_{\phi_\tau} = 10$ and $\sigma^2_{f^{rsd}_d} = 20$, and $\sigma^2_{\phi_\tau} = 5$, the EE performance gain of the proposed robust algorithm is improved by 9.8% and 6.7%, respectively, compared with the traditional nonrobust algorithm. The system EE of the proposed joint user scheduling and hybrid beamforming design algorithm is higher than that of the decoupling design algorithm, and the more preselected users, the higher performance improvement.

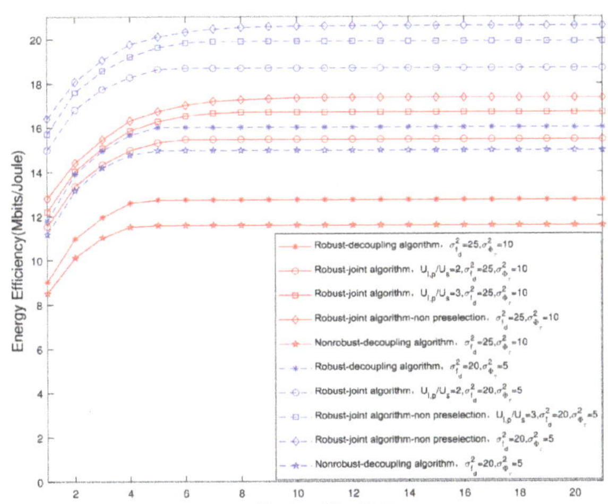

Figure 7. Convergence trajectory of system EE relative to different CSI errors, different number of preselected users and different scheduling algorithms.

Figure 8 compares the EE performance of the proposed algorithm and the traditional algorithm under different system parameters, versus different transmission power thresholds P_T. It can be seen that with the increase in transmission power, the EE performance shows a trend of first rising and then falling. The reason is that the growth rate of the system rate is lower than that of the power consumption. Meanwhile, we can see that the

EE of the joint user scheduling and hybrid beamforming design algorithm is higher than that of the decoupling design algorithm versus different transmission power thresholds P_T. Under the conditions of $\sigma^2_{f_d^{rsd}} = 25$, $\sigma^2_{\phi_\tau} = 10$ and $P_T = 15$ W, when $U_{l,p}/U_s = 2, U_{l,p} = 4$ and $U_{l,p}/U_s = 3, U_{l,p} = 6$, the EE performance gain of the proposed joint design algorithm is 28.41% and 45.19% higher than that of the traditional decoupling design algorithm.

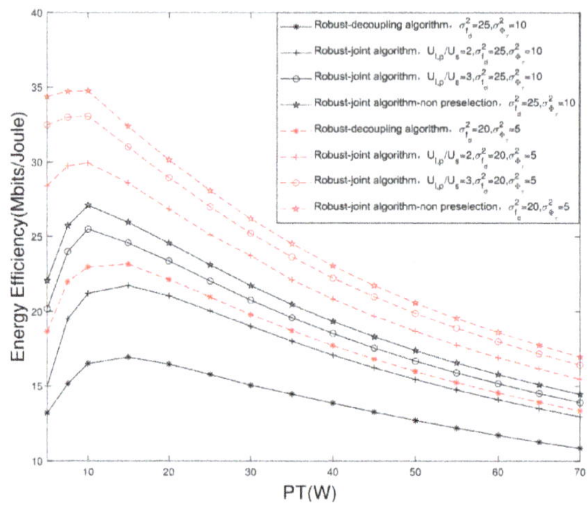

Figure 8. Comparison of system EE of different algorithms with different transmission power thresholds P_T.

Figure 9 indicates the change trend of the SE of the massive MIMO LEO satellite multigroup multicast communication system versus different transmission power thresholds P_T. It can be seen that the SE increases with the increase in the transmission power. Meanwhile, we can see that the SE of the joint user scheduling and hybrid beamforming design algorithm is higher than that of the decoupling design algorithm. In addition, the more preselected users, the higher the system SE. Compared with the traditional algorithm, the proposed robust joint design algorithm can obtain higher system SE at the same transmission power, and thus can improve the system EE. Under the conditions of $\sigma^2_{f_d^{rsd}} = 25$, $\sigma^2_{\phi_\tau} = 10$ and $P_T = 30$ W, when $U_{l,p}/U_s = 2, U_{l,p} = 4$ and $U_{l,p}/U_s = 3, U_{l,p} = 6$, with the improvement of system SE performance, the EE performance gain of the proposed joint design algorithm is 26.16% and 37.85% higher than that of the traditional decoupling design algorithm.

Figure 10 shows the SE comparison of different multicast groups. It can be seen that the SE of each multicast group of the proposed robust algorithm is higher than that of the nonrobust algorithm. Meanwhile, with the increase in the number of preselected users, the diversity of users increases, and the performance of the proposed joint user scheduling and hybrid beamforming design algorithm also improves. This is because with the increase in the number of preselected users, the range of users that can be scheduled and selected increases. By scheduling different users in each multicast group, the SE can be further improved. Meanwhile, with the improvement of the system SE, the system EE performance gain also increases, which verifies the effectiveness of the joint user scheduling and hybrid beamforming design algorithm.

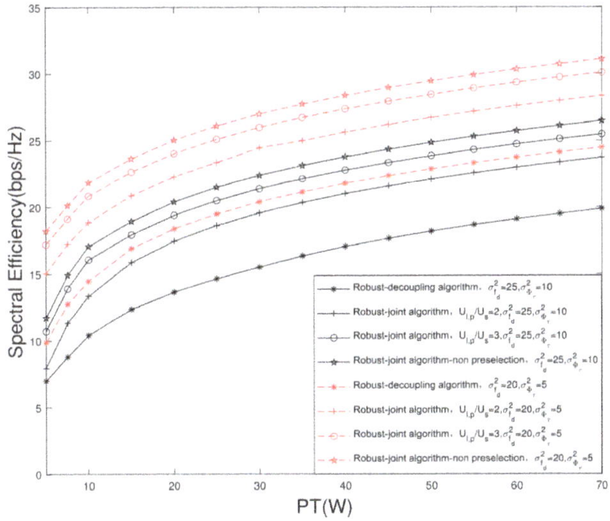

Figure 9. Change trajectory of system SE versus different transmission power thresholds P_T.

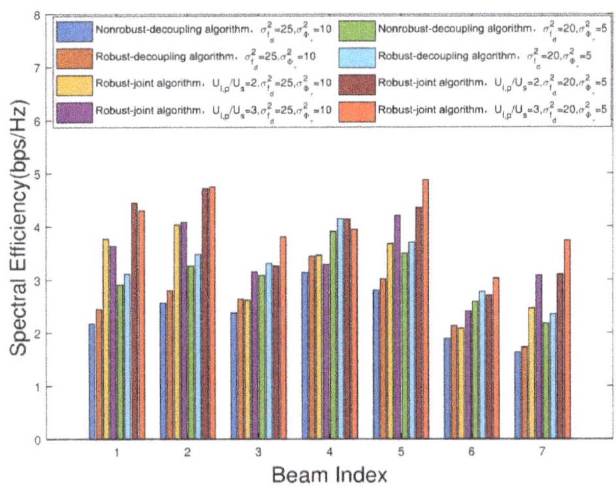

Figure 10. Comparison of SE of different multicast groups.

Figure 11 shows the change trajectory of the system EE and SE versus the different U_s. It can be seen that with the increase in U_s, the system EE and SE show a downward trend. This is because the communication rate of each multicast group is constrained by the user with the worst SINR in the multicast group. With the increase in U_s, if the users' channel vectors in the multicast group remain collinear, the EE and SE would remain unchanged. However, according to the rules of the user preselection and scheduling, with the increase in U_s, the collinearity among users in the multicast group would decrease, which can result in the increase in interference and the decrease in the worst SINR in each multicast group. In other words, with the decrease in U_s, the users' SINR will be improved. Therefore, with the improvement of SINR, the system EE performance gain also increases, as shown in Figure 11a. Under the condition of $U_{l,p}/U_s = 2, U_{l,p} = 4$, when $U_s = 2$, the EE performance gain is 21.05% higher than when $U_s = 4$.

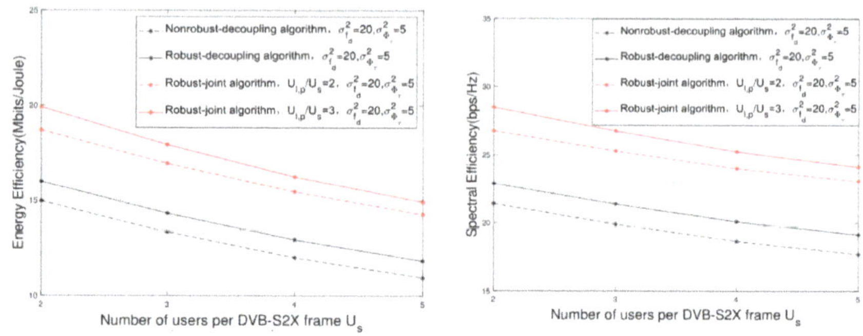

(a) (b)

Figure 11. Change trajectory of system performance versus different U_s. (**a**) Change trajectory of system EE versus different U_s; (**b**) change trajectory of system SE versus different U_s.

Figure 12 shows the performance comparison of different algorithms for solving $\boldsymbol{F}_{BB}$ and $\boldsymbol{F}_{RF}$. In this simulation, we set three comparison algorithms, i.e., the optimal design algorithm, the alternating minimization algorithm based on the phase extraction (PE-Altmin) algorithm and the orthogonal matching pursuit (OMP) algorithm. The optimal design algorithm refers to the numerical simulation result of the hybrid beamforming matrix $\boldsymbol{F}$. It can be seen that the system performance of the MM-AltOpt algorithm is slightly lower than that of the optimal design algorithm. Meanwhile, in Figure 12a, when $P_T = 10$ W, we can see that the system EE performance gain of the proposed MM-AltOpt algorithm is improved by about 2% and 5%, respectively, compared with the PE-Altmin algorithm and the OMP algorithm. In addition, from the perspective of algorithm complexity, the complexities of the MM-AltOpt algorithm, PE-Altmin algorithm and OMP algorithm are $O\big(I_{MM}\big(N^3 + I_{Inner}2NL^2 + NL\big)\big)$, $O\big(I_{PE}\big(N^3 + L^3 + NL\big)\big)$ and $O\big(I_{OMP}\big(L^4N + L^2 + N^2L^2 + 2L^3\big)\big)$, respectively, where I_{MM}, I_{Inner}, I_{PE} and I_{OMP} are the number of iterations of the corresponding algorithm. In conclusion, we can see that the complexity of the proposed MM-AltOpt algorithm is close to that of the other two algorithms, however, the system EE performance is higher, which can verify the effectiveness of the proposed algorithm.

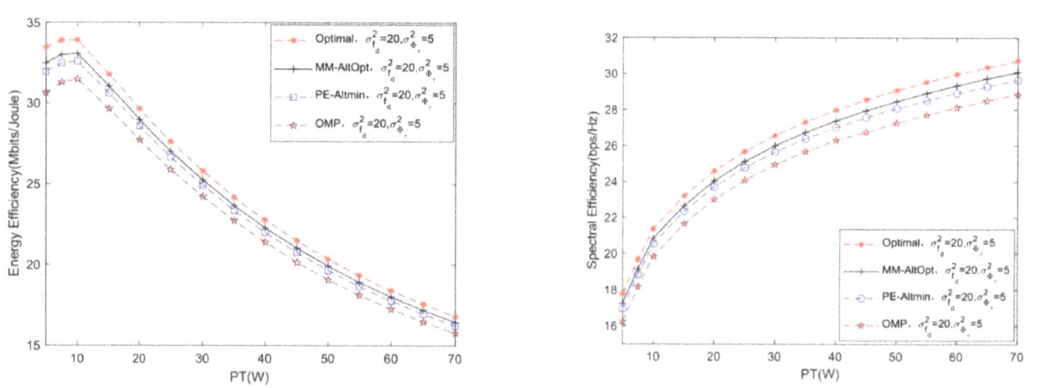

(a) (b)

Figure 12. Comparison of system performance of different algorithms for solving $\boldsymbol{F}_{BB}$ and $\boldsymbol{F}_{RF}$. (**a**) Comparison of system EE of different algorithms for solving $\boldsymbol{F}_{BB}$ and $\boldsymbol{F}_{RF}$; (**b**) comparison of system SE of different algorithms for solving $\boldsymbol{F}_{BB}$ and $\boldsymbol{F}_{RF}$.

9. Conclusions

In this paper, we investigated the robust joint user scheduling and hybrid beamforming design scheme for the massive MIMO LEO satellite multigroup multicast communication system. The scheme design considered the limited transmission power of the LEO satellite and the requirement of QoS and analyzed the influence of residual Doppler shift and phase disturbance on CSI errors. On this basis, taking the system EE as the optimization objective, we focused on the robust joint user scheduling and hybrid beamforming design. To reduce the complexity of the algorithm, we proposed the user preselection step, which can significantly reduce the system complexity while ensuring the system performance. For the nonconvex problem of the objective function, we adopted the CCCP framework after transforming the optimization problem into the DC programming problem. For the rank-one constraint, we proposed the penalty iterative algorithm. Finally, to obtain the digital and analog beamforming matrices, we adopted the MM-AltOpt algorithm.

Numerical results indicated that the proposed algorithm can effectively improve the system EE. The EE performance gain of the proposed robust algorithm was improved by nearly 10% compared with the traditional nonrobust algorithm. Meanwhile, the EE performance gain of the proposed joint user scheduling and hybrid beamforming design algorithm was improved by nearly 40% compared with the traditional decoupling design algorithm. In conclusion, the robust joint user scheduling and hybrid beamforming design algorithm proposed in this paper can significantly improve the system EE performance.

Author Contributions: Conceptualization, Y.L. and C.L.; methodology, Y.L. and J.L.; software, Y.L. and L.F.; validation, Y.L., C.L. and J.L.; formal analysis, Y.L. and C.L.; investigation, Y.L.; writing—original draft preparation, Y.L. and L.F.; writing—review and editing, C.L. and J.L. All authors have read and agreed to the published version of the manuscript.

Funding: This research was funded by National Natural Science Foundation of China, grant number 62001516.

Institutional Review Board Statement: Not applicable.

Informed Consent Statement: Not applicable.

Data Availability Statement: The data presented in this study are available in the manuscript.

Conflicts of Interest: The authors declare no conflict of interest.

Abbreviations

MIMO	multi-input multi-output
LEO	low earth orbit
UPA	uniform planar array
EE	energy efficiency
SDP	semidefinite programming
CCCP	concave convex process
MM	majorization-minimization
AltOpt	alternative optimization
SE	spectrum efficiency
CSI	channel state information
GEO	geosynchronous earth orbit
QoS	quality of service
BFP	Boolean fractional programming
QCQP	quadratic constraint quadratic programming
DC	difference of convex
FEC	forward error correction
SINR	signal to interference plus noise ratio
SNR	signal-to-noise ratio
CRLB	Cramer–Rao lower bound
EVD	eigenvalue decomposition

GRA	Gaussian randomization algorithm
PE-Altmin	alternating minimization algorithm based on the phase extraction
OMP	orthogonal matching pursuit
OFDMA	orthogonal frequency division multiple access
WMMSE	weighted minimum mean-square error

References

1. Jin, L.; Wang, L.; Jin, X.; Zhu, J.; Duan, K.; Li, Z. Research on the Application of LEO Satellite in IOT. In Proceedings of the 2022 IEEE 2nd International Conference on Electronic Technology, Communication and Information (ICETCI), Changchun, China, 27–29 May 2022; pp. 739–741.
2. Chen, S.; Sun, S.; Kang, S. System integration of terrestrial mobile communication and satellite communication—the trends, challenges and key technologies in B5G and 6G. *China Commun.* **2020**, *17*, 156–171. [CrossRef]
3. de Figueiredo, F.A.P. An Overview of Massive MIMO for 5G and 6G. *IEEE Lat. Am. Trans.* **2022**, *20*, 931–940. [CrossRef]
4. Lorincz, J.; Begusic, D. Adaptive beamforming structure with STBC for IEEE 802.11n WLAN systems. In Proceedings of the 2008 16th International Conference on Software, Telecommunications and Computer Networks, Split, Croatia, 25–27 September 2008; pp. 258–263. [CrossRef]
5. You, L.; Li, K.X.; Wang, J.; Gao, X.; Xia, X.G.; Ottersten, B. Massive MIMO transmission for LEO satellite communications. *IEEE J. Sel. Areas Commun.* **2020**, *38*, 1851–1865. [CrossRef]
6. Morello, A.; Mignone, V. DVB-S2X: The new extensions to the second generation DVB satellite standard DVB-S2. *Int. J. Satell. Commun.* **2016**, *34*, 323–325. [CrossRef]
7. Qiang, X.; You, L.; Li, K.X.; Tsinos, C.G.; Wang, W.; Gao, X.; Ottersten, B. Hybrid A/D Precoding for Downlink Massive MIMO in LEO Satellite Communications. In Proceedings of the 2021 IEEE International Conference on Communications Workshops (ICC Workshops), Montreal, QC, Canada, 14–23 June 2021.
8. Li, K.X.; You, L.; Wang, J.; Gao, X.; Tsinos, C.G.; Chatzinotas, S.; Ottersten, B. Downlink Transmit Design in Massive MIMO LEO Satellite Communications. *IEEE Trans. Commun.* **2020**, *70*, 1014–1028. [CrossRef]
9. Gupta, N.; Bitragunta, S. Green Satellite Communication Link Design, Optimization, and Performance Analysis. In Proceedings of the 2020 IEEE 7th Uttar Pradesh Section International Conference on Electrical, Electronics and Computer Engineering (UPCON), Prayagraj, India, 27–29 November 2020.
10. Jain, S.; Markan, A.; Markan, C. Performance Evaluation of a Millimeter Wave MIMO Hybrid Beamforming System. In Proceedings of the 2020 IEEE Latin-American Conference on Communications (LATINCOM), Santo Domingo, Dominican Republic, 18–20 November 2020; pp. 1–5.
11. Christopoulos, D.; Chatzinotas, S.; Ottersten, B. Multicast Multigroup Precoding and User Scheduling for Frame-Based Satellite Communications. *IEEE Trans. Wirel. Commun.* **2015**, *14*, 4695–4707. [CrossRef]
12. Wang, W.; Liu, A.; Zhang, Q.; You, L.; Gao, X.; Zheng, G. Robust Multigroup Multicast Transmission for Frame-Based Multi-Beam Satellite Systems. *IEEE Access* **2018**, *6*, 46074–46083. [CrossRef]
13. Bandi, A.; Chatzinotas, S.; Ottersten, B. Joint Scheduling and Precoding for Frame-Based Multigroup Multicasting in Satellite Communications. In Proceedings of the 2019 IEEE Global Communications Conference (GLOBECOM), Waikoloa, HI, USA, 9–13 December 2019; pp. 1–6.
14. Zhang, S.; Jia, M.; Wei, Y.; Guo, Q. User scheduling for multicast transmission in high throughput satellite systems. *EURASIP J. Wirel. Commun. Netw.* **2020**, *2020*, 133. [CrossRef]
15. Guidotti, A.; Vanelli-Coralli, A. Clustering Strategies for Multicast Precoding in Multi-Beam Satellite Systems. *Int. J. Satell. Commun. Netw.* **2018**, *38*, 85–104. [CrossRef]
16. Bandi, A.; Chatzinotas, S.; Ottersten, B. A Joint Solution for Scheduling and Precoding in Multiuser MISO Downlink Channels. *IEEE Trans. Wirel. Commun.* **2020**, *19*, 475–490. [CrossRef]
17. Jiang, J.; Kong, D. Joint User Scheduling and MU-MIMO Hybrid Beamforming Algorithm for mmWave FDMA Massive MIMO System. *Int. J. Antennas Propag.* **2016**, *2016*, 4341068. [CrossRef]
18. Bogale, T.E.; Le, L.B.; Haghighat, A. User scheduling for massive MIMO OFDMA systems with hybrid analog-digital beamforming. In Proceedings of the 2015 IEEE International Conference on Communications (ICC), London, UK, 8–12 June 2015; pp. 1757–1762. [CrossRef]
19. Yu, X.; Shen, J.C.; Zhang, J.; Letaief, K.B. Alternating Minimization Algorithms for Hybrid Precoding in Millimeter Wave MIMO Systems. *IEEE J. Sel. Top. Signal Process.* **2016**, *10*, 485–500. [CrossRef]
20. You, L.; Li, K.X.; Wang, J.; Gao, X.; Xia, X.G.; Otterstenx, B. LEO Satellite Communications with Massive MIMO. In Proceedings of the ICC 2020–2020 IEEE International Conference on Communications (ICC), Dublin, Ireland, 7–11 June 2020.
21. Palacios, J.; Gonzalez-Prelcic, N.; Mosquera, C.; Shimizu, T.; Wang, C.H. A hybrid beamforming design for massive MIMO LEO satellite communications. *arXiv* **2021**, arXiv:2104.11158. [CrossRef]
22. Khan, T.A.; Afshang, M.A. Stochastic Geometry Approach to Doppler Characterization in a LEO Satellite Network. In Proceedings of the ICC 2020–2020 IEEE International Conference on Communications (ICC), Dublin, Ireland, 7–11 June 2020; pp. 1–6.
23. Wei, Q.; Chen, X.; Zhan, Y.F. Exploring Implicit Pilots for Precise Estimation of LEO Satellite Downlink Doppler Frequency. *IEEE Commun. Lett.* **2020**, *24*, 2270–2274. [CrossRef]

24. Liu, Y.; Li, C.; Li, J.; Feng, L. Robust Energy-Efficient Hybrid Beamforming Design for Massive MIMO LEO Satellite Communication Systems. *IEEE Access* **2022**, *10*, 63085–63099. [CrossRef]
25. Vilnrotter, V.A.; Hinedi, S.; Kumar, R. Frequency estimation techniques for high dynamic trajectories. *IEEE Trans. Aerosp. Electron. Syst.* **1989**, *25*, 559–577. [CrossRef]
26. Mangum, J.G.; Wallace, P. Atmospheric Refractive Electromagnetic Wave Bending and Propagation Delay. *Publ. Astron. Soc. Pac.* **2015**, *127*, 500–501. [CrossRef]
27. Hopfield, H.S. Two-Quartic tropheriere fractivity profile for correcting satellite data. *J. Geophys. Res.* **1969**, *74*, 4487–4499. [CrossRef]
28. Saastaminen, J. Contribution to the Theory of Atmospheric regraction. *Bull. Géodésique* **1973**, *107*, 13–14.
29. Wang, W.; Gao, L.; Ding, R.; Lei, J.; You, L.; Chan, C.A.; Gao, X. Resource Efficiency Optimization for Robust Beamforming in Multi-Beam Satellite Communications. *IEEE Trans. Veh. Technol.* **2021**, *70*, 6958–6968. [CrossRef]
30. Sun, X.; Gao, X.; Li, G.Y.; Han, W. Agglomerative user clustering and downlink group scheduling for FDD massive MIMO systems. In Proceedings of the 2017 IEEE International Conference on Communications (ICC), Paris, France, 21–25 May 2017; pp. 1–6.
31. Gao, L.; Wang, W.; Ding, R.; You, L.; Gao, X. Resource efficiency optimization for robust multigroup multicast satellite communications. In Proceedings of the IEEE ICC Workshops, Montreal, QC, Canada, 14–23 June 2021; pp. 1–6.
32. Bengtsson, M.; Ottersten, B. Optimal and suboptimal transmit beamforming. In *Handbook of Antennas in Wireless Communications*; CRC Press: Boca Raton, FL, USA, 2001.
33. Sun, Y.; Babu, P.; Palomar, D.P. Majorization-Minimization Algorithms in Signal Processing, Communications, and Machine Learning. *IEEE Trans. Signal Process.* **2017**, *65*, 794–816.
34. Shen, K.; Yu, W. Fractional Programming for Communication Systems—Part I: Power Control and Beamforming. *IEEE Trans. Signal Process.* **2018**, *66*, 2616–2630. [CrossRef]
35. Kachouh, A.; Nasser, Y.; Artail, H.A. On the capacity optimization of D2D underlying cellular communications. In Proceedings of the 2016 23rd International Conference on Telecommunications (ICT), 16–18 May 2016; pp. 1–5.
36. Luo, Z.Q.; Ma, W.K.; So, A.M.C.; Ye, Y.; Zhang, S. Semidefinite relaxation of quadratic optimization problems. *IEEE Signal Process. Mag.* **2010**, *27*, 20–34. [CrossRef]
37. Lipp, T.; Boyd, S. Variations and extension of the convex–concave procedure. *Optim. Eng.* **2016**, *17*, 263–287. [CrossRef]
38. Gahinet, P.; Nemirovski, A.; Laub, A.J.; Chilali, M. *LMI Control Toolbox Users Guide*; MathWorks: Natick, MA, USA, 1995.
39. Shao, M.; Li, Q.; Ma, W.K.; So, A.M.C. A framework for one-bit and constant-envelope precoding over multiuser massive MISO channels. *IEEE Trans. Signal Process.* **2019**, *67*, 5309–5324. [CrossRef]

sensors

MDPI

Article

Binary PSO with Classification Trees Algorithm for Enhancing Power Efficiency in 5G Networks

Mayada Osama [1,2,*], Salwa El Ramly [2] and Bassant Abdelhamid [2]

[1] Electronics and Communications Department, Faculty of Engineering Science and Arts, Misr International University, Cairo 11828, Egypt
[2] Electronics and Communications Department, Faculty of Engineering, Ain Shams University, Cairo 11517, Egypt
* Correspondence: mayada.osama@miuegypt.edu.eg

Abstract: The dense deployment of small cells (SCs) in the 5G heterogeneous networks (HetNets) fulfills the demand for vast connectivity and larger data rates. Unfortunately, the power efficiency (PE) of the network is reduced because of the elevated power consumption of the densely deployed SCs and the interference that arise between them. An approach to ameliorate the PE is proposed by switching off the redundant SCs using machine learning (ML) techniques while sustaining the quality of service (QoS) for each user. In this paper, a linearly increasing inertia weight–binary particle swarm optimization (IW-BPSO) algorithm for SC on/off switching is proposed to minimize the power consumption of the network. Moreover, a soft frequency reuse (SFR) algorithm is proposed using classification trees (CTs) to alleviate the interference and elevate the system throughput. The results show that the proposed algorithms outperform the other conventional algorithms, as they reduce the power consumption of the network and the interference among the SCs, ameliorating the total throughput and the PE of the system.

Keywords: 5G HetNets; BPSO; classification trees (CTs); soft frequency reuse (SFR); small cells (SCs)

Citation: Osama, M.; El Ramly, S.; Abdelhamid, B. Binary PSO with Classification Trees Algorithm for Enhancing Power Efficiency in 5G Networks. *Sensors* **2022**, *22*, 8570. https://doi.org/10.3390/s22218570

Academic Editor: Josip Lorincz

Received: 5 October 2022
Accepted: 4 November 2022
Published: 7 November 2022

1. Introduction

Recently, the exponential growth of the numerous wireless devices and the data-hungry applications have earned huge significance. This required imperious expansion of the 5G network to support the forthcoming 5G use cases, such as video live-streaming, conferencing, online gaming, etc. [1,2]. Moreover, the 5G cellular network is planned to elevate the capacity 1000 times and the spectrum efficiency by 5–15 times with respect to 4G [3,4]. This can be achieved by utilizing heterogeneous networks (HetNets), which can enhance the system data rates and the quality of service (QoS) of the users as the small cells (SCs) are deployed within the macro cells (MCs) coverage area. Furthermore, SCs offer the benefit of providing service to previously uncovered regions and in the network regions demanding larger capacity [2,3,5–7]. Figure 1 shows a general representation of the HetNet scenario with the MCs underlaid by the densely deployed SCs.

It is foreseen that the massive growth in the SC deployment will be continued in the coming years [8,9], leading to various challenges such as the interference among the SCs [10], their elevated power consumption [11,12], and the elevated operating expenses [9]. Thus, it is crucial to face these challenges to ameliorate the performance of 5G HetNets.

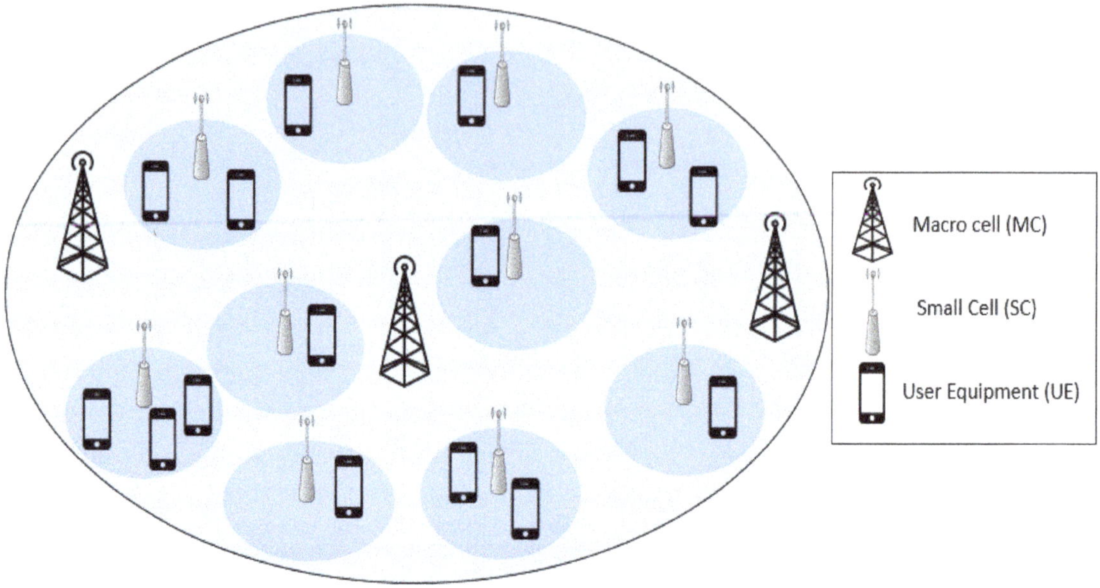

Figure 1. General representation of HetNet scenario with densely deployed small cells.

The main objective of our paper is to propose a new approach for the irregular nature of the 5G HetNets that merge the usage of both binary particle swarm optimization (BPSO) algorithm with linear increasing inertia weight (IW) and soft frequency reuse (SFR) to maximize the power efficiency (PE) of the SCs and minimize the number of active SCs while guaranteeing the QoS for the UEs. In SFR, every SC is split into center and edge regions where one of the unutilized sub-bands by the edge regions of the adjoining SCs is allocated to the edge region of the SC, while the remaining sub-bands are utilized in the center region of the SC with reduced transmission power to diminish the interference to the adjoining SCs. Unlike the prior works [9,13,14] that will be discussed later in Section 2, the proposed algorithm utilizes the linear increasing IW-BPSO algorithm for selecting the SCs to be switched on/off, and we propose the SFR utilizes classification trees (CTs) to enhance the network PE, taking into consideration the irregular nature of the 5G HetNets. The on/off switching of SCs using BPSO algorithm is carried out first; then the SFR is applied for sub-band allocation to minimize the number of operations required for allocating the sub-bands to the SCs.

The main contributions of this paper can be summarized as follow:

1. Propose an algorithm for irregular 5G HetNets based on BPSO algorithm for SC on/off switching to ameliorate PE of the system, and, using a linearly increasing IW approach where the IW is linearly increasing in each iteration, to enhance the convergence of the BPSO algorithm.
2. Propose a novel frequency allocation algorithm for SFR based on the CTs as it is simple and accurate machine learning (ML) technique to mitigate the interference among the irregularly shaped SCs.

Results demonstrate that the proposed algorithms have superior performance over the conventional algorithms (always on, random 10%, and BPSO only), as it has higher total system throughput and PE, and lower system power consumption and outage probability.

The remainder of the paper is organized as follows: the literature review is presented in Section 2. Section 3 demonstrates the system model, and Section 4 explains the proposed algorithms. Then, Section 5 shows the simulation results. Eventually, Section 6 concludes the paper.

2. Literature Review

Recently, immense research has been carried out to ameliorate the PE of the mod-ern communication networks, such as satellite and terrestrial networks [15,16], massive MIMO systems [17] and SCs networks [18–23].

SC on/off switching is an auspicious approach to minimize the power consumption and enhance the PE of the system [18–23]. The authors in [18,19] studied the elevated the energy consumption of WLAN. They proposed solving the problem by on/off switching and power adjustment of the access stations. The authors in [20] proposed a load-aware strategy, where SCs in HetNets are switched to sleep mode according to their load level. In [21], every SC independently switches off upon the decrease in the number of user equipments (UEs) and activates using one of three approaches. The first approach is the sleeping SC keeps sensing the interference plus noise levels and switches on when a new UE is sensed in its coverage area. The second approach is that the MC sends a wake-up request to all SCs upon increasing the number of UEs associated with the MC, then switches off the SCs with no UEs later. In the third approach, a time advance indicator is sent to the MC by the UEs and the SCs and is utilized by the MC to determine the nearest SC to the UE to switch on. In [22], pre-sleeping SCs at the same zone create a sleeping cluster. Then, the SCs in the sleeping cluster are randomly selected to be switched off leaving only one active SC to guarantee the coverage. On the other hand, the authors in [23] proposed switching off the SCs and handing over the UEs to the MC. However, the unplanned SC off switching may increase the unnecessary handovers and underutilize the SCs. Thus, novel SC on/off switching techniques are required to enhance the performance of the 5G HetNets and to reduce the power consumption of the system.

PSO is a prevalent meta-heuristic algorithm used in solving optimization problems [9]; thus, the authors in [9,24,25] utilized the PSO algorithm for SC switching to enhance the performance of the system. The authors in [9] proposed an efficient cell modeling (ECM) algorithm to set up the connection initially between the UEs and the SCs by selecting the strongest received signals. Then, the BPSO algorithm is utilized to turn off the excessive SCs. The authors in [24] proposed first utilization of BPSO algorithm to choose the MCs' optimum locations not only to achieve minimum overlap but also to guarantee a reasonable coverage for the UEs. Then, a multi-stage PSO (MS-PSO) algorithm consisting of two interactive loops are utilized. The outer loop is utilized to switch the SCs (on or off), while the inner loop is utilized to optimize the active power of the SC and to elevate the power of the SC if the data rate rises. On the other hand, a combined optimal frequency and power allocation (COFPA) scheme is proposed in [25]. First, using the BPSO algorithm, the MCs are switched on and off until the lead interferer is abolished with minimum cost function and with a reasonable coverage to the UEs. Then, the MS-PSO algorithm is utilized to control the SC switching to mitigate the interference and minimize the power consumption. However, the convergence of the PSO can be enhanced by adjusting its IW, leading to improve the system performance.

IW has a significant role in the process of offering a trade-off among diversification and intensification skills of PSO algorithm. Reducing the IW facilitates exploring the search space (global search), although raising the IW aids exploiting the search space (local search) [26] to find the solution (particle). Numerous approaches are presented to adjust the IW such as the constant IW [9,27] and the random IW [28]. However, the constant IW approach can fail to balance exploration and exploitation because of the lack of adjustment of IW [29,30]. On the other hand, the authors in [31] propose a linearly decreasing IW technique where the IW is initialized at a larger value; then it is linearly reduced to a smaller value. However, in this technique, the tendency of the particles to local search is constantly increasing. The authors in [32,33] demonstrated that the increase in IW surpasses the decrease in IW for PSO on all their tested benchmarks. Thus, the linearly increasing IW is adopted in our paper.

Various interference mitigation techniques for the modern networks are presented in the literature such as advanced multiple access techniques [34,35] and frequency reuse (FR) [13,36–39]. FR is an auspicious approach aiming to mitigate the interference in the modern HetNets [37]. SFR is presented to minimize the interference in HetNets [13,36,38,39]. The authors in [36] present SFR in HetNets, where the MC organizes the assignment of the resource plans to the SCs, and the SCs choose the resource plan. However, this algorithm cannot be implemented in the absence of the MC. A multi-level SFR (MSFR) for HetNets is demonstrated in [38], where every cell is split into three zones (central, intermediate, and edge), utilizing various frequency segments and transmission power levels. The authors in [13] proposed a novel SFR algorithm to mitigate the interference by splitting the SC to the center and edge regions. Moreover, the on/off switching of the SCs depends on their interference contribution rate (ICR) values. The authors in [39] presented an MSFR scheme where every MC is split into various circular areas, and different spectrum and power are allocated to every area. Unfortunately, in this scheme, SFR is not applied to the SCs but only the MCs. New SFR approaches are essential to alleviate the interference and improve the network throughput in 5G HetNets.

Moreover, the performance of 5G HetNets can be enhanced using ML techniques [40]. Decision trees (DTs) is one of the propitious supervised learning (SL) techniques, where every training example must be fed with its label to train a learning model, then utilize this model to predict the output for any new data. The authors in [41] proposed using supervised ML to enhance both the classification of services and the distribution of network resources in 5G networks. Moreover, the results reveal that DTs and random forests are the best approaches. DTs are split to classification trees (CTs) and regression trees [40]. The authors in [42] studied several anomaly detection techniques in 5G traffic. The performance of these techniques is analyzed based on multiple factors such as the probability of identifying anomalies and the probability of detecting a false positive. The results demonstrated that the CTs technique outperforms the other techniques. The problem of monitoring and predicting the quality of experience of cellular networks is studied in [43]. The authors compared various SL techniques and trained them utilizing training data based on traffic measurements of the UEs from a field trial study. The CTs are the ultimately chosen model because of their superb prediction accuracy and their prediction speed. In our paper, CT is utilized for the first time to allocate the frequencies in the SFR 5G network.

Finally, to match real-life scenarios, the irregular nature of the 5G HetNets was modeled utilizing Voronoi cells [44–47]. Because of the immense deployment of SCs, Voronoi cells are considered more practical than traditional hexagonal grids [48].

3. System Model

Consider 5G HetNet with densely deployed Voronoi SCs, where the SCs are deployed within the MCs coverage area. In this scenario, the MCs and the SCs utilize different frequency bands, mitigating the cross-tier interference between them. The MCs stay active to maintain the coverage of the network when the SCs are turned off. On the other hand, the SCs are either active (on) or asleep (off). When the SC is turned off, regular discovery signals are sent by the SC to be detected by any potential user. Each UE reports its channel state information and its reference signal received power (RSRP) to its SC. The SC sub-band allocation and the SC switching is organized by a main controller, or the MC if the main controller is absent, to collect the data from the SCs, allocate the sub-bands to the SCs, and determine the on/off switching decisions of the SCs. Furthermore, it is assumed that all UEs in every SC are located inside the coverage area of the SC. In case of the existence of any coverage holes in the SCs, relay nodes can be utilized to cover these holes [49]. However, this is not considered in our paper. Nomenclature lists the described symbols utilized in this paper.

An SC on/off indicator γ is defined, where $\gamma_m = 1$ when the SC m is active; otherwise, $\gamma_m = 0$. The number of SCs is denoted by "M" and the number of UEs in SC m is denoted by "U_m". The UE association indicator $\varphi_{m,k} = 1$ if UE k is associated with SC m; otherwise, $\varphi_{m,k} = 0$. The signal to interference and noise ratio (SINR) of UE k in SC m can be calculated as [14]:

$$SINR_{k,m} = \frac{\gamma_m P_m G_{m,k}}{\sum_{n \neq m,\ n \in M} \gamma_n P_n G_{n,k} + N} \tag{1}$$

where P_m and P_n are the transmission powers of the serving SC m and the interfering SC n, respectively. The channel gain between UE k and serving SC m is $G_{m,k} = {d_{m,k}}^{-\alpha}$, where $d_{m,k}$ is the distance between SC m and UE k and α is the path loss exponent [50]; the channel gain between UE k and interfering SC n is $G_{n,k}$ and N is the noise power. The data rate $R_{k,m}$ of UE k in SC m is also calculated by Shannon's formula as [13]:

$$R_{k,m} = B_k \log_2(1 + SINR_{k,m}) \tag{2}$$

where $B_k = B_{RB} L_{k,m}$ is the bandwidth allocated to UE k, while the resource block (RB) bandwidth is B_{RB} and $L_{k,m}$ is the number of requisite RBs for UE k in SC m to achieve the minimum data rate [14].

While the total throughput of the system is given by [9]:

$$C_{sys} = \sum_{m=1}^{M} \gamma_m \sum_{k=1}^{U_m} R_{k,m} \tag{3}$$

Due to the dense deployment of the SCs, some SCs can be turned off without affecting the QoS of the UEs. Thus, the total power consumption of SC m can be calculated as [13]:

$$P_{m_{tot}} = \beta P_{m_{on}} + (1 - \beta) P_{m_{on}} \gamma_m + \theta_m P_{m_{tx}} \gamma_m \tag{4}$$

where $P_{m_{on}}$ and $P_{m_{tx}}$ are the baseline and the transmission power consumption, respectively, while β is the inactive level of the SC, such that $P_{m_{off}} = \beta P_{m_{on}}$, and θ_m is the portion of power consumption that is due to the feeder losses and power amplifier of SC m [51]. The total power consumption of the system (P_{sys}) is the sum of the power consumption of all SCs. The PE of the system is given by [9]:

$$PE_{sys} = \frac{C_{sys}}{P_{sys}} \tag{5}$$

To improve the PE, the SCs on/off switching decisions using BPSO is proposed in this paper. The PSO is an iterative population-based search algorithm inspired by the hunting behavior of a flock of flying birds [52–54]. In PSO, every particle is considered a bird of the flock and represents a possible solution to the problem [30,55–57]. The search begins with an initial set of particles and attempts to find the best solution by searching around the solution space. The motion of the particle is based on its local best position, and the best-known position of all the other particles [9,52]. The fitness value of every particle is calculated using a fitness function that is optimized in every iteration [9,52].

The set of particles X = $\{x_1, x_2, \ldots, x_{N_{par}}\}$ is defined, where $x_{N_{par}}$ represents one possible status for the SCs, while N_{par} is the swarm size. In BPSO algorithm, the population is randomly initialized as binary values. For every particle, the population binary value of 1 signifies the active SC, while 0 signifies the sleeping SC. The velocity of the particle j is initialized as [9,58]:

$$v_j = v_{min} + (v_{max} - v_{min})\, a_1 \tag{6}$$

where v_{min} and v_{max} denote the minimum and the maximum velocity of the particle, respectively, while a_1 is a random number uniformly distributed between 0 and 1 [9]. The velocity and the position of the particles are updated in each iteration. The velocity of the particle j in iteration $(z+1)$ is updated as [9,26]:

$$v_j^{\,z+1} = \omega_z\, v_j^{\,z} + c_1\, a_2\Big(P_{best_j} - x_j^{\,z}\Big) + c_2\, a_3\Big(G_{best} - x_j^{\,z}\Big) \tag{7}$$

where ω_z is the IW in the zth iteration, while v_j^z and x_j^z are the velocity and position of the particle j in the zth iteration, respectively. The best position of the particle j is denoted as P_{best_j}, while the global best position of all the particles is denoted as G_{best}. Additionally, c_1 and c_2 denote the acceleration parameters [9]. Moreover, a_2 and a_3 are two random numbers uniformly distributed between 0 and 1. It is noted that the particle x_j is a binary vector and the velocity v_j is also a vector. The sigmoid function $(Sig(v_j(m)^z))$ is given as [52,55]:

$$Sig(v_j(m)^z) = \frac{1}{1+e^{-v_j(m)^z}} \tag{8}$$

Thus, the on/off state of the SC m in particle j in the z-th iteration is calculated as [52,55]:

$$x_j(m)^z = \begin{cases} 1, & a_4 < Sig(v_j(m)^z) \\ 0, & otherwise \end{cases} \tag{9}$$

where a_4 is a random number uniformly distributed between 0 and 1.

Since the IW is the pivotal factor in the convergence of the PSO, it should be carefully adjusted. Thus, linearly increasing IW is utilized in this paper, where the IW linearly increases every iteration from ω_{min} to ω_{max} . The IW in the zth iteration is given as [33]:

$$\omega_z = (\omega_{max} - \omega_{min})\left(\frac{z-1}{Z_{max} - 1}\right) + \omega_{min} \tag{10}$$

where ω_{max} and ω_{min} denote the maximum and minimum IW, respectively [33], and Z_{max} is the maximum number of iterations [59].

4. Proposed Algorithms

To alleviate the number of active SCs and enhance the PE of the system, a linearly increasing IW-BPSO algorithm for SC on/off switching is proposed in this paper. Moreover, a novel SFR technique using CTs is proposed for SC sub-band allocation. The BPSO algorithm is applied first while the linearly increasing IW enhances the convergence of the algorithm. Then, the sub-bands are allocated to the active SCs using the novel SFR technique. It is worth noting that the new SFR technique is applied after the SC switching to allocate the sub-bands to the active SCs, only aiming to reduce the number of operations needed in the sub-band allocation to the SCs

4.1. SC on/off Switching Using Linearly Increasing IW-BPSO Algorithm

In this paper, SC switching utilizing a linearly increasing IW-BPSO algorithm is proposed. The UE can associate with an SC m if $SINR_{k,m}$ exceeds a certain threshold $(SINR_{thr})$. At the beginning of our proposed algorithm, each UE k calculates the SINR from all SCs to determine all SCs that it can possibly associate with, then sorts the received SINR from these SCs in a descending order. The UE associates with the SC with the highest received SINR. If the number of UEs in this SC is larger than the maximum number of UEs in the SC (U_{max}), the UE connects with the SC having the next highest SINR. This continues until every UE is associated with one SC.

To minimize the number of active SCs and to elevate the PE, it is required to switch off the excessive SCs, taking into consideration the QoS of the UEs, since the SINR of every

UE exceeds $SINR_{thr}$. Thus, a multi-objective optimization problem had to be solved; this problem can be written as:

$$Objective1:\ \min\left(\sum_{m=1}^{M}\gamma_m\right)Objective2:\ \max\left(PE_{sys}\right) \tag{11}$$

subject to:

$$\gamma_m, \in \{0,1\},\ \forall\, m \in M \tag{11a}$$

$$\varphi_{m,k} \in \{0,1\},\ \forall\, m \in M,\ k \in U_m \tag{11b}$$

$$\sum_{m=1}^{M}\varphi_{m,k}\,\gamma_m = 1,\ \forall\, m \in M,\ k \in U_m \tag{11c}$$

$$U_m \le U_{max},\ \forall\, m \in M \tag{11d}$$

where constraint (11a) is the on/off state indicator of the SC. Constraint (11b) is the UE association indicator. Constraint (11c) states that the UE is associated with only one active SC. Constraint (11d) indicates that the number of UEs in an SC cannot exceed the maximum number of UEs in an SC (U_{max}).

Algorithm 1 summarizes the proposed BPSO-based on/off SC switching algorithm. In line 1, after the random binary initialization of the population of every particle, the velocity of the particles is initialized as Equation (6). Then, for each iteration till the maximum number of iterations (Z_{max}) is reached, the IW (ω_z) is computed (line 3) and the position (x_j) and velocity (v_j) of every particle j are updated using Equations (7) and (9), respectively (lines 5–6). Next, the fitness value of every particle j (F (x_j)) is computed using Equation (11) (line 7). If $F\left(x_j\right) < F\left(P_{best_j}\right)$, then $P_{best_j} = x_j$, and if $F\left(x_j\right) < F(G_{best})$, then $G_{best} = x_j$ (lines 8–13).

Algorithm 1: Proposed linearly increasing IW-BPSO-based on/off SC switching.

Inputs: Locations of UEs, locations of SCs, swarm size (N_{par}), maximum number of iterations (Z_{max})
Output: SC on/off indicator

```
1: Initialize the position (x_j) randomly and velocity (v_j) of every particle j as Equation (6).
2: For z = 1 to Z_max
3:     Calculate ω_z using Equation (10)
4:     For each particle j
5:         Update v_j using Equation (7)
6:         Update x_j using Equation (9)
7:         Calculate new fitness value F (x_j) as Equation (5)
8:         if F (x_j) < F(P_best_j)
9:             P_best_j ← x_j
10:        end if
11:        if F (x_j) < F(G_best)
12:            G_best ← x_j
13:        end if
14:    end For
15: end For
```

4.2. SC Sub-Band Allotment Using Classification Trees (CTs)

After the on/off switching decisions for all SCs are taken, the second phase is initialized, which is the sub-band allocation for the active SCs based on the SFR, which is illustrated in Figure 2, over three cells without loss of generality. In the shown example, if we have three hexagonal-shaped cells, each split into center and edge regions, the frequency band is divided to three (N_{sub} = 3) sub-bands: X, Y, and Z. The edge region of cells 1, 2, and 3 are allocated sub-bands X, Y, and Z, respectively. Consequently, cell 1 is allocated sub-bands Y and Z in the center region. Similarly, cell 2 is allocated X and Z, and cell 3 is allocated X and Y.

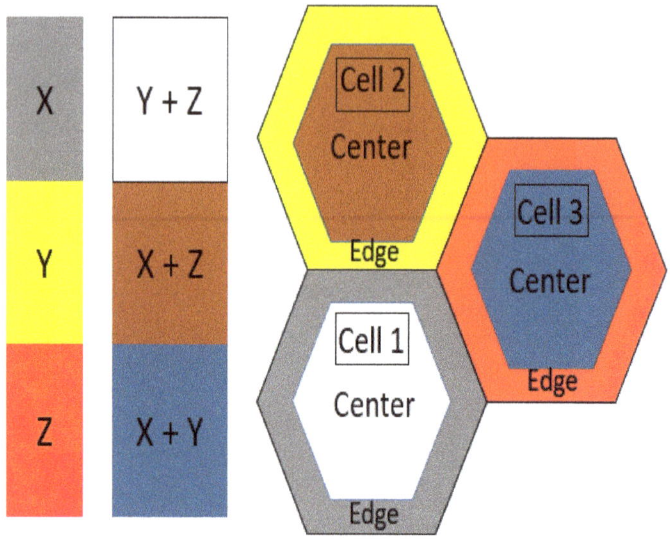

Figure 2. SFR example for hexagonal shaped cells with N_{sub} = 3.

In this technique, every SC is divided into center and edge regions. One of the N_{sub} sub-bands can be used in the edge region of every SC, on condition that it is not used by the edge regions of the adjoin SCs. The center region of every SC can use the remaining sub-bands with reduced transmission power. This alleviates not only the interference to the adjoin SCs but also the power consumption of the whole network. To determine the vertices of the center region, the distance between the center of the SC and every vertex of the SC is computed. Then the distance between the center of the SC and the nearest SC vertex (the smallest distance) is determined and is regarded as the SC radius. The radius of the center zone is chosen as 50% of the SC radius as it maximizes the throughput of the system [13]. A real example is demonstrated in Figure 3a, displaying an SC (the purple SC) and its adjoining SCs. Figure 3b displays the seven sub-bands. Every SC uses one of the seven sub-bands in its edge region. While the center region of this SC (the grey region) can use the remaining six sub-bands.

First, the SC senses the signals of the edge region of the adjoin SCs upon switching to obviate utilizing them. A binary indicator, $\tau_m = 1$, indicates the presence of unused sub-bands by the edge region of the adjoin SCs; otherwise, $\tau_m = 0$. If there are unused sub-bands by the edge region of the adjoined SCs ($\tau_m = 1$), then for every unused sub-band f, the distance between the SC and the closest nonadjacent SC using sub-band f in its edge region ($D_{Non-adf}$) is measured, and this is repeated for the rest of the remaining unused sub-bands. If all the sub-bands are used ($\tau_m = 0$), then for every sub-band q, the distance between the SC and the adjoin SC using sub-band q in its edge region (D_{Adq}) is measured. Then, the edge region of the SC is allocated the sub-band used in the farthest adjoined SC.

(a) (b)

Figure 3. (**a**) A real example demonstrating a SC (the purple SC) with the center region (the grey region); (**b**) the seven used sub-bands.

According to τ_m and $D_{Non-adf}$ or D_{Adq}, the CT takes the decision to assign which sub-band to the edge region of the SC. A CT example of 3 sub-bands is shown in Figure 4. In case of the presence of unused sub-bands in the edge region of the adjoin SCs ($\tau_m = 1$), then if the distance between the SC and the closest SC using the sub-band *1* in its edge region ($D_{Non-adf} = D_{Non-ad1}$) is larger than $D_{Non-ad2}$; $D_{Non-ad1}$ and $D_{Non-ad3}$ are checked. If $D_{Non-ad1}$ is larger than $D_{Non-ad3}$, then sub-band *1* is selected. In case all the sub-bands are utilized ($\tau_m = 0$), then if the distance between the SC and the adjacent SC using sub-band *2* ($D_{Adq} = D_{Ad2}$) is larger than D_{Ad1}, then D_{Ad2} and D_{Ad3} are checked. If D_{Ad3} is larger than D_{Ad2}, then sub-band *3* is selected. Afterward, the center region can utilize the remaining sub-bands with lower transmission power.

Unlike the computational complexity of the conventional SFR algorithm $O(S^2 . {N_{sub}}^2)$, which depends on the network size (S) and the number of sub-bands (N_{sub}) [60], the computational complexity of the proposed CTs algorithm is much lower. Since the computational complexity of the DTs is *O(1)* [61], as no multiplication process is done and only a sequence of branching operations are performed, then computational complexity of the CTs is also *O(1)* for every SC. Thus, the computational complexity of the system is S x *O(1)*, since only a sequence of branching operations are performed on moving along the CT according to the binary indicator (τ_m) and the distance between the SC and the closest nonadjacent SC ($D_{Non-adf}$)/farthest adjoining SC (D_{Ad_q}) using the sub-band in its edge region. Hence, applying SFR after the SC switching greatly reduces the computational complexity as SFR will be applied to the active SCs only (smaller S), while the complexity of the BPSO algorithm $O(Z_{max} . P)$ depends on the maximum number of iterations and the population size (P) [62,63]. On the other hand, the average computational time of the proposed algorithm is less than one minute (54.857 s). In modern HetNets, the traffic load of the network is monitored for longer time periods in the order of 5–15 min [6,64–66], while the on-off switching decisions of the SCs are usually taken every 15–60 min [64,66]. Thus, the proposed algorithm is suitable for practical implementation in real time.

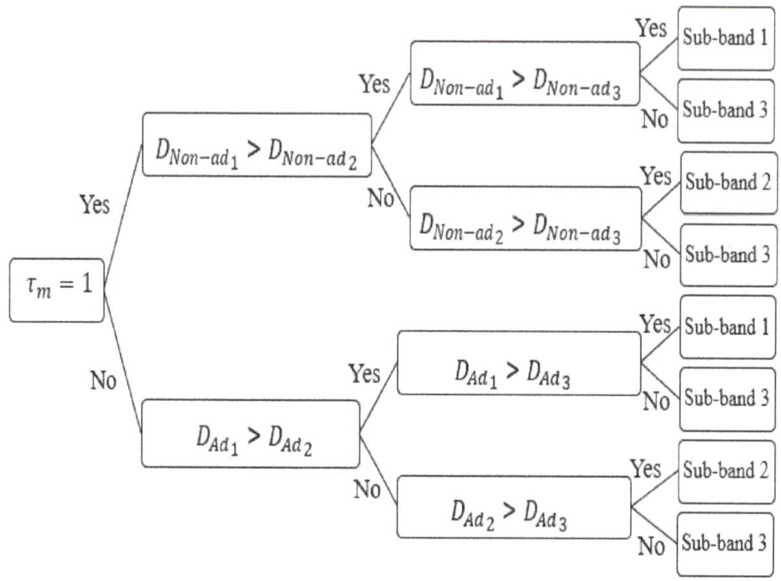

Figure 4. Classification tree (CT) example for three sub-bands.

5. Numerical Results

The simulation parameters are presented in Table 1. Various Voronoi SCs are allocated in the coverage area of the MCs. The UEs are randomly deployed within the entire network. The simulations are performed utilizing MATLAB R2018. The swarm size is chosen in the range of 20–50 particles [9,67–71]. It is shown in [71] that 25 is the optimum size over 16 different sizes on 3 different cases. The IW is linearly increased from 0.4 to 0.9, which is the range recommended in [33,72,73]. The velocity of the particles is chosen in the range of [−0.6, 0.6] [9,74,75]. Adjusting these parameters is essential for the convergence of the BPSO algorithm.

The performance of several algorithms is assessed, they are as follow:

- Always on: SFR is not utilized and all SCs are active.
- BPSO only: SFR is not utilized, but SC on/off switching is done via BPSO algorithm.
- Random 10%: SFR is not utilized, but 10% of the SCs are randomly chosen to be switched off. The remaining SCs are kept active.
- Proposed: the SC on/off switching is decided first using the BPSO algorithm, then the SFR is carried out using CTs.

The number of active SCs for various number of UEs is demonstrated in Figure 5. The number of active SCs increases with increasing the number of UEs, as more SCs are being activated to guarantee the minimum required SINR of the UEs allowing the UEs to associate with the SC having the better SINR, enhancing their data rates and improving the system performance. The "ω = 0.4" has better convergence than the "ω = 0.9" for the 200, 500, and 900 UEs, respectively, since reducing the IW facilitates in exploring the search space (global search), while raising the IW aids in exploiting (local search) the search space. The "linearly increasing ω" shows better convergence than the fixed "ω = 0.9" and "ω = 0.4", as in the "linearly increasing ω" case, the IW is lower at the beginning, allowing better exploration of the search space, and then it linearly increases, enhancing the local search. The fixed IW can get trapped during the search and is unable to find the global minimum number of active SCs. Therefore, the linearly increasing IW is chosen to be utilized in the proposed algorithm for the rest of the simulations in this paper.

Table 1. Simulation Parameters [9,13,14,33,59].

Parameters	Value
SC transmission power [13]	SFR: 20 dBm(center), 22 dBm(edge) No SFR: 22 dBm
SC baseline power ($P_{m_{on}}$) [14]	6.8 W
Maximum number of UEs in the SC (U_{max}) [9]	30
Swarm size (N_{par}) [9]	25
Maximum IW (ω_{max}) [33]	0.9
Minimum IW (ω_{min}) [33]	0.4
Maximum velocity of the particle (v_{max}) [9]	0.6
Minimum velocity of the particle (v_{min}) [9]	−0.6
Maximum number of iterations (Z_{max}) [59]	500
Total bandwidth [13]	20 MHz
RB bandwidth [13]	180 KHz
Maximum number of RBs [13]	106
Number of sub-bands (N_{sub}) [13]	7
Noise power spectral density [14]	−174 dBm/Hz
SINR threshold ($SINR_{thr}$) [9]	−5 dB
SC inactive level (β) [14]	0.63
Portion of power consumption due to the feeder losses and power amplifier (θ) [14]	4

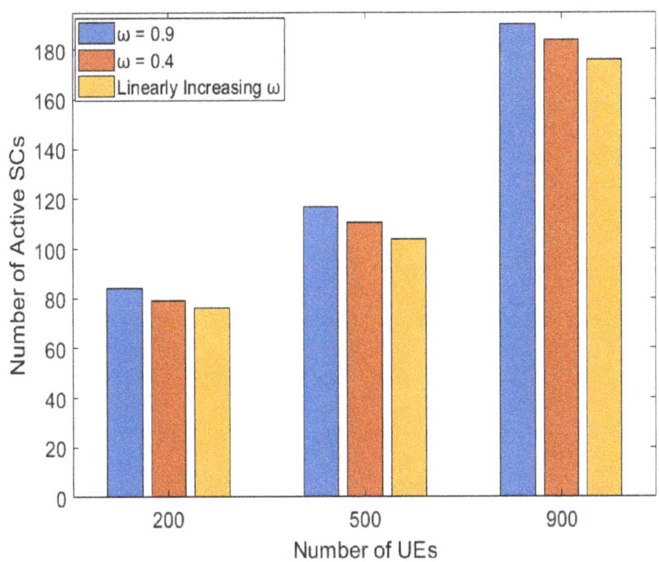

Figure 5. Number of active SCs for various numbers of UEs for different values of IW.

The number of active SCs for various numbers of UEs is presented in Figure 6. The "always on" algorithm has the highest number of active SCs, since all the SCs are kept active. The "random 10%" algorithm has a lower number of active SCs than the "always on" algorithm, since in the "random 10%" algorithm, 10% of the SCs are randomly turned off. However, the number of active SCs is constant with regard to the number of UEs. The proposed and "BPSO only" algorithms have the least number of active SCs, as the SCs are turned off utilizing the BPSO algorithm. In both algorithms (BPSO only and proposed), BPSO is used for SC switching. That is why the number of active SCs of the two algorithms is very close. Moreover, the number of active SCs increases on increasing the number of UEs, since more SCs are switched on to sustain the minimum SINR of the UEs.

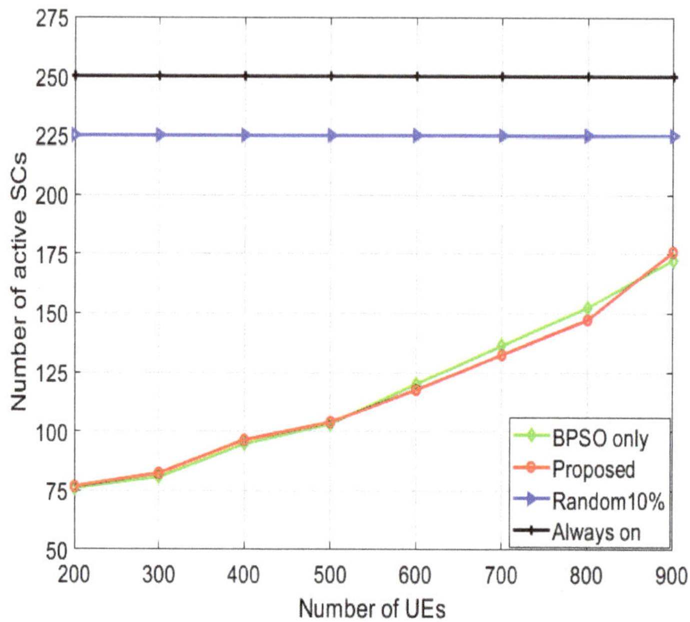

Figure 6. Number of active SCs for various numbers of UEs in case of linearly increasing IW.

The total system throughput for various numbers of UEs is shown in Figure 7. The proposed algorithm has the largest system throughput, since the proposed algorithm alleviates the interference levels, as it utilizes the SFR and switches the SCs on/off using the BPSO algorithm. The "BPSO only" algorithm has lower system throughput than the proposed algorithm because of its larger interference levels for not using SFR. The "random 10%" algorithm has lower system throughput than the "BPSO only" algorithm because of the random selection of the switched-off SCs, which is not the optimum one. The "always on" algorithm has the minimum system throughput because it has the largest interference levels as all the SCs are continuously active and it does not utilize the SFR.

The total power consumption of the system for various numbers of UEs is presented in Figure 8. The "always on" algorithm has the largest power consumption since no switching-off techniques are used. The "random 10%" algorithm has less power consumption than the "always on" algorithm because of the power savings from the switched-off SCs and higher power consumption than the "BPSO only" algorithm. The proposed algorithm consumes the least power because of the turning-off of the SCs utilizing BPSO, then applying the SFR minimizing the power consumption furthermore, because of the reduced transmission power for the UEs in the center region. In the proposed algorithm, the power consumption increases with the increment of the number of UEs, as more SCs are kept active to ensure the QoS of the UEs.

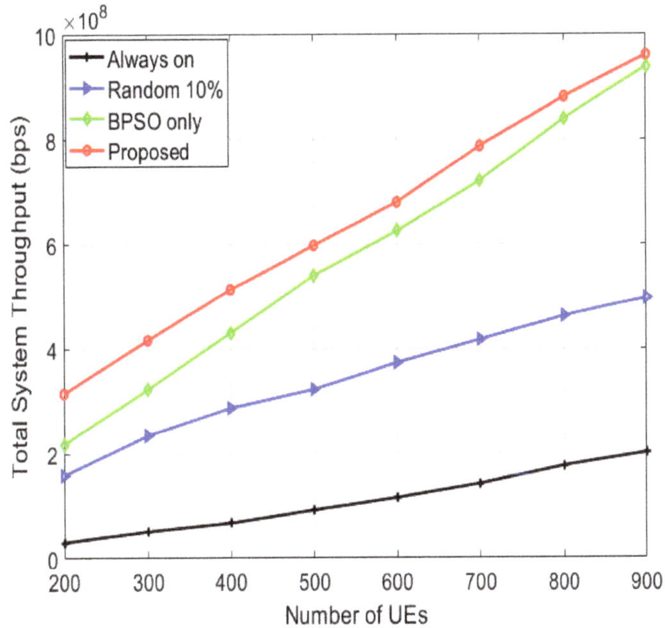

Figure 7. Total system throughput for various numbers of UEs.

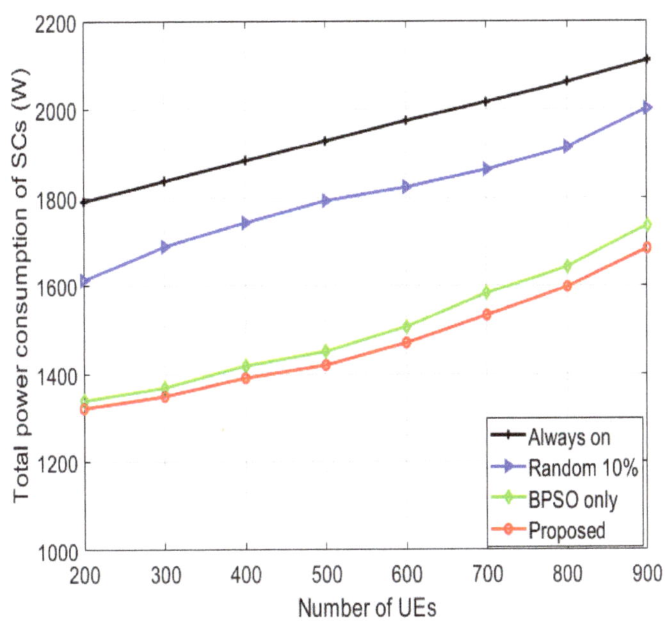

Figure 8. Total system power consumption for various numbers of UEs.

The PE of the system for various numbers of UEs is demonstrated in Figure 9. The proposed algorithm has the highest PE because of the reduced power consumption and the elevated system throughput. The "BPSO only" algorithm has lower PE because of its alleviated throughput and higher system power consumption compared to the proposed algorithm. The "random 10%" has lower PE than the "BPSO only" algorithm because of its alleviated system throughput caused by the random choice of the switched-off SCs. The "always on" algorithm has the least PE, since it has the largest power consumption and the least data rates because all the SCs are continuously active.

The outage probability for various SINR thresholds in the case of 900 UEs is depicted in Figure 10. The outage probability is defined as the percentage of UEs that are unable to attain a certain SINR threshold. The proposed algorithm has the minimum outage probability because of the enhanced SINR of the users, which resulted from minimizing the interference. The "BPSO only" algorithm has larger outage probability than the proposed algorithm, since it has larger interference levels. Both algorithms consider the QoS of the users, as all the users have SINR larger than $SINR_{thr}$ in both algorithms. The "random 10%" algorithm has larger outage probability than the "BPSO only" because of the deteriorated system performance. The "always on" algorithm has the largest outage probability because of the huge interference levels, as the SFR principle is not utilized and no on/off switching techniques are used, diminishing the total performance of the system.

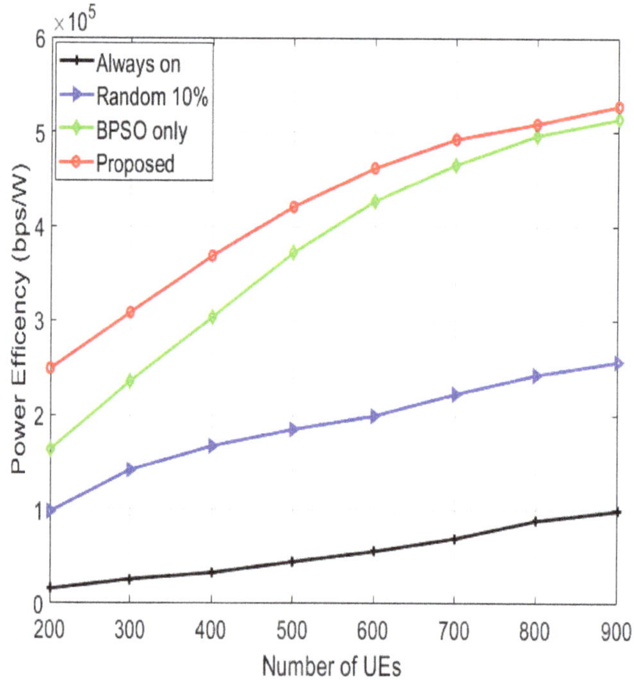

Figure 9. Power efficiency for various numbers of UEs.

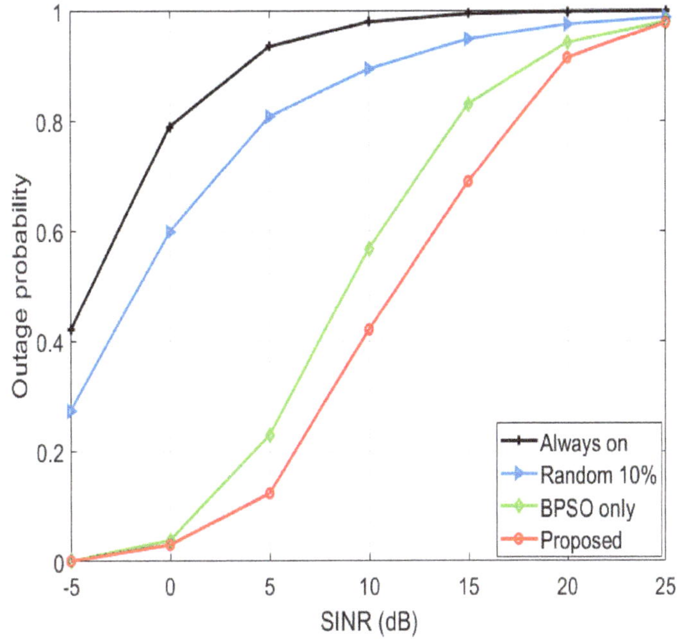

Figure 10. Outage probability for various SINR thresholds in the case of 900 UEs.

6. Conclusions

Minimizing the power consumption of the SCs and elevating the PE of the network are huge challenges facing the 5G HetNets. In this paper, to tackle these challenges, novel algorithms are proposed based on linear increasing IW-BPSO and SFR. The BPSO algorithm is used for SC on/off switching reducing the power consumption of the system without deteriorating the QoS of the UEs. Moreover, the linearly increasing IW is exploited to enhance the convergence of the BPSO algorithm to find the minimum number of active SCs. Furthermore, the CT-based SFR is proposed, where the SCs are divided into center and edge regions and different sub-bands are allocated to the edge regions of the adjoining SCs, minimizing the interference among the SCs. The results demonstrate that the proposed algorithms surpass the other conventional algorithms with regard to the total system power consumption, the total system throughput, the PE, and the outage probability. Additional work can be done in the future to address the coverage hole problem in Voronoi cells and to enhance the accuracy of the PSO algorithm using dynamic inertia weight.

Author Contributions: Conceptualization, M.O., B.A. and S.E.R.; methodology, M.O., B.A. and S.E.R.; software M.O.; writing—original draft preparation, M.O.; writing—review and editing, B.A. and S.E.R.; All authors have read and agreed to the published version of the manuscript.

Funding: This research received no external funding.

Conflicts of Interest: The authors declare no conflict of interest.

Nomenclature

Symbol	Description
a_1, a_2, a_3 and a_4	Random numbers uniformly distributed between 0 and 1
B_k	Bandwidth allocated to UE k
B_{RB}	Resource block (RB) bandwidth
c_1 and c_2	Acceleration parameters
C_{sys}	Total throughput of the system
D_f	Distance between the SC and the closest SC using sub-band f in its edge region
$d_{m,k}$	Distance between SC m and UE k
$G_{m,k}$	Channel gain between UE k and SC m
G_{best}	Global best position of all the particles
$L_{k,m}$	Number of requisite RBs for UE k in SC m to achieve the minimum data rate
M	Number of SCs
N	Noise power
P_{best_j}	Best position of the particle j
P_m	Transmission power of SC m
$P_{m_{tot}}$	Total power consumption of SC m
$P_{m_{tx}}$	Transmission power consumption of SC m
P_{sys}	Total power consumption of the system
PE_{sys}	PE (power efficency) of the system
$R_{k,m}$	Data rate of UE k in SC m
$SINR_{k,m}$	Signal to interference noise ratio of UE k in SC m
U_m	Number of UEs in SC m
$v_j{}^z$	Velocity of the particle j in the zth iteration
ω_z	Inertia Weight (IW) in the zth iteration
$x_j{}^z$	Position of the particle j in the zth iteration
α	Path loss exponent
γ_m	SC m on/off indicator
$\varphi_{m,k}$	UE k association indicator with SC m

References

1. Tanveer, J.; Haider, A.; Ali, R.; Kim, A. An Overview of Reinforcement Learning Algorithms for Handover Management in 5G Ultra-Dense Small Cell Networks. *Appl. Sci.* **2022**, *12*, 426. [CrossRef]
2. Salahdine, F.; Opadere, J.; Liu, Q.; Han, T.; Zhang, N.; Wu, S. A survey on sleep mode techniques for ultra-dense networks in 5G and beyond. *Comput. Netw.* **2021**, *201*, 108567. [CrossRef]
3. Osseiran, A.; Boccardi, F.; Braun, V.; Kusume, K.; Marsch, P.; Maternia, M.; Queseth, O.; Schellmann, M.; Schotten, H.; Taoka, H.; et al. Scenarios for 5G mobile and wireless communications: The vision of the METIS project. *IEEE Commun. Mag.* **2014**, *52*, 26–35. [CrossRef]
4. Boccardi, F.; Heath, R.W.; Lozano, A.; Marzetta, T.L.; Popovski, P. Five disruptive technology directions for 5G. *IEEE Commun. Mag.* **2014**, *52*, 74–80. [CrossRef]
5. Xu, Y.; Gui, G.; Gacanin, H.; Adachi, F. A survey on resource allocation for 5G heterogeneous networks: Current research, future trends and challenges. *IEEE Commun. Surv. Tutor.* **2021**, *23*, 668–695. [CrossRef]
6. Saeed, A.; Katranaras, E.; Zoha, A.; Imran, A.; Imran, M.A.; Dianati, M. Energy efficient resource allocation for 5G heterogeneous networks. In Proceedings of the IEEE International Workshop on Computer Aided Modelling and Design of Communication Links and Networks (CAMAD), Guildford, UK, 7–9 September 2015; pp. 119–123.
7. Lorincz, J.; Matijevic, T. Energy-efficiency analyses of heterogeneous macro and micro base station sites. *Comput. Electr. Eng.* **2014**, *40*, 330–349. [CrossRef]
8. Hashim, M.F.; Abdul Razak, N.I. Ultra-dense networks: Integration with device to device (D2D) communication. *Wireless Pers. Commun.* **2019**, *106*, 911–925. [CrossRef]
9. Venkateswararao, K.; Swain, P. Binary-PSO-based energy-efficient small cell deployment in 5G ultra-dense network. *J. Supercomput.* **2022**, *78*, 1071–1092. [CrossRef]
10. Kamel, M.I.; Hamouda, W.; Youssef, A.M. Ultra-Dense Networks: A Survey. *IEEE Commun. Surv. Tutor.* **2016**, *18*, 2522–2545. [CrossRef]
11. Liu, C.; Natarajan, B.; Xia, H. Small cell base station sleep strategies for energy efficiency. *IEEE Trans. Veh. Technol.* **2015**, *65*, 1652–1661. [CrossRef]

12. Ge, X.; Yang, J.; Gharavi, H.; Sun, Y. Energy efficiency challenges of 5G small cell networks. *IEEE Commun. Mag.* **2017**, *55*, 184–191. [CrossRef] [PubMed]
13. Osama, M.; El Ramly, S.; Abdelhamid, B. Interference Mitigation and Power Minimization in 5G Heterogeneous Networks. *Electronics* **2021**, *10*, 1723. [CrossRef]
14. Shen, B.; Lei, Z.; Huang, X.; Chen, Q. An interference contribution rate based small cells on/off switching algorithm for 5G dense heterogeneous networks. *IEEE Access* **2018**, *6*, 29757–29769. [CrossRef]
15. Lin, Z.; Niu, H.; An, K.; Wang, Y.; Zheng, G.; Chatzinotas, S.; Hu, Y. Refracting RIS aided hybrid satellite-terrestrial relay networks: Joint beamforming design and optimization. *IEEE Trans. Aerosp. Electron. Syst.* **2022**, *58*, 3717–3724. [CrossRef]
16. Lin, Z.; An, K.; Niu, H.; Hu, Y.; Chatzinotas, S.; Zheng, G.; Wang, J. SLNR-based Secure Energy Efficient Beamforming in Multibeam Satellite Systems. *IEEE Trans. Aerosp. Electron. Syst.* **2022**, 1–4. [CrossRef]
17. Isabona, J.; Srivastava, V.M. Downlink Massive MIMO Systems: Achievable Sum Rates and Energy Efficiency Perspective for Future 5G Systems. *Wirel. Pers. Commun.* **2017**, *96*, 2779–2796. [CrossRef]
18. Lorincz, J.; Bogarelli, M.; Capone, A.; Begusic, D. Heuristic approach for optimized energy savings in wireless access networks. In Proceedings of the International Conference on Software, Telecommunications and Computer Networks (SoftCOM), Split, Croatia, 23–25 September 2010; pp. 60–65.
19. Lorincz, J.; Capone, A.; Begusic, D. Heuristic Algorithms for Optimization of Energy Consumption in Wireless Access Networks. *KSII Trans. Internet Inf. Syst.* **2011**, *5*, 626–648. [CrossRef]
20. Xiao, Z.; Li, H.; Li, Z.; Wang, D. Load-awareness energy saving strategy via success probability constraint for heterogeneous small cell networks. In Proceedings of the IEEE Annual International Symposium on Personal, Indoor, and Mobile Radio Communications (PIMRC), Hong Kong, China, 30 August–2 September 2015; pp. 743–747.
21. Manssour, J.; Frenger, P.; Falconetti, L.; Moon, S.; Na, M. Smart small cell wake-up field trial: Enhancing end-user throughput and network energy performance. In Proceedings of the IEEE Vehicular Technology Conference (VTC Spring), Glasgow, UK, 11–14 May 2015; pp. 1–5.
22. Tang, L.; He, Y.; Zhou, Z.; Ren, Y.; Mumtaz, S.; Rodriguez, J. A distance-sensitive distributed repulsive sleeping approach for dependable coverage in heterogeneous cellular networks. *Trans. Emerg. Tel. Tech.* **2019**, *30*, e3784. [CrossRef]
23. Tao, R.; Liu, W.; Chu, X.; Zhang, J. An energy saving small cell sleeping mechanism with cell range expansion in heterogeneous networks. *IEEE Trans. Wirel. Commun.* **2019**, *18*, 2451–2463. [CrossRef]
24. AL-Samarrie, A.K.; Alyasiri, H.; AL-Nakkash, A.H. Proposed multi-stage PSO scheme for LTE network planning and operation. *Int. J. Appl. Eng. Res.* **2016**, *11*, 10199–10210.
25. Alyasiri, H.; AL-Samarrie, A.K.; AL-Nakkash, A.H. Interference Mitigation of Heterogeneous Networks by Proposed Combined Optimal Frequency and Power Allocations Scheme. *Int. J. Appl. Eng. Res.* **2016**, *11*, 11925–11934.
26. Rathore, A.; Sharma, H. Review on inertia weight strategies for particle swarm optimization. In Proceedings of the Sixth International Conference on Soft Computing for Problem Solving, Singapore; 2017; pp. 76–86.
27. Hsieh, S.T.; Sun, T.Y.; Liu, C.C.; Tsai, S.J. Efficient population utilization strategy for particle swarm optimizer. *IEEE Trans. Syst. Man Cybern. Part B* **2009**, *39*, 444–456. [CrossRef] [PubMed]
28. Eberhart, R.C.; Shi, Y. Tracking and optimizing dynamic systems with particle swarms. In Proceedings of the Congress on Evolutionary Computation (CEC), Seoul, Korea, 27–30 May 2001; pp. 94–100.
29. Tian, D.; Shi, Z. MPSO: Modified particle swarm optimization and its applications. *Swarm Evol. Comput.* **2018**, *41*, 49–68. [CrossRef]
30. Taherkhani, M.; Safabakhsh, R. A novel stability-based adaptive inertia weight for particle swarm optimization. *Appl. Soft Comput.* **2016**, *38*, 281–295. [CrossRef]
31. Xin, J.; Chen, G.; Hai, Y. A particle swarm optimizer with multi-stage linearly-decreasing inertia weight. In Proceedings of the International Joint Conference on Computational Sciences and Optimization, Sanya, China, 24–26 April 2009; pp. 505–508.
32. Zheng, Y.; Ma, L.; Zhang, L.; Qian, J. Empirical study of particle swarm optimizer with an increasing inertia weight. In Proceedings of the 2003 Congress on Evolutionary Computation, Canberra, ACT, Australia, 8–12 December 2003; Volume 1, pp. 221–226.
33. Lee, C.Y.; Cheng, Y.H. Motor Fault Detection Using Wavelet Transform and Improved PSO-BP Neural Network. *Processes* **2020**, *8*, 1322. [CrossRef]
34. Lin, Z.; Lin, M.; Wang, J.B.; De Cola, T.; Wang, J. Joint beamforming and power allocation for satellite-terrestrial integrated networks with non-orthogonal multiple access. *IEEE J. Sel. Top. Signal Process.* **2019**, *13*, 657–670. [CrossRef]
35. Lin, Z.; Lin, M.; De Cola, T.; Wang, J.B.; Zhu, W.P.; Cheng, J. Supporting IoT with rate-splitting multiple access in satellite and aerial-integrated networks. *IEEE Internet Things J.* **2021**, *8*, 11123–11134. [CrossRef]
36. Giambene, G.; Le, V.A.; Bourgeau, T.; Chaouchi, H. Soft frequency reuse schemes for heterogeneous LTE systems. In Proceedings of the IEEE International Conference on Communications (ICC), London, UK, 8–12 June 2015; pp. 3161–3166.
37. Mohamed, M.O.; Abdelhamid, B.; El Ramly, S. Interference mitigation in heterogeneous networks using Fractional Frequency Reuse. In Proceedings of the International Conference on Wireless Networks and Mobile Communications (WINCOM), Fez, Morocco, 26–29 October 2016; pp. 154–159.
38. Giambene, G.; Le, V.A.; Bourgeau, T.; Chaouchi, H. Iterative multi-level soft frequency reuse with load balancing for heterogeneous lte-a systems. *IEEE Trans. Wirel. Commun.* **2016**, *16*, 924–938. [CrossRef]

39. Hossain, M.S.; Tariq, F.; Safdar, G.A.; Mahmood, N.H.; Khandaker, M.R. Multi-layer soft frequency reuse scheme for 5G heterogeneous cellular networks. In Proceedings of the IEEE Globecom Workshops (GC Wkshps), Singapore, 4–8 December 2017; pp. 1–6.
40. Moysen, J.; Giupponi, L. From 4G to 5G: Self-organized network management meets machine learning. *Comput. Commun.* **2018**, *129*, 248–268. [CrossRef]
41. Preciado-Velasco, J.E.; Gonzalez-Franco, J.D.; Anias-Calderon, C.E.; Nieto-Hipolito, J.I.; Rivera-Rodriguez, R. 5G/B5G Service Classification Using Supervised Learning. *Appl. Sci.* **2021**, *11*, 4942. [CrossRef]
42. Radivilova, T.; Kirichenko, L.; Lemeshko, O.; Ageyev, D.; Mulesa, O.; Ilkov, A. Analysis of anomaly detection and identification methods in 5G traffic. In Proceedings of the Eleventhth IEEE International Conference on Intelligent Data Acquisition and Advanced Computing Systems: Technology and Applications (IDAACS), Cracow, Poland, 22–25 September 2021; pp. 1108–1113.
43. Casas, P.; D'Alconzo, A.; Wamser, F.; Seufert, M.; Gardlo, B.; Schwind, A.; Tran-Gia, P.; Schatz, R. Predicting QoE in cellular networks using machine learning and in-smartphone measurements. In Proceedings of the Ninth International Conference on Quality of Multimedia Experience (QoMEX), Erfurt, Germany, 29 May–2 June 2017; pp. 1–6.
44. Galinina, O.; Pyattaev, A.; Andreev, S.; Dohler, M.; Koucheryavy, Y. 5G Multi-RAT LTE-WiFi Ultra-Dense Small Cells: Performance Dynamics, Architecture, and Trends. *IEEE J. Sel. Areas Commun.* **2015**, *33*, 1224–1240. [CrossRef]
45. Niu, C.; Li, Y.; Hu, R.Q.; Ye, F. Fast and Efficient Radio Resource Allocation in Dynamic Ultra-Dense Heterogeneous Networks. *IEEE Access* **2017**, *5*, 1911–1924. [CrossRef]
46. Shabbir, A.; Khan, H.R.; Ali, S.A. Traffic Load Aware Approach for Optimum Throughput in 5G Heterogeneous Cellular Networks. In Proceedings of the Fourth International Conference on Computer and Information Sciences (ICCOINS), Kuala Lumpur, Malaysia, 13–14 August 2018; pp. 1–6.
47. Su, G.; Chen, B.; Lin, X.; Wang, H.; Li, L. User Association and Base Station Sleep Management in Dense Heterogeneous Cellular Networks. *KSII Trans. Int. Inf. Sys.* **2017**, *11*, 2058–2074.
48. Huang, X.; Zhang, D.; Tang, S.; Chen, Q.; Zhang, J. Fairness-based distributed resource allocation in two-tier heterogeneous networks. *IEEE Access* **2019**, *7*, 40000–40012. [CrossRef]
49. Mendis, H.V.K.; Balapuwaduge, I.A.M.; Li, F.Y. Dependability-based reliability analysis in URC networks: Availability in the space domain. *IEEE ACM Trans. Netw.* **2019**, *27*, 1915–1930. [CrossRef]
50. Xu, Y.; Yang, P.; Gong, J.; Niu, K. A self-organizing base station sleeping strategy in small cell networks using local stable matching games. In Proceedings of the International Conference on Wireless Algorithms, Systems, and Applications, Berlin, Germany; 2018; pp. 545–556.
51. Ghazzai, A.H.; Farooq, M.J.; Alsharoa, A.; Yaacoub, E.; Kadri, A.; Alouini, M.S. Green networking in cellular HetNets: A unified radio resource management framework with base station ON/OFF switching. *IEEE Trans. Veh. Technol.* **2016**, *66*, 5879–5893. [CrossRef]
52. Akram, M.R.; Al-Nakkash, A.H.; Salim, O.N.M.; AlAbdullah, A.A. Proposed APs Distribution Optimization Algorithm: Indoor Coverage Solution. *J. Phys. Conf. Ser.* **2021**, *1804*, 012134. [CrossRef]
53. Cai, H.; Li, X.; Xie, C.; Guo, K.; Liu, H.; Liu, C. Area-to-point heat conduction enhancement using binary particle swarm optimization. *Appl. Therm. Eng.* **2019**, *155*, 449–460. [CrossRef]
54. Anuradha, J.; Tripathy, B.K. Improved intelligent dynamic swarm PSO algorithm and rough set for feature selection. In Proceedings of the International Conference on Computing and Communication Systems. Berlin; 2011; pp. 110–119.
55. Zhang, H.; Wang, Y.; Ji, H.; Li, X. A sleeping mechanism for cache-enabled small cell networks with energy harvesting function. *IEEE Trans. Green Commun. Net.* **2020**, *4*, 497–505. [CrossRef]
56. Shariatmadar, H.; Meshkat Razavi, H. Seismic control response of structures using an ATMD with fuzzy logic controller and PSO method. *Struct. Eng. Mech.* **2014**, *51*, 547–564. [CrossRef]
57. Izquierdo, J.; Montalvo, I.; Pérez, R.; Fuertes, V.S. Design optimization of wastewater collection networks by PSO. *Comput. Math. Appl.* **2008**, *56*, 777–784. [CrossRef]
58. Tasetiren, M.F.; Lian, Y.C. A Binary Particle Swarm Optimization Algorithm for Lot Sizing Problem. *J. Econ. Soc. Res.* **2003**, *5*, 1–20.
59. Kumar, N.; Sharma, S.K. Inertia Weight Controlled PSO for Task Scheduling in Cloud Computing. In Proceedings of the International Conference on Computing, Power and Communication Technologies (GUCON), Greater Noida, UP, India, 28–29 September 2018; pp. 155–160.
60. Qian, M.; Hardjawana, W.; Li, Y.; Vucetic, B.; Yang, X.; Shi, J. Adaptive Soft Frequency Reuse Scheme for Wireless Cellular Networks. *IEEE Trans. Veh. Technol.* **2014**, *64*, 118–131. [CrossRef]
61. Sánchez-Rodríguez, D.; Hernández-Morera, P.; Quinteiro, J.M.; Alonso-González, I. A Low Complexity System Based on Multiple Weighted Decision Trees for Indoor Localization. *Sensors* **2015**, *15*, 14809–14829. [CrossRef]
62. Ji, B.; Lu, X.; Sun, G.; Zhang, W.; Li, J.; Xiao, Y. Bio-Inspired Feature Selection: An Improved Binary Particle Swarm Optimization Approach. *IEEE Access.* **2020**, *8*, 85989–86002. [CrossRef]
63. Aghdam, M.H.; Sharifi, A.A. PAPR reduction in OFDM systems: An efficient PTS approach based on particle swarm optimization. *ICT Express.* **2019**, *5*, 178–181. [CrossRef]
64. Donevski, I.; Vallero, G.; Marsan, M.A. Neural networks for cellular base station switching. In Proceedings of the IEEE Conference on Computer Communications Workshops (INFOCOM WKSHPS), Paris, France, 29 April–2 May 2019; pp. 738–743.

65. Qin, M.; Yang, Q.; Cheng, N.; Li, J.; Wu, W.; Rao, R.R.; Shen, X. Learning-Aided Multiple Time-Scale SON Function Coordination in Ultra-Dense Small-Cell Networks. *IEEE Trans. Wirel. Commun.* **2019**, *18*, 2080–2092. [CrossRef]
66. Sesto-Castilla, D.; Garcia-Villegas, E.; Lyberopoulos, G.; Theodoropoulou, E. Use of Machine Learning for energy efficiency in present and future mobile networks. In Proceedings of the IEEE Wireless Communications and Networking Conference, Marrakesh, Morocco, 15–18 April 2019; pp. 1–6.
67. Chen, W.N.; Zhang, J.; Lin, Y.; Chen, N.; Zhan, Z.H.; Chung, H.S.H.; Li, Y.; Shi, Y.H. Particle Swarm Optimization with an aging leader and challengers. *IEEE Trans. Evol. Comput.* **2012**, *17*, 241–258. [CrossRef]
68. Gong, Y.J.; Li, J.J.; Zhou, Y.; Li, Y.; Chung, H.S.H.; Shi, Y.H.; Zhang, J. Genetic learning particle swarm optimization. *IEEE Trans. Cybern.* **2015**, *46*, 2277–2290. [CrossRef]
69. Liang, J.J.; Qin, A.K.; Suganthan, P.N.; Baskar, S. Comprehensive learning particle swarm optimizer for global optimization of multimodal functions. *IEEE Trans. Evol. Comput.* **2006**, *10*, 281–295. [CrossRef]
70. Lin, A.; Sun, W.; Yu, H.; Wu, G.; Tang, H. Global genetic learning Particle Swarm Optimization with diversity enhancement by ring topology. *Swarm Evol. Comput.* **2019**, *44*, 571–583. [CrossRef]
71. Hashim, N.; Ismail, N.F.N.; Johari, D.; Musirin, I.; Rahman, A.A. Optimal population size of particle swarm optimization for photovoltaic systems under partial shading condition. *Int. J. Elec. & Comp. Eng.* **2022**, *12*, 4599–4613.
72. Shi, Y.; Eberhart, R.C. Empirical study of particle swarm optimization. In Proceedings of the Congress on Evolutionary Computation (CEC), Washington, DC, USA, 6–9 July 1999; pp. 1945–1950.
73. Agrawal, A.; Tripathi, S. Particle swarm optimization with adaptive inertia weight based on cumulative binomial probability. *Evol. Intell.* **2021**, *14*, 305–313. [CrossRef]
74. Yalcin, N.; Tezel, G.; Karakuzu, C. Epilepsy diagnosis using artificial neural network learned by PSO. *Turk. J. Elec. Eng. Comp. Sci.* **2015**, *23*, 421–432.
75. Wan, R.; Zhu, L.; Li, T.; Bai, L.A. NOMA-PSO Based Cooperative Transmission Method in Satellite Communication Systems. In Proceedings of the Nineth International Conference on Wireless Communications and Signal Processing (WCSP), Nanjing, China, 11–13 October 2017; pp. 1–6.

sensors

MDPI

Article

Energy-Efficient Resource Allocation Based on Deep Q-Network in V2V Communications

Donghee Han and Jaewoo So *

Department of Electronic Engineering, Sogang University, Seoul 04107, Republic of Korea
* Correspondence: jwso@sogang.ac.kr; Tel.: +82-2-705-8464

Abstract: Recently, with the development of autonomous driving technology, vehicle-to-everything (V2X) communication technology that provides a wireless connection between vehicles, pedestrians, and roadside base stations has gained significant attention. Vehicle-to-vehicle (V2V) communication should provide low-latency and highly reliable services through direct communication between vehicles, improving safety. In particular, as the number of vehicles increases, efficient radio resource management becomes more important. In this paper, we propose a deep reinforcement learning (DRL)-based decentralized resource allocation scheme in the V2X communication network in which the radio resources are shared between the V2V and vehicle-to-infrastructure (V2I) networks. Here, a deep Q-network (DQN) is utilized to find the resource blocks and transmit power of vehicles in the V2V network to maximize the sum rate of the V2I and V2V links while reducing the power consumption and latency of V2V links. The DQN also uses the channel state information, the signal-to-interference-plus-noise ratio (SINR) of V2I and V2V links, and the latency constraints of vehicles to find the optimal resource allocation scheme. The proposed DQN-based resource allocation scheme ensures energy-efficient transmissions that satisfy the latency constraints for V2V links while reducing the interference of the V2V network to the V2I network. We evaluate the performance of the proposed scheme in terms of the sum rate of the V2X network, the average power consumption of V2V links, and the average outage probability of V2V links using a case study in Manhattan with nine blocks of 3GPP TR 36.885. The simulation results show that the proposed scheme greatly reduces the transmit power of V2V links when compared to the conventional reinforcement learning-based resource allocation scheme without sacrificing the sum rate of the V2X network or the outage probability of V2V links.

Keywords: vehicular communications; deep reinforcement learning; deep Q-network; resource allocation; energy efficiency

Citation: Han, D.; So, J. Energy-Efficient Resource Allocation Based on Deep Q-Network in V2V Communications. *Sensors* **2023**, *23*, 1295. https://doi.org/10.3390/s23031295

Academic Editor: Josip Lorincz

Received: 29 December 2022
Revised: 18 January 2023
Accepted: 19 January 2023
Published: 23 January 2023

1. Introduction

Today, with the development of autonomous driving technologies, vehicular communication technologies are receiving significant attention from both the industry and academia [1,2]. The 3GPP has recently designed a new radio (NR) sidelink to support direct vehicle-to-vehicle (V2V) communication without the help of a base station (BS) in a low-latency, high-throughput, and high-connection-density network [1,3,4]. V2V networks require ultra-reliable and low-latency communication (URLLC) services for use cases that demand certain safety features, such as autonomous driving systems that send and receive warning messages to and from nearby vehicles, even as the number of vehicles increases [5]. Therefore, it is important to manage radio resources efficiently to satisfy the quality of service (QoS) of vehicles in the V2V network.

Radio resource management is often formulated as a set of combined optimizations used to find the optimal solution of an objective problem, which is generally an NP-hard problem. In recent years, machine learning has been successfully applied in a wide range of areas, resulting in significant performance improvements. In particular, reinforcement learning (RL) has shown its superiority in solving the resource allocation problems in communications [6,7]. Resource allocation can be dived into three categories according to which layer of the OSI 7-layer model performs [8–10]. The first category is bandwidth allocation at the network layer, which aims to provide call-level QoS guarantees. The second category is the allocation of resource blocks (RBs) at the link layer. The link layer determines which RB the transmitter will use, on the basis of the channel state measured in the physical layer. The third category is the joint RB and power allocation at the cross-layer between the link layer and the physical layer [9,10]. In this paper, we focus on the resource allocation of the RB and the transmit power of V2V links at both the link layer and the physical layer. Resource allocation is based on the channel state information (CSI), i.e., the signal-to-interference-plus-noise ratio (SINR). We propose a deep Q-network (DQN)-based spectrum and power allocation scheme for energy-efficient V2V communications while maximizing the sum rate of the V2I and V2V links. The proposed Q-network uses the CSI of V2I and V2V links and the latency constraints of vehicles to find the optimal RB and transmit power of the V2V links. The contributions of this paper are as follows: First, we developed a decentralized resource allocation problem that incorporates the power consumption as well as the latency of V2V links while increasing the sum rate of the V2I and V2V links. Second, we developed a DQN model to solve the resource allocation problem, where the reward function includes the power consumption and latency conditions represented as penalties. Third, the simulation results show that the DQN-based energy-efficient resource allocation scheme greatly reduces overall power consumption in comparison with the conventional RL-based scheme without sacrificing the sum rate and latency requirements of V2V links.

The rest of the paper is organized as follows. Section 2 describes the system model. Section 3 presents a DQN-based resource allocation scheme, where the state, action, and reward functions of reinforcement learning (RL) are described in detail. Section 4 shows the simulation results in a case study of Manhattan. Finally, Section 5 concludes the paper.

2. Related Work

The resource allocation mechanism in vehicle-to-everything (V2X) communications has been studied in various ways. The authors of [11] introduced the deep reinforcement learning (DRL)-based resource allocation scheme and showed experimental results for both unicast and broadcast scenarios. They designed a reward function to ensure the latency constraints of the V2V links were satisfied. The authors of [12] proposed a QoS-aware resource allocation scheme based on the DRL framework in V2X communications, where they took QoS parameters such as the priority of V2V messages into consideration. The proposed scheme of [12] aims to maximize the sum rate of vehicle-to-infrastructure (V2I) links while satisfying the latency constraints of V2V links. The authors of [13] developed a power allocation problem in the cellular device-to-device (D2D)-based V2X communication network and mathematically solved the problem. They showed that the proposed power allocation scheme outperforms the existing algorithms in terms of power consumption. The authors of [14] developed a multi-agent RL (MARL)-based resource allocation for V2V links in the spectrum-sharing V2X network. They aimed to maximize the capacity of V2I links while also improving the reliability of the payload delivery in V2V links. They showed the MADRL-based resource allocation is efficient for the V2I and V2V network collaboration although decisions are made locally and distributed at each V2V transmitter. The authors of [15] proposed a MARL-based resource allocation scheme in order to maximize the sum rate of V2I links while satisfying the latency and reliability requirements of V2V links. In this work, they developed individual double-dueling deep recurrent Q-networks (D3RQN), where they used interference power measurements instead of the conventional CSI under the assumption that it is difficult to acquire the perfect CSI in the vehicular network. They showed that the proposed MARL-based resource allocation jointly adjusts the sub-channel and transmit power using only local interference measurements without inter-agent communication. Some studies have focused on the energy consumption in V2X networks.

The authors of [16] developed an energy efficiency problem in an NR V2X network, where energy efficiency is defined as the ratio of the sum rate to power consumption. They proposed a heuristic algorithm of traffic-density-based random selection to solve the developed mixed-integer problem. The authors of [17] also developed an energy efficiency problem of vehicle users while considering the QoS requirement of cellular users in the cellular network underlying V2V communications. They transformed the latency constraint into the constraint of the queue length and solved the virtual queue problem based on the Lyapunov optimization. In V2X networks, the energy consumption of V2I links as well as V2V links is important. Some studies have focused on optimizing energy consumption across the entire wireless access network [18–20]. The authors of [18] proposed an energy-efficient resource management scheme based on the transmit power scaling and on/off switching of base stations. The authors of [19] formulated an optimization problem for the energy consumption of a wireless location area network (WLAN) by adjusting the transmit power and turning access stations on and off based on realistic traffic patterns. They proposed integer linear programming (ILP) optimization models and heuristic algorithms to minimize the energy consumption of the network. The authors of [20] developed an ILP model for energy saving of wireless access networks, and also developed a heuristic algorithm based on a greedy method to cope with the computational complexity of the ILP model.

Recently, graph-based deep learning solutions have been proposed for resource allocation in communication networks [7,21–24]. Graph neural networks (GNNs) have achieved some success in solving resource allocation problems in various communication networks, e.g., wireless networks, wired networks, and software-defined networks, because of their abilities to learn to capture the dependencies of graphs and to learn non-Euclidean structure data [21]. The authors of [22] presented a comprehensive review and analysis of graph-based resource allocation methods in cellular, device-to-device, and cognitive radio networks. Here, they classified the graph-based resource allocation methods in terms of graph models, tasks solved via graphs, graph formulation, and optimization methods. The authors of [23] proposed a heterogeneous bipartite GNN (HBGNN) to solve the joint user association and power allocation problem in heterogeneous ultra-dense networks (HUDNs). They modeled the downlink of the HUDN as a heterogeneous bipartite graph and compared the performance of the proposed HBGNN with the fully connected neural network and the convolutional neural network (CNN). However, the HBGNN requires supervised learning, unlike RL. The authors of [24] developed a graph convolutional network (GCN)-based DRL framework to perform joint channel selection and power adaptation in the underlying cognitive radio networks, maximizing the data rate of secondary users while maintaining the level of interference to primary users. They modeled the environment of the cognitive radio network as a dynamic graph and adopted a DRL to explore the optimal resource allocation strategy. However, the work of [24] did not take the energy efficiency and the latency constraints of the secondary users into account. The authors of [7] presented a GNN-augmented RL method to perform spectrum allocation for vehicular networks. They expressed the V2V network as a graph and exploited RL to perform resource allocation. The deep Q-network was developed to select the spectrum for each V2V pair.

Deep learning technologies for Internet of vehicle (IoV) networks have been studied previously [25–29]. The authors of [25] discussed deep learning applications for security and collision prediction in the internet of vehicle (IoV) networks, and they proposed a DRL-based resource allocation method to enhance multiple QoS requirements, such as latency and suitable data rate requirements. They introduced an actor–critic framework to achieve an intelligent resource allocation in the IoV network. The authors of [26] discussed deep learning techniques to enhance the performance of the overall IoV system. They addressed various learning networks, e.g., CNN, recurrent neural networks, DRL, classification, clustering, and regression. The authors of [27] presented a comprehensive review and analysis of machine learning technologies for IoV applications, e.g., energy- and buffer-aware optimization, edge caching, intelligent decisions for network scheduling and adaptation, intelligent autonomous driving, etc. The authors of [28] presented a comprehensive review of resource allocation and management for the IoV over 5G radio access networks. They described learning-based resource allocation approaches to improve the QoS and quality of experience in distributed and cloud-computing resource allocation schemes, along with big data resource allocation. The authors of [29] conducted a critical review and analysis of machine learning models used to resolve the challenge in IoV applications. Moreover, they proposed a Markov decision-process-based, edge-computing offloading model and evaluated its performance in terms of its power consumption and task latency.

Moreover, vehicular edge computing (VEC) technologies have been studied to dynamically manage computing resources, caching, and networking [30]. The authors of [31] proposed a generic approach to improve the performance of application outsourcing in the caching-assisted VEC. They mathematically showed that application caching can optimize the average response time while satisfying the long-term energy consumption constraint. The authors of [32] addressed route planning in a navigation system that finds an optimal route from the source to the target location. They proposed a real-time cache-aided route planning system based on mobile edge computing with the aim of reducing the communication delay between the access network and the remote central server and the computational time of route planning queries. The authors of [33] proposed a caching-enabled VEC scheme for jointly optimizing task caching and computation offloading in a VEC system; task caching was shown to reduce response latency but increase energy consumption. They then formulated an optimization problem that minimizes the weighted sum of the service time and energy consumption in the caching-assisted VEC system and used a genetic algorithm to solve the problem. The authors of [34] presented a comprehensive review and analysis of the vehicle routing problem (VRP). They mainly reviewed machine learning-assisted VRP modeling and optimization approaches.

3. System Model

We consider a V2X network consisting of a V2I network and a V2V network as shown in Figure 1. We focus on the uplink in the V2I network, where there are L V2I links denoted by $\mathcal{L} = \{1, 2, \ldots, L\}$. In the V2V network, there are K V2V links denoted by $\mathcal{K} = \{1, 2, \ldots, K\}$. In the V2I network, the spectrum is orthogonally allocated to the vehicles, where the number of orthogonal RBs is N_{RB}. However, the V2V links share the resources, N_{RB}, of the V2I network.

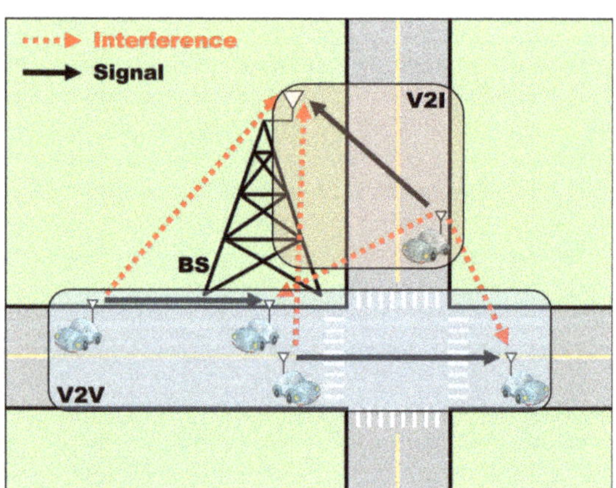

Figure 1. A system model.

In the V2I network, the received SINR and capacity of the lth V2I link are represented as follows:

$$\text{SINR}_l = \frac{P_l^{\text{V2I}} h_l}{\sigma^2 + \sum_{k \in \mathcal{K}} \mu_{k,l} p_k h_k} \quad (1)$$

$$C_l = W \log(1 + \text{SINR}_l) \quad \text{[bits/second]}, \quad (2)$$

where P_l^{V2I} is the transmit power of the vehicle and h_l is the channel power gain in the lth V2I link. Additionally, p_k is the transmit power of the kth V2V link, h_k is the channel power gain from the transmitter of the k V2V link to the base station, σ^2 is the noise power, and W is the bandwidth. The indicator function, $\mu_{k,l}$, denotes whether the resource is shared between the kth V2V link and the lth V2I link. That is, if the kth V2V link shares the RB of the lth V2I link, $\mu_{k,l} = 1$; otherwise, $\mu_{k,l} = 0$.

In the V2V network, the received SINR and capacity of the kth V2V link are represented as follows:

$$\text{SINR}_k = \frac{p_k g_k}{\sigma^2 + I_k^{\text{V2V}} + I_k^{\text{V2I}}} \quad (3)$$

$$I_k^{\text{V2I}} = \sum_{l \in \mathcal{L}} \mu_{k,l} P_l^{\text{V2I}} g_{l,k} \quad (4)$$

$$I_k^{\text{V2V}} = \sum_{l \in \mathcal{L}} \sum_{j \in \mathcal{K}, j \neq k} \mu_{k,l} \mu_{j,l} p_j g_{j,k} \quad (5)$$

$$C_k = W \log(1 + \text{SINR}_k) \quad \text{[bits/second]}, \quad (6)$$

where p_k is the transmit power of the vehicle and g_k is the channel power gain in the kth V2V link. Additionally, I_k^{V2I} is the interference from the V2I link sharing the RB of the kth V2V link, I_k^{V2V} is the interference from the V2V link sharing the RB of the kth V2V link, and $g_{l,k}$ is the channel power gain from the transmitter of the lth V2I link to the receiver of the kth V2V link. The indicator function, $\mu_{j,l}$, denotes whether the resource is shared between the jth V2V link and the lth V2I link. That is, if the jth V2V link shares the RB of the lth V2I link, $\mu_{j,l} = 1$; otherwise, $\mu_{j,l} = 0$.

In order for the BS to know the channel state of the V2V links, each receiver of the V2V link reports its CSI to the BS, which results in a large signaling overhead. Hence, we assume that the BS does not know the CSI of the V2V links. The BS independently controls the resource allocation of the V2I links without considering the channel state of the V2V links. Consequently, vehicles on the V2V link individually select the RB and determine the transmit power based on the locally observed channel information. Here, the locally observed channel information in the V2V link consists of the following: the CSI of the V2I link, the interference power observed in the previous time slot, the instantaneous CSI of the V2V link, and the information on the RB selected by nearby vehicles.

Our objective is to maximize the sum rate of the V2I links while increasing the probability of meeting the latency constraint of the V2V links by controlling the selection of RB and the transmit power of each V2V link. However, finding the optimal allocations of the RB and transmit power is an NP-hard problem. Hence, we propose a DQN-based approach to solve the resource allocation problem.

4. Deep Q-Network for Energy-Efficient Resource Allocation

4.1. Reinforcement Learning

In RL, an agent observes a state in an environment that satisfies the Markov decision process (MDP). Then, the optimal action is selected according to the given policy. Depending on the selected action, the agent interacts with the environment, receives a reward from the environment, and transitions to the next state.

The goal of RL is to maximize the expected return value after the episode ends. The return formula is given as follows:

$$R_{t:T} = r_t + \gamma r_{t+1} + \gamma^2 r_{t+2} + \ldots + \gamma^T r_{t+T-1}, \tag{7}$$

where r_t denotes the reward obtained immediately at time t, T is the time step, and γ denotes the discount factor. The structure of our RL is shown in Figure 2. The agent observes the state of the environment at the time (t) and selects the best action according to the given policy. When the agent selects an action for all V2V links, the actions are stored in the joint action group and interact with the environment at the same time, and the agent receives a reward. In our system model, we develop the RL with the following parameters:

1. *State space*: We use the following state, similar to the unicast scenario of [11].

$$\mathbf{s}_t = \{\mathbf{H}_t, \mathbf{I}_{t-1}, \mathbf{G}_t, \mathbf{N}_{t-1}, U_t, L_t\}, \tag{8}$$

   where $\mathbf{H}_t$ is the CSI of V2I links at time t; $\mathbf{I}_{t-1}$ is the interference power to the link at time $t-1$; $\mathbf{G}_t$ is the instantaneous CSI of the corresponding V2V link at time t; $\mathbf{N}_{t-1}$ is the information of RBs selected by surrounding vehicles at time $t-1$; U_t is the time remaining to satisfy the latency constraints at time t; and L_t is the remaining data to be received from the transmitter of the V2V link at time t. $\mathbf{H}_t$, $\mathbf{I}_{t-1}$, $\mathbf{G}_t$, and $\mathbf{N}_{t-1}$ are vectors containing the state information of the corresponding RBs, and U_t and L_t are scalar values that are the time remaining to satisfy the latency constraints and the remaining data, respectively. Therefore, the dimension of the state space is given by $D_{\text{state}} = (4 \times N_{\text{RB}}) + 2$.
2. *Action space*: The action determines the transmit power and the allocation of RBs. Hence, the dimensions of the action space are given by $D_{\text{action}} = N_{\text{pwr}} \times N_{\text{RB}}$, where N_{pwr} is the number of transmit power levels in the V2V link and N_{RB} is the number of RBs.
3. *Reward*: We formulate the following reward function taking two penalties into account, the transmission time and the power consumption:

$$r_t = \lambda_{\text{V2I}} \sum_{l \in \mathcal{L}} C_l + \lambda_{\text{V2V}} \sum_{k \in \mathcal{K}} C_k - \lambda_{\text{latency}}(T_0 - U_t) - \lambda_{\text{pwr}} \frac{1}{K} \sum_{k \in \mathcal{K}} \frac{p_k}{p_{\max}}, \tag{9}$$

   where T_0 is the maximum tolerable latency, and therefore, $(T_0 - U_t)$ means the transmission time. Moreover, $p_{\max}$ is the maximum transmit power in the V2V link. λ_{V2I} and λ_{V2V} represent the weight for the sum rate of the V2I links and the sum rate of the V2V links, respectively. λ_{latency} and λ_{pwr} represent the weight of the penalty according to an increase in the transmission time and the penalty according to an increase in the transmit power, respectively. As the sum rate of the V2I or V2V links increases, a positive factor is added. However, as the transmission time or power consumption increases, a negative factor is added.

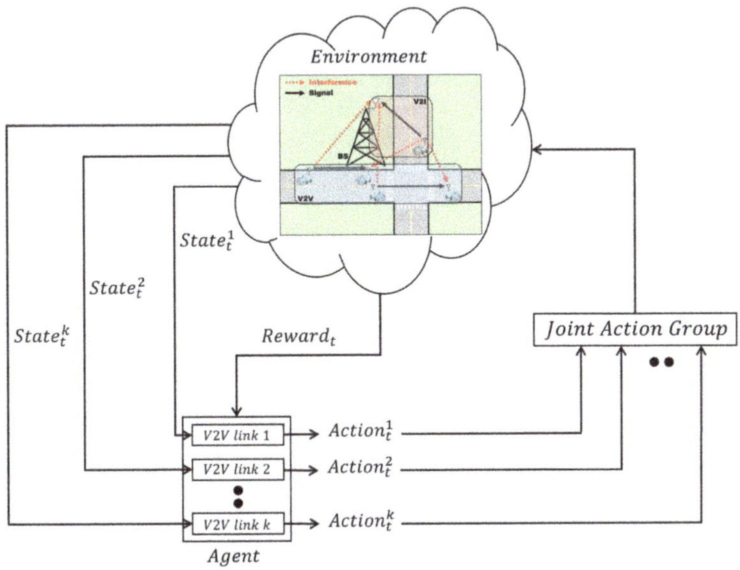

Figure 2. The structure of the RL for the vehicular network.

4.2. Deep Q-Network

A frequently used framework in the RL is a DQN [35–38]. The DQN framework is a structure that includes a Q-network consisting of a deep neural network (DNN) in the Q-learning structure. In order to train the Q-network in the DQN framework, several learning methods need to be applied [39].

In Q-learning, the Q-value means the expected return when reaching the terminal state from the state observed in time t, as follows:

$$Q(\mathbf{s},a) = \mathbb{E}[R_{t:T}|\mathbf{s}_t = \mathbf{s}, A_t = a]. \tag{10}$$

The Q-value is updated as follows:

$$Q(\mathbf{s}_t,a_t) = Q(\mathbf{s}_t,a_t) + \alpha[r_t + \gamma \max_a Q(\mathbf{s}_{t+1},a) - Q(\mathbf{s}_t,a_t)], \tag{11}$$

where α denotes the learning rate. The agent's behavior is determined based on the ϵ-greedy policy. The ϵ-greedy policy is a method of randomly selecting an action if the randomly sampled value is lower than the value of ϵ and selecting the action with the highest Q-value is greedy if it is high. However, because Q-learning uses a lookup table called a Q-table that stores Q-values in order to find the state and action pairs, it has several disadvantages: First, the probability of visiting the same state is very low. Second, a very large storage device is required to store an exponentially increasing number of state and action pairs. A DQN framework has been developed to overcome these disadvantages.

As shown in Figure 3, the DQN framework calculates Q-values using a Q-network in which weights and biases are stored. Therefore, when an agent needs a Q-value that matches a state, the agent puts the state as input to the Q-network and obtains the appropriate Q-value as output. The loss function for training the Q-network is as follows:

$$Loss(\boldsymbol{\theta},\boldsymbol{\beta}) = \sum_{\mathbf{s}_t,a_t \in E} (y - Q(\mathbf{s}_t,a_t,\boldsymbol{\theta},\boldsymbol{\beta}))^2 \tag{12}$$

$$y = r_t + \gamma \max_a Q(\mathbf{s}_{t+1},a,\boldsymbol{\theta},\boldsymbol{\beta}), \tag{13}$$

where $\boldsymbol{\theta}$ and $\boldsymbol{\beta}$ mean the weights and biases in the Q-network, respectively. $Q(\mathbf{s}_t, a_t, \boldsymbol{\theta}, \boldsymbol{\beta})$ means the Q-value. E is a mini-batch sampled from the experience replay memory that stores the state, action, and reward of the next-state tuples collected while the agent interacts with the environment. However, in order for the Q-network to perform an approximation function in the DQN framework, a training process that adjusts the weights and biases stored in the Q-network is required.

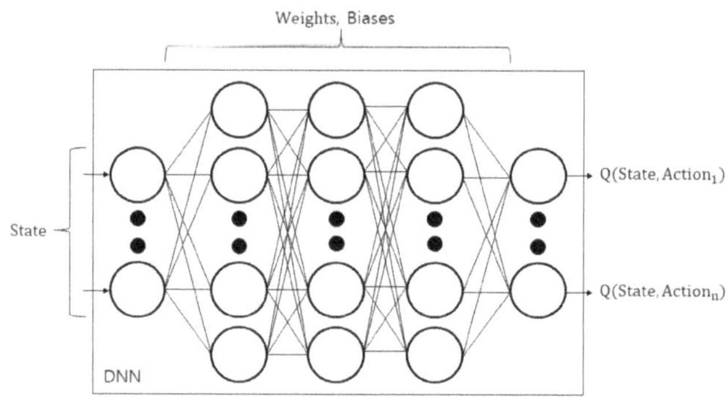

Figure 3. A Q-network of the DQN framework.

4.3. *Training and Testing Algorithm*

We train the DQN with the following methods: First, we use a data sampling method with experience replay memory. Data sampling is used to remove the temporal relationships between the used data to learn the Q-network. Here, the experience replay memory is a data storage technique in which the agent collects data while interacting with the environment. The data consist of tuples of the state, action, reward, and next state. The Q-network is trained by randomly sampling data tuples from experience replay memory. In this paper, the experience replay memory is denoted by $\mathcal{D}$. Second, we use a fixed target network method that includes two Q-networks, the target network and the online network, in the training process. $Q(\mathbf{s}_{t+1}, a, \boldsymbol{\theta}, \boldsymbol{\beta})$ of (13) is calculated as the target network, and $Q(\mathbf{s}_t, a_t, \boldsymbol{\theta}, \boldsymbol{\beta})$ of (12) is calculated as the online network. Additionally, the weights of the online network are periodically copied to those of the target network.

The training process is described in Algorithm 1. Parameters of the online and target networks are initialized (lines 1–3). The agent observes the state in the environment and selects the action according to the ϵ-greedy policy (lines 10–11). That is, the agent performs a random selection with the probability of ϵ, inputs the current state to the online network with the probability of $\epsilon - 1$, and selects the largest value among the observed Q-values as output. The selected action is saved in the joint action group (line 12). If the agent selects the actions for all V2V links, the joint action group interacts with the environment and acquires a reward (line 14). The data tuples collected through the above process are stored in the experience replay memory (lines 15–16). The sampled data tuples are used to update the online network. When the online network repeatedly updates the weights and biases, the weights and biases of the online network are copied to those of the target network (lines 17–25).

The testing process is described in Algorithm 2. Unlike the training process, the testing process greedily selects an action based on the Q-network learned by the training process (lines 8–9). After that, the action is stored in the joint action group in the same way as the training process (line 10). If the agent selects the action for all V2V links, the joint action group interacts with the environment (line 12). When the time step t reaches the simulation end time, the performances are evaluated in terms of the sum rate of the V2I and V2V links, the outage probability of V2V links, and the average power consumption of V2V links (lines 14–17).

Algorithm 1 Training algorithm

Initialize the online Q-network with random weights $\boldsymbol{\theta}$ and random biases $\boldsymbol{\beta}$;
Initialize the target Q-network with random weights $\boldsymbol{\theta}^{target}$ and random biases $\boldsymbol{\beta}^{target}$;
Generate Experience replay memory $\mathcal{D}$;
for each episode e **do**
    Initialize environment;
    Generate V2V and V2I networks;
    for each time step t **do**
        Generate a joint action group A;
        for each V2V links **do**
            Get state $\mathbf{s}_t$ from the environment
            Choose an action a_t based on the ϵ-greedy policy;
            Append the a_t to A;
        end for
        Interact with the environment based on A and Calculate reward r_t;
        Get all V2V links state $\mathbf{s}_{t+1}$ from the environment;
        Append the $\mathbf{s}_t$, a_t, r_t, $\mathbf{s}_{t+1}$ to $\mathcal{D}$;
        for each update step i **do**
            Sample a mini-batch of experience set E from the $\mathcal{D}$;
            Calculate the loss:
            $y = r_t + \gamma \max_a(Q(\mathbf{s}_{t+1}, a, \boldsymbol{\theta}^{target}, \boldsymbol{\beta}^{target}))$;
            $\text{Loss}(\boldsymbol{\theta},\boldsymbol{\beta}) = \sum_{\mathbf{s}_t, a_t \in E}(y - Q(\mathbf{s}_t, a_t, \boldsymbol{\theta}, \boldsymbol{\beta}))^2$;
            Update the online Q-network with $\boldsymbol{\theta}$, $\boldsymbol{\beta}$;
        end for
        Update weights, $\boldsymbol{\theta}^{target} \leftarrow \boldsymbol{\theta}$
        Update biases with $\boldsymbol{\beta}^{target} \leftarrow \boldsymbol{\beta}$
    end for
end for

Algorithm 2 Testing algorithm

Load the Q-network with trained weights $\boldsymbol{\theta}$ and biases $\boldsymbol{\beta}$;
for each episode e **do**
    Initialize environment;
    Generate V2V and V2I networks;
    for each time step t **do**
        Generate a joint action group A;
        for each V2V link **do**
            Get state $\mathbf{s}_t$ from the environment
            Choose the a_t with the maximum value among the estimated Q-values by inputting the $\mathbf{s}_t$ into the Q-network;
            Append the a_t to A;
        end for
        Interact with the environment based on A;
    end for
    Calculate the sum rate of V2I links;
    Calculate the sum rate of V2V links;
    Calculate the outage probability of V2V links;
    Calculate the average transmit power of V2V links;
end for

5. Simulation Results

We consider a single-cell system with one base station and 20 V2I links. We follow the simulation setup for the urban case study of Manhattan with 9 blocks of 3GPP TR 36.885 [11,40]. The models of vehicle drops, mobility, and channels all follow the evaluation scenario of 3GPP TR 36.885. Vehicles are dropped on the road according to a spatial Poisson process, and the vehicle locations are updated every one time slot in the simulation. A vehicle moves at a constant speed defined in Table 1. The vehicle changes its direction at the intersection to go straight with a probability of 0.5, to turn left with a probability of 0.25, and to turn right with a probability of 0.25. Figure 4 shows the movement of vehicles for 20 s, where there are eight vehicles and one BS. The V2V channel model and V2I channel model are both described in Table 1, according to 3GPP TR 36.885. Each vehicle communicates with a vehicle nearby. We perform a time-driven simulation, where the simulation clock is advanced in increments of time slot units and the state variables are updated for every time slot. For each slot in the simulation, we calculate the CSI of the V2I and V2V links and the interference power, which results in the state of the DQN. The simulation parameters are summarized in Table 1.

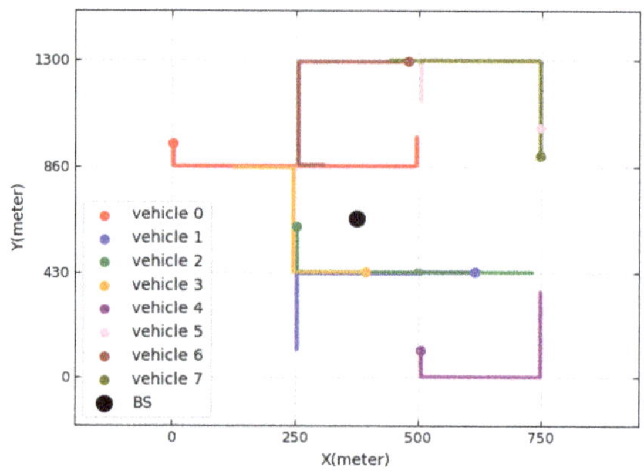

Figure 4. Vehicle movements.

Table 1. Simulation parameters.

Parameter	Value
Road intersection size	430 m × 250 m
Simulation area size	1300 m × 750 m
Absolute vehicle speed	36 km/h
Vehicle drop and mobility model	Urban case of A.12 in 3GPP TR 36.885 [40]
V2V path loss model	WINNER + B1 Manhattan [41]
V2V shadowing	Log-normal with $\sigma^2 = 3$ dB
V2I path loss model	$128.1 + 37.6 \log(R)$, where R in kilometers
V2I shadowing	Log-normal with $\sigma^2 = 8$ dB
V2V and V2I fast fading	Rayleigh fading
Noise power	−114 dBm
Carrier frequency, f_c	2 GHz
Sub-carrier frequency	1.5 MHz
Number of V2I links, L	20
Number of V2V links, K,	[60, 120, 180, 240, 300]
Antenna height of eNode B type RSU	25 m
Antenna gain of RSU	8 dBi
Noise figure of RSU's antenna	5 dB
Antenna height of vehicles	1.5 m
Antenna gain of vehicles	3 dBi
Noise figure of vehicle's antenna	9 dB
Latency constraints for V2V link	100 ms
V2V payload size	30 Mbits
Update time slot duration	2 ms
Simulation time	400 ms
transmit power level of V2V links	[5, 10, 23] dBm

In the proposed Q-network, the number of neurons in the input layer is set to 82, the number of neurons in the hidden layers is set to [500, 250 120], and the number of neurons in the output layer is set to 60. The activation function of the hidden layers uses the ReLU function. The optimizer for training the Q-network uses RMSProp. The detailed parameters of the DQN framework are summarized in Table 2. The DQN is trained for 20,000 episodes, where an episode means 1 simulation time and new vehicles are dropped each time an episode starts. After training the DQN, the simulation is run 1000 times, and the 1000 results are averaged.

Table 2. DQN framework parameters.

Parameter	Value
Number of neurons in the input layer	82
Number of neurons in each hidden layer	500, 250, 120
Number of neurons in the output layer	60
Reward discount factor	0.99
Hidden layer activation function	ReLU
Optimizer	RMSProp
Learning rate α	0.001
Values of λ_{V2I}, λ_{V2V}, $\lambda_{Latency}$, and λ_{pwr}	0.1, 0.9, 1, and 0.2, respectively

The proposed resource allocation is compared with the random resource allocation and the conventional RL-based resource allocation of [11] in terms of the average transmit power of the V2V links, the average outage probability of the V2V links, and the average sum rates of the V2V and V2I networks. In the random resource allocation, the transmitter of the V2V link transmits data with randomly selected transmit power through a randomly selected RB.

Figure 5 shows the average transmit power of V2V links according to the number of V2V links. As the number of V2V links increases, the average transmit power of vehicles increases in the proposed scheme and the conventional RL-based scheme, but the average transmit power in the random allocation scheme is fixed. Because the interference increases according to the increase in the number of V2V links, the transmit power of vehicles increases in order to overcome the interference, in the proposed scheme and the conventional RL-based scheme. The proposed scheme significantly reduces the power consumption of V2V links in comparison with the conventional RL-based scheme because of the penalty function of the transmit power in the reward. In the random allocation scheme, because the transmitter randomly selects the transmit power, the transmit power of the V2V link is fixed on average, regardless of the amount of interference caused by other V2V links. In the simulation environment of this paper, the random allocation scheme shows a low power consumption due to the low transmit power but shows an outage probability that is too high.

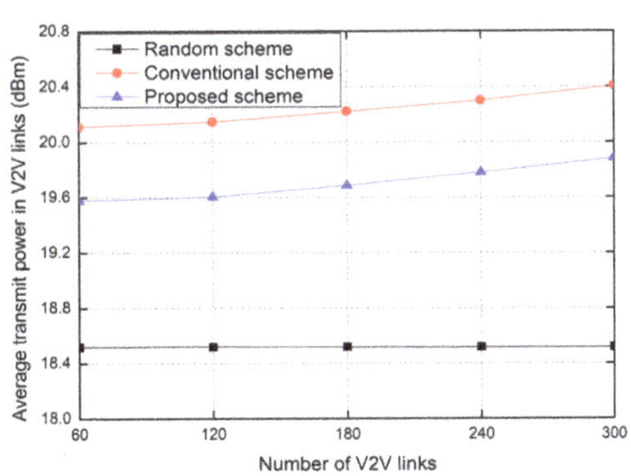

Figure 5. Average transmit power of V2V links.

Figure 6 shows the sum rate of V2V links as the number of V2V links increases. As the number of V2V links increases, the sum rate of V2V links increases in all the resource allocation schemes. The sum rate of the conventional RL-based scheme is slightly higher than that of the proposed scheme because the proposed scheme suppresses the transmit power of vehicles for the purpose of energy efficiency. The sum rate of the random allocation scheme is the worst because it randomly selects the RBs regardless of the interference to others.

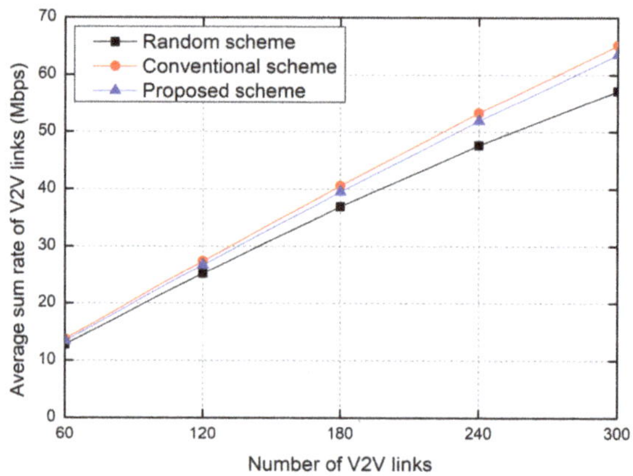

Figure 6. Average sum rate of V2V links.

Figure 7 shows the sum rate of V2I links as the number of V2V links increases. Because the number of V2I links is fixed at 20, the interference from the V2V links increases according to the number of V2V links, and thus the sum rate of V2I links decreases with the increase in the V2V links. In particular, the performance of the proposed scheme is slightly better than that of the conventional RL-based scheme. Moreover, because the interference from the V2V links to the V2I link increases according to the number of V2V links, the sum rate of the random allocation greatly decreases with the number of V2V links.

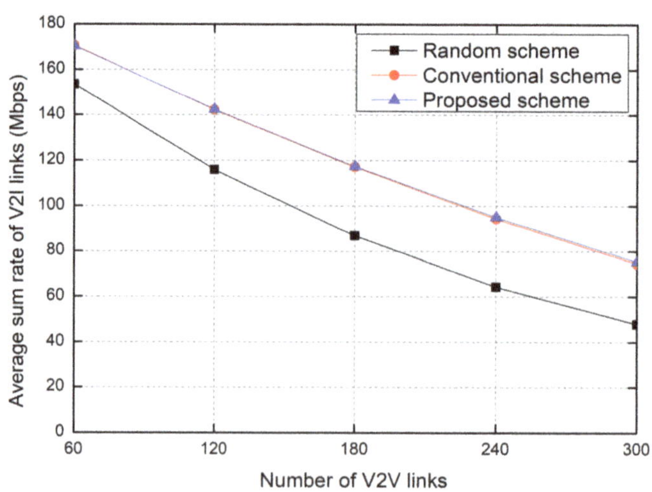

Figure 7. Average sum rate of V2I links.

Figure 8 shows the outage probability as the number of V2V links increases. Here, the outage probability is defined as the probability that a transmitter on the V2V link fails to transmit data within the maximum allowable latency, T_0. The outage probability is inversely proportional to the sum rate. Hence, the outage probability increases according to the number of V2V links. Moreover, the outage probability of the random allocation scheme is much higher than that of other schemes. That is, in order to efficiently allocate resources, RL-based resource allocation is required.

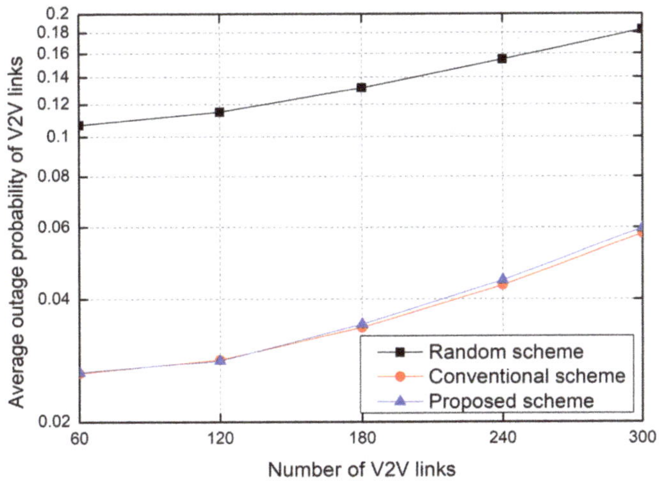

Figure 8. Average outage probability of V2V links.

The major concerns with deep-learning-based approaches are the computational complexity and the memory space, which depend on the number of parameters to be stored and to be computed. In the proposed RL, from (8), the dimension of the state space is $D_{\text{state}} = (4 \times N_{\text{RB}}) + 2$ (=82 in our simulation) and the dimension of the action space is $D_{\text{action}} = N_{\text{pwr}} \times N_{\text{RB}}$ (=60 in our simulation), where N_{RB} is the number of RBs and N_{pwr} is the number of the transmit power levels. Consider a feed-forward network with l layers, where layer 0 is the input layer and layer $l-1$ is the output layer. Let the number of neurons of each layer be $n_0, n_1, \ldots, n_{l-1}$. Then, the number of parameters (weights) of the network, including biases, is given by $N_{\text{DNN}} = \sum_{i=0}^{l-2} n_i n_{i+1} + \sum_{i=1}^{l-1} n_i$ (=204,130 in our simulation). Because the DQN framework calculates Q-values using a DNN, the total number of parameters becomes $D_{\text{state}} + D_{\text{action}} + N_{\text{DNN}}$. Moreover, because of the use of two Q-networks, the online network and the target network, the total number of parameters to be processed doubles, and the replay memory is required to store a collection of experience tuples, i.e., the parameters of the online network. In our simulation, we set the replay buffer size to 100 tuples. The computational complexity is similar to [11]. In our implementation, each selection takes less than 10^{-4} s using GPU 2080 Ti. The computational speed is acceptable for vehicles thanks to the power of the GPU. The computational complexity of the DNN can be reduced by using lightweight DNNs [42,43].

6. Conclusions

Vehicular communications or V2X are key to the development of autonomous vehicles. In the V2X network, it is important to manage radio resources efficiently to provide low-latency and energy-efficient services. In this paper, we developed a DQN-based energy-efficient resource allocation scheme in a V2X communication network in which V2I and V2V networks share resource blocks. We formulated the reward of the DQN model by using two penalties and two positives. Here, the two penalties are the transmission time and the transmit power, and the two positives are the sum rate of the V2I and V2V networks. The proposed scheme significantly reduces the power consumption of vehicles in the V2V network without sacrificing the sum rate and outage probability. The results show that an energy-efficient resource allocation scheme is crucial in order to meet the latency and power consumption requirements of mission critical V2V applications.

Author Contributions: Conceptualization, D.H. and J.S.; software, D.H.; validation, D.H. and J.S.; investigation, D.H. and J.S.; writing—original draft preparation, D.H. and J.S.; writing—review and editing, J.S.; supervision, J.S.; project administration, J.S.; funding acquisition, J.S. All authors have read and agreed to the published version of the manuscript.

Funding: This research was supported by the National Research Foundation of Korea (NRF) grant funded by the Korea government (MSIT) (nos. 2020R1F1A1065109 and 2022R1F1A1062696).

Conflicts of Interest: The authors declare no conflict of interest.

References

1. Garcia, M.H.C.; Molina-Galan, A.; Boban, M.; Gozalvez, J.; Coll-Perales, B.; Şahin, T.; Kousaridas, A. A tutorial on 5G NR V2X communications. *IEEE Commun. Surv. Tutor.* **2021**, *23*, 1972–2026. [CrossRef]
2. Molina-Masegosa, R.; Gozalvez, J. LTE-V for sidelink 5G V2X vehicular communications: A new 5G technology for short-range vehicle-to-everything communications. *IEEE Veh. Technol. Mag.* **2017**, *12*, 30–39. [CrossRef]
3. Lien, S.Y.; Deng, D.J.; Lin, C.C.; Tsai, H.L.; Chen, T.; Guo, C.; Cheng, S.M. 3GPP NR sidelink transmissions toward 5G V2X. *IEEE Access* **2020**, *8*, 35368–35382. [CrossRef]
4. Nam, Y.; So, J. One-bit signaling-based interference management for MIMO V2V sidelink. *ICT Express* **2022**, *8*, 595–599. [CrossRef]
5. Hong, C.; Shan, H.; Song, M.; Zhuang, W.; Xiang, Z.; Wu, Y.; Yu, X. A joint design of platoon communication and control based on LTE-V2V. *IEEE Trans. Veh. Technol.* **2020**, *69*, 15893–15907. [CrossRef]
6. Liang, L.; Ye, H.; Yu, G.; Li, G. Y. Deep-learning-based wireless resource allocation with application to vehicular networks. *Proc. IEEE* **2020**, *108*, 341 – 356. [CrossRef]
7. He, Z.; Wang, L.; Ye, H.; Li, G.Y.; Juang, B.H.F. Resource allocation based on graph neural networks in vehicular communications. In Proceedings of the 2020 IEEE Global Communications Conference (GLOBECOM), Taipei, Taiwan, 7–11 December 2020; pp. 1–5.
8. Xu, L.; Zhuang, W. Energy-efficient cross-layer resource allocation for heterogeneous wireless access. *IEEE Tran. Wirel. Commun.* **2018**, *17*, 4819–4829. [CrossRef]
9. Ismail, M.; Gamage, A.T.; Zhuang, W.; Shen, X.; Serpedin, E.; Qaraqe, K. Uplink decentralized joint bandwidth and power allocation for energy-efficient operation in a heterogeneous wireless medium. *IEEE Trans. Commun.* **2015**, *63*, 1483–1495. [CrossRef]
10. Zhu, X.; Yang, B.; Chen, C.; Xue, L.; Guan, X.; Wu, F. Cross-layer scheduling for OFDMA-based cognitive radio systems with delay and security constraints. *IEEE Trans. Veh. Technol.* **2015**, *64*, 5919–5934. [CrossRef]
11. Ye, H.; Li, G.Y.; Juang, B.H.F. Deep reinforcement learning based resource allocation for V2V communications. *IEEE Trans. Veh. Technol.* **2019**, *68*, 3163–3173. [CrossRef]
12. Bhadauria, S.; Shabbir, Z.; Roth-Mandutz, E.; Fischer, G. QoS based deep reinforcement learning for V2X resource allocation. In Proceedings of the 2020 IEEE International Black Sea Conference on Communications and Networking (BlackSeaCom), Odessa, Ukraine, 26–29 May 2020; pp. 1–6.
13. Xiao, H.; Zhu, D.; Chronopoulos, A.T. Power allocation With energy efficiency optimization in cellular D2D-based V2X communication network. *IEEE Trans. Intell. Transp. Syst.* **2019**, *21*, 4947–4957. [CrossRef]
14. Liang, L.; Ye, H.; Li, G.Y. Spectrum sharing in vehicular networks based on multi-agent reinforcement learning. *IEEE J. Sel. Areas Commun.* **2019**, *37*, 2282–2292. [CrossRef]
15. Xiang, P.; Shan, H.; Wang, M.; Xiang, Z; Zhu, Z. Multi-agent RL enables decentralized spectrum access in vehicular networks. *IEEE Trans. Veh. Technol.* **2021**, *70*, 10750–10762. [CrossRef]
16. Soleymani, D.M.; Ravichandran, L.; Gholami, M.R.; Del Galdo, G.; Harounabadi, M. Energy-efficient autonomous resource selection for power-saving users in NR V2X. In Proceedings of the 2021 IEEE 32nd Annual International Symposium on Personal, Indoor and Mobile Radio Communications (PIMRC), Helsinki, Finland, 13–16 September 2021; pp. 972–978.
17. Gao, L.; Hou, Y.; Tao, X.; Zhu, M. Energy-efficient power control and resource allocation for V2V communication. In Proceedings of the 2020 IEEE Wireless Communications and Networking Conference (WCNC), Seoul, Republic of Korea, 25–28 May 2020; pp. 1–6.
18. Lorincz, J.; Matijevic, T.; Petrovic, G. On interdependence among transmit and consumed power of macro base station technologies. *Comput. Commun.* **2014**, *50*, 10–28. [CrossRef]
19. Lorincz, J.; Capone, A.; Begušić, D. Heuristic algorithms for optimization of energy consumption in wireless access networks. *KSII Trans. Internet Inf. Syst.* **2011**, *5*, 626–648. [CrossRef]

20. Lorincz, J.; Bogarelli, M.; Capone, A.; Begušić, D. Heuristic approach for optimized energy savings in wireless Access Networks. In Proceedings of the 18thInternational Conference on Software, Telecommunications and Computer Networks, Dalmatia, Croatia, 23-25 September 2010; pp. 1–6.
21. Jiang, W. Graph-based deep learning for communication networks: A survey. *Comput. Commun.* **2022**, *185*, 40–54. [CrossRef]
22. Ivanov, A.; Tonchev, K.; Poulkov, V.; Manolova, A.; Neshov, N.N. Graph-based resource allocation for integrated space and terrestrial communications. *Sensors* **2022**, *22*, 5778. [CrossRef]
23. Zhang, X.; Zhang, Z.; Yang, L. Learning-based resource allocation in heterogeneous ultradense network. *IEEE Internet Things J.* **2022**, *9*, 20229–20242. [CrossRef]
24. Zhao, D.; Qin, H.; Song, B.; Han, B.; Du, X.; Guizani, M. A graph convolutional network-based deep reinforcement learning approach for resource allocation in a cognitive radio network. *Sensors* **2020**, *20*, 5216. [CrossRef]
25. Alatabani, L.E.; Ali, E.S.; Mokhtar, R.A.; Saeed, R.A.; Alhumyani, H.; Hasan, M.K. Deep and reinforcement learning technologies on internet of vehicle (IoV) applications: Current issues and future trends. *J. Adv. Transp.* **2022**, *2022*, 1947886. [CrossRef]
26. Alatabani, L.E.; Ali, E.S.; Saeed, R.A. Deep learning approaches for IoV applications and services. In *Intelligent Technologies for Internet of Vehicles*; Springer: Berlin/Heidelberg, Germany, 2021; pp. 253–291.
27. Ali, E.S.; Hassan, M.B.; Saeed, R.A. Machine learning technologies on internet of vehicles. In *Intelligent Technologies for Internet of Vehicles*; Springer: Berlin/Heidelberg, Germany, 2021; pp. 225–252.
28. Elfatih, N.M.; Hasan, M.K.; Kamal, Z.; Gupta, D.; Saeed, R.A.; Ali, E.S.; Hosain, M.S. Internet of vehicle's resource management in 5G networks using AI technologies: Current status and trends. *IET Commun.* **2021**, *16*, 400–420. [CrossRef]
29. Ali, E.S.; Hasan, M.K.; Hassan, R.; Saeed, R.A.; Hassan, M.B.; Islam, S.; Nafi, N.S.; Bevinakoppa, S. Machine learning technologies for secure vehicular communication in internet of vehicles: Recent advances and applications. *Secur. Commun. Netw.* **2021**, *2021*, 8868355. [CrossRef]
30. He, Y.; Zhao, N.; Yin, H. Integrated networking, caching, and computing for connected vehicles: A deep reinforcement learning approach. *IEEE Trans. Veh. Technol.* **2018**, *67*, 44–55. [CrossRef]
31. Tang, C.; Zhu, C.; Wu, H.; Li, Q.; Rodrigues, J.J.P.C. Toward response time minimization considering energy consumption in caching-assisted vehicular edge computing. *IEEE Internet Things J.* **2022**, *9*, 5051–5064. [CrossRef]
32. Yao, Y.; Xiao, B.; Wang, W.; Yang, G.; Zhou, X.; Peng, Z. Real-time cache-aided route planning based on mobile edge computing. *IEEE Wirel. Commun.* **2020**, *27*, 151–161. [CrossRef]
33. Tang, C.; Wu, H. Joint optimization of task caching and computation offloading in vehicular edge computing. *Peer-to-Peer Netw. Appl.* **2022**, *15*, 854–869. [CrossRef]
34. Bai, R.; Chen, X.; Chen, Z.L.; Cui, T.; Gong, S.; He, W.; Jiang, X.; Jin, H.; Jin, J.; Kendall, G.; et al. Analytics and machine learning in vehicle routing research. *Int. J. Prod. Res.* **2023**, *61*, 4–30. [CrossRef]
35. Zhao, Y.; Niemegeers, I.G.; De Groot, S.M.H. Dynamic power allocation for cell-free massive MIMO: Deep reinforcement learning methods. *IEEE Access* **2021**, *9*, 102953–102965. [CrossRef]
36. Zhang, X.; Peng, M.; Yan, S.; Sun, Y. Deep-reinforcement-learning-based mode selection and resource allocation for cellular V2X communications. *IEEE Internet Things J.* **2020**, *7*, 6380–6391. [CrossRef]
37. Cao, Y.; Zhang, G.; Li, G.; Zhang, J. A deep Q-network based-resource allocation scheme for massive MIMO-NOMA. *IEEE Commun. Lett.* **2021**, *25*, 1544–1548. [CrossRef]
38. Choi, J.Y.; Jo, H.S.; Mun, C.; Yook, J.G. Deep reinforcement learning-based distributed congestion control in cellular V2X networks. *IEEE Wirel. Commun. Lett.* **2021**, *10*, 2582–2586. [CrossRef]
39. Mnih, V.; Kavukcuoglu, K.; Silver, D.; Rusu, A.A.; Veness, J.; Bellemare, M.G.; Graves, A.; Riedmiller, M.; Fidjeland, A.K.; Ostrovski, G.; et al. Human-level control through deep reinforcement learning. *Nature* **2015**, *518*, 529–533. [CrossRef] [PubMed]
40. Technical Specification Group Radio Access Network; Study LTE-Based V2X Services (Release 14), Document 3GPP TR 36.885 V14.0.0, 3rd Generation Partnership Project, June 2016. Available online: https://www.3gpp.org/ftp/Specs/archive/36_series/36.885/36885-e00.zip (accessed on 1 October 2022).
41. Kyösti, P.; Meinilä, J.; Hentila, L.; Zhao, X.; Jämsä, T.; Schneider, C.; Narandzic, M.; Milojević, M.; Hong, A.; Ylitalo, J.; et al. WINNER II Channel Models. 02 2008, iST-4-027756 WINNER II D1.1.2 V1.2. Available online: http://www.ero.dk/93F2FC5C-0C4B-4E44-8931-00A5B05A331B (accessed on 1 October 2022).
42. Wang, C.-H.; Huang, K.-Y.; Yao, Y.; Chen, J.-C.; Shuai, H.-H.; Cheng, W.-H. Lightweight deep learning: An overview. *IEEE Consum. Electron. Mag.* **2022**, 1–12. [CrossRef]
43. Rastegari, M.; Ordonez, V.; Redmon, J.; Farhadi, A. Xnor-net: Imagenet classification using binary convolutional neural networks. In Proceedings of the 2016 European Conference on Computing Vision (ECCV), Amsterdam, The Netherlands, 8–16 October 2016; pp. 525–542.

sensors MDPI

Article

A Novel Traffic Prediction Method Using Machine Learning for Energy Efficiency in Service Provider Networks

Francisco Rau [1,*], Ismael Soto [1,*], David Zabala-Blanco [2], Cesar Azurdia-Meza [3], Muhammad Ijaz [4], Sunday Ekpo [4] and Sebastian Gutierrez [5]

1 CIMTT, Department of Electrical Engineering, Universidad de Santiago de Chile, Santiago 9170124, Chile
2 Department of Computer Science and Industry, Universidad Católica del Maule, Talca 3480112, Chile; dzabala@ucm.cl
3 Department of Electrical Engineering, Universidad de Chile, Santiago 8370451, Chile; cazurdia@ing.uchile.cl
4 Department of Engineering, Faculty of Science and Engineering, Manchester Metropolitan University, Manchester M1 5GD, UK; m.ijaz@mmu.ac.uk (M.I.); s.ekpo@mmu.ac.uk (S.E.)
5 Faculty of Engineering, Universidad Autónoma de Chile, Santiago 7500912, Chile; sebastian.gutierrezl@usach.cl
* Correspondence: francisco.rau@usach.cl (F.R.); ismael.soto@usach.cl (I.S.)

Abstract: This paper presents a systematic approach for solving complex prediction problems with a focus on energy efficiency. The approach involves using neural networks, specifically recurrent and sequential networks, as the main tool for prediction. In order to test the methodology, a case study was conducted in the telecommunications industry to address the problem of energy efficiency in data centers. The case study involved comparing four recurrent and sequential neural networks, including recurrent neural networks (RNNs), long short-term memory (LSTM), gated recurrent units (GRUs), and online sequential extreme learning machine (OS-ELM), to determine the best network in terms of prediction accuracy and computational time. The results show that OS-ELM outperformed the other networks in both accuracy and computational efficiency. The simulation was applied to real traffic data and showed potential energy savings of up to 12.2% in a single day. This highlights the importance of energy efficiency and the potential for the methodology to be applied to other industries. The methodology can be further developed as technology and data continue to advance, making it a promising solution for a wide range of prediction problems.

Keywords: energy efficiency; machine learning; telecom services operator; traffic prediction

Citation: Rau, F.; Soto, I.; Zabala-Blanco, D.; Azurdia-Meza, C.; Ijaz, M.; Ekpo S.; Gutierrez S. A Novel Traffic Prediction Method Using Machine Learning for Energy Efficiency in Service Provider Networks. *Sensors* **2023**, *23*, 4997. https://doi.org/10.3390/s23114997

Academic Editor: Josip Lorincz

Received: 8 April 2023
Revised: 15 May 2023
Accepted: 18 May 2023
Published: 23 May 2023

1. Introduction

The increasing demand for energy in contemporary industry, global warming, and the development of new communication technologies such as the Internet of Things (IOT), 5G, and B5G have necessitated research into energy-saving strategies in the telecommunications sector, especially among telecommunications service operators (TSOs). From 2010 to 2018, the global energy demand for data centers increased from 194 TWh to 205 TWh, according to research [1]. According to [2], by the year 2030, telecommunications networks will consume up to 51% of the world's electricity if their energy efficiency is not significantly improved. Consequently, energy efficiency is becoming essential for existing and future 5G and beyond networks.

Some TSOs have designed networks with redundant links to avoid congestion in high-availability schemes (active/passive configuration) and load balancing (active/active configuration) [3]. From the perspective of energy efficiency, these designs can be considered as inefficient energy expenditure, because these links are always active [4]. The studies in [5,6] show that links are underutilized by only 40% in the central network, commonly called core network (CN), and that during off-peak hours, it is much lower. Moreover, there is not much difference in energy consumption between equipment at full load and standby mode [5]. To increase bandwidth capacity, TSOs connect the routers through multiple physical cables that form a single grouped logical link. The grouped logical links are also called link aggregation groups (LAG) or bundle Ethernet (BE). Link aggregation in the Ethernet was standardized in the IEEE 802.3ad standard and later renamed as IEEE 802.1ax to maintain consistency with other 802.3 standards [7].

The current CN of a TSO is composed of multiple networks, and trunk lines are often formed by many grouped optical fibers to increase capacity and add resilience. Connections with the number of sublinks within a BE or LAG range from 2 to approximately 20 in a typical TSO [6]. Additionally, there is the problem of the energy consumption of these links, which is even more when they are grouped. In recent years, several works have been published and proposed to reduce the energy consumption of simple links (unaggregated) [8]. One of the standards that helps solve this problem is IEEE 802.3az, which specifies Ethernet energy efficiency (EEE), which is a method for reducing the energy used by an Ethernet device during low link utilization periods [9]. The premise of EEE is that Ethernet links have idle time and therefore the opportunity to save energy during that period of time. The method is called low power idle (LPI), but it is only for copper interfaces [10]. Regarding LAGs, several energy efficiency studies have been carried out using different optimization methods [6]. Additionally, there are works that propose threshold point algorithms, as in [11], and others [12] that work with software-defined network (SDN) controllers. Moreover, [13] has used predictive techniques to reduce LAG energy, using a simple moving average (SMA).

On the other hand, neural networks have begun to be used to predict time series, because time series are the optimal method to describe network traffic behavior, in particular regarding recurrent neural networks (RNNs) and their variants, long short-term memory (LSTM) and gated recurrent unit (GRU) [14–17]. There is also a neural network that stands out for its speed called online sequential extreme learning machine (OS-ELM), which has been shown to be efficient and especially fast compared with gradient-based networks [18–20]. This paper's primary objective is to predict traffic in a short period of time to activate and deactivate the ports of the link aggregation (LAG or BE) between two nodes within a TSO network in order to save energy on fiber optic links while maintaining a high quality of service (QoS) for clients; the main contributions of this work are presented below:

1. A novel method is proposed and developed to compare different types of neural networks in terms of their ability to process time series data, specifically in real-time traffic analysis. This methodology aimed to evaluate the performance of various neural network models and identify the most suitable option for the task.
2. A novel bundle Ethernet energy efficiency methodology was designed. This algorithm was based on the expected traffic and used the best-performing neural network, selected by the methodology outlined in point 1.
3. The proposed traffic prediction method and energy-saving Ethernet bundle methodology were evaluated. The performance of the traffic prediction methodology was compared between neural networks. The energy-saving Ethernet bundle was evaluated in terms of energy savings by comparing the performance algorithms proposed in point 2. The results were analyzed to determine the feasibility and effectiveness of the proposed solution.

The remainder of this paper is structured as follows: Section 2 presents related works in LAG or BE energy efficiency, as well as the most popular machine learning models and neural networks for traffic prediction. Section 3 details the methodology involved in the selection of a neural network. Section 4 outlines the stages used in the methodology for the development of two energy-efficient algorithms. Section 5 describes the use case, including network topology, traffic description, and network equipment specifications. In Section 6, the results of training a neural network using training and testing data, along with various evaluation metrics, are presented for the use case. In Section 7, the performance of the proposed energy efficiency optimization algorithms is compared with the base case, and the results are presented for the use case. Finally, relevant conclusions and future works are presented in Section 8.

2. Related Works

In this Section, we review the related works on traffic prediction based on neural networks and energy efficiency in grouped links (LAG or BE).

2.1. Works Related to Methods of Traffic Prediction Based on Machine Learning and Neural Networks

Analyzing historical traffic is a critical challenge for generating an accurate model that reflects the structure of the series in order to allow for prediction and classification of future events [21]. Time series consist of a succession of ordered numerical data points. The problem of time series prediction is the forecast of future activity from past values and the related patterns [21,22].

For traffic prediction, several time series forecasting techniques can be used, grouped into two types of linear and nonlinear methods. Nonlinear methods are more suitable for predicting traffic due to the existing noise and the complex nature that traffic presents [23]. Neural networks are being widely used to predict time series [24].

According to the study in [25], convolutional neural networks (CNNs) and RNNs are the most widely used deep learning models for short-term traffic prediction. CNNs are good at capturing spatial characteristics, and RNNs are good at capturing the temporal characteristics of traffic data. Within deep learning and recurrent networks, there are variations such as the LSTM and GRU neural networks. Currently, predicting traffic with neural networks has made the analysis of time series an essential part of data modeling in a wide range of industries, including finance, health, transportation, and the environment [26–31].

In [22], the authors used a GRU neural network, a variant of LSTM, to predict the traffic flow of ships within an area of wind parks. The results of GRU were compared with the autoregressive integrated moving average (ARIMA) model, as well as support vector machine (SVM) and LSTM deep learning models, with GRU being the winner.

In the logistics sector, the prediction of passenger flow in metro stations has been carried out by [32], as well as the use of parking modeling in [33] and urbanization planning based on vehicle traffic prediction [34]. In the health field, the prediction of driver stress and drowsiness for accident prevention [35] and the prediction of the monthly progression of Alzheimer's disease [36] have also been demonstrated. With regard to the telecommunications sector, the prediction of TSO network traffic load has been used to avoid overloads, minimize response time, and optimize resource use [15,19,37,38].

With the arrival of the 5G standard, traffic prediction will be more difficult due to its heterogeneous nature. The coexistence of different networks and significantly different characteristics make traffic prediction, management, and optimization a difficult task. Therefore, the use and adaptation of neural networks will increasingly be used to analyze and manage network traffic based on data [39]. The diverse use of machine learning models in telecommunications operators and neural network research have led to more and more comparisons between them.

In [40], the use of cellular network traffic prediction for dynamic resource optimization in wireless backhaul networks is discussed. In [14], the authors state that the LSTM neural network is a type of recurrent neural network architecture, which is trained with a gradient-based learning algorithm. The framework proposed by the authors utilizes real traces from a TIER-1 TSO. With these traces, predictions were made at different time spans and compared with a combination of the classic method, such as ARIMA, and RNN. The results obtained in comparison with the ARIMA and RNN models show that the LSTM model performs well with a low normalized RMSE for the entire dataset and also generates predictions at very short time scales (less than thirty seconds). In the work in [41], the RNN, LSTM, GRU, deep neural network (DNN), and bidirectional LSTM (BLSTM) networks were compared, and the result was that LSTM was one of the networks that performed the best in terms of traffic prediction, due to having lower error metrics. In [42], the authors show the prediction of internet traffic in Telecom Italia, comparing deep learning models to conventional machine learning models. The performance of LSTM, GRU, and two conventional machine learning architectures, random forest (RF) and decision tree (DT), was compared for the prediction of mobile Internet traffic. The predictive quality of the models was evaluated using the root mean squared error (RMSE) and mean absolute error (MAE). Both deep learning algorithms were effective in modeling Internet activity and seasonality, both within days and over 2 months. The deep learning models outperformed the conventional machine learning models, placing the LSTM network as the winner over the GRU network in the experiments.

In the study in [15], it is shown that 5G networks can face network traffic peaks due to their numerous connections, so they focus on predicting these traffic peaks through deep learning techniques such as RNN, LSTM, and GRU from a real network. In terms of prediction, LSTM and GRU outperform RNN by 4.98% and 4.56%, respectively, and in terms of computational complexity, GRU is the worst compared with RNN and LSTM by 12.16% and 0.13%, respectively. Finally, since the times between GRU and LSTM are similar, when seeking greater precision for traffic peaks, the LSTM model performed the best. In [43], the prediction performance of the recurrent models RNN, LSTM, and GRU was compared with nonrecurrent models, such as XGBoost and RF, LASSO linear regression, and prediction models based on moving averages. The results indicate that RNN and its variants outperformed the other methods. The best nonrecurrent model was XGBoost. In comparison with XGBoost, GRU and RNN reduced 15% in the RMSE metric and 8% in MAPE. In another study, by [44], recurrent neural networks and their variants are compared with traditional statistical methods, such as ARIMA, seasonal ARIMA (SARIMA), and CNN. Three different congestion scenarios are tested: full day, morning peak hour, and afternoon peak hour. The results indicate that for the RMSE metric, the LSTM neural network had performances of 5.8, 7.9, and 10.2 and LSTM of 6.7, 8.6, and 10.9, depending on the scenario tested.

As previously mentioned, RNN, LSTM, and GRU have gained ground in traffic prediction compared with machine learning models (XGBoost, RF, SVM, and DT), conventional methods such as LASSO linear regression, ARIMA, and SARIMA, and other CNNs. One of the factors that hinders the RNN, LSTM, and GRU deep learning networks from being faster in prediction is the factor of seeking their optimization based on minimizing the gradient; thus, alternatives to this, without losing predictive performance, have been sought.

One of the neural networks that has attracted interest in recent times is the extreme learning machine (ELM) network, due to its good results in prediction and extremely fast training algorithm [45]. ELM is a particular type of feed-forward neural network. The learning mechanism allows for significantly faster training speed compared with classic neural networks in a variety of scenarios. A variation of this neural network is called online sequential extreme learning machine (OS-ELM). The advantage of OS-ELM over ELM is that it allows the algorithm to learn sequential data online, part by part, using the recursive least squares method [46]. In the study by [19], the authors compare the LSTM recurrent neural network with the OS-ELM neural network. The results show that OS-ELM outperforms LSTM in terms of computational cost by a factor of 2300, which is extremely high, and in terms of network prediction, OS-ELM was similar to LSTM. The OS-ELM neural network is simpler in architecture than LSTM, which makes it faster in terms of complexity than recurrent RNNs and their variants. The latter neural networks may have better accuracy in certain cases, but it depends on the input data, the architecture of the time series, and the hyperparameters that are configured.

For prediction models, the goal of evaluation metrics is to minimize error. For regression and prediction models, MAE and RMSE metrics are most commonly used to evaluate the performance of the model. For example, in studies such as [15,19,37,47–50], the RMSE metric has been used, and in others such as [36,37,47,51], the MAE metric has been used. The mean absolute percentage error (MAPE) is also used in some cases to evaluate regression and prediction [52,53]. RMSE and MAE are metrics that are dependent on the scale. MAPE is a percentage error metric. The RMSE metric is very sensitive to outliers, while the impact of these is reduced with MAE. On the other hand, MAE cannot indicate the bias of predictions in terms of overfitting or underfitting. MAPE can differentiate this type of bias and imposes a penalty on predicted values above the real ones.

In this study, the highest possible accuracy in prediction is required, with the processing time being relaxed to the maximum extent possible to the time of obtaining system data. As this depends on the data of the time series, this work makes a comparison between OS-ELM and recurrent neural networks and their variants (RNN, LSTM, and GRU) to determine which of these networks is better in terms of accuracy (RMSE, MAE, and MAPE) and computational time.

2.2. Works Related to Energy Efficiency in Link Aggregation Groups or Bundle Ethernet

In the field of energy-efficient link aggregation, several methods and techniques have been proposed to reduce the energy consumption of LAGs or BEs in telecommunications networks, which can be divided into four categories: optimization methods, threshold points, SDN controllers, and predictive techniques.

In [6], an integer linear programming (ILP) formulation was proposed to optimize the energy of LAGs. The energy savings were significant, reducing energy consumption by 79% compared with a normal network. Another study, in [54], investigated energy reduction in clustered links using a mixed-integer linear programming (MILP) model, showing a 50% reduction compared with shortest-path routing. In [55], the use of link aggregation activation and deactivation was studied using linear programming (LP) optimization, and it was shown how to find the most energy-efficient link configuration for all links of a BE. Energy savings ranged between 10% and 30% depending on the chosen configuration. In [56], convex optimization was used with the water-filling algorithm, reducing energy consumption by up to 50% in EEE links, an IEEE 802.3az standard that reduces the energy consumption of physical layer (PHY) devices during periods of low link utilization. EEE saves energy by switching part of the transmission circuit to low-power mode when the link is inactive.

An Ethernet link consumes energy even when the link is inactive. EEE provides a method for using energy so that Ethernet links only use it during data transmission. EEE uses a signaling LPI protocol to achieve energy savings when an Ethernet link is inactive. EEE allows PHYs to exchange LPI indications to signal the transition to low-power mode when there is no traffic. LPI indicates when a link can be inactive and when it should resume after a predefined delay, without affecting data transmission. The following copper PHYs are standardized by IEEE 802.3az: 100BASE-T, 1000BASE-T, and 10GBASE-T. However, the optical fiber standards are not.

In other works by [11], algorithms such as the fixed local heuristic threshold (FLHT) and the dynamic local heuristic threshold (DLHT), which are two locally optimized distributed algorithms, can dynamically adjust the number of active sublinks to save energy consumption by approximately 80% in the CN, for both bin packing and load balancing cases. In telecommunications networks with a higher intelligence using SDN controllers, the sleep port algorithm (SPA) and the two-queue algorithm (TQA) are applied, achieving an energy efficiency of up to 50% for grouped links [12,57]. In [13], the algorithm proposed by the author was able to reduce the average number of active links to 25.4%, using a mechanism based on the SMA prediction technique.

Although researchers have proposed many schemes to improve the energy efficiency of the BEs or LAGs, there are still issues with the scalability of the solution, because they are primarily based on a snapshot of the network, without taking into account the scalability and dynamic nature of a TSO network, and most of them are associated with copper standards.

3. Traffic Prediction Methodology

The methodology for traffic prediction involves several steps to obtain the best prediction based on the configuration and the model used. The process is illustrated in Figure 1. The diagram starts with raw data input, obtained from an online monitoring platform. The data are then processed for cleaning and adjustments to be used in simulations. Next, the data are divided for validation processes. Then, simulations are run using different neural networks, and their performance is evaluated using metrics such as prediction performance in terms of RMSE, MAE, MAPE, and computational time. Finally, the best prediction model is chosen.

The following section provides more detail on each of these steps.

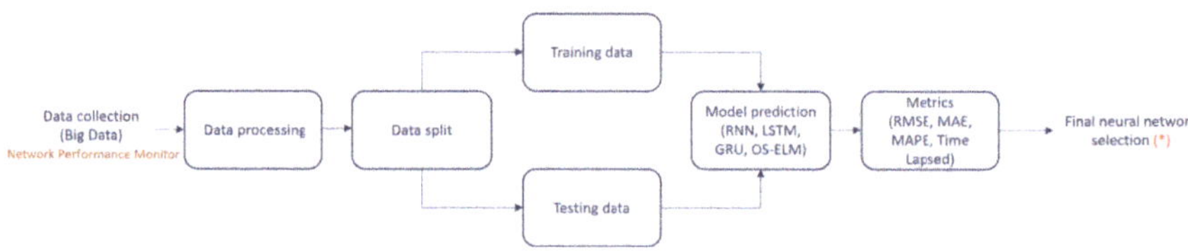

Figure 1. Traffic prediction methodology. * This output will be the input for Bundle Ethernet energy efficiency methodology shown in Section 4.

3.1. Data Collection

The utilization of a network monitoring platform is necessary in order to continuously monitor the traffic on routers and provide real-time statistics on various time scales, including but not limited to minutes, hours, days, and months. This results in the accumulation of a significant amount of data, which can be categorized as big data and is structured in the form of a time series. The platform stores this information for a minimum duration of 1 year, and the monitoring system updates the data every 5 min, with the precise time span being contingent on the system in use. The collected data serve as inputs for prediction models. It is imperative to note that for this particular use case, the effective monitoring of the traffic on TSO network routers can only be achieved through the utilization of the network performance monitor platform in conjunction with the simple network management protocol (SNMP).

3.2. Structure of RNN, LSTM, GRU, and OS-ELM

The basic architectures used by neural networks have been extensively studied and discussed in the literature; for example, ref. [58] discusses RNNs, ref. [23] discusses LSTM, ref. [59] discusses GRU, and refs. [60,61] discuss OS-ELM.

3.3. Data Processing for RNN, LSTM, GRU, and OS-ELM

In this stage, the network performance monitor has provided a database that will serve as the data source for training. However, it is crucial to ensure that this database is suitable for the purpose it is intended for, which is searching. To achieve this, the data will be preprocessed and transformed to align with the specifications of the neural network that will be utilized. The neural network model will play a crucial role in determining the final form of the data and shaping it to meet the necessary criteria for successful searching. In essence, the database will be tailored to the requirements of the neural network, ensuring that it can effectively extract meaningful information from the data and deliver accurate results.

3.3.1. Data Processing for RNN, LSTM, and GRU

The transformation of these three steps will be necessary to better process the data in the implementation of RNN, LSTM, and GRU.

1. Transform the data into a supervised learning problem. In the time series problem, the data are modified as follows: The observation at the last time step $(t-1)$ as the

input and the observation at the current time step (t) as the output. This represents the single-step sliding window. It is mainly the only variable to compare, so it is a univariate problem [41].

2. Time-dependent time series data. The trend can be removed from the observations and then returned to the original prediction scale. A standard way to remove a trend is to differentiate the data.
3. Normalize the observations. The default activation function of the RNN, LSTM, and GRU models is the hyperbolic tangent $(tanh)$, which has values between -1 and 1. The observations will be normalized in the same way, that is, between -1 and 1. This regularization helps to avoid corrupting the experimental set with information from the test dataset.

3.3.2. Data Processing for OS-ELM

The transformation of these three steps is necessary to better process the data in the implementation of OS-ELM.

1. Transform the data into a supervised learning problem. In preparing the data, the model is instructed in the same way as recurrent neural networks, that is, the sliding window or prediction step will be one step.
2. Activation function. The activation function of the OS-ELM neural network is the sigmoid function [62]. The rectified linear unit (RELU) activation function, also known as ramp function, is tested in [19], with poor results compared with the sigmoid function.
3. Normalize the observations. In the OS-ELM model, the recommended scale is to normalize the data by subtracting the mean and dividing by the standard deviation.

3.4. *Training and Testing Data*

Training and testing datasets make up the two sections of the data collection. The model is constructed and validated using the training and testing datasets, respectively. The forward-chaining method is generally used to compare and validate the models. Forward chaining is a technique used in machine learning to evaluate the performance of a predictive model. In this method, the training data are divided into two parts: a training set and a validation set. The model is trained on the training set and then tested on the validation set. The testing is performed in a forward direction, meaning that the model is tested on data that comes after the training data. Forward chaining is particularly useful when working with time series data, where the goal is to predict future values based on past observations. It is a form of cross-validation that ensures the model is not overfitted on the training data and can generalize well to new data [63,64]. Traditional cross-validation is not suitable for time series data due to temporal dependencies and the arbitrary nature of test set selection, among other factors. It is essential to prevent data leakage when partitioning time series data [65]. In conventional cross-validation, the test set selection is often arbitrary, which may result in the test set error being an unreliable estimator of the error in an independent test set. This problem can be addressed using a technique called stacked cross-validation, as described by [66].

For this case, we applied the same testing set for all simulations, so it will be possible to draw conclusions based on variations of the four neural networks and different hyperparameters. In the case of the training set, data will be added to verify if adding more training data improves the error metrics. As the time series data are seasonal, previous days will be added (*t-xdays*), as shown in Figure 2. The amount of data to be added depends on the time interval of the monitoring system.

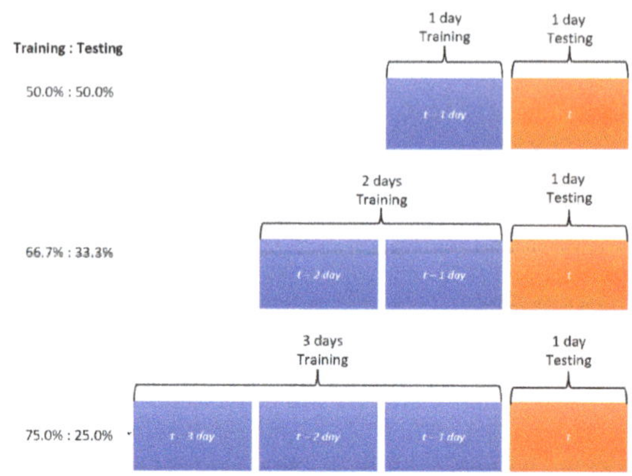

Figure 2. Training and testing dataset.

Thus, three sets of data in different portions for the simulations are obtained, which are *Training* : *Testing*. Therefore, the first set = 50.0%:50.0%, the second set = 66.7%:33.3%, and the third set = 75.0%:25.0%.

3.5. Hyperparameters of RNN, LSTM, GRU, and OS-ELM

Hyperparameters in neural networks are the parameters that are set before training a model, unlike the model parameters, which are learned during the training process. These hyperparameters control various aspects of the model's training, such as the number of neurons in each layer, the learning rate, the type of activation function, the type of optimization algorithm, epochs, and time steps. The optimal values of hyperparameters greatly affect the performance of the neural network, such as its ability to learn from the data and generalize well to new examples. Finding the best hyperparameters for a specific problem is usually done through a process called hyperparameter tuning, where different values are tried and the performance of each set of hyperparameters is evaluated.

3.5.1. RNN, LSTM, and GRU Hyperparameters

These steps explain the fundamental hyperparameters of the RNN, LSTM, and GRU model that will be implemented.

1. Number of neurons: It is the number of hidden layers added to the RNN, LSTM, and GRU cell.
2. Epochs: It is the number of times each training dataset will pass through the neural network.
3. Time steps: The number of time steps specified determines the number of input variables x used to predict the next time step h, as shown in Figure 3. In recurrent neural networks, time steps (also known as lags) refer to the number of previous time steps that are used as input to predict the next time step. For example, if the time steps are set to 3, the network will use the previous 3 time steps of the data as input to predict the next time step. The number of time steps can have a significant impact on the performance of the network, as it determines the amount of context that the network has access to when making predictions.
4. Adam optimizer: The Adam algorithm [67] is one that combines RMSProp with momentum. To date, there is no algorithm that has superior performance over others in different scenarios [68], so it is recommended to use the optimization algorithm with which the user feels the most comfortable when adjusting the hyperparameters. For running the simulations, the Adam-based optimization algorithm will be configured for RNN, LSTM, and GRU. Ref. [69] indicates that the Adam optimization algorithm has been a very popular optimizer in deep learning networks in recent years.

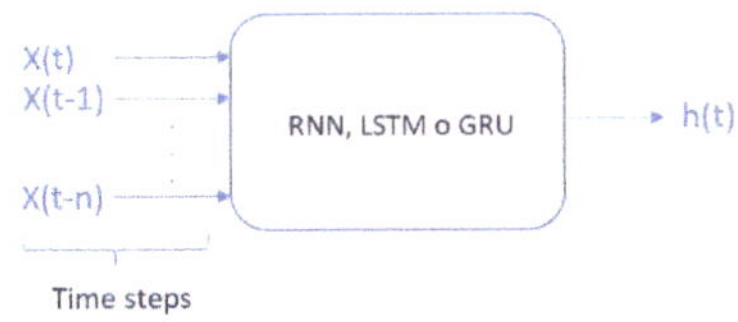

Figure 3. Time steps (lags) in recurrent neural networks.

3.5.2. OS-ELM Hyperparameters

These steps explain the fundamental hyperparameters of the OS-ELM model that will be implemented.

1. Number of neurons: It is the number of hidden layers added to the OS-ELM cell.
2. Forgetting factor: The forgetting factor allows the OS-ELM neural network to continuously forget obsolete input data in the training process, in order to reduce its negative effect on subsequent learning. If the forgetting factor equals 1, it means that the OS-ELM neural network does not forget anything. If the forgetting factor is less than 1, it starts to forget data.

3.6. Metrics

The root mean square error (RSME), the mean absolute error (MAE), and the mean absolute percentage error (MAPE) metrics are used to evaluate the performance of prediction models. These metrics provide a systematic approach for evaluating the accuracy of a model's predictions and facilitate the comparison of the performance of different models. Furthermore, it is important to consider computational time, namely how long it will take to process the prediction.

3.6.1. Root Mean Squared Error (RMSE)

RMSE is a commonly used measure of the difference between the predicted and actual values of a model. It is calculated by taking the square root of the mean of the squared differences between the predicted and actual values and is given by Equation (1):

$$\text{RMSE} = \sqrt{\frac{1}{n}\sum_{i=1}^{n}(y_i - \hat{y}_i)^2}, \quad (1)$$

where y_i is the actual value, $\hat{y}_i$ is the predicted value produced by the model, and n is the total number of samples.

3.6.2. Mean Absolute Error (MAE)

MAE is a measure of the difference between the predicted and actual values of a model. It is calculated as the average of the absolute differences between the predicted and actual values. Similar to RMSE, the lower the MAE, the better the fit of the model to the data. MAE is commonly used in time series forecasting; it is less sensitive to outliers than RMSE, and it is defined by Equation (2):

$$\text{MAE} = \frac{1}{n}\sum_{i=1}^{n}|y_i - \hat{y}_i|, \quad (2)$$

where y_i is the actual value, $\hat{y}_i$ is the predicted value produced by the model, and n is the total number of samples.

3.6.3. Mean Absolute Percentage Error (MAPE)

MAPE is a measure of the difference between the predicted and actual values of a model. It is calculated as the average of the absolute percentage differences between the predicted and actual values. It expresses the error as a percentage of the actual value, which can be useful for comparing the error of models that make predictions for different scales of values, and it is defined by Equation (3):

$$\text{MAPE} = \frac{1}{n}\sum_{i=1}^{n}\frac{|y_i - \hat{y}_i|}{max(\epsilon, |y_i|)}, \tag{3}$$

where y_i is the actual value, $\hat{y}_i$ is the predicted value produced by the model, n is the total number of samples, and ϵ is an arbitrary small but strictly positive number to avoid undefined results when y is zero. It is important to note that when the actual value is zero, MAPE is not defined, which could be a limitation of this measure.

For all neural networks, 100 runs are performed and the mean for each metric is obtained, as in the study by [70], which indicated that in optimization problems, which are heuristic, more than 100 tests should be carried out to find a true or optimal value of the solution.

3.6.4. Computational Time

Computational time or lapsed time refers to the amount of time required to perform a specific computation or task on a computer. It can include the time required to input data, process it, and output the results. It is measured in seconds (s). In the context of neural networks, computational time includes the time required to train and test the network, as well as any other computations that are necessary as part of the model's implementation.

The hardware and software that will be used in the simulations regarding the neural networks are specified in Table 1.

Table 1. System specifications.

	Hardware
CPU	Intel(R) Core(TM) 8600 K at 5.1 Ghz
RAM	32 Gb
Graphics card	NVIDIA GeForce(R) RTX 2080.
	Software
Python	3.7.10
Tensorflow	2.2.0
Keras	2.3.0
Pandas	1.2.4
Scikit-Learn	0.24.1

4. Bundle Ethernet Energy Efficiency Methodology

To understand the methodology for energy efficiency in bundle Ethernet, Figure 4 shows how the three cases will be compared. The first case corresponds to the base (without energy efficiency), which will be when the system is in current conditions, that is, how the system is currently functioning. The second and third case will be the development of algorithms that will be responsible for turning on or turning off the corresponding ports on the side that provides services to the router, in order to reduce the energy consumption of the network while meeting traffic demand.

The first algorithm will be based on the past time of the raw values (rv at $t-1$) of the network performance monitor, in which a threshold or safety factor of 100% will be added, that is, if the traffic at time $t-1$ was 100 Gbps, a capacity of at least 200 Gbps will be applied to time t, and based on that capacity, the required BE will be arranged. This algorithm is called the "threshold-based algorithm".

The second algorithm will have as input the prediction model that will be selected in the traffic prediction methodology, seen in Figure 1. With this prediction model, the predicted values (*pv at t*) will be obtained for each point in time t, with this, the capacity in the BE will be activated. This algorithm is called the prediction-based algorithm.

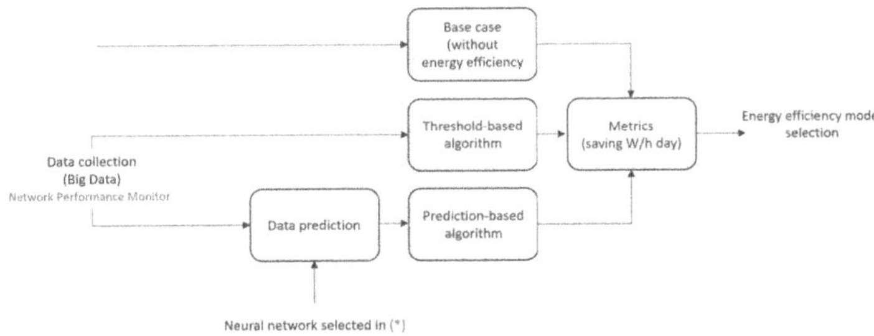

Figure 4. Bundle Ethernet energy efficiency methodology. * This input for Bundle Ethernet energy efficiency methodology is the neural network selected in traffic prediction methodology, according to Figure 1.

4.1. *Threshold-Based and Prediction-Based Algorithms*

The logic of both algorithms written in pseudocode are presented below.

4.1.1. Threshold-Based Algorithm

The threshold-based algorithm, which is shown in Algorithm 1, requires the following inputs in order to be executed:

- rv_{t-1}: Raw value of the link speed in Gbps at a previous timestamp, i.e., (rv *at* $t-1$). This value is obtained from the Network Performance Monitor platform and is a continuous variable.
- nl: Number of links initially possessed by the LAG or BE; it is a discrete integer variable and dimensionless.
- pb: Port bandwidth measured in Gbps, and it is a continuous variable.

Algorithm 1 Threshold-based Algorithm with raw value in $t-1$

Require: raw value in t-1: rv_{t-1}; numbers of links in LAG: nl; port bandwidth: pb
  $x \leftarrow rv_{t-1}/pb$ ▷ x is defined as a ratio variable
  $pa \leftarrow 0$ ▷ pa is defined as ports active
  $pd \leftarrow nl - pa$ ▷ pd is defined as ports deactivate
  if $x > 0$ **then**
    $pa_u = (integer(x) + 1) * 2$ ▷ pa_u is defined as ports active update
    if $pa_u > nl$ **then**
      $pa = nl$
      $pd = nl - pa$
      activate pa ports ▷ set in router activate ports
      deactivate pd ports ▷ set in router deactivate ports
    else
      $pd = nl - pa_u$
      $pa = pa_u$
      activate pa ports ▷ set in router activate ports
      deactivate pd ports ▷ set in router deactivate ports
    end if
  end if

As variables that initialize the algorithm, x is defined as a ratio between the variables rv_{t-1} and pb. In addition, pa represents the active ports of the link, which is initialized at 0, and it is a discrete-integer variable. pd is defined as the difference between nl and pa. This last variable is obtained from the ports that are deactivated at the time of executing the algorithm.

During the execution of the algorithm, while the variable x is greater than 0, meaning the router is present with traffic, the variable pa_u is executed, which corresponds to giving a threshold of 100% more than the value obtained from the previous traffic, because the past values are being used for this action. In other words, pa_u corresponds to the ports that should be active, according to the previous traffic plus the safety factor defined as double, due to the uncertainty of the future traffic.

After this, if the number of active ports exceeds the number of links of the LAG or BE, defined as nl, it must be limited to the maximum defined by this port channel. Otherwise, it calculates the difference between ports to be activated and deactivated. In both conditions, the number of ports to be activated and deactivated on the router for the next timestamp is executed.

4.1.2. Prediction-Based Algorithm

For this algorithm, the same logic as the previous one (threshold-based) is used; the only and important difference is that it has as input the future traffic value obtained in the prediction. As shown in Algorithm 2, the required inputs are the prediction value pv, the number of links in the LAG nl, and the port bandwidth pb.

Algorithm 2 Prediction-based algorithm

```
Require: prediction value: pv; numbers of links in LAG: nl; port bandwidth: pb
  x ← pv/pb                                  ▷ x is defined as a ratio variable
  pa ← 0                                     ▷ pa is defined as ports active
  pd ← nl − pa                               ▷ pd is defined as ports deactivate
  if x > 0 then
      pa_u = integer(x) + 1                  ▷ pa_u is defined as ports active update
      if pa_u > nl then
          pa = nl
          pd = nl − pa
          activate pa ports                  ▷ set in router activate ports
          deactivate pd ports                ▷ set in router deactivate ports
      else
          pd = nl − pa_u
          pa = pa_u
          activate pa ports                  ▷ set in router activate ports
          deactivate pd ports                ▷ set in router deactivate ports
      end if
  end if
```

For both algorithms to work correctly, the units of measure of the variables pv and pb must be the same.

4.2. Metrics

In order to compare both algorithms in addition to the base case, the unit of measure of energy watt-hour (Wh) will be utilized. Watt-hour serves as a means of measuring the amount of generated or performed work. The savings will be reflected in the difference in consumption from the base case per day in relation to each algorithm. It stands to reason that the chosen algorithm will be the one that produces the most savings in comparison with the base case.

The methodology being referred to is a general solution for prediction problems and can be adapted for use in various industries and fields. It involves the use of statistical and machine learning techniques to make accurate predictions about future events. For example, in the energy sector, this methodology can be used to forecast energy demand, anticipate disconnections in power generation centers, and make other predictions that are relevant to the energy industry. Having this information helps energy providers to better manage the supply and demand of energy, which is essential for the efficient operation of their businesses. The use of this methodology in the energy sector can improve the overall reliability of energy services and help to reduce the likelihood of blackouts or other disruptions.

5. Case Study

This section describes the network topology to be used in this case study, the traffic description to choose the best neural network in terms of prediction and computational time, and the characteristics of the equipment to be used in order to model the required energy savings as a target.

5.1. Network Topology

As previously mentioned, the LAG or BE allow for the logical grouping of multiple Ethernet physical links; this aggregation is treated as a single link and allows for the sum of the nominal speed of each Ethernet physical port used to obtain a high-speed trunk link. In a TSO, different routers can be found connecting to each other. As shown in Figure 5, it can be seen that the LAG connects to n optical fiber interfaces generating a LAG of n links. Over time, the interfaces of the optical transport network (OTN) and the interfaces of the routers have been growing in the optical IP Core networks of a TSO, reaching up to 400 GE interfaces in other operators [71], thus increasing as traffic demand grows and in accordance with the development of this technology.

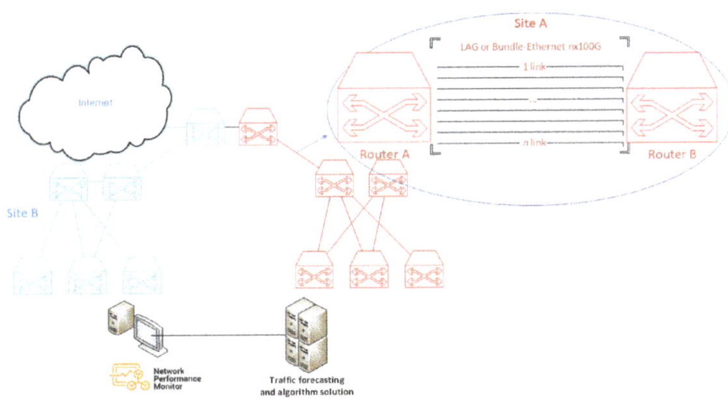

Figure 5. Link aggregation in a TSO.

In addition, many of the implementations in TSOs are geographically redundant as part of design of reliable communication networks (DRCN) [72], which in this case, as shown in Figure 5, means that Site A is equal to Site B; this means that if the router of Site A goes down (fails), the network traffic will switch to Site B via the interplane link (which connects both sites). The same applies vice versa.

In Figure 6, the architecture of a content distribution network (CDN) is shown. The CDN router aggregates different content providers, such as Google, Netflix, Facebook, Microsoft, and Akamai, among others. The CDN is a group of servers that are distributed geographically and interconnected. They provide cached Internet content from the closest network location to the user to accelerate information delivery.

The Internet output is provided by the router, commonly called the Internet gateway router (IGR). The IGR is the node that aggregates all the traffic of a TSO network and communicates with the Internet. Both the IGR and the CDN router are connected to 8 optical fiber with 100GE interfaces that generate a LAG or BE of 8 links, which is 800 Gbps of capacity. The actual traffic between both routers is approximately 400 Gbps, leaving the same capacity for backup to switch to another site (mirror) in case of failure.

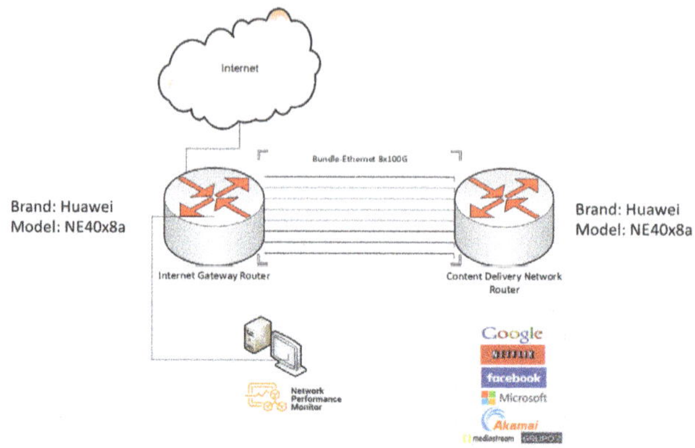

Figure 6. CDN architecture in a TSO.

As for the protocol that regulates the hardware providers and provides guidance on the practice of link aggregation for data connections, the IEEE defined the link aggregation control protocol (LACP) within IEEE 802.3ad [73], which is a standard-based method for controlling the aggregation of physical network links. Active LACP mode is the protocol generally configured on equipment to manage aggregated links. This means that the interface is in a permanently active negotiation state. LACP runs on any link that is configured to be in active state. The active port also automatically initiates negotiations with other ports by initiating LACP packets. Static LACP is configured, which increases the interface's bandwidth and provides reliability. When an Eth-Trunk or BE member link fails or is not active, traffic is automatically distributed to other available links, thereby avoiding traffic interruption. In addition, Eth-Trunk interfaces operating in LACP static mode can implement load balancing. When a BE or LAG is present, all links are active. As a result, energy consumption occurs on all optical links, as the LAG mechanism constantly sends test packets to check if the link or member is active or inactive. On an energy level, it is a waste of energy, because the link capacity should be projected to the network's peak traffic demand and in other cases to geographically established redundancy.

To model the energy consumption of a router, one must first know the power. In Figure 7, the service side considered in the present study is shown.

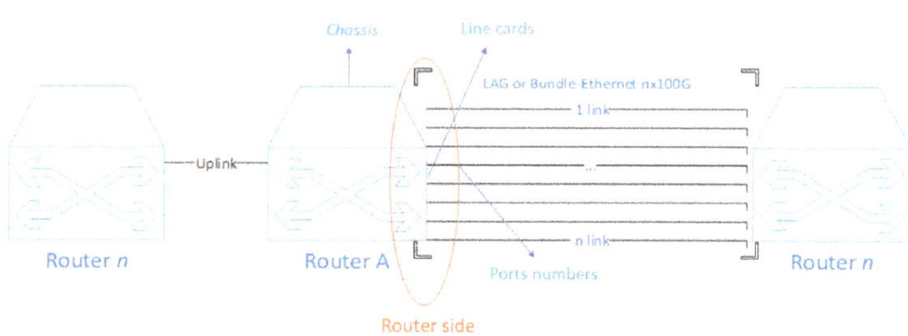

Figure 7. Router side to be considered.

According to the study in [5], calculating the equipment consumption mainly consists of a chassis, a number of line cards, and a number of ports. The energy consumption of the chassis and line cards is fixed, regardless of the traffic load, because they will always be active, so the power of a router P_r can be expressed by Equation (4):

$$P_r = P_{ch} + N_c P_c + \sum_{i=0}^{n} N_{p_i} P_{p_i} * fu_i, \tag{4}$$

where P_{ch} is the power of the chassis, which is the base power of the equipment. N_c is the number of line cards, and P_c is the power of the line card. N_p is the number of ports, P_p is the power of the ports, and fu is the port utilization factor. The port utilization factor is the percentage between the used traffic and the port capacity.

When talking about 100GE interfaces, there are standards that establish different technical norms for different purposes. When referring to connections within the CN, the 100GBASE-LR4 standard is the most commonly used [74], as it corresponds to the IEEE 802.3 physical layer specification for 100 Gb/s with 100GBASE-R encoding over four wavelength division multiplexing (WDM) lanes on single-mode fiber, with a reach of at least 10 km [75]. In the 100GBASE-LR4, 100GBASE-SR10, 100GBASE-SR4, and 100GBASE-ER4 standards, energy efficiency is not available [76]. In all of these standards, the common medium through which data are transmitted is optical fiber. In the IEEE 802.3 bm standard, it is indicated that EEE in 100GBASE-LR4 is optional. Additionally, many providers have not yet integrated this norm into their manufacturing. When EEE is not active, Ethernet standards operate at full power all the time, consuming 100% of the energy, regardless of the traffic load [77]. Therefore, for our case, we can assume that the utilization factor will always be equal to 1.

5.2. Traffic Description

For the purpose of the study, real traffic of the architecture shown in Figure 6 is used. The dataset represents the activity from 13 November 2021 to 16 November 2021 (4 days), consisting of traffic from the CDN router located in Santiago, Metropolitan Region, Chile (−33.444285499124504, −70.65611679943314). Historical data were captured from the monitoring system, with raw data that contained the time, average incoming traffic, average outgoing traffic, maximum incoming traffic, maximum outgoing traffic, etc. In the dataset, the output traffic peak is the variable used, as it is the maximum traffic that the network can have. In the monitoring system, the minimum data collection time is every 5 min. As a result, the dataset will consist of a traffic variable measured in Gbps every five minutes.

For the purpose of training, and as shown in the methodology, three groups will be divided according to the number of days to train the neural network. As shown in Figure 8, each day corresponds to 288 observations, and each point represents five minutes, with its respective value in Gbps, because this dataset works as a univariate time series. Therefore, the division of training groups is as follows, maintaining the proportion indicated in Figure 2.

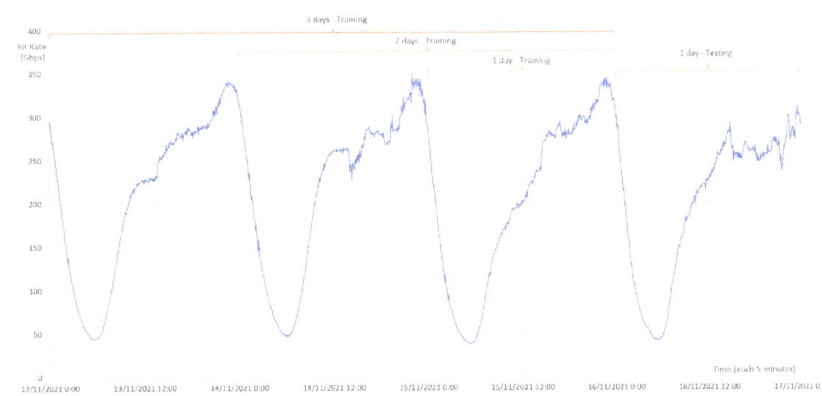

Figure 8. CDN traffic—final dataset.

- First set: 576 observations:
  - Training observations: 288;
  - Testing observations: 288.
- Second set: 864 observations:
  - Training observations: 576;
  - Testing observations: 288.
- Third set: 1152 observations:
  - Training observations: 864;
  - Testing observations: 288.

As the CN architecture adds traffic and delivers it to various clients through the CDN router, the traffic type is seasonal due to the nighttime usage feature of major CDNs such as Netflix, Facebook, and Google. Seasonal fluctuations in telecommunications traffic are due to changes in consumer behavior, such as increased usage during holidays and winter months. The adoption of 5G and B5G technologies may lead to increased usage and traffic, but it is still unclear how much impact they will have, as they are in their early stages of development and other factors such as infrastructure, regulations, and competition may also influence their adoption and usage. Ultimately, the impact of these technologies on seasonal traffic will depend on consumer behavior and market demand. Other factors such as the availability of infrastructure, government regulations, and competition from other technologies may also play a role.

5.3. Equipment Characteristics

In the network architecture shown in Figure 6, there is a configured BE or LAG of 8 links with 100GE interface, typically indicated as 8 × 100GE. According to the LACP protocol, these are configured with the same weight. This means that the load distribution is equal for all links when traffic is assigned, and therefore, in terms of energy consumption, all links are active. The studied equipment is a Huawei brand NE40E-X8A model, with a base configuration that consumes 784 W (typical power at 25 °C), without adding service cards or uplink links. If only the consumption of the BE on one side of the NE40E-x8A equipment is analyzed, as shown in Figure 7, the following configuration and consumption are shown in Table 2.

Table 2. Line card power of CDN router.

Slot	Board Info	Typical Power at 25 °C (W)
Slot1	LPUF-480-E	290
Slot1-PIC0	PIC-2*100GBase-QSFP28	73
Slot1-PIC1	PIC-2*100GBase-QSFP28	73
Slot2	LPUF-480-E	290
Slot2-PIC0	PIC-2*100GBase-QSFP28	73
Slot2-PIC1	PIC-2*100GBase-QSFP28	73

The base configuration data, as well as the details of each of the cards that make up the equipment, were obtained by the current configuration of the router in the network. Therefore, as all ports being active, the utilization factor (fu_i) will be equal to 1. The base power is calculated with Equation (4), as follows:

$$P_r = 784 + 2290 + \sum_{i=1}^{8} 36.51 = 1656W, \quad (5)$$

Therefore, the total base power is 1656 W when all ports are active.

6. Traffic Forecasting Results in Case Study

The results of both simulations are presented for RNN, LSTM, GRU, and OS-ELM.

6.1. Simulation Results of RNN, LSTM, and GRU

Simulations are performed by varying three hyper-parameters: Time-steps (number of inputs) or more commonly known as lags, number of neurons in the hidden layer and epochs. The variations of each hyper-parameter will be as follows:

- Time steps (lags): 1, 4, 8, 16, and 32.
- Number of neurons: 1, 10, and 50.
- Epochs: 1, 10, and 100.

Once each hyperparameter is varied, the results of the RMSE, MAE, MAPE, and computational time metrics will be obtained. Note that the RMSE, MAE, and MAPE metrics are errors, so they should be close to zero. For a better understanding of these results, graphs are made for each metric, showing each neural network in the different training sets or days. Remember that the training days are 1, 2, and 3 days, with 1 day of testing, which is already explained in the previous point.

The tabulation of the results is presented in Appendix A. The best performance values are shown in bold. The criterion was to have at least two metrics with a lower value in the configuration of the hyperparameters lags, number of neurons, and epochs for each group and deep learning neural network. To better interpret the graphs, Table 3 shows the number of hyperparameter configuration indicated on the x-axis. Each hyperparameter has the following description: lags, number of neurons, and epochs). That is, the hyperparameter number 1 has the configuration of lags: 1, number of neurons: 1, epochs: 1, and so on.

Table 3. Hyperparameters configuration.

Number	Hyperparameters Setting	Number	Hyperparameters Setting	Number	Hyperparameters Setting
1	1,1,1	16	1,1,10	31	1,1,100
2	4,1,1	17	4,1,10	32	4,1,100
3	8,1,1	18	8,1,10	33	8,1,100
4	16,1,1	19	16,1,10	34	16,1,100
5	32,1,1	20	32,1,10	35	32,1,100
6	1,10,1	21	1,10,10	36	1,10,100
7	4,10,1	22	4,10,10	37	4,10,100
8	8,10,1	23	8,10,10	38	8,10,100
9	16,10,1	24	16,10,10	39	16,10,100
10	32,10,1	25	32,10,10	40	32,10,100
11	1,50,1	26	1,50,10	41	1,50,100
12	4,50,1	27	4,50,10	42	4,50,100
13	8,50,1	28	8,50,10	43	8,50,100
14	16,50,1	29	16,50,10	44	16,50,100
15	32,50,1	30	32,50,10	45	32,50,100

The results of the RMSE metric are shown in Figure 9. The x-axis represents the number of hyperparameter configurations, indicated in Table 3. From number 16, the number of epochs changes to 10, and at number 31, to 100. On the y-axis, RMSE is represented. The data are in Mbps, so the range of these errors is from 5.5 to 9.0 Gbps.

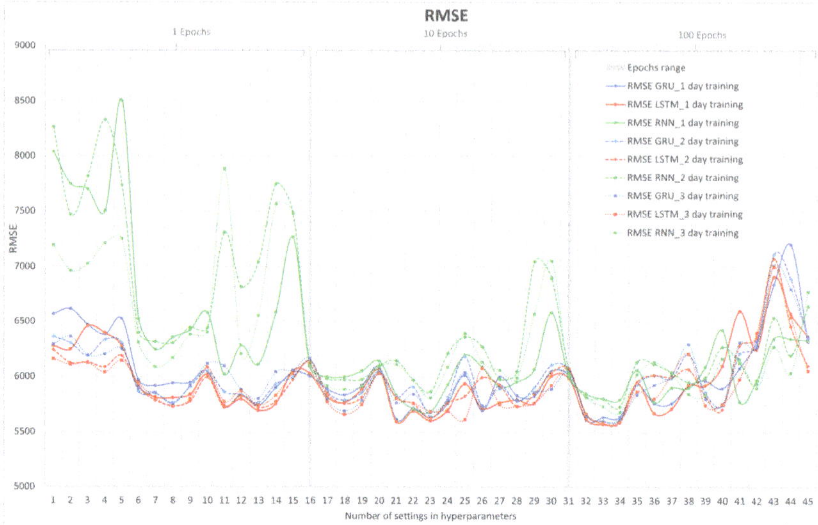

Figure 9. RMSE metrics for RNN, LSTM, and GRU.

It can be observed that in general, the RNN for the epochs at 1 and 10 behaves with a high degree of error compared with the LSTM and GRU neural networks for 1, 2, and 3 days of training. This is because they have a lower number of epochs. That is why, at 100 epochs, the RNN improves, and it is even better than other networks in some cases. Remember that each cycle of backpropagation and forward correction to reduce loss is called an epoch. Backpropagation consists of determining the best input weights and biases to obtain a more accurate result or minimize losses.

Another conclusion from this Figure 9 is that because the RNN neural network does not have a memory effect, and that in 1 day of training it has a lower amount of data, it has worse performance than in 2 and 3 days of training. This is minimized in other neural networks such as LSTM and GRU by having a greater effect on memory than RNN. Furthermore, we can observe that from hyperparameter 40 to 45, in which the number of neurons is set to 50 and the epochs to 100, varying the number of lags makes the LSTM and GRU networks worse compared with the RNN. It can be inferred that many epochs and inputs may be influencing in some way the memory effect of these two networks, which would lead to the impoverishment of these two neural networks due to an excess of data.

Finally, in the RMSE metric, as can be seen, the lowest point is the hyperparameter 33 of the LSTM neural network for 1, 2, and 3 days of training. Within this network, the configuration of the hyperparameter number 33 is 8,1,100, whose value of RMSE closest to zero is on 1 day of training (first group), which can be found in Appendix A.

The MAE error metric is calculated as an average of absolute differences between the target values and the predictions. MAE is a linear score, which means that all individual differences are weighted equally in the average. From the point of view of interpretation, the MAE metric is preferable, since RMSE has the advantage of penalizing larger errors (outlier values) more, so focusing on the upper limit, which means that the RMSE number tends to be increasingly larger than that of MAE as the test sample size increases. In other words, since the data being analyzed represent a time series, it usually does not show outlier values, so it is preferred to plot this metric.

Figure 10 shows the results of the MAE metric for the RNN, LSTM, and GRU neural networks in their 1, 2, and 3 days of training. On the x-axis, the number of hyperparameter configurations is represented, which is already known from the previous graph, and on the y-axis, MAE is represented.

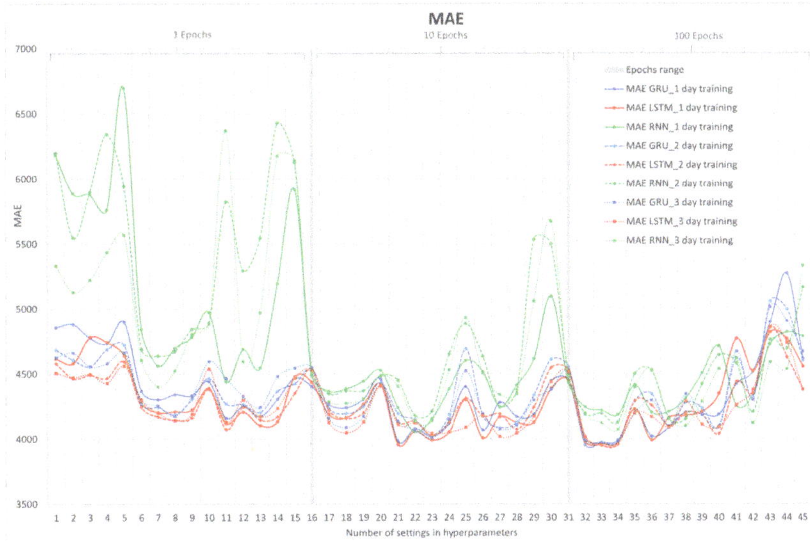

Figure 10. MAE metrics for RNN, LSTM, and GRU.

Note that the curves in Figure 10 are similar to the result of the RMSE metric with some small differences. It is emphasized that the RNN after 3 days of training does not have good performance, as evidenced by the RMSE metric. Finally, the closest RMSE error to zero is achieved by the LSTM neural network after 2 days of training (second group) with a hyperparameter configuration (8,1,100) that can be found in Appendix A.

Since MAPE is a further development of the MAE calculation, there is similarity between both metrics. Both are not sensitive to outliers since they use the absolute difference. MAPE is more understandable than MAE for the final user, because the error value is in terms of percentage.

In Figure 11, the results of the MAPE metric for RNN, LSTM, and GRU in their 1, 2, and 3 days of training are shown. On the x-axis, the number of hyperparameter configurations is represented, and on the y-axis, MAPE is represented.

Here, the error is represented in percentage terms, which leads us to believe that we are working with minimal errors close to 1.9%, which translates to the prediction of the three neural networks having good performances. Now, in Figure 11, it can be seen that at 100 epochs is where the three networks make predictions with the least error and that the neural network with the best results continues to be LSTM.

In Figure 12, the results of the computational metric or lapsed time for RNN, LSTM, and GRU in their 1, 2, and 3 days of training are shown. The x-axis represents the configuration number of the hyperparameters, and the y-axis represents the elapsed time.

One of the predominant factors is the amount of time it takes for the neural network to predict the next value. This time must be less than the input time of the monitoring system, which was established at 5 min (300 s). In Figure 12, it can be clearly seen that as the number of epochs increases, they are determinant in terms of computational calculation. Note that these values vary depending on the computer system. Table 1 shows the equipment used. In this case, a GPU is used to accelerate the vectorial calculations of the gradient of each of the neural networks.

In addition to the number of epochs, the number of days of training increases the elapsed time. This is clearly visible where the results of the RMSE, MAE, and MAPE metrics are closer to zero, that is, centered on the value of 100 epochs.

Another point to note is the stability in terms of computational time possessed by the LSTM and GRU neural networks, which does not vary significantly if the number of lags is increased as the RNN does, which increases the time exponentially.

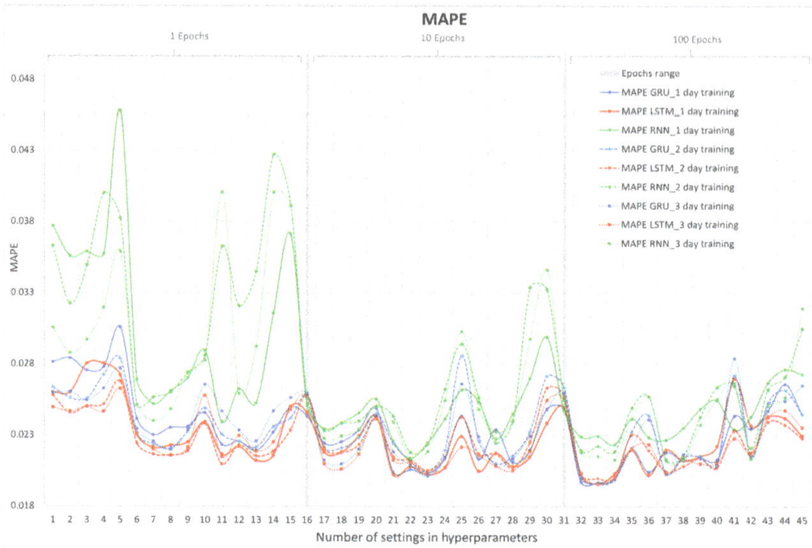

Figure 11. MAPE metrics for RNN, LSTM, and GRU.

Finally, in the computational time metric, it would be incorrect to choose the value closest to zero, as it must be accompanied by the metrics seen earlier. Now, it is known that computational times greater than 300 s cannot be selected.

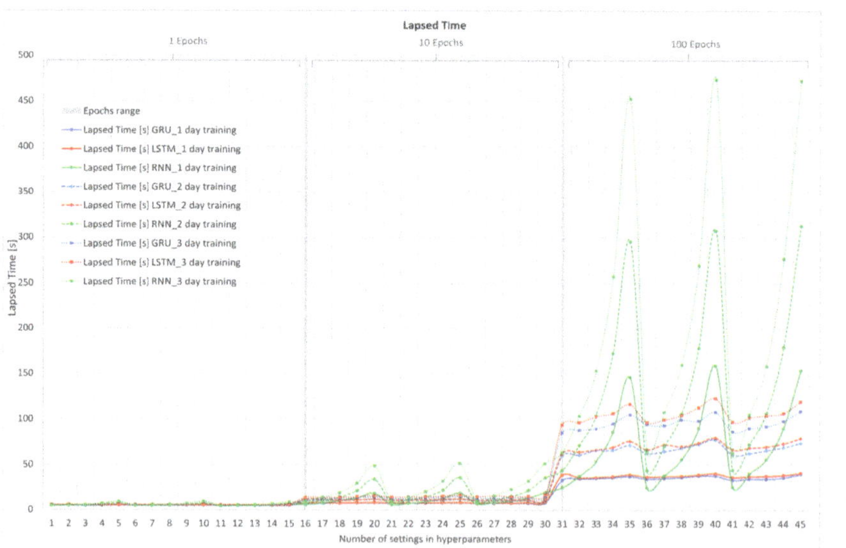

Figure 12. Computational time or lapsed time metric for RNN, LSTM, and GRU.

Table 4 shows a summary of the results containing the best metrics of the selected neural networks.

Table 4. Summary of results for RNN, LSTM, and GRU.

Type	Group	Training	Time Step (Lags)	Number Neurons	Epochs	RMSE	MAE	MAPE	Lapsed Time [s]
RNN	First	1 day	4	10	10	5711.161	4054.292	0.02137	7.177
RNN	Second	2 days	16	1	100	5728.889	4126.839	0.02184	172.208
RNN	Third	3 days	16	1	100	5679.286	4070.981	0.02126	257.146
LSTM	**First**	**1 day**	**8**	**1**	**100**	**5573.399**	**3961.516**	**0.01959**	**35.745**
LSTM	Second	2 days	8	1	100	5581.399	3947.681	0.01998	66.765
LSTM	Third	3 days	8	1	100	5585.884	3950.845	0.01958	103.195
GRU	First	1 day	4	1	100	5612.749	3953.799	0.01970	34.093
GRU	Second	2 days	8	1	100	5600.504	3972.234	0.01962	66.179
GRU	Third	3 days	8	1	100	5600.595	3971.039	0.01962	89.822

The results indicate that the predictions of the LSTM neural network are considerably more accurate than RNN and GRU. If the number of training days that is being given to each neural network is visualized, in RNN, when these days are increased, the computation time increases considerably. In the case of LSTM, there is a deterioration of the RMSE, MAE, and MAPE metrics, and at the same time, an increase in computational time. For GRU, it behaves similarly to LSTM.

On the other hand, it can be observed that the results of the RMSE, MAE, and MAPE metrics in LSTM are more consistent than in the RNN and GRU networks, because they present the same configuration of hyperparameters (8 inputs, 1 hidden layer, and 100 epochs) in different training days. The LSTM and GRU neural networks are surprising, because even with a complex structure on the RNN, they present low computational times when adjusting the hyperparameters to high values. The best combination of the neural network and training days to choose is the LSTM network with 1 day of training with a configuration of 8 lags, 1 neuron, and 100 epochs, as shown in Table 4.

6.2. *Simulation Results of the OS-ELM Neural Network*

For the OS-ELM network, the same method as for RNN, LSTM, and GRU is used, which consists of varying the value of the hyperparameters. Simulations are carried out by varying two parameters: the number of neurons in the hidden layer and forgetting factor. The value that each hyperparameter will take is as follows:

- Number of neurons: 10, 110, 210, 310, 410, 510, 610, 710, 810, 910, 1010, 1110, 1210, 1310, 1410, 1510, 1610, 1710, 1810, and 1910.
- Forgetting factor: 0.9, 0.95, 0.99, and 1.00.

Once each hyperparameter is varied, the results of the RMSE, MAE, and MAPE metrics and computational time will be obtained. In Figure 13, the results of four graphs of the RMSE metric are shown, varying the forgetting factor. On the x-axis, the number of neurons is represented. On the y-axis, RMSE is represented.

It can be seen that for different training days of the OS-ELM neural network, the RMSE metric can vary between values of 50,000 to 2700, indicating that the behavior of this network is more sensitive to the variation of its hyperparameters compared with RNN, LSTM, and GRU. The tabulation of the results is presented in Appendix A. Remember that if the forgetting factor is 1.00, it means that the OS-ELM network does not forget anything. The forgetting factor allows for continuously forgetting obsolete input data during the training process in order to reduce its negative effect on future learning.

When the forgetting factor is 0.90, the RMSE is highest when 110 neurons are configured. If the forgetting factor is 0.95, the error decreases as the number of hidden nodes increases and converges after 410 neurons. When the forgetting factors are 0.99 and 1.00, the neural network presents a huge error with 10 neurons, which rapidly decreases as the number of neurons increases. Additionally, it can be seen that as the forgetting factor and number of neurons increase, the OS-ELM network does not show much difference for the different days (1, 2, and 3) that are used as input for training data.

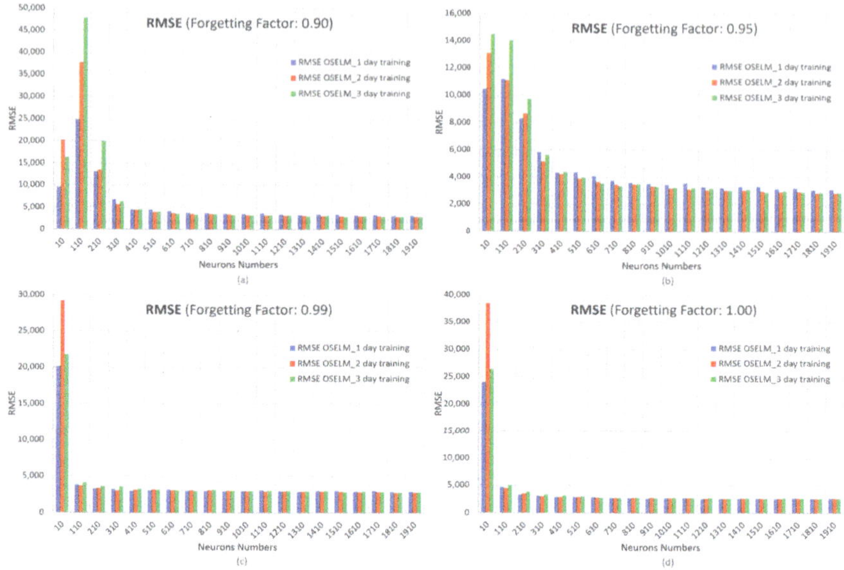

Figure 13. RMSE metric for the OS-ELM network with forgetting factors at (**a**) 0.90, (**b**) 0.95, (**c**) 0.99, and (**d**) 1.00.

In Figure 14, the results of the MAE metric with varying the forgetting factor are shown. The x-axis represents the number of neurons and the y-axis represents MAE.

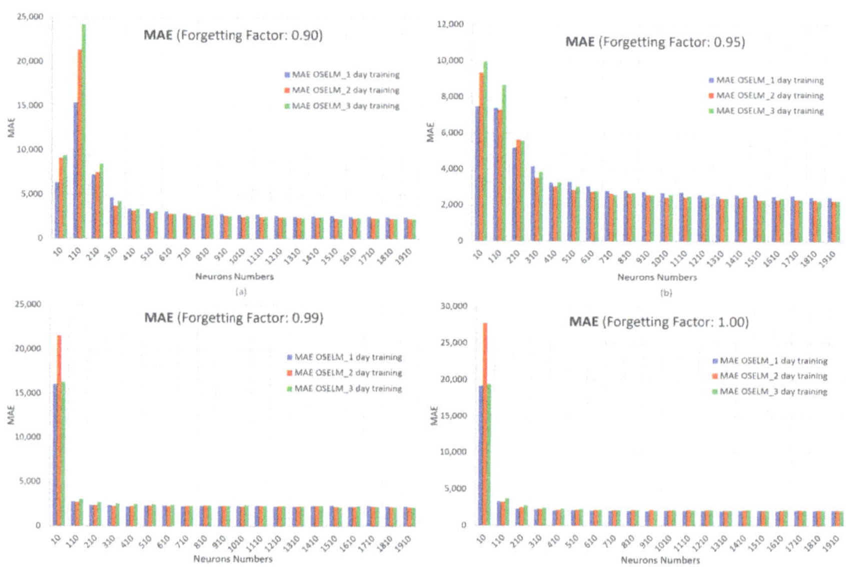

Figure 14. MAE metric for the OS-ELM network with forgetting factors at (**a**) 0.90, (**b**) 0.95, (**c**) 0.99, and (**d**) 1.00

These results show the same behavior already indicated in the RMSE metric. Remember that MAE is more robust to outliers and does not penalize errors as severely as RMSE.

This is why the OS-ELM network trained for 3 days (green bar) is visualized in Figure 14 with a lower error than in Figure 13. This behavior is due to the presence of extreme values.

Figure 15 shows the MAPE metric by varying the forgetting factor. The x-axis represents the number of neurons and the y-axis represents MAPE.

Figure 15 shows the influence of the data input to the OS-ELM network in percentage terms, with the training of 1 day standing out compared with the other days, as it presents the lowest error. Moreover, the convergence of this network is highlighted after 410 neurons with a forgetting factor of 0.95. However, in the OS-ELM network, overfitting can be observed when the number of neurons is increased and the forgetting factor is greater than 0.95. Overfitting is an undesired behavior of neural networks that occurs when the machine learning model provides accurate predictions for the training data but not for new data. If the forgetting factor is close to 1.00, the network may not forget the previous data, and this can cause the network to provide inaccurate predictions when there is a new behavior.

In Figure 16, the results of the computational time metric for the OS-ELM neural network in its 1, 2, and 3 days of training are shown. The x-axis represents the number of neurons and the y-axis represents the lapsed time or computational time.

It is very clear that the computational times at the forgetting factors of 0.90, 0.95, 0.99, and 1.00 increase considerably with a greater number of neurons. This is why the criterion for choosing the best hyperparameter configuration and training days for the OS-ELM network will be based on time and take into account a forgetting factor of 0.95 that allows for possible traffic variations in the network, in order to avoid overfitting, as seen in Figure 15.

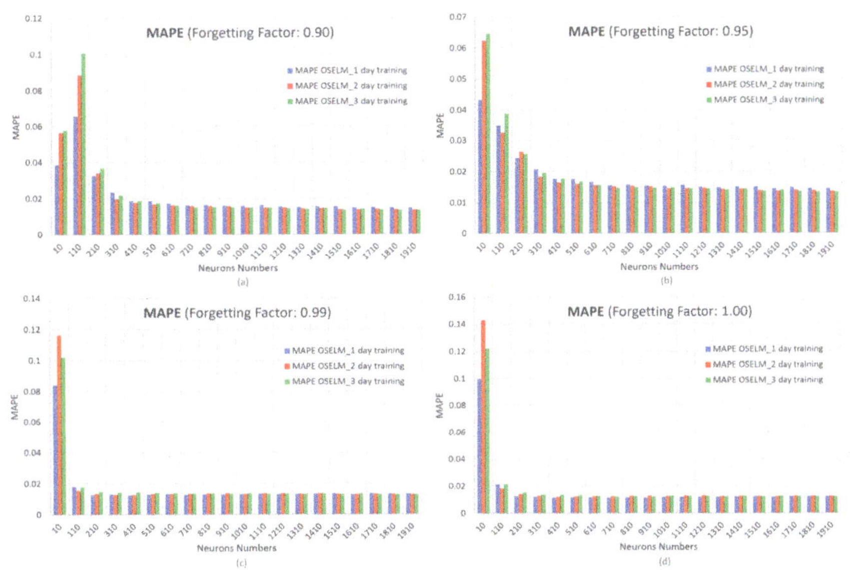

Figure 15. MAPE metric for the OS-ELM network with forgetting factors at (**a**) 0.90, (**b**) 0.95, (**c**) 0.99, and (**d**) 1.00.

Table 5 presents a summary of the best results for each day of OS-ELM network training with a forgetting factor of 0.95 and 410 as the number of neurons in terms of computational time and convergence for that number of neurons.

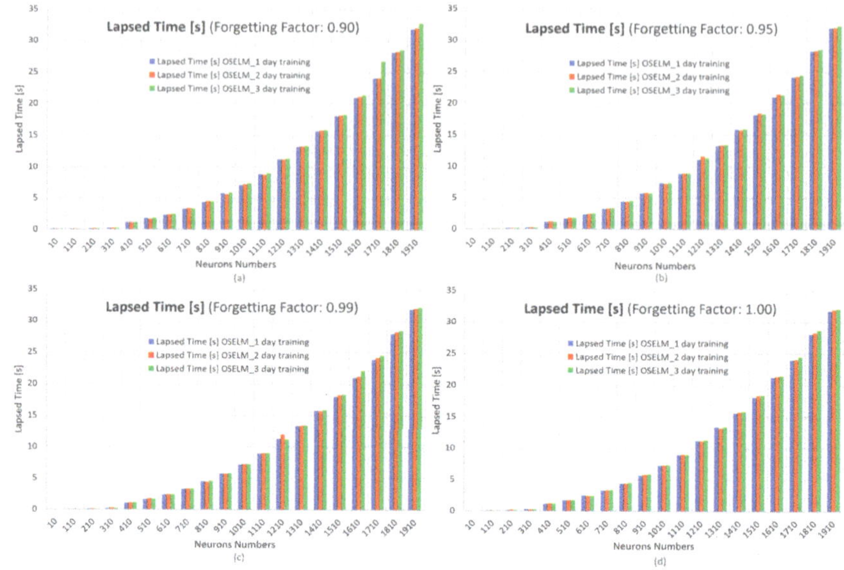

Figure 16. Computational time metric for the OS-ELM network with forgetting factors at (**a**) 0.90, (**b**) 0.95, (**c**) 0.99, and (**d**) 1.00.

Table 5. Summary of results for OS-ELM.

Type	Group	Training	Forgetting Factor	Number Neurons	RMSE	MAE	MAPE	Lapsed Time [s]
OS-ELM	First	1 day	0.95	410	4336.068	3273.084	0.01761	1.128
OS-ELM	**Second**	**2 days**	**0.95**	**410**	**4221.912**	**3037.816**	**0.01642**	**1.182**
OS-ELM	Third	3 days	0.95	410	4384.103	3276.136	0.01778	1.113

When comparing the RMSE, MAE, and MAPE metrics, it can be seen that the network with 2 days of training performs the best. With regard to computational times, they are all around 1.2 s, so the final choice is the OS-ELM neural network with 2 days of input.

6.3. Final Neural Network Selection

Table 6 shows a comparison of the four best-rated neural networks according to the RMSE, MAE, and MAPE metrics in each of the training groups. For comparative purposes, the worst of the corresponding RNNs is used as a reference. The results are decisive in terms of prediction. Starting with the RMSE metric, the OS-ELM, LSTM, and GRU neural networks surpass the RNN by 26%, 2%, and 1%, respectively. For the MAE metric, the OS-ELM, LSTM, and GRU networks surpass the RNN by 25%, 3%, and 2%, respectively, and for the MAPE measure, the OS-ELM, LSTM, and GRU networks surpass the RNN by 23%, 6%, and 8%, respectively.

Finally, in terms of computational time, the OS-ELM, LSTM, and GRU networks surpass the RNN by factors of 217.6, 7.2, and 3.9 times, respectively.

For the purpose of reviewing the comparison of the two best networks and differentiating their optimization technique, Table 7 shows LSTM, which is based on the best gradient descent optimization network, and OS-ELM, which uses the Moore–Penrose pseudoinverse. It shows the percentages of the RMSE, MAE, and MAPE metrics, using the worst of them as a reference. In terms of prediction, OS-ELM outperforms LSTM by 24% in RMSE, 23% in MAE, and 16% in MAPE. In terms of computational time, OS-ELM is 30.2 times faster than LSTM.

Table 6. Traffic comparison of prediction of neural networks.

Type	Training	RMSE	% RMSE	MAE	% MAE	MAPE	% MAPE	Lapsed Time [s]	Times
OS-ELM	2 days	4221.912	26%	3037.816	25%	0.01642	23%	1.182	217.6x
LSTM	1 day	5573.399	2%	3961.516	3%	0.01999	6%	35.745	7.2x
GRU	2 days	5600.504	1%	3972.234	2%	0.01962	8%	66.179	3.9x
RNN	3 days	5679.286	-	4070.981	-	0.02126	-	257.146	-

Table 7. Traffic prediction metrics for the best LSTM and OS-ELM networks.

Type	Training	RMSE	% RMSE	MAE	% MAE	MAPE	% MAPE	Lapsed Time [s]	Times
OS-ELM	2 days	4221.912	24%	3037.816	23%	0.01642	18%	1.182	30.2x
LSTM	1 day	5573.399	-	3961.516	-	0.01999	-	35.745	-

In Figure 17, the two neural networks (OS-ELM and LSTM) indicated in Table 7 are shown. These present the prediction of the traffic load of the CDN router every five minutes in one day. On the x-axis, the bit rate is represented in Gbps, and on the y-axis, the time is indicated in hours, with each point representing 5 min (300 s).

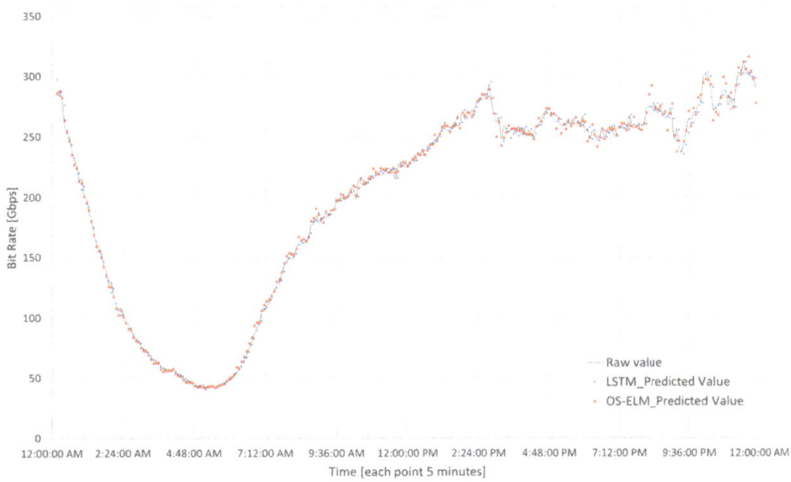

Figure 17. Traffic prediction comparison of the best networks: LSTM and OS-ELM.

The final choice of the sequential neural network OS-ELM is clear given the results obtained.

7. Results of Energy Efficiency Algorithms in the Case Study

The performance of the proposed energy efficiency optimization algorithms are shown below and indicated in Figure 4. These algorithms are compared with the base case, that is, the system operating under current conditions. Remember that the algorithms are responsible for turning on or off the corresponding ports on one side of the CDN router and the link connection to reduce the energy consumption of the network, while satisfying the traffic demand. The first algorithm is based on past traffic with a threshold value of 100%, called the threshold-based algorithm. The second algorithm is based on traffic prediction performed by the neural network selected in the previous chapter, called the prediction-based algorithm.

The results of the simulations are presented in Figures 18–21, which were carried out for the test day presented in Figure 8.

In Figure 18, the results of the accumulated savings on the test day are shown for the base case, threshold-based, and prediction-based algorithms. On the x-axis, the time is represented in hours. On the y-axis, the accumulated savings are represented, measured in W/h.

As can be observed in Figure 18, the savings achieved by the prediction-based algorithm is 4829.58 W/h per day, or 4.83 kW/h per day. On the other hand, when considering the threshold-based algorithm, the savings amount to 2.68 kW/h per day. Clearly, there are no energy savings in the base case. The power of the router is 1656 W, as calculated by Equation (5); thus, the energy consumed is 39.6 kW/h per day. Consequently, the daily savings are 12.2%, only taking into account one side of the equipment (that pertaining to services), the consumption of the chassis, and that of the cards.

Furthermore, the maximum savings that can be achieved by not utilizing any port of the router is 7 kW/h per day, considering 36.5 W of power per port, multiplied by the 8 ports present in the BE. If we compare the savings of prediction (4.83 kW/h per day) to the total consumption of the ports (7 kW/h per day), the savings are almost 70%.

In Figure 19, the number of active ports based on turning on or off during the testing day is shown for the base case, threshold-based algorithm, and prediction-based algorithm. The x-axis represents time in hours, while the y-axis represents the number of active ports.

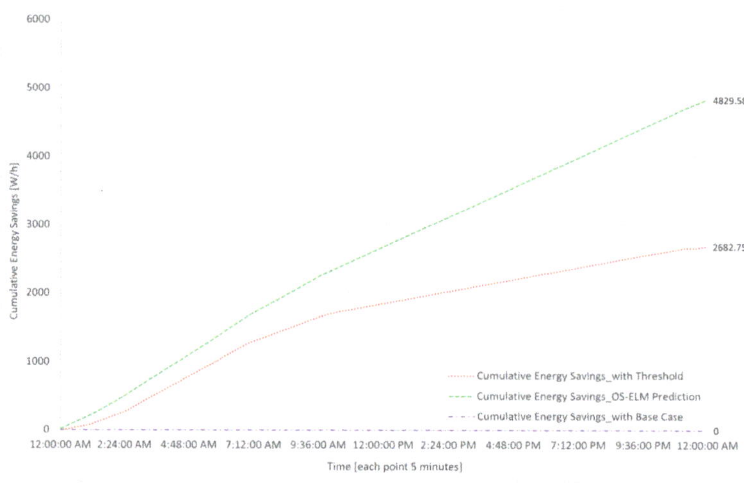

Figure 18. Cumulative one-day savings for base, threshold, and forecast cases.

Regarding the available capacity of the BE, Figure 20 shows that in the base case, there is a high availability of capacity or, in other words, a low utilization; this is due to the network being configured in such a way that it can absorb a site with similar characteristics in case of failure for backup purposes. In the case of the threshold-based algorithm, there is still available capacity in any situation or event. However, in the prediction-based algorithm, this backup capacity is lost to some extent in exchange for the benefit of the energy savings presented in this study.

In this simulation, the prediction-based algorithm exhibits points where traffic is lost due to less accurate predictions, specifically −0.3 Gbps at 12:05 a.m. and −0.79 Gbps at 10:10 PM, as shown in Figure 21.

It can be seen in Figure 19 that for the base case, all ports are active at all times, and for both algorithms, the use of ports begins to decrease in a stepped manner starting at 2:00 a.m., until reaching the lowest point within the range of 3:00 a.m. to 7:00 a.m.; this is due to low traffic usage during nighttime hours. We should note that in the case of the threshold-based algorithm, there is a high variation at one point in the morning not present in the prediction algorithm, which is observed to be more stable.

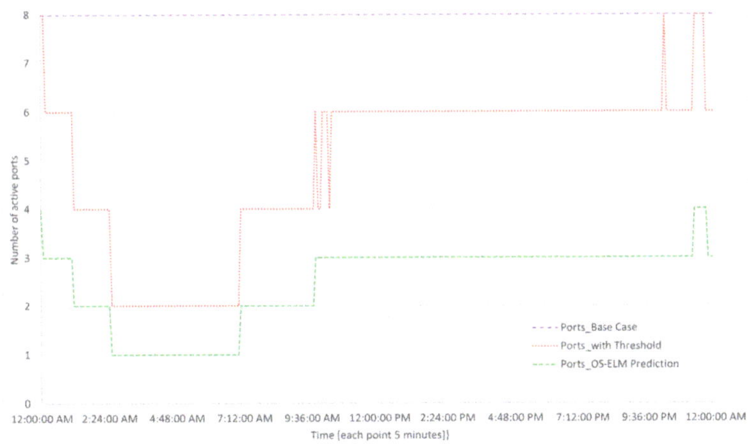

Figure 19. Active number of ports in a day for the base case, threshold, and prediction.

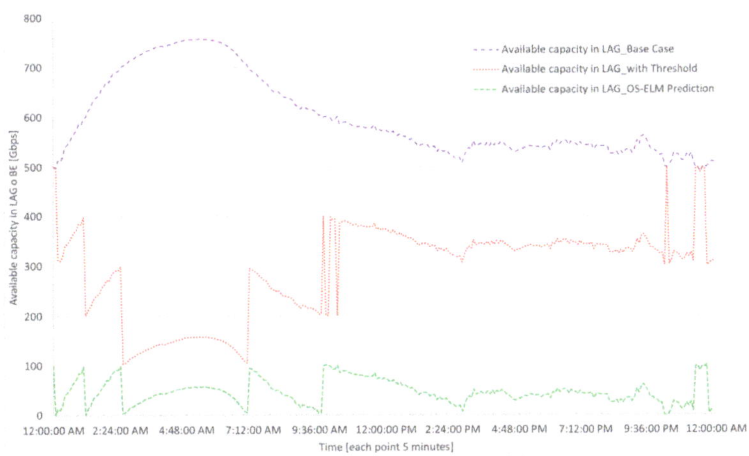

Figure 20. Capacity available in the BE or LAG in a day for base case, threshold, and prediction.

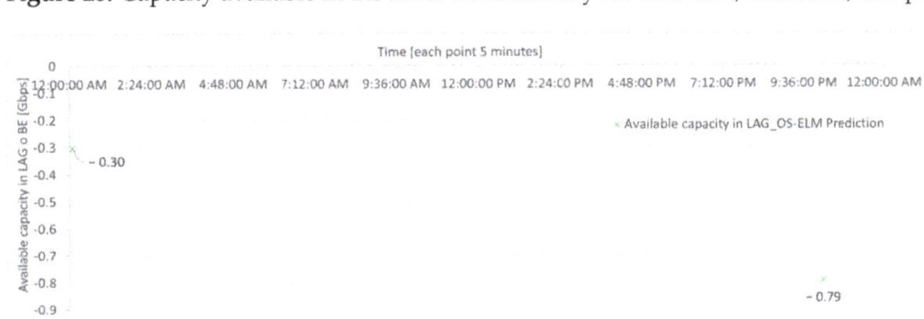

Figure 21. Capacity loss in BE or LAG in one day for base case, threshold, and prediction.

For the purposes of proportionality, it is considered low, given that it represents approximately 0.2% of total traffic (400 Gbps); however, for a TSO this is critical, as it would result in packet loss at a specific moment on the network, with subsequent attempts at reconnection. A solution to this would be to consider a safety factor in the prediction.

8. Conclusions and Future Works

This paper outlines a systematic method for resolving complex problems requiring precise predictions. Utilizing a neural network as the primary tool for prediction enables high accuracy and adaptability to different data types. In addition, the emphasis on energy efficiency emphasizes the significance of reducing energy consumption and discovering ways to optimize resource utilization.

In the case study, a solution was implemented to address the issue of energy efficiency in the data centers of telecommunications service providers. In order to accomplish this, four recurrent and sequential neural networks were compared, allowing predictions to be made every 5 min using sliding windows and hyperparameters of varying values. OS-ELM is the best high-precision network. In terms of prediction, the OS-ELM, LSTM, and GRU networks outperform the RNN by 26%, 2%, and 1% on the RMSE metric; 25%, 3%, and 2% on the MAE metric; and 23%, 6%, and 8% on the MAPE metric, respectively OS-ELM, LSTM, and GRU outperform RNN in terms of computational time by factors of 217.5, 7.2, and 3.9, respectively. For each one, prediction execution times are shorter than the time required for the system to collect data (less than 300 s). The OS execution of ELM's time for this effect is approximately 1.2 s due to its simple structure and absence of gradient minimization in its search for the optimal solution.

The simulations were applied to real traffic from a telecommunications service provider, which offers a real solution for energy efficiency and energy savings that can be applied not only to the core part but also to the aggregation networks, where there are a large number of BEs or LAGs and significant energy savings can be achieved. Regarding the base case (current conditions), the threshold-based algorithm yielded 6.8% and the prediction-based algorithm yielded 12.2% energy savings per day. It should be noted that only one side of the equipment was considered for energy savings in this simulation (customer or service side). If extrapolated to a large quantity of equipment, it would represent substantial cost savings.

As mentioned previously, the methodology presented in this paper can be expanded and applied to other industries in future research. The approach's versatility and adaptability make it a promising solution for a wide variety of prediction issues. As technology and data continue to advance, this methodology can be further developed and improved to provide even more accurate predictions and drive innovation across numerous industries. The potential for future applications and its impact in a variety of fields highlight the significance of this methodology and the need for additional research in this field.

The proposed online sequential extreme learning machine (OS-ELM) scheme holds great potential for addressing energy efficiency in telecoms networks as a whole system challenge. For instance, the deterministic, causal, and universality dimensions of the OS-ELM consider the impact of creating inefficiencies elsewhere in the telecoms and/or other systems while driving networks more efficiently. The prediction scheme can provide near-real-time trade-offs to enhance the flexibility of telecoms networks and demands that are susceptible to efficiency measures. This capability of the proposed scheme serves to augment the energy efficiency networks (EENs) enhancements occasioned with architecture and technologies. Consequently, the current OS-ELM approach will be advanced to enable integrated system-level energy efficiency prediction and optimization across communications and energy systems.

The future direction of research will also explore telecoms equivalents of energy ideas around self-generation, storage, flexibility, and demand reduction. For the considered data centers' case study, our proposed scheme can enable the segmentation of use/users to predict uses that are wasteful, normal, important, and critical. This four-level classification of data centers will consider the temporally and the spatially complex system-level constraints of the EEN of the telecoms system. Holistic, system-level OS-ELM prediction can provide a ubiquitous, seamless, and deeper understanding of the embedded carbon footprints of telecoms network equipment in data centers. This has the potential to predict and utilize feasible EEN solutions via edge computing and Open-RAN equipment. The proposed scheme will be extended by incorporating the EEN trade-offs to encompass end-to-end costs of energy, centralized cloud, end-to-end shared infrastructure, and radio access technologies (5G/6G). This will enable consumer behavior, end-to-end power consumption budget, and user equipment energy use challenges to be predicted for near-real-time EEN optimization for cost-effective data center operation.

Author Contributions: conceptualization, F.R.; methodology, F.R., I.S. and D.Z.-B.; software, F.R.; validation, I.S., D.Z.-B. and C.A.-M.; investigation, F.R.; resources, F.R.; data curation, F.R.; writing—original draft preparation, F.R., I.S., D.Z.-B. and C.A.-M.; writing—review and editing, F.R., I.S., D.Z.-B., C.A.-M., M.I., S.E. and S.G.; supervision, I.S., D.Z.-B., C.A.-M., M.I. and S.E.; funding acquisition, I.S. and S.G. All authors have read and agreed to the published version of the manuscript.

Funding: This research received funding in Chile from USACH, Proyecto Dicyt 062117SG, Vicerrectoría de Investigación, Desarrollo e Innovación, FONDEF No. ID21 I 10191, FONDECYT Regular No. 1211132, and STIC-AmSud AMSUD220026.

Institutional Review Board Statement: Not applicable.

Informed Consent Statement: Not applicable.

Data Availability Statement: Not applicable.

Acknowledgments: We acknowledge the support of the University of Santiago's Postgraduate Vicerectory and Master's Program in Engineering Sciences in Electrical Engineering.

Conflicts of Interest: The authors declare no conflict of interest.

Abbreviations

The following abbreviations are used in this manuscript:

ARIMA	Autoregressive Integrated Moving Average
BE	Bundle Ethernet
BLSTM	Bidirectional Long Short-Term Memory
CDN	Content Delivery Network
CNN	Convolutional Neural Networks
DLHT	Dynamic Local Heuristic Threshold
DNN	Deep Neural Network
DRCN	Design of Reliable Communication Networks
DT	Decision Tree
EEE	Energy Efficient Ethernet
ELM	Extreme Learning Machine
EPC	Evolved Packet Core
FLHT	Fixed Local Heuristic Threshold
GPU	Graphics Processing Unit
GRU	Gated Recurrent Units
IEEE	Institute of Electrical and Electronics Engineers
IGR	Internet Gateway Router
ILP	Integer Linear Programming
IOT	Internet of Things
IP	Internet Protocol
LACP	Link Aggregation Control Protocol
LAG	Link Aggregation Groups
LP	Linear Programming
LPI	Low Power Idle
LSTM	Long Short-Term Memory
MAE	Mean Absolute Error
MAPD	Mean Absolute Percentage Deviation
MAPE	Mean Absolute Percentage Error
MILP	Mixed-Integer Linear Problem

OS-ELM	Online Sequential Extreme Learning Machine
OTN	Optical Transport Network
PHY	Physical Layer
RF	Random Forest
RMSE	Root Mean Squared Error
RNN	Recurrent Neural Network
SARIMA	Seasonal Autoregressive Integrated Moving Average
SDN	Software-Defined Network
SMA	Simple Moving Average
SNMP	Simple Network Management Protocol
SPA	Standby Port Algorithm
SVM	Support Vector Machine
TQA	Two-Queuing Algorithm
TSO	Telecom Service Operator
WDM	Wavelength Division Multiplexing

Appendix A

The link mentions a table that displays the results obtained from for RNN, LSTM, GRU, and OS-ELM in the applied use case. The table shows different performance metrics, such as RMSE, MAE, MAPE, and computational time. Link: simulation results of neural networks RNN, LSTM, GRU, and OS-ELM. https://github.com/frzrau/database_energyefficiencypaper/blob/5a7b8859c5c6275924da31320fd646019f11fd1a/Simulation%20Results.pdf (accessed on 17 May 2023).

References

1. Ahmed, K.M.U.; Bollen, M.H.J.; Alvarez, M. A Review of Data Centers Energy Consumption and Reliability Modeling. *IEEE Access* **2021**, *9*, 152536–152563. [CrossRef]
2. Andrae, A.; Edler, T. On Global Electricity Usage of Communication Technology: Trends to 2030. *Challenges* **2015**, *6*, 117–157. [CrossRef]
3. Junior, R.R.R.; Vieira, M.A.M.; Vieira, L.F.M.; Loureiro, A.A.F. Intra and inter-flow link aggregation in SDN. *Telecommun. Syst.* **2022**, *79*, 95–107. [CrossRef]
4. Addis, B.; Capone, A.; Carello, G.; Gianoli, L.G.; Sanso, B. Energy Management Through Optimized Routing and Device Powering for Greener Communication Networks. *IEEE/ACM Trans. Netw.* **2014**, *22*, 313–325. [CrossRef]
5. Mahadevan, P.; Sharma, P.; Banerjee, S.; Ranganathan, P. A power benchmarking framework for network devices. In *Lecture Notes in Computer Science (including Subseries Lecture Notes in Artificial Intelligence and Lecture Notes in Bioinformatics), Proceedings of the 8th International IFIP-TC 6 Networking Conference, Aachen, Germany, 11–15 May 2009*; Springer: Berlin/Heidelberg, Germany, 2009; pp. 795–808. [CrossRef]
6. Fisher, W.; Suchara, M.; Rexford, J. Greening backbone networks. In *Proceedings of the first ACM SIGCOMM Workshop on Green Networking*; ACM: New York, NY, USA, 2010; pp. 29–34. [CrossRef]
7. IEEE Standards Association. *IEEE Std 802.1AX-2020 (Revision of IEEE Std 802.1AS-2014)*; IEEE Std 802.1AX™-2020, IEEE Standard for Local and Metropolitan Area Networks—Link Aggregation. IEEE Standards Association: Piscataway, NJ, USA, 2020; Volume 2020, pp. 1–421.
8. Bianzino, A.P.; Chaudet, C.; Rossi, D.; Rougier, J.L. A Survey of Green Networking Research. *IEEE Commun. Surv. Tutor.* **2012**, *14*, 3–20. [CrossRef]
9. *IEEE 802.3az-2010*; Energy Efficient Ethernet. IEEE: Piscataway, NJ, USA, 2010.
10. Reviriego, P.; Christensen, K.; Bennett, M.; Nordman, B.; Maestro, J.A. Energy Efficiency in Ethernet. In *Green Communications*; John Wiley & Sons, Ltd.: Chichester, UK, 2015; pp. 277–290. [CrossRef]
11. Liu, L.; Ramamurthy, B. A dynamic local method for bandwidth adaptation in bundle links to conserve energy in core networks. *Opt. Switch. Netw.* **2013**, *10*, 481–490. [CrossRef]
12. Fondo-Ferreiro, P.; Rodríguez-Pérez, M.; Fernández-Veiga, M.; Herrería-Alonso, S. Matching SDN and Legacy Networking Hardware for Energy Efficiency and Bounded Delay. *Sensors* **2018**, *18*, 3915. [CrossRef]
13. Imaizumi, H.; Nagata, T.; Kunito, G.; Yamazaki, K.; Morikawa, H. Power Saving Mechanism Based on Simple Moving Average for 802.3ad Link Aggregation. In Proceedings of the 2009 IEEE Globecom Workshops, Honolulu, HI, USA, 30 November–4 December 2009 ; IEEE: Piscataway, NJ, USA, 2009; pp. 1–6. [CrossRef]

14. Nihale, S.; Sharma, S.; Parashar, L.; Singh, U. Network Traffic Prediction Using Long Short-Term Memory. In Proceedings of the 2020 International Conference on Electronics and Sustainable Communication Systems (ICESC), Coimbatore, India, 2–4 July 2020; IEEE: Piscataway, NJ, USA, 2020; pp. 338–343. [CrossRef]
15. Rau, F.; Soto, I.; Zabala-Blanco, D. Forescating Mobile Network Traffic based on Deep Learning Networks. In Proceedings of the 2021 IEEE Latin-American Conference on Communications (LATINCOM), Santo Domingo, Dominican Republic, 17–19 November 2021; IEEE: Piscataway, NJ, USA, 2021; pp. 1–6. [CrossRef]
16. Andreoletti, D.; Troia, S.; Musumeci, F.; Giordano, S.; Maier, G.; Tornatore, M. Network Traffic Prediction based on Diffusion Convolutional recurrent neural networks. In Proceedings of the IEEE INFOCOM 2019—IEEE Conference on Computer Communications Workshops (INFOCOM WKSHPS), Paris, France, 29 April–2 May 2019; IEEE: Piscataway, NJ, USA, 2019; pp. 246–251. [CrossRef]
17. Wang, W.; Zhou, C.; He, H.; Wu, W.; Zhuang, W.; Shen, X. Cellular Traffic Load Prediction with LSTM and Gaussian Process Regression. In Proceedings of the ICC 2020–2020 IEEE International Conference on Communications (ICC), Dublin, Ireland, 7–11 June 2020; IEEE: Piscataway, NJ, USA, 2020; pp. 1–6. [CrossRef]
18. Liang, N.-Y.; Huang, G.-B.; Saratchandran, P.; Sundararajan, N. A Fast and Accurate Online Sequential Learning Algorithm for Feedforward Networks. *IEEE Trans. Neural Netw.* **2006**, *17*, 1411–1423. [CrossRef]
19. Rau, F.; Soto, I.; Adasme, P.; Zabala-Blanco, D.; Azurdia-Meza, C.A. Network Traffic Prediction Using Online-Sequential Extreme Learning Machine. In Proceedings of the 2021 Third South American Colloquium on Visible Light Communications (SACVLC), Toledo, Brazil, 11–12 November 2021; IEEE: Piscataway, NJ, USA, 2021; pp. 1–6. [CrossRef]
20. Singh, R.; Kumar, H.; Singla, R. An intrusion detection system using network traffic profiling and online sequential extreme learning machine. *Expert Syst. Appl.* **2015**, *42*, 8609–8624. [CrossRef]
21. Liu, Z.; Zhu, Z.; Gao, J.; Xu, C. Forecast Methods for Time Series Data: A Survey. *IEEE Access* **2021**, *9*, 91896–91912. [CrossRef]
22. Wu, J.; He, Y. Prediction of GDP in Time Series Data Based on Neural Network Model. In Proceedings of the 2021 IEEE International Conference on Artificial Intelligence and Industrial Design (AIID), Guangzhou, China, 28–30 May 2021; IEEE: Piscataway, NJ, USA, 2021; pp. 20–23. [CrossRef]
23. Do, Q.H.; Doan, T.T.H.; Nguyen, T.V.A.; Duong, N.T.; Linh, V.V. Prediction of Data Traffic in Telecom Networks based on Deep Neural Networks. *J. Comput. Sci.* **2020**, *16*, 1268–1277. [CrossRef]
24. Mao, Q.; Hu, F.; Hao, Q. Deep Learning for Intelligent Wireless Networks: A Comprehensive Survey. *IEEE Commun. Surv. Tutor.* **2018**, *20*, 2595–2621. [CrossRef]
25. Hou, Y.; Zheng, X.; Han, C.; Wei, W.; Scherer, R.; Połap, D. Deep Learning Methods in Short-Term Traffic Prediction: A Survey. *Inf. Technol. Control* **2022**, *51*, 139–157. [CrossRef]
26. Huang, S.C.; Wu, C.F. Energy Commodity Price Forecasting with Deep Multiple Kernel Learning. *Energies* **2018**, *11*, 3029. [CrossRef]
27. Xiao, C.; Choi, E.; Sun, J. Opportunities and challenges in developing deep learning models using electronic health records data: A systematic review. *J. Am. Med. Inform. Assoc.* **2018**, *25*, 1419–1428. [CrossRef] [PubMed]
28. Lepot, M.; Aubin, J.B.; Clemens, F. Interpolation in Time Series: An Introductive Overview of Existing Methods, Their Performance Criteria and Uncertainty Assessment. *Water* **2017**, *9*, 796. [CrossRef]
29. Zhang, X.; Kuehnelt, H.; De Roeck, W. Traffic Noise Prediction Applying Multivariate Bi-Directional Recurrent Neural Network. *Appl. Sci.* **2021**, *11*, 2714. [CrossRef]
30. Shin, J.; Yeon, K.; Kim, S.; Sunwoo, M.; Han, M. Comparative Study of Markov Chain With Recurrent Neural Network for Short Term Velocity Prediction Implemented on an Embedded System. *IEEE Access* **2021**, *9*, 24755–24767. [CrossRef]
31. Impedovo, D.; Dentamaro, V.; Pirlo, G.; Sarcinella, L. TrafficWave: Generative Deep Learning Architecture for Vehicular Traffic Flow Prediction. *Appl. Sci.* **2019**, *9*, 5504. [CrossRef]
32. Sha, S.; Li, J.; Zhang, K.; Yang, Z.; Wei, Z.; Li, X.; Zhu, X. RNN-Based Subway Passenger Flow Rolling Prediction. *IEEE Access* **2020**, *8*, 15232–15240. [CrossRef]
33. Zeng, C.; Ma, C.; Wang, K.; Cui, Z. Parking Occupancy Prediction Method Based on Multi Factors and Stacked GRU-LSTM. *IEEE Access* **2022**, *10*, 47361–47370. [CrossRef]
34. Khan, Z.; Khan, S.M.; Dey, K.; Chowdhury, M. Development and Evaluation of Recurrent Neural Network-Based Models for Hourly Traffic Volume and Annual Average Daily Traffic Prediction. *Transp. Res. Rec. J. Transp. Res. Board* **2019**, *2673*, 489–503. [CrossRef]
35. Chui, K.T.; Gupta, B.B.; Liu, R.W.; Zhang, X.; Vasant, P.; Thomas, J.J. Extended-Range Prediction Model Using NSGA-III Optimized RNN-GRU-LSTM for Driver Stress and Drowsiness. *Sensors* **2021**, *21*, 6412. [CrossRef] [PubMed]
36. Nguyen, M.; Sun, N.; Alexander, D.C.; Feng, J.; Yeo, B.T. Modeling Alzheimer's disease progression using deep recurrent neural networks. In Proceedings of the 2018 International Workshop on Pattern Recognition in Neuroimaging (PRNI), Singapore, 12–14 June 2018; IEEE: Piscataway, NJ, USA, 2018; pp. 1–4. [CrossRef]
37. Li, P.; Shi, Y.; Xing, Y.; Liao, C.; Yu, M.; Guo, C.; Feng, L. Intra-Cluster Federated Learning-Based Model Transfer Framework for Traffic Prediction in Core Network. *Electronics* **2022**, *11*, 3793. [CrossRef]
38. Zhang, C.; Zhang, H.; Yuan, D.; Zhang, M. Citywide Cellular Traffic Prediction Based on Densely Connected Convolutional Neural Networks. *IEEE Commun. Lett.* **2018**, *22*, 1656–1659. [CrossRef]

39. Fu, Y.; Wang, S.; Wang, C.X.; Hong, X.; McLaughlin, S. Artificial Intelligence to Manage Network Traffic of 5G Wireless Networks. *IEEE Netw.* **2018**, *32*, 58–64. [CrossRef]
40. Zhang, D.; Liu, L.; Xie, C.; Yang, B.; Liu, Q. Citywide Cellular Traffic Prediction Based on a Hybrid Spatiotemporal Network. *Algorithms* **2020**, *13*, 20. [CrossRef]
41. Kao, C.C.; Chang, C.W.; Cho, C.P.; Shun, J.Y. Deep Learning and Ensemble Learning for Traffic Load Prediction in Real Network. In Proceedings of the 2020 IEEE Eurasia Conference on IOT, Communication and Engineering (ECICE), Yunlin, Taiwan, 23–25 October 2020; IEEE: Piscataway, NJ, USA, 2020; pp. 36–39. [CrossRef]
42. Santos, G.L.; Rosati, P.; Lynn, T.; Kelner, J.; Sadok, D.; Endo, P.T. Predicting short-term mobile Internet traffic from Internet activity using recurrent neural networks. *Int. J. Netw. Manag.* **2022**, *32*, e2191. [CrossRef]
43. Nejadettehad, A.; Mahini, H.; Bahrak, B. Short-term Demand Forecasting for Online Car-hailing Services Using recurrent neural networks. *Appl. Artif. Intell.* **2020**, *34*, 674–689. [CrossRef]
44. Kumar, B.P.; Hariharan, K.; Shanmugam, R.; Shriram, S.; Sridhar, J. Enabling internet of things in road traffic forecasting with deep learning models. *J. Intell. Fuzzy Syst.* **2022**, *43*, 6265–6276. [CrossRef]
45. Huang, G.B.; Zhu, Q.Y.; Siew, C.K. Extreme learning machine: Theory and applications. *Neurocomputing* **2006**, *70*, 489–501. [CrossRef]
46. Park, J.M.; Kim, J.H. Online recurrent extreme learning machine and its application to time-series prediction. In Proceedings of the 2017 International Joint Conference on Neural Networks (IJCNN), Anchorage, AK, USA, 14–19 May 2017; IEEE: Piscataway, NJ, USA, 2017; Volume 2017-May; pp. 1983–1990. [CrossRef]
47. Tian, Y.; Zhang, K.; Li, J.; Lin, X.; Yang, B. LSTM-based traffic flow prediction with missing data. *Neurocomputing* **2018**, *318*, 297–305. [CrossRef]
48. Baytas, I.M.; Xiao, C.; Zhang, X.; Wang, F.; Jain, A.K.; Zhou, J. Patient Subtyping via Time-Aware LSTM Networks. In *KDD '17, Proceedings of the 23rd ACM SIGKDD International Conference on Knowledge Discovery and Data Mining, Halifax, NS, Canada, 13–17 August 2017*; ACM: New York, NY, USA, 2017; pp. 65–74. [CrossRef]
49. Rubanova, Y.; Chen, R.T.; Duvenaud, D. Latent odes for irregularly-sampled time series. *arXiv* **2019**, arXiv:1907.03907.
50. Vecoven, N.; Ernst, D.; Drion, G. A bio-inspired bistable recurrent cell allows for long-lasting memory. *PLoS ONE* **2021**, *16*, e0252676. [CrossRef]
51. Zhou, J.; Huang, Z. Recover missing sensor data with iterative imputing network. In Proceedings of the Workshops at the Thirty-Second AAAI Conference on Artificial Intelligence, New Orleans, LA, USA, 2–7 February 2018. [CrossRef]
52. Weerakody, P.B.; Wong, K.W.; Wang, G.; Ela, W. A review of irregular time series data handling with gated recurrent neural networks. *Neurocomputing* **2021**, *441*, 161–178. [CrossRef]
53. Mahmood, A.; Mat Kiah, M.L.; Reza Z'Aba, M.; Qureshi, A.N.; Kassim, M.S.S.; Azizul Hasan, Z.H.; Kakarla, J.; Sadegh Amiri, I.; Azzuhri, S.R. Capacity and Frequency Optimization of Wireless Backhaul Network Using Traffic Forecasting. *IEEE Access* **2020**, *8*, 23264–23276. [CrossRef]
54. Ba, S.; Ouédraogo, I.A.; Oki, E. A power consumption reduction scheme in hose-model networks with bundled links. In Proceedings of the 2013 IEEE International Conference on Green Computing and Communications and IEEE Internet of Things and IEEE Cyber, Physical and Social Computing, GreenCom-iThings-CPSCom 2013, Beijing, China, 20–23 August 2013; pp. 40–45. [CrossRef]
55. Galán-Jiménez, J.; Gazo-Cervero, A. Designing energy-efficient link aggregation groups. *Ad. Hoc. Netw.* **2015**, *25*, 595–605. [CrossRef]
56. Rodriguez-Perez, M.; Fernandez-Veiga, M.; Herreria-Alonso, S.; Hmila, M.; Lopez-Garcia, C. Optimum Traffic Allocation in Bundled Energy-Efficient Ethernet Links. *IEEE Syst. J.* **2018**, *12*, 593–603. [CrossRef]
57. Fondo-Ferreiro, P.; Rodriguez-Perez, M.; Fernandez-Veiga, M. Implementing energy saving algorithms for ethernet link aggregates with ONOS. In Proceedings of the 2018 5th International Conference on Software Defined Systems, SDS 2018, Barcelona, Spain, 23–26 April 2018; pp. 118–125. [CrossRef]
58. Ramakrishnan, N.; Soni, T. Network Traffic Prediction Using recurrent neural networks. In Proceedings of the 2018 17th IEEE International Conference on Machine Learning and Applications (ICMLA), Orlando, FL, USA, 17–20 December 2018; IEEE: Piscataway, NJ, USA, 2018; pp. 187–193. [CrossRef]
59. Chung, J.; Gulcehre, C.; Cho, K.; Bengio, Y. Empirical Evaluation of Gated recurrent neural networks on Sequence Modeling. In Proceedings of the NIPS 2014 Deep Learning and Representation Learning Workshop, Montreal, QC, Canada, 12 December 2014. [CrossRef]
60. Zhang, J.; Xiao, W.; Li, Y.; Zhang, S. Residual compensation extreme learning machine for regression. *Neurocomputing* **2018**, *311*, 126–136. [CrossRef]
61. seok Lim, J.; Lee, S.; Pang, H.S. Low complexity adaptive forgetting factor for online sequential extreme learning machine (OS-ELM) for application to nonstationary system estimations. *Neural Comput. Appl.* **2013**, *22*, 569–576. [CrossRef]
62. Jian, L.; Gao, F.; Ren, P.; Song, Y.; Luo, S. A Noise-Resilient Online Learning Algorithm for Scene Classification. *Remote Sens.* **2018**, *10*, 1836. [CrossRef]
63. Shrivastava, S. Cross Validation in Time Series, 2020.
64. Bergmeir, C.; Benítez, J.M. On the use of cross-validation for time series predictor evaluation. *Inf. Sci.* **2012**, *191*, 192–213. [CrossRef]

65. Tashman, L.J. Out of Sample Tests of Forecasting Accuracy: An Analysis and Review. *Int. J. Forecast.* **2000**, *16*, 437–450. [CrossRef]
66. Varma, S.; Simon, R. Bias in error estimation when using cross-validation for model selection. *BMC Bioinform.* **2006**, *7*, 91. [CrossRef] [PubMed]
67. Kingma, D.P.; Ba, J. Adam: A Method for Stochastic Optimization. In Proceedings of the 3rd International Conference on Learning Representations, ICLR 2015, San Diego, CA, USA, 7–9 May 2015.
68. Schaul, T.; Antonoglou, I.; Silver, D. Unit Tests for Stochastic Optimization. In Proceedings of the 2nd International Conference on Learning Representations, ICLR 2014, Banff, AB, Canada, 14–16 April 2014.
69. Bock, S.; Goppold, J.; Weiß, M. An improvement of the convergence proof of the ADAM-Optimizer. In Proceedings of the OTH CLUSTERKONFERENZ, Weiden, Germany, 13 April 2018.
70. Carling, K.; Meng, X. Confidence in Heuristic Solutions. *J. Glob. Optim.* **2015**, *63*, 381–399. [CrossRef]
71. Yue, Y.; Wang, Q.; Yao, J.; O'Neil, J.; Pudvay, D.; Anderson, J. 400GbE Technology Demonstration Using CFP8 Pluggable Modules. *Appl. Sci.* **2018**, *8*, 2055. [CrossRef]
72. Zhang, W.; Bathula, B.G.; Sinha, R.K.; Doverspike, R.; Magill, P.; Raghuram, A.; Choudhury, G. Cost Comparison of Alternative Architectures for IP-over-Optical Core Networks. *J. Netw. Syst. Manag.* **2016**, *24*, 607–628. [CrossRef]
73. IEEE Standards Association. *IEEE Std 802.3ad-2000*; IEEE Standard for Information Technology—Local and Metropolitan Area Networks—Part 3: Carrier Sense Multiple Access with Collision Detection (CSMA/CD) Access Method and Physical Layer Specifications-Aggregation of Multiple Link Segments. IEEE Standards Association: Piscataway, NJ, USA, 2000; pp. 1–184. [CrossRef]
74. Braun, R.P. 100Gbit/s IP-Router and DWDM Transmission Interoperability Field Tests. In Proceedings of the Photonic Networks, 12. ITG Symposium, Leipzig, Germany, 2–3 May 2011 ; pp. 1–3.
75. IEEE. *IEEE Std 802.3-2015*; IEEE Std 802.3-2015 (Revision of IEEE Std 802.3-2012). IEEE Standards Association: Piscataway, NJ, USA, 2016; pp. 1–4017. [CrossRef]
76. IEEE Standards Association. *IEEE Std 802.3bm-2015*; IEEE Standard for Ethernet Amendment 3: Physical Layer Specifications and Management Parameters for 40 Gb/s and 100 Gb/s Operation over Fiber Optic Cables. IEEE: Piscataway, NJ, USA, 2015.
77. Reviriego, P.; Hernadez, J.A.; Larrabeiti, D.; Maestro, J.A. Burst Transmission in Energy Efficient Ethernet. *IEEE Internet Comput.* **2010**, *14*, 50–57. [CrossRef]

sensors

MDPI

Article

Scheduling Sparse LEO Satellite Transmissions for Remote Water Level Monitoring

Garrett Kinman [1], Željko Žilić [2,*] and David Purnell [2,3]

[1] Octasic Inc., 2901 Rachel, Montréal, QC H1W 4A4, Canada; garrett.kinman@mail.mcgill.ca
[2] Department of Electrical and Computer Engineering, McGill University, 3480 University, Montréal, QC H3A 0E9, Canada; david.purnell.1@ulaval.ca
[3] Department of Civil and Water Engineering, Laval University, pavillon Adrien-Pouliot 1065, av. de la Médecine, Québec, QC G1V 0A6, Canada
* Correspondence: zeljko.zilic@mcgill.ca

Abstract: This paper explores the use of low earth orbit (LEO) satellite links in long-term monitoring of water levels across remote areas. Emerging sparse LEO satellite constellations maintain sporadic connection to the ground station, and transmissions need to be scheduled for satellite overfly periods. For remote sensing, the energy consumption optimization is critical, and we develop a learning approach for scheduling the transmission times from the sensors. Our online learning-based approach combines Monte Carlo and modified k-armed bandit approaches, to produce an inexpensive scheme that is applicable to scheduling any LEO satellite transmissions. We demonstrate its ability to adapt in three common scenarios, to save the transmission energy 20-fold, and provide the means to explore the parameters. The presented study is applicable to wide range of IoT applications in areas with no existing wireless coverages.

Keywords: Internet of Remote Things; sparse LEO satellite transmission; water-level monitoring

1. Introduction

The Internet of Things (IoT) has made huge advances in smart homes, industrial and other settings with numerous networking options already present. To achieve progress in the Internet of Remote Things (IoRT) for environmental monitoring in wilderness, connectivity solutions are needed that are widespread, energy-efficient and cost-efficient. This paper presents the exploration of satellite-based connectivity in the context of environmental water-level monitoring.

Citation: Kinman, G.; Žilić, Ž.; Purnell, D. Scheduling Sparse LEO Satellite Transmissions for Remote Water Level Monitoring. *Sensors* **2023**, *23*, 5581. https://doi.org/10.3390/s23125581

Academic Editor: Josip Lorincz

Received: 8 May 2023
Revised: 10 June 2023
Accepted: 12 June 2023
Published: 14 June 2023

1.1. Water-Level Monitoring and Its Role in Climate

Global water-level monitoring is critical in hydrology and climate change tracking. The polar regions are arguably at the center of the climate crisis, because these regions are experiencing the most rapid changes and the largest current and future contribution to sea level rise is predicted to be from ice sheets losing mass to the ocean [1]. Predicting how the polar regions will change in the future requires field measurements, for example, from sensors that monitor changes in the atmosphere (weather stations), coastal water-level sensors, ocean buoys, or from Global Navigation Satellite System (GNSS) stations (for monitoring solid earth deformation) [1]. Despite the ever-expanding capabilities of remote sensing satellites, such measurements cannot yet be obtained from space with the same accuracy or temporal resolution as from ground-based sensors.

Climate model predictions become more reliable with an increased density of sensors, hence low-cost environmental sensor networks are emerging as a powerful tool for climate monitoring [2]. One recent innovation repurposes mass-market GNSS technology for water-level monitoring, using a technique called GNSS Interferometric Reflectometry (GNSS-IR), and has the potential to be used to increase the density of coastal water-level stations [3,4].

In remote regions such as Greenland or Antarctica, where sea level information is critical for climate monitoring [5,6], field campaigns are expensive and it may be prohibitively expensive to maintain a dense network of sensors. Wireless connectivity should reduce the maintenance cost of remote sensor networks by reducing the frequency of expensive site visits to collect data or check the status of instruments. This paper focuses on a low-cost and energy-efficient wireless communication technique using low earth orbit (LEO) satellites that is suitable for remote water-level sensor networks, notably GNSS-IR ones, Figure 1.

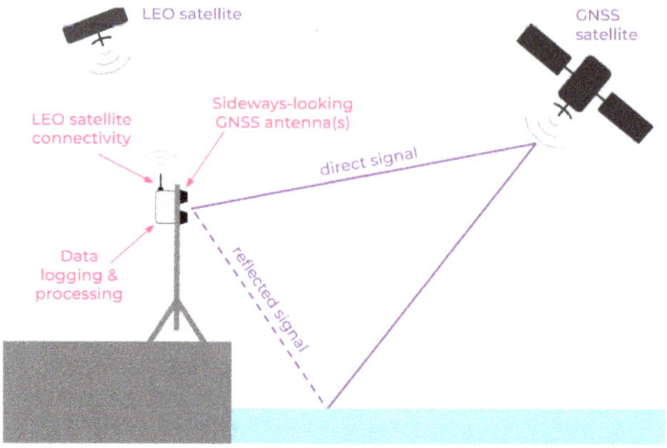

Figure 1. Schematic showing a GNSS-IR water-level sensor with integrated LEO satellite connectivity.

1.2. *Connectivity for Internet of Remote Things*

This paper addresses the problem of providing inexpensive and energy-efficient satellite IoT links in the context of GNSS-IR monitoring. Such a water-level sensor must be affordable and widely deployable. Wide geographic reach imposes the challenge of data uplink from remote locations to where that data is needed [3]. This limitation of IoT has spawned a subdomain dedicated to solving the issues of bringing IoT to the remote corners of the globe, the IoRT [7].

There is an abundance of connectivity options for IoT around populated areas. Several standard IoT connectivity options range from cellular technologies to LoRa. Connectivity options for IoRT range from low-power wide area networks (LPWANs) to low-power cellular network standards to geostationary and LEO satellites [8]. Additionally, there have been efforts into unmanned aerial vehicles supporting IoRT [8,9]. Satellite options are the only proven connectivity options for truly global coverage [8,10,11], but only if the cost and energy consumption are kept low enough. Traditional geostationary satellites are always overhead for a fixed earth location, but they are costly, require higher transmission power, and incur around 70 times longer latency than LEO satellites [7,12].

1.3. *LEO Satellite Communications for Internet of Remote Things*

For IoT applications, LEO satellites are practical for the most remote regions where terrestrial infrastructure is out of reach [7,12]. LEO communications are categorized by:

- Communication directness;
- LEO orbit configuration;
- By satellite service type.

Regarding communication directness, individual sensor nodes can communicate directly to a satellite (known fittingly as "direct-to-satellite") or indirectly via a local network (often an LPWAN such as Bluetooth) centered around a satellite *gateway* [8,10]. The latter case is impractical in our case, as a gateway adds another independent hardware part not under the IoRT node control, which needs to be designed and dimensioned for multiple unknown nodes, and requires a critical mass of nodes to be useful.

Among satellite services, there are those provided by companies from the pre-IoT era (who often offer satellite internet and phone coverage as well), and the emerging sparse constellation [13] networks, such as Swarm, Lacuna, Enxaneta, Kepler or Astrocast provided by independent LEO satellite companies [8,10], which are more suited for IoT. The independent LEO satellite services use CubeSats, which are small and modular picosatellites [8]. Since IoT can tolerate intermittent connectivity better than satellite phones, their satellites can use polar orbits to provide global, but intermittent, coverage [10] by sparse constellations. In contrast, traditional LEO providers deliver continuous or near-continuous coverage using a combination of polar and non-polar orbits [10].

Using emerging *sparse LEO constellations* for IoRT has the primary benefit of requiring fewer satellites and less cost [10], but requires waiting until a satellite passes overhead [10]. To achieve low-energy LEO networking, a suitable algorithm must be devised to schedule sensing and transmissions at appropriate times such that the data is transmitted at (near-)minimal cost in energy [10].

1.4. Relation to Previous Work

To the best our knowledge, one previous paper [11] has examined this problem and proposed an online learning algorithm, which can learn sample-by-sample in the field, as opposed to offline in batched datasets. That work on indirect-to-satellite communications is not applicable here, as it assumes perfect knowledge of uplink availability, unpredictable multiple streams of data that can easily overflow the buffers, and is thus posed as a queue scheduling problem [11]. They then propose an online learning algorithm based on Lyupanov optimization, which is a common approach for similar queue optimization problems [11,14].

In contrast, our paper deals with direct-to-satellite communications with a known data production rate. The algorithm presented in this paper is derived from reinforcement learning, specifically Monte Carlo learning and the k-armed bandit problem. Because of the relative youth of the LEO satellite service (provided commercially by Swarm Technologies), an integrative approach is taken to the design from requirements, to communications technology selection, to hardware, and finally to software and algorithm design. In doing this, this paper aims to highlight key design considerations for creating a low-cost, low-power communications scheme for an IoRT device. The key contribution of this paper is the online learning-based direct-to-satellite scheduling, and associated energy model.

2. Materials and Methods

We implemented GNSS-IR water-level detection system on a printed circuit board (PCB), Figure 2 and deployed on Swarm network by help of our software. The PCB includes from right to left: a Swarm M138 LEO Modem, a Raspberry Pi Pico, and pads for four GNSS modules for GNSS-IR water-level measuring (together with GNSS antennas). The proposed scheduling algorithm was implemented on a dual-core ARM Cortex M processor of the Raspberry Pi Pico, where each processor core executes one process of the code. The board is sending one 128-bit message per hour, to fit within a single Swarm data plan for USD 60/year.

Figure 3 presents the schematic for the final prototype PCB design produced in this project. The left-hand side displays spaces for four GPS receivers and four GPS antenna connectors, which are the project-specific sensing components for GNSS-R water level sensing. The remaining two-thirds of the schematic are generalizeable to other projects that use the Swarm M138 LEO modem, including an mPCI-e connector, decoupling and feed-through capacitors, and headers for the microcontroller.

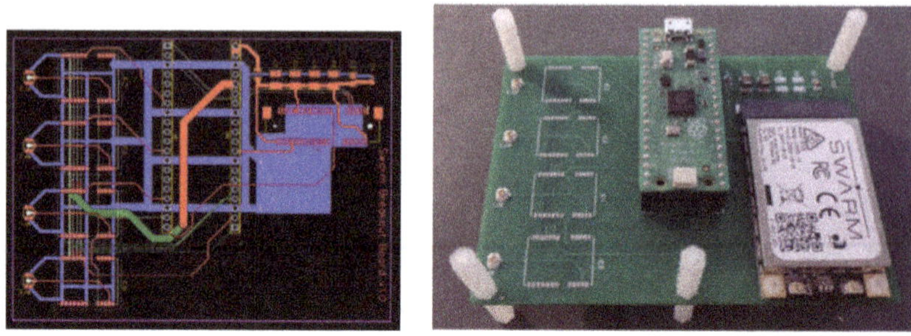

Figure 2. Board design: PCB layers and board populated with Raspberry and Swarm modules.

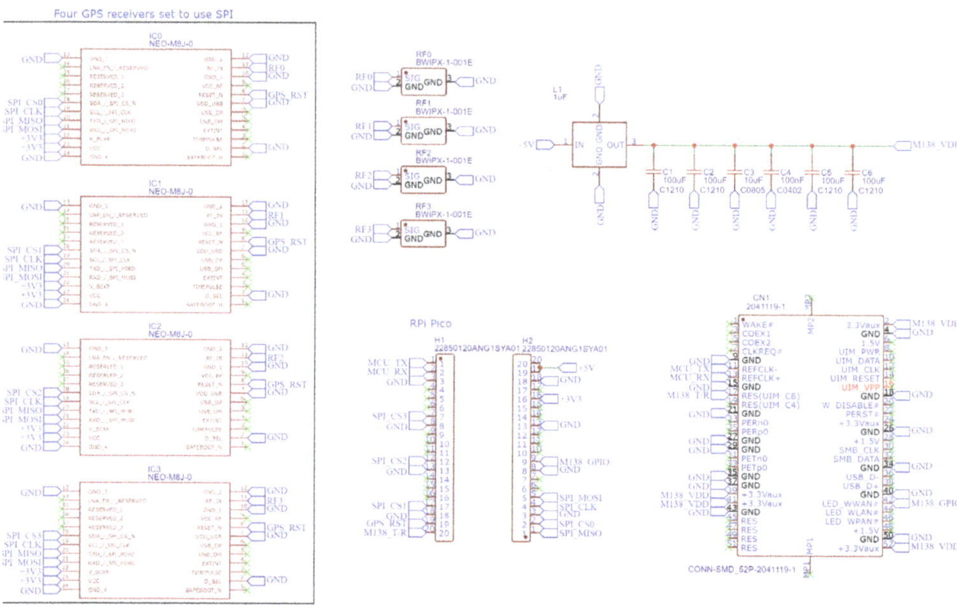

Figure 3. Board schematic with GNSS receivers, Raspberry Pico microcontroller and Swarm modem.

2.1. Satellite Modem Operating Specifications

For remote GNSS-IR sensors, we use the independent LEO satellite provider Swarm Technologies. The energy consumption and, consequently, transmission scheduling will depend on Swarm's service specification and operation of the Swarm M138 modem built into our board. The Swarm modem has four operating states: (1) Sleep Mode, (2) GPS Acquisition Mode, (3) Receive Mode, and (4) Transmit Mode, Figure 4.

When the modem powers on, it enters GPS acquisition mode to determine the time and location. The modem will also re-enter GPS Acquisition Mode every 4 h or when awoken. Once a GPS fix has been acquired (30 s typical duration), the modem enters the Receive Mode, wherein it listens for a packet from any satellites passing overhead. This mode lasts until either a packet is received from a satellite (at which point it enters Transmit Mode), the modem is instructed to enter Sleep Mode, or enough time elapses that the modem automatically re-enters GPS Acquisition Mode. Robust operation and enhanced availability [15] is built into the M138 modem, as well as ensured by handling the exceptions, such as those caused by lost signals (Swarm or GNSS) or power.

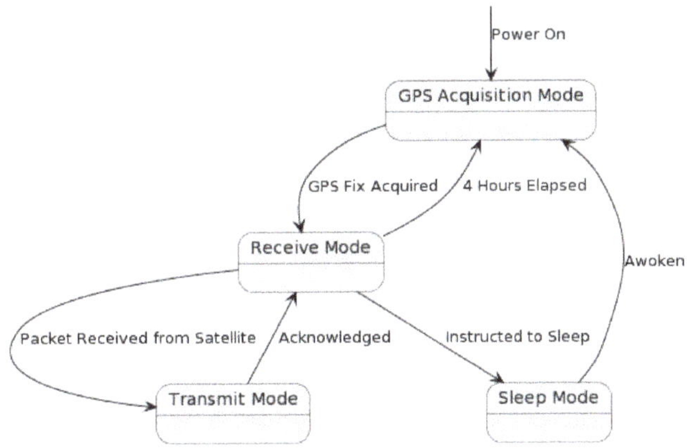

Figure 4. State machine of the Swarm M138 operating modes.

If a packet is received from a satellite, the modem enters Transmit Mode, attempting to transmit queued packets and receive an acknowledgement. If successful, it will return to Receive Mode, unless put into Sleep Mode. Table 1 shows that transmission is 1–2 orders of magnitude more costly, while Sleep Mode uses 2 to 3 orders of magnitude less energy. Communication incurs a dominant part of energy consumption in IoT nodes [16]—even more so for satellite access. For instance, Swarm reports that sending a maximum-length 192-byte packet at $P = 2.8$ W takes $\Delta T = 3.7$ s and $E_{total} = 12.24$ J, while in comparison Raspberry Pico benchmark for embedded code, `hello_sleep` runs at 1.5 mW, as per the datasheet.

Table 1. DC power characteristics of 4 modes of operation for 5V/3.3V power supply.

Mode	Typical Current at 5 V/3.3 V	Typical Power at 5 V/3.3 V
Transmit	550 mA/850 mA	2.8 W/2.8 W
GPS Acquisition	45 mA/45 mA	230 mW/150 mW
Receive	26 mA/26 mA	130 mW/86 mW
Sleep	<110 µA/80 µA	<550 µW/260 µW

For Swarm modem's operating modes, the energy-saving strategy includes:

1. Keep the modem in Sleep Mode as much as possible. When not in Sleep Mode, its default state is Receive Mode, which uses much more power.
2. Being awake dominates energy usage, either from the actual transmission energy or the GPS Acquisition and Receive Modes.
3. Failure to transmit will waste considerable energy. Thus, one should schedule transmission to when there is a high probability of successful communication.

2.2. *Swarm Satellite Transmission*

To minimize transmission power consumption requires understanding how the satellites, transmission, and data plans work. There is not always a satellite overhead, nor are the elevation angle and environmental conditions (e.g., background RF noise) always suitable. Data rate is limited, and frequent transmissions consume energy. These factors critically impact how we orchestrate transmissions.

The nature of Swarm satellite passages is disclosed by their Web-based tool that lists upcoming satellite passes, their times, durations and max elevation angles for a given location. Elevation angles observed in Montreal, Quebec, Canada range between 15 and 85 degrees, and pass durations typically range between 10 min and an hour. In reality, even with a satellite pass, the modem might not always be able to transmit. There are many factors impacting this: satellite pass "quality", RF background noise, environmental conditions, antenna setup, and many others.

The first factor, satellite pass quality, is due to the pass duration and maximum elevation angle. Swarm gives no guidance on what factors impair successful transmission, and one objective of this paper is for each sensor to construct an empirical model for quantifying the likelihood of a pass leading to successful transmission. The second factor, RF background noise, does have guidance provided by Swarm, Table 2, by which noise intensity of -93 dBm or lower is expected for successful transmission.

Table 2. Background noise intensity required for likelihood of transmission.

Background Noise RSSI (dBm)	Quality (for Transmission)
−90 and higher	Bad (unlikely to work)
−93 and lower	Marginal
−97 and lower	OK
−100 and lower	Good
−105 and lower	Great

There are also the constraints imposed by Swarm data plans, priced at USD 60 per year per data plan, with up to four data plans stackable onto a single modem. Each data plan permits up to 750 packets per month, or about 25 packets per day, or about one per hour. These constraints imply that for finer temporal resolution (e.g., every 15 min), one must either bundle measurements, or pay to stack multiple data plans. The later, costly option also reduces the battery life, while bundling reduces the number of packets and possibly the cost.

The high-level view of the two main processes is shown in Figure 5. The process on the left produces and inserts the data into a circular queue. Since the Swarm modem's internal queue can drop packets after 48 h, the circular queue needs to contain 48 h of data. Each cycle of waking from sleep, acquiring GPS, listening for a satellite, and transmitting uses a lot of extra energy. In addition, due to environmental variables, there is inherent uncertainty as to how long one can expect the modem to be awake before transmitting successfully. This precise question is examined in the rest of the paper.

2.3. Efficient Packet Data Bundling

Note that there are a few important functions in Figure 5, such as the data bundling, as the transmission duration directly causes energy consumption. Table 3 shows the format of data. Each datum includes a timestamp, expressed in minutes since 1 January 1970. Due to the nature of the GNSS-IR, we omit seconds, which allows the re-purposing of 4 bits for 16 status codes. For completeness, using 28 bits allows timekeeping for 510 years. This format allows the whole datum to fit in 16 bytes, which divides evenly into 192 bytes per packet, such that each packet will be maximally utilized with 12 data points per packet.

Table 3. Format for each datum within the software.

Name	Type	Bits
Water level	Floating-point	32 bits (4 bytes)
Error	Floating-point	32 bits (4 bytes)
Roughness	Floating-point	32 bits (4 bytes)
Minutes since 1 January 1970	Positive integer	28 bits (<4 bytes)
Status	Positive integer	4 bits (<1 byte)
Total		128 bits (16 bytes)

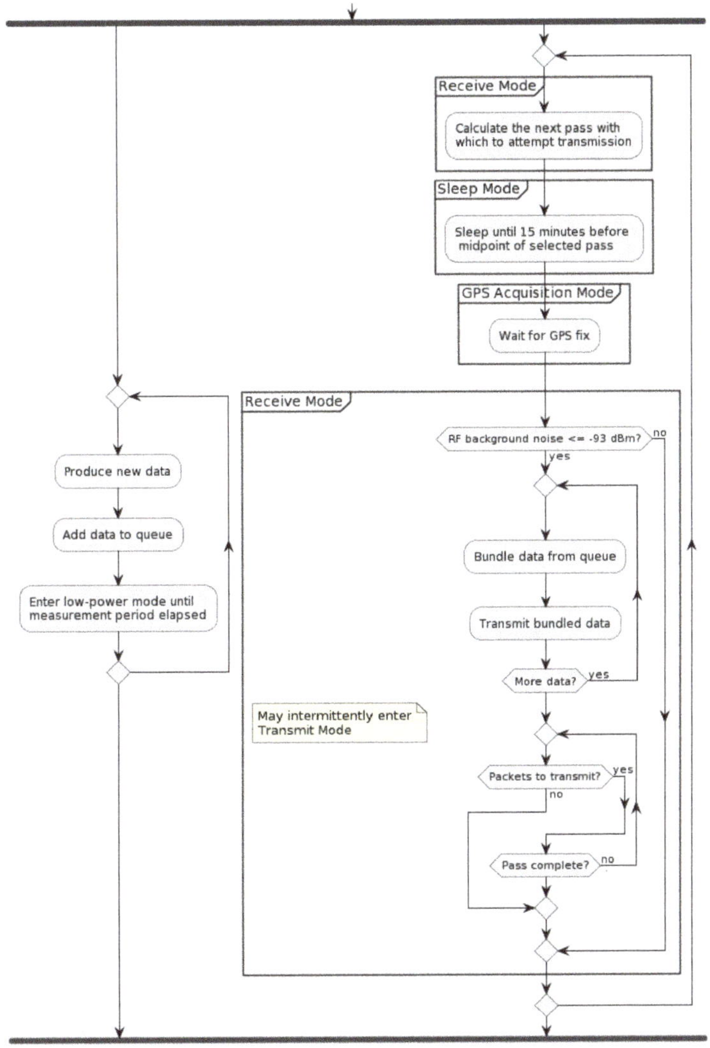

Figure 5. Activity diagram for the host device with two processes

Second implicit function within the high-level processes shown in Figure 5 is that of good satellite pass selection, while the algorithm for actually predicting satellite passes—at least from the user perspective—is made fairly simple with the help of an open-source SGP4 satellite pass prediction Arduino library, quantifying what satellite passes are "good" depends much on environmental conditions, setup details, and empirical observations, as described next.

2.4. Online Learning Direct-to-Satellite Packet Scheduling

Transmitting to the LEO satellites can be unreliable due to minute changes in equipment setup or environmental factors. For example, severely cloudy days lead to too high RF background noise (i.e., higher than −93 dBm). Further, unshielded microcontroller within 10 to 20 cm of the antenna could increase measured RF background noise by as much as 5 to 10 dB. Further, slightly angling the antenna towards or away from a cell tower a few kilometers away could vary the RF background noise by several dB. With all these factors, creating a generalizable pass model is intractable.

Previous work with indirect-to-satellite scheduling shows that online learning is a successful strategy [11]. Thus, each individual sensor should learn for itself and for its exact site conditions and hardware setup via online learning. Previous work in indirect-to-satellite scheduling uses a Lyupanov optimization problem for network queuing to avoid making assumptions about when new data would become available [11] while the perfect knowledge of satellite overpasses is assumed. However, the data production rate is constant in our case, so rather than treating it as a network queuing problem, we ought to predict the uplink availability. Thus, a novel approach will be used.

2.4.1. Algorithmic Problem Statement

For a sensor placed in a remote location, a simple and *interpretable model* is needed to be trusted to perform as expected [17]. To achieve this, a relatively simple algorithm inspired from reinforcement learning has been devised. The goal is to learn the probability of successful transmission, given three input variables: (1) the satellite pass duration (in minutes), (2) the maximum elevation angle of the satellite pass (in degrees), and (3) the RF background noise (in dBm).

Borrowing the notation from reinforcement learning, the state space S is the set of all possible input variable combinations, and the action space A is the set of all possible actions [18]. In this case, A consists of the actions to transmit or not to transmit for each satellite pass with pass characteristics $s \in S$. Let the function v be the mapping of S to a probability of successful transmission, $v : S \mapsto [0,1]$, and let the policy π represent the conditional probability of choosing a particular action $a \in A$ given a state $s \in S$. Hence, the policy is the mechanism for choosing which satellite pass to select, given a set of passes and their characteristics.

$$\pi(a|s) = P(A_t = a|S_t = s) \tag{1}$$

In Equation (1), A_t represents the action at time step t, and S_t represents the state at time step t. Regarding the probability success mapping V, a natural objective is thus to approximate it with collected experience: as the system runs and has successes and failures transmitting with different states $s \in S$, it will converge to true probabilities of successful transmission for a given state, i.e., the value function v [18,19].

2.4.2. Modified Monte Carlo Learning

Monte Carlo learning methods approximate a value function in a simple and interpretable way by taking the value of a state to be the average return at the end of a training episode [18,19]. In the direct-to-satellite packet scheduling, the episodes are of length one, i.e., there is no sequential decision-making, simplifying the problem. If the reward is taken to be 1 for a successful transmission and 0 for an unsuccessful transmission, then the value function can be taken to be the average rate of successful transmission from a given state. If for a given state of satellite pass characteristics and RF noise, transmission is successful 50% of the time, then the value function is 0.5.

However, Monte Carlo learning requires a discrete state space, whereas the state space for this problem is continuous, so we discretize the state space. Using the Swarm pass checker, it is known that all satellite passes shown are between 15 and 90 degrees and almost all between 10 and 60 min. Additionally, while RF noise is technically continuous, the modems only report whole numbers, e.g., −95 dBm. If only integers within the range −93 dBm (the highest noise Swarm reports success transmitting with) to −106 dBm (the lowest noise measured in this project) are considered, this is naturally discretized. Table 4 shows how the state space has been chosen to be discretized. With 5 buckets for each state variable, some simple combinatorics gives 125 unique combinations, where the total set of 125 combinations represents the discretized state space.

Table 4. State space bucketing for each state variable.

Bucket Number	Max Elevation Angle (°)	Pass Duration (minutes)	RF Background Noise (dBm)
1	15 to 30	10 to 20	−93 to −95
2	31 to 45	21 to 30	−96 to −98
3	46 to 60	31 to 40	−99 to −101
4	61 to 75	41 to 50	−102 to −104
5	76 to 90	51 and higher	−105 and lower

The remaining question is that of the policy π. Clearly, once a good approximation of the true value function is made, the policy π should exploit that knowledge to select the most promising satellite passes. At the beginning, the system will not know about a good satellite pass, and it will thus have to explore with passes of different characteristics. This is an example of the *exploration–exploitation* problem in reinforcement learning [18,19]. A common approach is to explore early on and gradually exploit more with time.

2.4.3. Modified *k*-Armed Bandit

Regarding the policy for packet scheduling, there is a similarity to the *k*-armed bandit problem, whereby an agent repeatedly plays the same one-step episode. In each game, the agent has a selection of options, which may give varying stochastic rewards. The goal is to learn over time which actions give the greatest expected reward [18,19]. A common approach to this problem involves softmax (Boltzmann) exploration, which derives a set of probabilities corresponding to each possible action [20]. The action with the highest expected reward has the highest probability of selection, plus all the choices are guaranteed to sum to 1 by the design.

Our problem is slightly different from the *k*-armed bandit problem in two important ways: (1) the set of actions available to the agent in each episode is different, and (2) expected reward is not only the probability of successful transmission, but its utility in the given application, most notably the timeliness. Regarding the first point, the agent is faced with a different selection of satellite passes each episode, each with their own set of pass characteristics and times at which they occur. This problem is solvable, as the modified Monte Carlo learning methods will allow keeping track of the estimated reward of each action.

2.4.4. Temporal Bounds for Packet Scheduling

Addressing the reward modeling, we apply the following reasoning. A good pass in an hour is not the same as an equally good pass occurring after 24 h because: (1) data needs to be transmitted regularly, (2) the circular queue holding data has a finite capacity, and (3) the Swarm modem will drop packets from its transmission queue after a timeout. Thus, to model he preference for more prompt transmissions, a *discount factor* λ is applied to reduce the value of later passes.

Under the data plan, each modem can transmit at most one packet per hour to remain within the budget. Since there are many satellite passes to consider, the rules are needed for the interval of packet scheduling. Such rules are shown in Figure 6, in which t_{min} and t_{max} are the minimum and maximum amount of time (in hours) for a satellite pass, respectively, $\mathbf{a}$ is a vector of satellite passes between t_{min} and t_{max} (a_i is the i-th element of $\mathbf{a}$). Further, $\mathbf{s}$ is a vector of states (i.e., pass characteristics and RF background noise) of satellite passes of $\mathbf{a}$. Let $\mathbf{t}$ be a vector of midpoint times of satellite passes of $\mathbf{a}$.

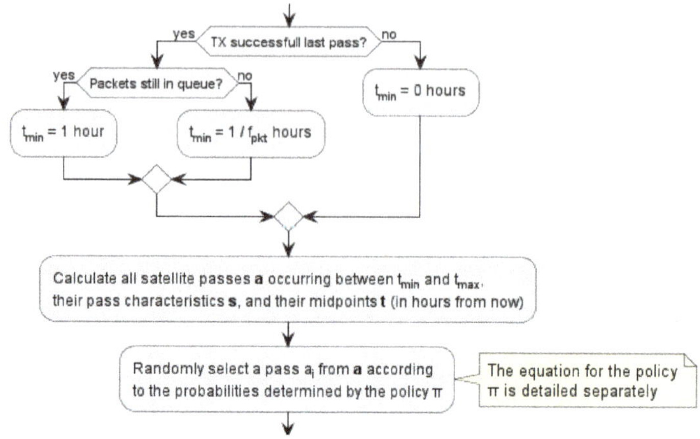

Figure 6. Routine for selecting the next satellite pass to attempt transmission.

2.4.5. Algorithmic Formulation

Since we are developing a learning approach to the LEO transmission scheduling, we will rely on the activation function for classification/learning, softmax. For the set of values $x_j, \{j, 1, N\}$, it is for each value x_i from as:

$$softmax(x_i) = \frac{exp(x_i)}{\sum_j exp(x_j)}. \tag{2}$$

Since softmax adds up to 1 across all inputs, it effectively creates a probability distribution function that disproportionally favors larger values of x_i.

Let r_{data} be the data point generation rate (in data points per hour), and *bundlesize* be the number of data points that comprise a full bundle. Then, the rate of full packet bundling r_{pkt} is: $r_{pkt} = \frac{r_{data}}{bundlesize}$. Let softmax($\mathbf{z}$) be the vectorized softmax function where softmax$(\mathbf{z})_i$ is the softmax of the i-th element of $\mathbf{z}$, and let $v(\mathbf{s})$ the vectorized value function. We express the policy π as:

$$\pi(a_i|s_i) = \text{softmax}(\lambda^{\mathbf{t} \ominus t_{min}} \odot v(\mathbf{s}))_i \tag{3}$$

where the $\ominus$ and $\odot$ symbols operating on vectors $\mathbf{t}$ and $\mathbf{s}$ are the element-wise subtraction and multiplication, respectively. Equation (3) expresses the probability of selecting a satellite pass a_i from interval $[t_{min}, t_{max}]$ as the softmax of the estimated transmission success probability for the pass, multiplied by a discount factor for future passes. Pass quality and promptness will be prioritized, while still giving a chance for exploration of passes currently predicted to be worse. This preference allows Monte Carlo learning to improve the value function estimates with time.

2.5. Uplink Transmission Energy Model

Energy consumption modeling of communication interfaces is a complex issue, and we have relied on the existing Iridium satellite communication model [21], as well as a model for long-range terrestrial network Sigfox [22], as the closest detailed model that similarly to us relies on the published energy consumption values from the datasheets. To determine the average power consumption, we introduce the unified uplink transmission energy model. Since the stochastic nature of transmission success prohibits the derivation of a deterministic model, a probabilistic model is created to give an estimate of average power consumption. There are two key causes of transmission non-determinism: (1) whether a transmission will succeed for a given pass, and (2) if it does succeed, how long the modem will be in Receive Mode before it is able to transmit.

To build the model, let t_{SL} be the mean time that the modem is in Sleep Mode, t_{GPS} be the mean time the modem is in GPS Acquisition Mode, and t_{RX} be the mean time the modem is in Receive Mode before transmission is successful. With typical modem power consumption values P_{SL}, P_{GPS}, and P_{RX}, the total energy usage in these modes over a single transmit attempt cycle, $E_{attempt}$ is:

$$E_{attempt} = P_{SL}t_{SL} + P_{GPS}t_{GPS} + P_{RX}t_{RX} + E_{TX}N_{pkt} \tag{4}$$

where N_{pkt} is the number of packets transmitted in a given pass. Depending on satellite pass selection and/or previous transmission attempt successes, N_{pkt} may be 1 or larger. In the case of an unsuccessful attempt, N_{pkt} is 0. An expression for non-zero N_{pkt} is:

$$N_{pkt} = \frac{r_{pkt}}{p_{success}r_{attempt}} \tag{5}$$

where $p_{success}$ is the transmission success probability, $r_{attempt}$ is the mean transmission attempt rate, and r_{pkt} is the rate at which fully bundled packets are generated. Since $r_{attempt}$ is smaller or equal to r_{pkt}, N_{pkt} is guaranteed to be 1 or greater because successful transmission of one packet entails successful transmission of all queued packets.

In Equation (4), also note that, while P_{SL}, P_{GPS}, t_{GPS}, P_{RX}, and E_{TX} (at least for full packets) are constant, t_{SL} and t_{RX} are variable. Here, t_{SL} represents the mean time in the Sleep Mode before making a transmission attempt, approximated as: $t_{SL} = \frac{1}{r_{attempt}}$.

The value of t_{RX} depends on how long the modem waits until it receives a packet and begins the transmission, or the pass is over. For a successful transmission, the quickest case is to transmit immediately after exiting GPS Acquisition Mode. The slowest success case is to transmit at the very end of the satellite pass. The worst failure case is the modem reaching the end of a given satellite pass in Receive Mode, with no transmission. In terms of t_{RX}, this case and successful transmission at the very end of the pass would be approximately equal. All three cases depend on the mean pass duration, denoted as t_{pass}.

$E_{attempt}$ can take two forms, depending on the transmission attempt success. A success is expressed in Equation (6), where ϵ_{pass} represents the proportion of a satellite pass spent in Receive Mode before receiving a packet from the satellite and is able to transmit. For pessimistic and optimistic models, ϵ_{pass} can be treated as either 1 or 0, as these serve as the upper and lower bounds of the time in Receive Mode for a given satellite pass.

$$E_{success} = P_{SL}\frac{1}{r_{attempt}} + P_{GPS}t_{GPS} + \epsilon_{pass}P_{RX}t_{pass} + E_{TX}\frac{r_{pkt}}{p_{success}r_{attempt}} \tag{6}$$

If the attempt is a failure, the model is represented by Equation (7). Note that there is no ϵ_{pass} value and no N_{pkt}, as the system will wait out a full pass without transmissions.

$$E_{fail} = P_{SL}\frac{1}{r_{attempt}} + P_{GPS}t_{GPS} + P_{RX}t_{pass} \tag{7}$$

The above two cases can be combined into a complete model: $E_{attempt} = p_{success}E_{success} + (1 - p_{success})E_{fail}$, which expands into the following expression:

$$\begin{aligned} E_{attempt} = \frac{P_{SL}}{r_{attempt}} + P_{GPS}t_{GPS} + p_{success}(\epsilon_{pass}P_{RX}t_{pass} + \frac{E_{TX}r_{pkt}}{p_{success}r_{attempt}}) \\ +(1 - p_{success})P_{RX}t_{pass} \end{aligned} \tag{8}$$

where P_{SL}, P_{GPS}, and P_{RX} are all constants and given by Swarm. Similarly, location fix time t_{GPS} is rather constant, reported to be about 30 s by Swarm. Furthermore, note that $r_{attempt}$ depends on site conditions, project requirements, and packet scheduling. Similarly, $p_{success}$ and t_{pass} depend heavily on site conditions and packet scheduling.

2.6. Simulation Model for Online Learning Evaluation

Setting up a number of sensors in the representative environment is expensive in time and money. Instead, the algorithm is tested with a simulated environment, similar to the methodology chosen in previous work on indirect-to-satellite scheduling [11]. Using simulations first can demonstrate the ability of the algorithm to learn underlying unknown patterns about satellite pass qualityand tune the discount factor λ parameter. To simulate the algorithm, two key components are needed: (1) virtual transmitters with an underlying probability model for which pass qualities are likely to result in transmission, and (2) randomly generated satellite pass characteristics and RF noise data. For virtual transmitters, three conceptual preference models were created, Table 5, to see how different transmitting obstacles would affect the algorithm. Note that the preferences in Table 5 refer to the conditions required for a high likelihood of success. For example, the first preference model requires high angles, long durations, and low noise for a high likelihood of success.

Table 5. Conceptual preference models for the virtual transmitters.

Model	Elevation Angle	Pass Duration	RF Background Noise
1	High angles	Long time	Low noise
2	Mid to high angles	Mid to long time	Low to mid noise
3	Low to high angles	Short to long time	Low to high noise

To create the virtual transmitter models, a function is constructed that outputs a transmission success probability by multiplying three stretched-and-shifted sigmoid (theshold activation) curves, one for each of three preference variables from Table 5. For example, the sigmoid to represent a preference for high angles would produce a value close to 1 for high angles (e.g., 70 degrees and higher) but a value close to 0 for low angles (e.g., 30 degrees and lower). The general form of the preference models is shown in Equation (9).

$$P(success) = \sigma(k_\theta(\theta - \theta_0)) \times \sigma(k_d(d - d_0)) \times \sigma(k_\gamma(\gamma - \gamma_0)) \tag{9}$$

where $\sigma(x)$ represents the sigmoid function, θ represents the max elevation angle, d represents the pass duration, and γ represents the RF background noise. Note that k_θ, k_d, k_γ, θ_0, d_0, and γ_0 represent configurable stretching and threshold shifting constants to represent the different conceptual preference models. The values of these constants used to create the three preference models by Equation 9 are shown in Table 6.

Table 6. Constants for the three preference models.

Preference Model	k_θ	θ_0	k_d	d_0	k_γ	γ_0
1	0.5	70	0.5	35	−1	−102
2	0.5	50	0.5	20	−1	−99
3	0.5	30	0.5	10	−1	−96

Simulated satellite passes are presented to virtual transmitters by agents imbued with a preference model and a value function approximator. The generated pass characteristics are randomly generated: each agent is exposed to random RF background noise, a vector **a** of satellite passes with corresponding random midpoint times **t**, and random pass characteristics **s** (except each $s_i \in \mathbf{s}$ also includes the RF noise value). The randomly generated pass characteristics are drawn from a uniform distribution, and the RF background noise values are drawn from two differing distributions:

1. Uniform across all buckets (−107 to −93 dBm).
2. Uniform within one bucket (−107 to −105 dBm).

to express that a given sensor may experience either a full range of RF noise, or (as expected in a remote location) a narrow sub-range of RF noise.

Each agent calculates the probabilities of selecting satellite passes from the discretized states, agents' value function approximators, and the pass midpoints. These probabilities are calculated from the policy π and satellite passes are chosen by these probabilities. Satellite pass and the RF noise characteristics are used by the agents' preference models for transmission success probabilities. Finally, transmission successes are determined according to the agent preference model outputs, and the process repeats.

3. Findings

This section summarizes the findings on the suitability of the proposed on-line learning to adapt in different scenarios, on tuning of the parameters, as well as the overall energy savings and the parameters tradeoffs. For evaluation of the transmission scheduling, the goal is to (1) demonstrate the ability of the algorithm to learn patterns behind transmission success probabilities, and (2) to make apparent how the performance gets affected by the main variables, such as the overpass duration, angle and the RF noise.

Simulations were conducted for the proposed algorithm under all three preference models in Table 5. The power data published by Swarm, as per Table 1 forms the basis of the transmitter energy consumption. The effects of random noise were expressed in two ways: noise contained in a single bucket, or across all buckets from Table 4, to model the remote and populated locations, respectively. Figures 7–12 show the transmission success and and average time to transmit for three preference models, as training epochs(i.e., the number of times that the learning steps are applied) progress, all parameterized by the discount factor λ. For their evaluation, notable is the response to overall RF noise distribution and transmission difficulty. We observe that wider RF noise distributions negatively impact the success, as well as when it is relatively hard (or easy) to transmit, there is less room for the algorithm to make huge improvements in success rate. For example, when there are few good passes, the algorithm often has to make a choice between a mediocre and a bad pass. This is seen in the lower (but still significant) improvement for the first preference model, and lower TX success rates, as there is simply no room for much improvement.

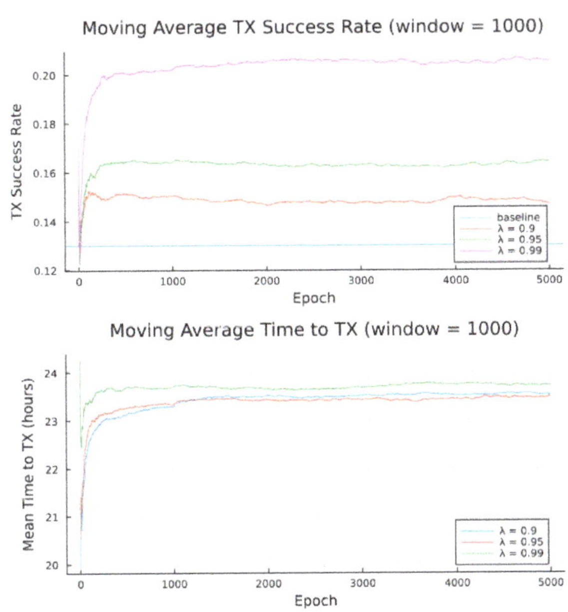

Figure 7. Simulated results for **preference model 1** and random noise within **1 bucket**.

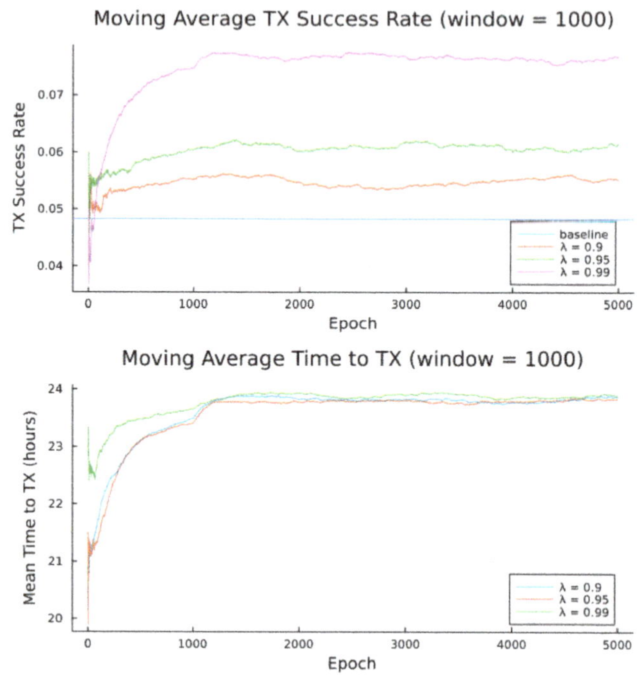

Figure 8. Simulated results for **preference model 1** and random noise across **all buckets**.

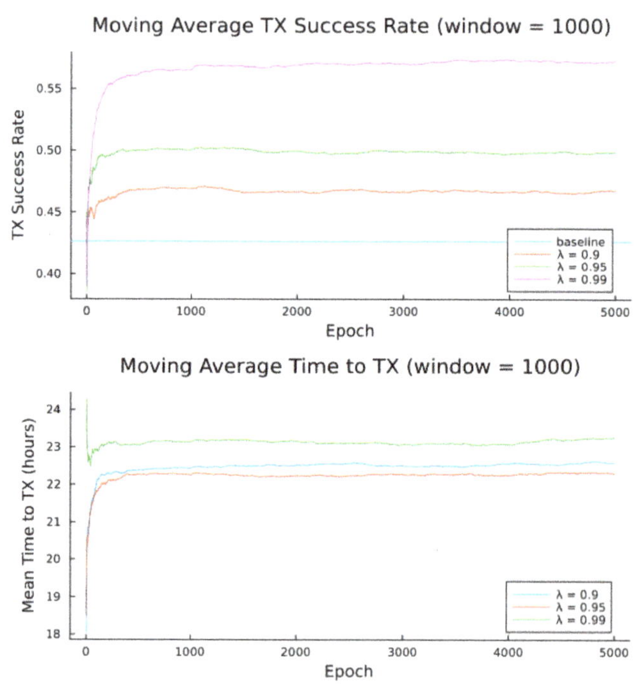

Figure 9. Simulated results for **preference model 2** and random noise within **1 bucket**.

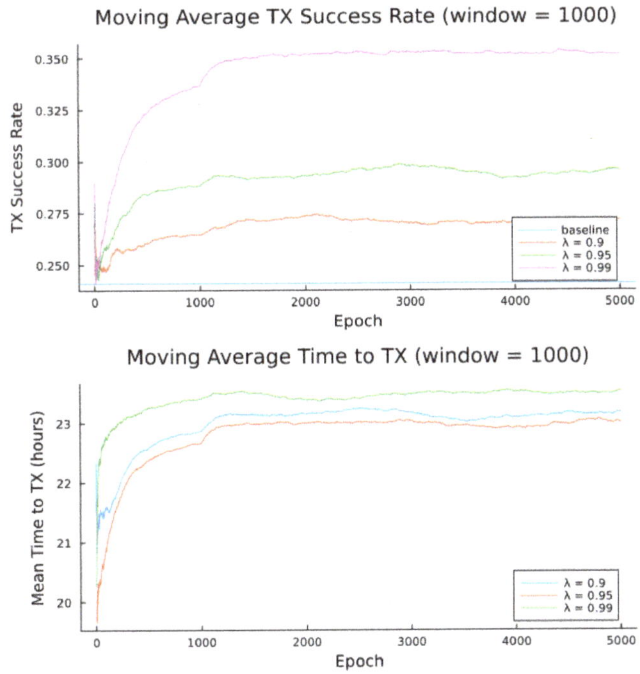

Figure 10. Simulated results for **preference model 2** and random noise across **all buckets**.

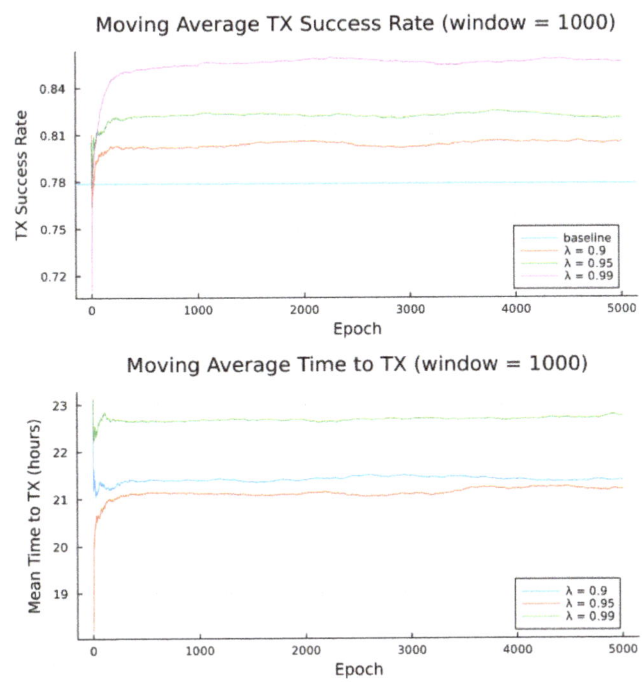

Figure 11. Simulated results for **preference model 3** and random noise within **1 bucket**.

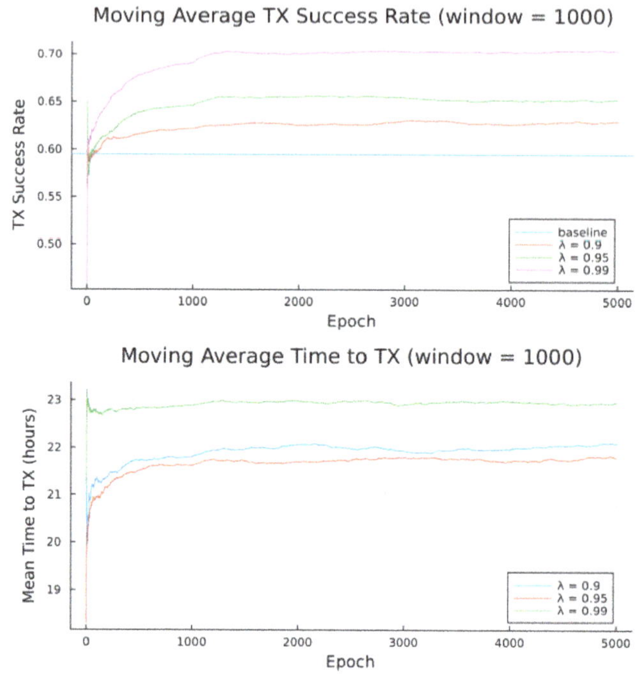

Figure 12. Simulated results for **preference model 3** and random noise across **all buckets**.

Second to note is the difference in response to discount factors λ. In the first preference model, less likely to succeed, differing discount factors made little difference to $r_{attempt}$, as the penalty for long wait for a decent pass outweigh the cost of other poorer options. For the other two preference models, however, the discount factors closer to 1 did see significantly higher average times to attempt transmission. For certain applications, an average time to transmit of 24 h may be unacceptable. If it is necessary to keep $r_{attempt}$ higher, using a lower value of t_{max} or a value of λ closer to 0 would lower the average time to transmit. A lower value of t_{max} in particular forces the algorithm to only consider satellite passes within a more constrained time frame.

Regarding noise distribution, Figures 7–12 also compare two noise models, where the RF noise values were drawn uniformly from all noise buckets or from one bucket only. This second model was explored as remote sites have largely consistent RF background noise levels. The primary impact is that the algorithm consistently learns faster and converges to higher success rates when exposed to RF noise associated with remote locations. For example, the algorithm learned in 1000 epochs for preference model 2 and fully random noise what it learned in under 300 epochs for same-bucket noise.

The most permissive conceptual preference model 3, Figures 11 and 12 shows that a high baseline success rate is rapidly improved through fewer epochs of training, especially as RF noise is constrained in a single bucket, as the expectation for remote, wilderness areas noise is not to occupy all buckets. Not only that the learning will be successfull, but the results are interpretable based on the understanding that the wireless channels with less contention perform better than those found in populated areas.

Power and Energy Consumption

We now evaluate expected transmission energy savings. Energy spent for each transmission attempt $E_{attempt}$, (Equation (8)) employs parameter values listed in Table 7. They are derived from Table 1 obtained from the datasheet values. While these values may depend on exact hardware setup (e.g., voltage supply), they are all constant, as opposed to variable terms, as detailed next.

Table 7. Constants used for transmission attempt energy model.

Constant	Meaning	Value
P_{SL}	Sleep Power	550 μ W
P_{GPS}	GPS Power	230 mW
t_{GPS}	Time to GPS Fix	30 s
P_{RX}	Receive Power	130 mW
E_{TX}	Transmit Energy	12.24 J
r_{pkt}	Packet Rate	$\frac{1}{3}$ h^{-1}

The ranges of the variables in Equation (8) are shown in Table 8. Note that "pessimistic" refers to the boundary of the interval that results in higher average consumption $E_{attempt}$, while "optimistic" refers to the boundary of the interval with lower average consumption. A value denoted as optimistic or pessimistic does not necessarily mean a value judgement for system operation. For example, a lower $r_{attempt}$ means less frequent transmission attempts, which is good for energy consumption, but causes high average time to transmit and lost data. Furthermore, note the difference between average energy consumption and the value for $E_{attempt}$ given by Equation (8); while a low $r_{attempt}$ will result in a higher $E_{attempt}$, it will result in lower average power, as shown by Equation (10) below. The term in the denominator is the average time elapsed during a complete cycle.

$$P_{avg} = \frac{E_{attempt}}{\frac{1}{r_{attempt}} + t_{GPS} + p_{success}\epsilon_{pass}t_{pass} + (1 - p_{success})t_{pass}} \quad (10)$$

The results of these four variables on average modem power consumption are shown in Figure 13. The two dominant factors in determining average power consumption are the success rate and average time between attempts, while the smaller proportion of time ϵ_{pass} idled in Receive mode helps to keep the power low irrespective of other variables.

Table 8. Sample variable ranges for transmission attempt energy model.

Parameter	Variable	Pessimistic Value	Optimistic Value
Attempt rate	$r_{attempt}$	1 h^{-1}	$\frac{1}{48}$ h^{-1}
Success probability	$p_{success}$	0.0	1.0
Portion of time in read mode	ϵ_{pass}	1.0	0.0
Overpass duration	t_{pass}	60 min	10 min

The impact of the success rates on power and battery requirements is shown in Table 9 for the three preference models. Note that the $p_{success}$ and $r_{attempt}$ values are taken from the simulations, and ϵ_{pass} and t_{pass} are taken as 0.5 and 25 min, respectively. For each preference model, three results are shown: (1) a baseline based on taking the earliest available passes and no scheduling, (2) another baseline based on the same average $r_{attempt}$ as the simulated results but no scheduling, and (3) the test case with scheduling and simulated average $r_{attempt}$.

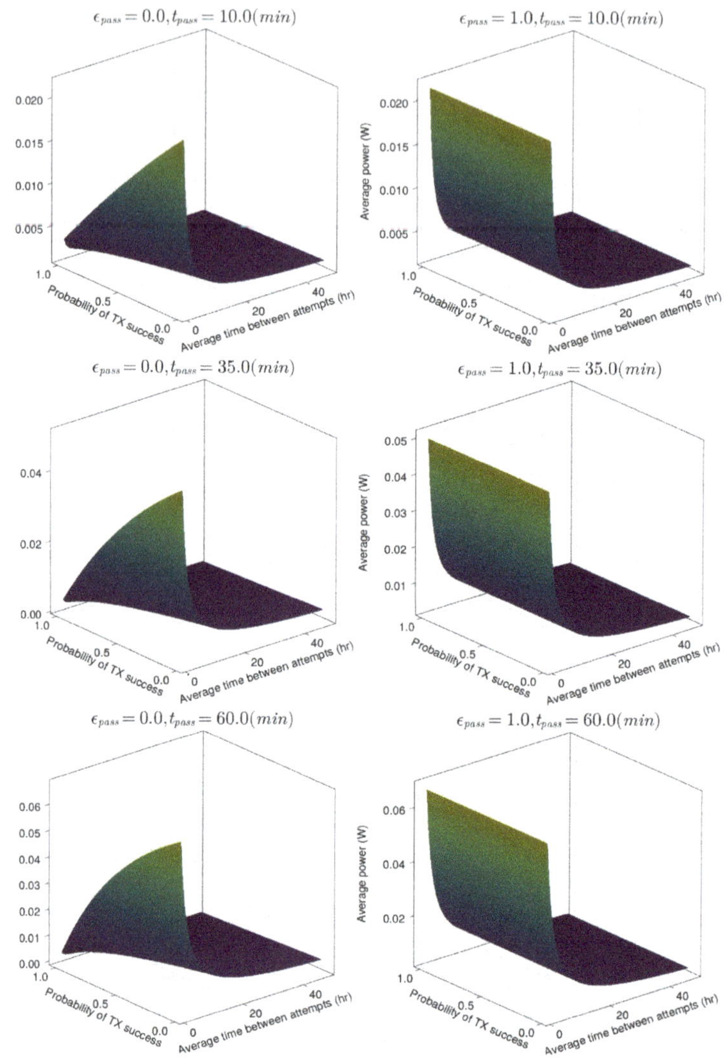

Figure 13. Average power of the Swarm modem under different pass variable values.

Table 9. Sample energy savings from simulated packet scheduling for a year of operation

Preference Model	Success Prob. $p_{success}$	Attempt Rate $r_{attempt}$	Average Power	Required Battery Capacity
1	0.13	2.564 h^{-1}	67.54 mW	592.1 Wh
	0.13	$\frac{1}{24}$ h^{-1}	3.810 mW	33.40 Wh
	0.20	$\frac{1}{24}$ h^{-1}	3.735 mW	32.74 Wh
2	0.42	0.7937 h^{-1}	29.31 mW	256.9 Wh
	0.42	$\frac{1}{23}$ h^{-1}	3.575 mW	31.34 Wh
	0.57	$\frac{1}{23}$ h^{-1}	3.405 mW	29.85 Wh
3	0.78	0.4274 h^{-1}	14.95 mW	131.1 Wh
	0.78	$\frac{1}{22}$ h^{-1}	3.234 mW	28.35 Wh
	0.85	$\frac{1}{22}$ h^{-1}	3.151 mW	27.62 Wh

For simulated baseline values, the average time to attempt is represented as the first available satellite pass, expressed by

$$t_{SL,baseline} = \frac{p_{success}}{r_{pkt}} \Rightarrow r_{attempt,baseline} = \frac{r_{pkt}}{p_{success}},$$

which represents the algorithm for determining t_{min}, where successful transmission lead to waiting a minimum of $\frac{1}{r_{pkt}}$ hours, while unsuccessful attempts lead to no minimum wait.

The results demonstrate that online learning direct-to-satellite packet scheduling is capable of reducing average power and battery requirements around 20 times. Note that the dominant energy saving comes from reducing average attempt frequency, although the improved success rate of the scheduling algorithm introduce significant power saving. Additionally, our direct-to-satellite packet scheduling scheme enabled us to lower the attempt frequency, as it provides a built-in mechanism for selecting future passes.

Tradeoff between Low Power and Learning Rate

The proposed algorithm in its current state exhibits low average attempt rate values $r_{attempt}$ (still sufficiently high for the intended application). The discount factor parameter λ is investigated in Figures 14–16, as it impacts the transmission attempt rate and thus the time a node takes to learn, as well as the time to transmit and the modem power.

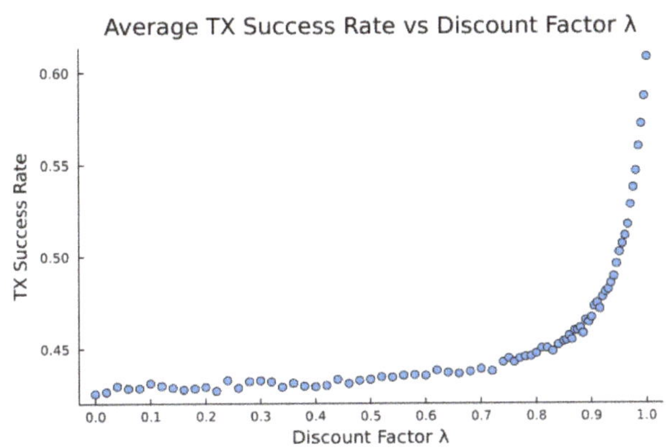

Figure 14. Average success rate versus discount factor λ for bucketed noise and moderate preference model 2.

Figures 14 and 15 show that with low values of λ, future candidate passes are all discounted to such a degree that the resultant probabilities from the softmax function show little to no preference for earlier passes. Furthermore, with high values of λ (very close to 1), we observe that future candidate passes are all so little discounted that the resultant probabilities from the softmax function show little to no preference for earlier passes. It is only for λ values from around 0.9 to 0.95 ("sweet spot") that the values of future candidate passes are discounted such that there is an apparent preference for earlier passes, such that the transmission success and time to transmit are more favorable.

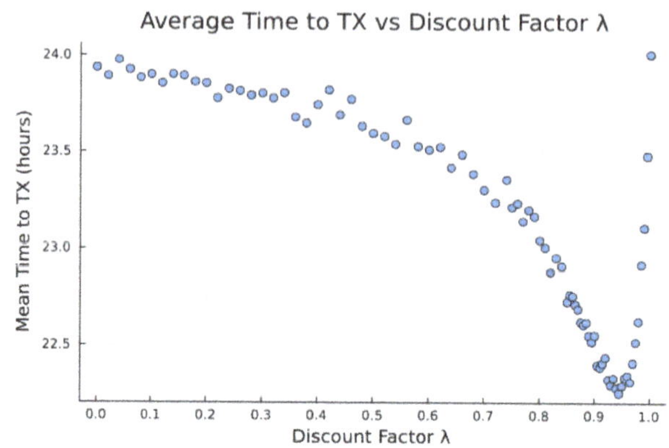

Figure 15. Average time to attempt transmission versus discount factor λ for bucketed noise and moderate preference model 2.

Earlier passes means more frequent transmission attempts. A downside to this is in smaller expected power savings, as more frequent transmissions require significantly more average power. There is thus a tradeoff between having a high learning rate and achieving low average power consumption, Figures 14–16, with the "sweet spot" area of λ selected for all earlier experiments, e.g., Table 9.

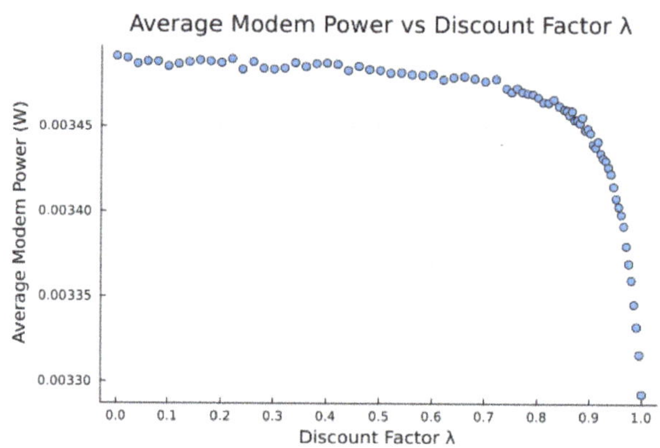

Figure 16. Average modem power versus discount factor λ for bucketed noise and moderate preference model 2, using $\epsilon_{pass} = 0.5$, $r_{attempt} = \frac{1}{24}$ hr^{-1}, and $t_{pass} = 25$ min.

4. Conclusions and Future Work

We have presented and evaluated the transmission scheduling for emerging sparse LEO satellite services suitable for IoT. This is the first published study that addresses the critical uplink availability issue with sparse LEO constellation. A detailed probabilistic energy consumption model was developed used to evaluate our on-line learning scheme for predicting transmission periods. Our learning proposal is inexpensive computationally, learns in small increments and in a modest number of training epochs, and is interpretable, unlike most modern machine learning approaches. For three common scenarios, we have observed up to 20-fold reduction in power and battery requirements of the transmissions attributable to this learning approach. Since the whole scheme is suitable for the smallest embedded microcontrollers, we have demonstrated its implementation on Raspberry Pi Pico that manages to pack all sensed data within the least expensive dataplan.

While the intended application is the water-level monitoring in remote locations, the proposed scheme is practical for other IoRT applications, as it incurs modest computing costs. The results demonstrate the ability of the proposal to facilitate energy-efficient data collection over prolonged periods of time.

The whole analysis was done by simulations to reduce the burden in cost, time and also to combat location-dependence, since RF noise in the city differs from the intended remote locations; while we demonstrate the ability of our proposed scheme to adapt appropriately, obtain interpretable learning and explore the impact of variables on the performance of the algorithm, they are nonetheless simulated results.

Complete code and data for modeling, learning and energy consumption evaluation is made available by GitHub (as reported below in "Data Availability" statement). The code and data can be readily used for further exploration of sparse LEO connectivity for IoT in general.

4.1. Future Work

Related to the learning approach, further research could examine the possibility of adaptive learning rates. One possibility for polar regions would be increasing the transmission attempt frequency (e.g., by decreasing the discount factor λ) during the summer when solar power is abundant, and lowering the transmission attempt frequency during the polar winters.

There are two natural steps that could be taken to improve the simulation setup in future: (1) the simulations could be based on extracted real-world data for a given application, or (2) multiple sensors could be placed in the field for weeks or months. The first option is natural, as the risk of needing to adjust the algorithm is high, making tuning in simulations and testing on hardware only when confident in good results is the best route.

4.1.1. Potential Simulation Model Improvements

The most apparent way to improve the simulations is to generate a more representative RF noise distribution and to use the Swarm satellite pass information, something which is not available in the remote locations. For the latter, there are the versions of the online satellite pass library, as used in the open source SGP4 Arduino library or in Python [23,24].

4.1.2. RF Noise Impact on Prototype System

One factor discovered in verification testing of a prototype is the sensitivity of the antenna to noise. For example, minor changes in the exact positioning of the PCB and microcontroller under the ground plane could vary the measured noise by as much as 3 dB. Clearly, the antenna receives RF interference from an unshielded device in the immediate vicinity (e.g., 10 to 20 cm). In fact, it was precisely this interference that played a role in the design of a online learning direct-to-satellite packet scheduling algorithm; since no method could possibly account for the range of possible housing, ground plane, and even power supply designs, as well as varied site conditions and lines of sight, there could be no single ultimate "good" pass model, so a learning-based approach is used.

Author Contributions: Conceptualization, G.K. and D.P.; methodology, G.K..; software, G.K..; validation, G.K.; formal analysis, G.K..; investigation, G.K., Ž.Ž. and D.P.; resources, Ž.Ž. and D.P.; writing—original draft preparation, G.K. and Ž.Ž.; writing—review and editing, G.K., Ž.Ž. and D.P.; visualization, G.K., Ž.Ž. and D.P.; supervision, Ž.Ž. and D.P. All authors have read and agreed to the published version of the manuscript.

Funding: This research was funded by NSERC Canada Discovery Grant.

Institutional Review Board Statement: Not applicable.

Informed Consent Statement: Not applicable.

Data Availability Statement: The data used in this study was generated by a simulation program placed in the repository: https://github.com/garrettkinman/Self-Learning-Satellite-Pass-Selection (accessed on 31 May 2023).

Conflicts of Interest: The authors declare no conflict of interest.

Abbreviations

GNSS	Global Navigational Satellite System
GNSS-IR	GNSS Interferometric Reflectometry
LEO	Low Earth Orbit
IoRT	Internet of Remote Things
PCB	Printed Circuit Board
MC	Monte Carlo
RSSI	Received Signal Strength Intensity

References

1. Pörtner, H.O.; Roberts, D.; Masson-Delmotte, V.; Zhai, P.; Tignor, M.; Poloczanska, E.; Mintenbeck, K.; Alegría, A.; Nicolai, M.; Okem, A.; et al. Summary for policymakers. In *IPCC Special Report on the Ocean and Cryosphere in a Changing Climate*; Intergovernmental Panel on Climate Change: Paris, France, 2019; Volume 7.
2. Carotenuto, F.; Brilli, L.; Gioli, B.; Gualtieri, G.; Vagnoli, C.; Mazzola, M.; Viola, A.P.; Vitale, V.; Severi, M.; Traversi, R.; et al. Long-Term Performance Assessment of Low-Cost Atmospheric Sensors in the Arctic Environment. *Sensors* **2020**, *20*, 1919. [CrossRef] [PubMed]
3. Purnell, D.J.; Gomez, N.; Minarik, W.; Porter, D.; Langston, G. Precise water level measurements using low-cost GNSS antenna arrays. *Earth Surf. Dyn.* **2021**, *9*, 673–685. [CrossRef]
4. Karegar, M.A.; Kusche, J.; Geremia-Nievinski, F.; Larson, K.M. Raspberry Pi Reflector (RPR): A Low-Cost Water-Level Monitoring System Based on GNSS Interferometric Reflectometry. *Water Resour. Res.* **2022**, *58*, e2021WR031713. [CrossRef]
5. Mitrovica, J.X.; Hay, C.C.; Kopp, R.E.; Harig, C.; Latychev, K. Quantifying the Sensitivity of Sea Level Change in Coastal Localities to the Geometry of Polar Ice Mass Flux. *J. Clim.* **2018**, *31*, 3701–3709. .: 10.1175/JCLI-D-17-0465.1. [CrossRef]
6. Coulson, S.; Dangendorf, S.; Mitrovica, J.X.; Tamisiea, M.E.; Pan, L.; Sandwell, D.T. A detection of the sea level fingerprint of Greenland Ice Sheet melt. *Science* **2022**, *377*, 1550–1554. [CrossRef] [PubMed]
7. Sanctis, M.D.; Cianca, E.; Araniti, G.; Bisio, I.; Prasad, R. Satellite Communications Supporting Internet of Remote Things. *IEEE Internet Things J.* **2016**, *3*, 113–123. [CrossRef]
8. Centenaro, M.; Costa, C.E.; Granelli, F.; Sacchi, C.; Vangelista, L. A Survey on Technologies, Standards and Open Challenges in Satellite IoT. *IEEE Commun. Surv. Tutorials* **2021**, *23*, 1693–1720. [CrossRef]
9. Jia, Z.; Sheng, M.; Li, J.; Niyato, D.T.; Han, Z. LEO-Satellite-Assisted UAV: Joint Trajectory and Data Collection for Internet of Remote Things in 6G Aerial Access Networks. *IEEE Internet Things J.* **2021**, *8*, 9814–9826. [CrossRef]
10. Fraire, J.A.; Umaña, S.C.; Accettura, N. Direct-To-Satellite IoT—A Survey of the State of the Art and Future Research Perspectives—Backhauling the IoT Through LEO Satellites. In Proceedings of the ADHOC-NOW, Luxembourg, 1–3 October 2019.
11. Huang, H.; Guo, S.; Liang, W.; Wang, K. Online Green Data Gathering from Geo-Distributed IoT Networks via LEO Satellites. In Proceedings of the 2018 IEEE International Conference on Communications (ICC), Kansas City, MO, USA, 20–24 May 2018; pp. 1–6.
12. Palma, D.; Birkeland, R. Enabling the Internet of Arctic Things With Freely-Drifting Small-Satellite Swarms. *IEEE Access* **2018**, *6*, 71435–71443. [CrossRef]
13. Capez, G.M.; Henh, S.; Fraire, J.A.; Garello, R. Sparse Satellite Constellation Design for Global and Regional Direct-to-Satellite IoT Services. *IEEE Trans. Aerosp. Electron. Syst.* **2022**, *58*, 3786–3801. [CrossRef]
14. Qi, Y.; Pan, L.; Liu, S. A Lyapunov optimization-based online scheduling algorithm for service provisioning in cloud computing. *Future Gener. Comput. Syst.* **2022**, *134*, 40–52. [CrossRef]
15. Chiang, M.W.; Zilic, Z.; Radecka, K.; Chenard, J.S. Architectures of increased availability wireless sensor network nodes. In Proceedings of the 2004 International Conference on Test, Charlotte, NC, USA, 26–28 October 2004; pp. 1232–1241.
16. Wielandt, S.; Dafflon, B. Minimizing Power Consumption in Networks of Environmental Sensor Arrays using TDD LoRa and Delta Encoding. In Proceedings of the 2021 55th Asilomar Conference on Signals, Systems, and Computers, Pacific Grove, CA, USA, 31 October–3 November 2021; pp. 318–323.
17. Marcinkevics, R.; Vogt, J.E. Interpretability and Explainability: A Machine Learning Zoo Mini-tour. *arXiv* **2020**, arXiv:2012.01805.
18. Silver, D. Lectures on Reinforcement Learning. 2015. Available online: https://www.davidsilver.uk/teaching/ (accessed on 31 May 2023).
19. Sutton, R.; Barto, A. *Reinforcement Learning, Second Edition: An Introduction*; Adaptive Computation and Machine Learning Series; MIT Press: Cambridge, MA, USA, 2018.
20. Kuleshov, V.; Precup, D. Algorithms for multi-armed bandit problems. *arXiv* **2014**, arXiv:1402.6028.
21. Gomez, C.; Darroudi, S.M.; Naranjo, H.; Paradells, J. On the Energy Performance of Iridium Satellite IoT Technology. *Sensors* **2021**, *21*, 7235. [CrossRef] [PubMed]
22. Gomez, C.; Veras, J.C.; Ferré, R.V.; Casals, L.; Aspas, J.P. A Sigfox Energy Consumption Model. *Sensors* **2019**, *19*, 681. [CrossRef] [PubMed]

23. Hopperpop. Library for Calculating Satellites Positions and Predicting overpasses. Available online: https://github.com/Hopperpop/Sgp4-Library (accessed on 31 May 2023).
24. Rhodes, B. Python Version of the SGP4 Satellite Position Library. Available online: https://pypi.org/project/passpredict/ (accessed on 31 May 2023).

sensors

MDPI

Article

Energy-Efficient Decentralized Broadcasting in Wireless Multi-Hop Networks †

Artur Sterz [1,*,‡], Robin Klose [2,‡], Markus Sommer [1], Jonas Höchst [1], Jakob Link [2], Bernd Simon [3], Anja Klein [3], Matthias Hollick [2] and Bernd Freisleben [1]

1 Department of Mathematics & Computer Science, Philipps-Universität Marburg, 35043 Marburg, Germany; msommer@informatik.uni-marburg.de (M.S.); hoechst@informatik.uni-marburg.de (J.H.); freisleb@informatik.uni-marburg.de (B.F.)
2 Department of Computer Science, Technical University of Darmstadt, 64289 Darmstadt, Germany; rklose@seemoo.de (R.K.); jlink@seemoo.de (J.L.); mhollick@seemoo.de (M.H.)
3 Department of Electrical Engineering & Information Technology, Technical University of Darmstadt, 64283 Darmstadt, Germany; b.simon@nt.tu-darmstadt.de (B.S.); a.klein@nt.tu-darmstadt.de (A.K.)
* Correspondence: sterz@informatik.uni-marburg.de

† This paper is an extended version of our paper published in the 48th IEEE Conference on Local Computer Networks, Daytona Beach, FL, USA, 1–5 October 2023 (accepted for publication).
‡ These authors contributed equally to this work.

Abstract: Several areas of wireless networking, such as wireless sensor networks or the Internet of Things, require application data to be distributed to multiple receivers in an area beyond the transmission range of a single node. This can be achieved by using the wireless medium's broadcast property when retransmitting data. Due to the energy constraints of typical wireless devices, a broadcasting scheme that consumes as little energy as possible is highly desirable. In this article, we present a novel multi-hop data dissemination protocol called BTP. It uses a game-theoretical model to construct a spanning tree in a decentralized manner to minimize the total energy consumption of a network by minimizing the transmission power of each node. Although BTP is based on a game-theoretical model, it neither requires information exchange between distant nodes nor time synchronization during its operation, and it inhibits graph cycles effectively. The protocol is evaluated in Matlab and NS-3 simulations and through real-world implementation on a testbed of 75 Raspberry Pis. The evaluation conducted shows that our proposed protocol can achieve a total energy reduction of up to 90% compared to a simple broadcast protocol in real-world experiments.

Keywords: wireless networks; data dissemination; broadcast tree

Citation: Sterz, A.; Klose, R.; Sommer, M.; Höchst, J.; Link, J.; Simon, B.; Klein, A.; Hollick, M.; Freisleben, B. Energy-Efficient Decentralized Broadcasting in Wireless Multi-Hop Networks. *Sensors* **2023**, *23*, 7419. https://doi.org/10.3390/s23177419

Academic Editor: Josip Lorincz

Received: 29 July 2023
Revised: 21 August 2023
Accepted: 22 August 2023
Published: 25 August 2023

1. Introduction

In several areas of wireless networking, such as wireless sensor networks (WSN) or the Internet of Things (IoT), application data need to be disseminated across multiple devices to reach their destination. In particular, there are applications in which one device needs to disseminate data to all other nodes in the network, e.g., in tasks such as network configuration [1], update dissemination [2], or network diagnosis. Other typical examples are announcements, notifications, and event distribution [3].

To enable these applications, the wireless medium's broadcast property can be utilized. This means that multiple devices in the vicinity of the sender can receive a transmitted packet. Since the possible transmission range of a wireless device is limited by its maximum transmission power, a multi-hop data dissemination scheme is required, i.e., nodes in different locations retransmit the received packets to distribute the data over an extended area. Furthermore, since wireless network devices usually have a limited energy budget, only selected nodes should actively participate in data dissemination through retransmission, while the other nodes should be passive consumers of the received packets. Thus, the active nodes form a spanning tree that connects all nodes of the network to the root node. There are several combinations of nodes that can form a spanning tree, but the construction of an energy-minimal spanning tree is NP-hard [4]. Furthermore, to minimize the total energy consumption of a network in a real setting, the practical applicability of a spanning tree construction algorithm is important. However, although there are several spanning tree approaches in the literature, virtually none of them can be implemented in a real-world environment due to different shortcomings. Many approaches assume global knowledge of parameters or states of nodes, but such knowledge is not available without additional effort in practice. Most works do not consider the potential occurrence of graph cycles. Some studies assume a pre-existing spanning tree and omit the initial spanning tree construction phase. Others do not optimize the transmission power levels of the individual nodes, resulting in missed potential. Centralized approaches are generally not suitable for wireless ad hoc and multi-hop networks.

In this article, we present the Broadcast Tree Protocol (BTP), a novel multi-hop data dissemination protocol that constructs a spanning tree in a decentralized manner such that the total energy consumption in a network is minimized. This is achieved by letting each node connect to the broadcast tree in a way that minimizes its own contribution to the total energy consumption. To realize this approach, we first model the broadcast tree using a game-theoretical model (based on the work conducted by Mousavi et al. [5]) that has been proven to converge to a Nash Equilibrium, yielding better results than other approaches from the literature. However, the original model makes assumptions that render a practical implementation impossible. First, it assumes that every node is aware of the transmission power of its neighbors at all times. However, this assumption translates to complicated and expensive status updates between neighboring nodes. Second, the original model assumes that all nodes make their decisions at the same time. Achieving this in practice would require the precise time synchronization of all nodes, which, in itself, is a challenging task. Third, the potential game proposed by Mousavi et al. [5] uses a weakly dominant strategy where a node can switch between different parent nodes even if the energy costs that it incurs stay the same. This strategy can result in a ping-pong effect, i.e., a node may permanently switch between multiple parent nodes, while the overall transmission power of the network is not reduced further. Fourth, the original model assumes that the constructed broadcast tree does not have cycles. However, due to the lack of time synchronization of all nodes and the lack of global knowledge at each node, graph cycles cannot be easily avoided in practice. Therefore, we present an approximation of the original model that retains the convergence properties of the original model and can be implemented in real-world settings.

In particular, we make the following contributions:

1. We present BTP, a practical protocol that approximates a game-theoretical model for constructing an energy-minimal broadcast tree while preserving both its convergence and Nash Equilibrium properties. To the best of our knowledge, BTP is the first protocol for energy-efficient data dissemination in wireless multi-hop networks based on a provably optimal game-theoretical model that is implemented on real hardware.
2. We design and implement a discovery mechanism that allows the nodes in a wireless multi-hop network to construct a broadcast tree in a decentralized fashion using locally available information from their direct neighbors.
3. We change the decision strategy of the algorithm from a *weakly* dominant strategy to a *strictly* dominant strategy to avoid ping-pong effects, in which a node may potentially continue changing its decision without further reducing the transmission power.
4. We implement and evaluate three different algorithms for inhibiting graph cycles. Specifically, (1) the Path-to-Source algorithm avoids cycles by letting each node keep track of the path from the root to itself so that each node can check the consistency of the spanning tree when making decisions. (2) The Mutex algorithm avoids cycles by letting each node lock its sub-tree when connecting to a different parent node,

ensuring consistency at all times. (3) The Ping-to-Source algorithm allows for cycles temporarily, but it detects and resolves such cycles immediately.

5. We evaluate BTP using different tools to assess its feasibility under various conditions. First, we use Matlab simulations to compare BTP against approaches from the literature. Second, we perform NS-3 simulations to investigate the scalability of BTP. Third, we present a real-world implementation of BTP, that is evaluated on a testbed of 75 Raspberry Pis deployed in one of our university buildings to explore its practical feasibility. The evaluations show that BTP can achieve an energy reduction of up to 90% in real-world experiments compared to a simple broadcast protocol.
6. The code of the NS-3 implementation and the real-world implementation has been released under a permissive open-source license. Furthermore, all code required to reproduce the experiments as well as the experimental artifacts are also been made available.

This article is organized as follows. Section 2 discusses related work. Section 3 introduces our system model. Section 4 presents the game-theoretical model, followed by design of the formal protocol. Section 5 presents comparisons of BTP against other algorithms in Matlab simulations, while Section 6 explores the scalability of BTP in NS-3 simulations. Section 7 presents the implementation of BTP on Raspberry Pis and an evaluation through practical experiments. Section 8 concludes this article and outlines areas for future work.

2. Related Work

Several approaches have been proposed for designing energy-efficient broadcast trees. In one approach, the nodes broadcasted beacon packets with increasing transmission power to iteratively build a set of all their neighbors [6]. In another work, each node was connected to the node for which the minimal additional energy usage was required [7]. In yet another work, each node computed its optimal parent under a certain optimization objective, while the parent was on the path to the source node [8]. Energy accumulation to minimize the nodes' transmission power was also used [9], while others attempted to find a spanning tree that minimizes the number of transmissions and the transmission delay at the same time [10]. Also, clusters of trees have been built using different methodologies [11–13]. Several approaches use game theory to construct a spanning tree that requires minimal transmission power [14,15]. Finally, some authors have attempted to minimize the path lengths from the source to all other nodes to minimize energy consumption [16,17].

Furthermore, several centralized approaches have been proposed. For example, there are approaches in which sensor nodes send a beacon packet to a central controller that constructs the tree [18] or minimizes the path lengths of the tree [19].

Several approaches do not rely on a tree structure to disseminate data in wireless ad hoc networks but still attempt to minimize total energy consumption by, for example, proposing an energy-efficient sensor placement algorithm [20] or by letting a controller decide which nodes should transmit the data [21,22]. Other approaches involve the formulation of a linear programming problem that finds the most energy-efficient unicast path to reach all nodes [23], the proposal of a data-forwarding scheme that utilizes nodes' contextual information [24], or the use of a k-coverage algorithm to distribute data energy-efficiently in wireless underwater sensor networks [25].

Additionally, some approaches do not minimize energy consumption but optimize throughput [26], delivery probability [27], end-to-end latency [28–30], or fairness [31].

Furthermore, practical approaches with an implementation were either aiming not to minimize energy consumption but other metrics such as path length or latency [32,33] or were not based on a proven theoretical model [34–36].

Finally, apart from approaches that rely on topology control or spatial placement, energy efficiency can also be optimized through techniques like satellite communication [37,38], beamforming [39–41], or MU-MIMO [41]. However, these approaches are beyond the scope of our work since devices used in multi-hop wireless sensor networks are typically not equipped with hardware capable of using such techniques.

Compared to BTP, the aforementioned decentralized approaches suffer from various problems. Many approaches require global knowledge of parameters or the states of neighboring nodes and do not explicitly address the inhibition of graph cycles. Some works do not consider the initial tree construction phase but assume that an already-constructed tree exists, where, afterwards, only the transmission power is adjusted. Other approaches also squander their potential by not leveraging the possibility of adjusting the transmission power of each node. Furthermore, centralized approaches are usually not suitable in the area of wireless ad hoc and multi-hop networks. Spanning trees offer the possibility to reach all nodes in a network with a minimum amount of energy, making approaches that do not rely on a tree structure questionable in terms of minimizing the energy consumption. Finally, the presented approaches either do not provide an implementation or a real-world evaluation using off-the-shelf Wi-Fi devices or are not based on a proven theoretical model. In fact, most of the presented works only propose a theoretical model without considering its applicability. Some of these models cannot be implemented under real-world constraints.

3. System Model

Our system model is based on the assumption that a given source node contains data to send to all other nodes in a network. The nodes are spatially distributed, and each node has a single antenna and a maximum transmission power of $p^{\max}$. Due to the path loss properties of wireless transmissions, the source node may not be able to reach all other nodes, even if transmitting with $p^{\max}$. Therefore, a multi-hop transmission scheme is required. Table 1 lists the mathematical notations of our system model described below.

Table 1. Mathematical notations.

Notation	Description	Notation	Description
T	Broadcast tree consisting of nodes V and edges E	$v \in V$	A node of the broadcast tree T
$S \in V$	Source node of the broadcast tree	$e \in E$	An edge of the broadcast tree T
$p^{\max}$	Maximum possible Tx power	p_i	Tx power of node $i \in V$
$\lvert h_{i,j} \rvert^2$	Channel gain of $(i,j) \in E$	σ^2	Noise power
γ_j	SNR at node j	γ^{min}	Minimum required SNR
C_i	All children of node $i \in V$	N_k	All neighbours of node $k \in V$
$p_{i,j}$	Tx power that node $i \in V$ must use to reach node $j \in V$	$p_i(C_i)$	Tx power that node $i \in V$ must use to reach all its children C_i
p	Sum of Tx powers of all nodes in V		
$\mathcal{G}$	Potential game	$\mathcal{P}$	Set of rational players $V \setminus \{S\}$
$a_j^{(t)}$	Selected parent of destination node j at iteration t	$\boldsymbol{a}^{(t)}$	Action profile at iteration t
$\varphi_j\left(a_j^{(t)}\right)$	Cost function of destination node j for action $a_j^{(t)}$ at iteration t	$\mathcal{A}_j^{(t)}$	Set of possible actions (parents) for node j at iteration t

3.1. Graph Representation

A rooted spanning tree is an appropriate graph for a multi-hop broadcast scheme. In the remainder of this article, we call such a tree a *broadcast tree*. A broadcast tree is defined as a graph $T = (V, E)$, where the vertices V correspond to all nodes of the network, the source node $S \in V$ is the root of the broadcast tree, and the edges E are the connections between the nodes. Each edge $e \in E$ has a weight $p_{i,j}$ corresponding to the transmission power required to successfully establish a wireless communication link between the two nodes $\{i,j\} \in V_e$ connected via e. Furthermore, a broadcast tree must not contain any cycles, i.e., for a given broadcast tree $T = (V, E)$, there is no path of edges $(e_1, e_2, \ldots, e_n)$ with a vertex sequence $(v_1, v_2, \ldots, v_n, v_1)$.

In a broadcast tree, each node $j \in V \setminus \{S\}$ has exactly one parent i, whereas each parent may have multiple children. The set of children of a parent i is denoted as C_i, as indicated by the blue box in Figure 1. Further, we assume that all communication links are bi-directional. Due to the broadcast property of wireless communications, a parent node i has to send data only once, while all its children C_i should be able to receive it.

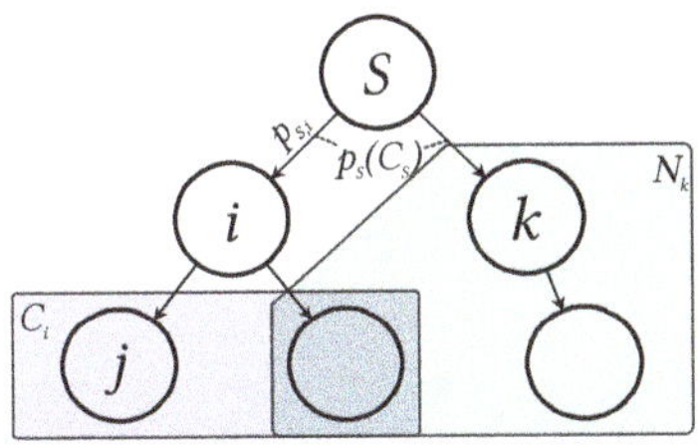

Figure 1. Broadcast tree overview.

3.2. Transmission Power Model

To establish a connection between a parent i and all its children C_i, node i has to transmit data with enough transmission power to ensure that the received signal strength exceeds a certain level at all of its children C_i. The signal-to-noise-ratio (SNR) γ_j at receiver $j \in C_i$ is:

$$\gamma_j = \frac{p_i |h_{i,j}|^2}{\sigma^2} \tag{1}$$

Here, p_i is the transmission power that parent i uses to send the data, $|h_{i,j}|^2$ is the channel gain, and σ^2 is the noise power. Furthermore, based on the minimally required SNR γ^{min}, the minimum transmission power that i must use to reach j is as follows:

$$p_{i,j} = \frac{\gamma^{min}\sigma^2}{|h_{i,j}|^2} \tag{2}$$

In Figure 1, this is visualized as $p_{S,i}$ on the connection between S and i. Furthermore, a node j that receives a signal from node i can calculate the required transmission power as follows:

$$p_{i,j} = \frac{p_i \gamma^{min}}{\gamma_j} \tag{3}$$

The transmission power that a parent node i must use to reach all its children C_i depends on its most distant child and is generally bound by $p^{\max}$:

$$p_i(C_i) = \max_{j \in C_i}(p_{i,j}) \leq p^{\max} \tag{4}$$

Note that both connections of S are marked with $p_S(C_S)$ in Figure 1 due to the broadcast property. The neighborhood of a parent k is defined in the following manner:

$$N_k = \{l | l \in V, p_{k,l} \leq p^{\max}\}, \tag{5}$$

i.e., it contains all nodes that can be reached by parent k with the transmission power $p^{\max}$ or less, as represented by the green box in Figure 1.

Our goal is to minimize the total transmission power in the network.

$$p = \sum_{i \in V} p_i(C_i) \tag{6}$$

The total transmission power is defined as the sum of the transmission powers $p_i(C_i)$ of all nodes $i \in V$ for transmitting to their respective child nodes C_i.

4. Broadcast Tree Protocol

This section presents the design of our Broadcast Tree Protocol (BTP) that is based on a game-theoretical model.

4.1. Potential Game

Our approach is based on a *potential game* [42], i.e., all nodes cooperate to minimize the total transmission power (Equation (6)) required to disseminate data over an entire broadcast tree T. The use of potential games to calculate energy-efficient broadcast trees in a decentralized manner has been proposed by Mousavi et al. [5].

4.1.1. Design of the Potential Game

Our potential game is designed as a *child-driven* game, meaning that the receiving nodes (children) decide which transmitting node they select (parents). The construction of the broadcast tree T is executed in iterations, and the current iteration number is denoted by the index t. The potential game $\mathcal{G}$ is described by a set of rational players $\mathcal{P}$ containing all destination nodes $j \in V \setminus \{S\}$, a set $\mathcal{A}_j$ of possible actions for each player j, and a player-specific local cost function φ_j for each player j.

Each node j individually decides from which parent node it should receive data. Therefore, the set $\mathcal{A}_j^{(t)}$ of possible actions for each player j is the set $\mathcal{A}_j^{(t)} = V \setminus \{j\}$ with its potential parents. In each iteration t, each node $j \in \mathcal{P}$ selects a parent node, which is denoted by the action $a_j^{(t)} \in \mathcal{A}_j^{(t)}$. The action profile of the game $\mathbf{a}^{(t)} = (a_1^{(t)}, \ldots, a_n^{(t)})$ is a vector containing the actions of all nodes in iteration t, and $\mathbf{a}_{-j}^{(t)}$ represents the actions of all nodes except the j-th node. Each node j has a player-specific local cost function $\varphi_j(a_j^{(t)}, \mathbf{a}_{-j}^{(t)})$, which depends on the node's action $a_j^{(t)}$ and the actions $\mathbf{a}_{-j}^{(t)}$ of all other nodes in the network.

This cost function, φ_j, is designed to be the marginal contribution of the transmission power required by its parent i to reach the considered node j:

$$\varphi_j\left(i, \mathbf{a}_{-j}^*\right) = p_i(C_i) - p_i(C_i \setminus j) \tag{7}$$

To describe the solutions of $\mathcal{G}$, we introduce the concept of a Nash Equilibrium (NE). An action profile $\mathbf{a}^*$ is an NE of $\mathcal{G}$ if

$$\varphi_j\left(a_j^*, \mathbf{a}_{-j}^*\right) \leq \varphi_j\left(a_j, \mathbf{a}_{-j}^*\right), \quad \forall j \in \mathcal{P}, a_j \in \mathcal{A}_j^{(t)} \tag{8}$$

holds, i.e., no node in $\mathcal{P}$ can reduce its local cost φ_j any further by changing its action a_j. We consider the best response of node j in iteration t to the other nodes' actions as follows:

$$a_j^{(t)} = \underset{a_j^{(t)} \in \mathcal{A}_j^{(t)}}{\operatorname{argmin}} \ \varphi_j\left(a_j^{(t)}, \mathbf{a}_{-j}^{(t)}\right). \tag{9}$$

Mousavi et al. [5] proved that if all players select their actions according to their best response (Equation (9)), the strategy profile will converge to an NE, which minimizes the transmission power over all nodes (Equation (6)).

Although the presented potential game contributes to the decentralized construction of energy-efficient broadcast trees, its practical implementation poses several challenges. First, in the potential game, it is assumed that every node j knows all potential parents in V at any given iteration t as well as the actions of all other players $k \in \mathcal{P}$. However, this information is not easily available in a real-world system; therefore, it must be exchanged and maintained. Second, the potential game assumes iterations with discrete time steps, while every node makes a decision at each time step. This is not feasible for a distributed algorithm in the real world since there is no mechanism to precisely synchronize all nodes. Therefore, each node must make a decision whenever it obtains information asynchronously from its neighborhood. Third, the potential game follows a weakly dominant strategy, where a child node j changes parents if the total transmission power p remains, at most, the same. An alternative is a strictly dominant strategy, where a child node j only switches to another parent node if the total transmission power p is strictly reduced. In a real-world implementation, a problem may arise wherein no stable broadcast tree T is found when a child node can reach two potential parent nodes with the same transmission power, causing the child node to repeatedly switch between the two parent nodes. Fourth, while the potential game assumes that there are no graph cycles, a real-world implementation needs a mechanism to ensure this property. Thus, we modify the potential game as follows.

4.1.2. Approximation of the Potential Game

In contrast to the original approach, (t) now denotes the iteration of a node j that is not synchronized with other nodes. Therefore, in any given iteration (t), a node j possesses the information about the transmission power $p_{Q_j,j}$ that j's current parent Q_j needs to reach node j. Furthermore, for a given potential parent i at iteration (t), j knows the transmission power required by i to reach all of its children C_i, i.e., $p_i(C_i)$, since node i shares this information through broadcast messages. Therefore, in iteration (t), j has two possible actions $\mathcal{A}'^{(t)}_j = \{Q_j, i\}$: either stay with the current parent Q_j or switch to i. Furthermore, j can also derive the transmission power $p_{i,j}$ that i needs to reach j, as shown in Equation (3). With the information received from Q_j and i, j can also derive how much Q_j can potentially reduce its transmission power if j is no longer a child of Q_j as well as how much i would need to increase its transmission power to reach all its children if j was a child of i. With this information in hand, j has two cost functions that express its marginal contribution to i and Q_j, respectively:

$$\varphi_j^{(t)}(i) = p_i(C_i \cup j) - p_i(C_i) \tag{10}$$

$$\varphi_j^{(t)}(Q_j) = p_{Q_j}(C_{Q_j}) - p_{Q_j}(C_{Q_j} \setminus j) \tag{11}$$

Using this information, and changing the weakly dominant strategy to a strictly dominant strategy, a node j switches to the potential parent i if the marginal contribution to i is lower than the marginal contribution to the current parent Q_j:

$$a_j^{(t)} = \underset{a_j^{(t)} \in \mathcal{A}'^{(t)}_j}{\operatorname{argmin}} \ \varphi_j\left(a_j^{(t)}\right) \tag{12}$$

This modification enables the child node to make decisions whenever a new potential parent node is discovered instead of having to exchange information with all other nodes. To solve the problem with our approximation in which a child node j does not know all the potential parents in V, each node needs more than one turn in the game to find the optimal parent. Therefore, our last modification is the introduction of a counter that tracks how often a potential parent has been discovered without j switching parents. Once a threshold is reached, we consider all parents of a node to be discovered and the node to be finished. With this last change, child j might not find the optimal parent in the first iteration, but, over time, it will switch parents until the best parent is found.

In the following, we show that the modified game still converges to an NE in finite time. We make the following assumptions: (1) nodes use discovery packets to identify potential parents, no discovery packets are lost, and all nodes receive discovery packets from their entire neighborhood N_k; (2) the obtained values $p_i(C_i \cup j)$, $p_i(C_i)$, $p_{Q_j}(C_{Q_j})$, and $p_{Q_j}(C_{Q_j} \setminus j)$ are undisturbed. The first assumption requires the counter mentioned above to be set to a reasonably high value, which is carried out empirically in the evaluation in Section 7.3. The second assumption might seem impractical, but our evaluation shows that noise only affects the obtained values marginally and does not cause frequent decision changes.

Mousavi et al. [5] have shown that an NE exists for the original game. We show that BTP converges to an NE of the modified game. For this purpose, we use the concept of an improvement path [43], which is defined as a sequence of action profiles $\{\mathbf{a}^{(0)}, \mathbf{a}^{(1)}, \dots\}$, where in $\mathbf{a}^{(t+1)}$, every node j that changes its action $a_j^{(t+1)} \neq a_j^{(t)}$ has a lower cost function $\varphi_j\left(a_j^{(t+1)}\right) < \varphi_j\left(a_j^{(t)}\right)$. Furthermore, at least one node needs to choose the same action in $\mathbf{a}^{(t)}$ and $\mathbf{a}^{(t+1)}$. Using BTP, the action profile $a_j^{(t)}$ of the nodes follows an improvement path. This can be directly seen by referring to the best response strategy (12), in which nodes only change their actions if their cost function decreases. For potential games, the finite improvement path property holds, leading to an NE in a finite amount of time [43]. Therefore, BTP converges to an NE in a finite number of steps.

4.2. BTP

BTP is based on the game-theoretical model presented in Section 4.1. BTP consists of two phases: (1) the broadcast tree construction phase, where the source node S, i.e., the node intending to disseminate data to all other nodes, initiates the construction of the broadcast tree T (this phase ends with a tree topology, where the source node S is the root), and (2) the data dissemination phase, where the actual data are sent from S to all other nodes.

The broadcast tree T consists of the source node S, which is the root of the broadcast tree, and parent and child nodes. Here, all parents except S are also children, and all children except the leaves are also parents. Since this approach is child-driven, parents only broadcast their own state information periodically, and children use this information to decide which node they will choose to be their parent. Furthermore, a child may choose a node as its parent, but a chosen parent node may or may not accept parenthood for the given child. In the latter case, the child must find another parent.

4.2.1. Broadcast Tree Construction Phase

To initialize the decentralized broadcast tree construction phase, the source node S sends a Neighbor Discovery packet using the maximum transmission power $p^{\max}$. Every node $i \in V$ that is already part of the broadcast tree T also periodically broadcasts Neighbor Discovery packets with maximum transmission power $p^{\max}$. Every receiver $j \in N_i$ of a Neighbor Discovery packet checks whether the sender i of the received packet is a suitable parent Q. Two cases may occur. In the first case, j is not connected to any parent, e.g., during the initial construction of the broadcast tree. In this case, j requests i to become i's child by sending it a Child Request packet, which it, in turn, may or may not accept, as described later. Second, j is already a child and is connected to parent Q_j, which is different from i. In this case, j checks if switching from Q_j to i would decrease p (see Equation (6)), as shown in Section 4.1.2. To this end, j must check whether switching to having node i as its parent would reduce the transmission power of Q_j to a greater extent than i would have to increase its transmission power to reach all its children. To do so, j needs four transmission power values (see Section 4.1.2):

1. $p_{Q_j}(C_{Q_j})$, i.e., the transmission power of Q_j needed to reach all its children;
2. $p_{Q_j}(C_{Q_j} \setminus \{j\})$, i.e., the transmission power of Q_j if j is no longer Q_j's child;
3. $p_i(C_i)$, i.e., the transmission power of i needed to reach all its current children;
4. $p_i(C_i \cup j)$, i.e., the transmission power of i if j becomes i's child.

To provide these values to all potential children of a node i, every BTP packet includes the transmission power $p_i(C_i)$, as well as the transmission power that would be required to reach the second-farthest child of i, along with its respective address. When node j receives a BTP packet from node i, it can additionally calculate the transmission power $p_{i,j}$ that i would need to reach j by means of Equation (3). Thus, j can also calculate $p_i(C_i \cup j)$. When j knows the four values, it switches to i if the following condition holds:

$$p_i(C_i \cup j) - p_i(C_i) < p_{Q_j}(C_{Q_j}) - p_{Q_j}(C_{Q_j} \setminus \{j\}) \quad (13)$$

i.e., if the current parent Q_j can reduce its transmission power to a greater extent than the new parent i must increase its own transmission power.

When node j decides to choose node i as its new parent, it sends a Child Request packet to i in order to become its child. Node i verifies that j is not i's parent, j is not already a child of i, and j is reachable from i, i.e., $p_{i,j} \leq p^{\max}$. If all these checks are successful, i accepts j as a child using a Child Confirmation packet or rejects it otherwise using a Child Rejection packet. When i accepts j, it adjusts its transmission power to $p_i(C_i \cup j)$, resulting in an increase if $p_{i,j}$ is greater than $p_i(C_i)$; otherwise, the value stays the same as before. Even if i must increase its transmission power, the total transmission power p is still lower because j only connects to i if the condition of Equation (13) holds. When i accepts j as its child, j disconnects from its old parent Q_j using a Child Revocation packet. Q_j, in turn, removes j from its child list C_{Q_j} and adjusts its transmission power accordingly to reach the farthest child $k \in C_{Q_j} \setminus \{j\}$. However, when i rejects j, j places i on a blocklist to avoid repetitive connection attempts and repeats the above processes to find a new parent.

Finally, since S does not know the global state of the network, it cannot know when the broadcast tree has reached its optimal state. Therefore, each node maintains a counter that tracks the iterations without any changes, i.e., without connecting to or disconnecting from parents or adding or removing children. As soon as this counter reaches a threshold, j considers itself finished and notifies its parent Q_j using an End of Construction packet. When Q_j has received such a packet from all its children $k \in C_{Q_j}$ and when it is itself finished, it notifies its own parent that it has finished its game. This procedure continues until S has received End of Construction packets from all its children $l \in C_S$. The broadcast tree construction phase is then finished, and the data dissemination phase starts. Once the broadcast tree is constructed, each node i sets its transmission power to $p_i(C_i)$ so that i just barely reaches all its children C_i. Thus, the total transmission power is minimized.

During the broadcast tree construction phase, graph cycles may potentially occur, which must either be avoided or detected and broken up. Graph cycles can occur in three cases. In the first case, the source node S may try to connect to another node as its parent. Since S is defined as the root of the broadcast tree, this eventuality must be avoided, which can easily be accomplished. In the second case, a parent node i may try to connect to one of its children $j \in C_i$. To circumvent this eventuality, i must check if $j \in C_i$ and refrain from connecting to children. In the third case, a parent node i may try to connect to a node k that is not its direct child but that includes i on the path from k to S $(S, \ldots, i, \ldots, Q_k, k)$. To handle this case, we propose two cycle avoidance algorithms and one cycle detection algorithm.

Path-to-Source

In the Path-to-Source cycle avoidance algorithm, every node j that successfully connects to a parent Q_j adds its own address to a list of addresses $(S, \ldots, Q_j, j)$ that represents the entire path from source S to node j. This list is included in each Neighbour Discovery packet. When a node k tries to connect to j, it first has to check if $k \in (S, \ldots, Q_j, j)$. If it is part of the path, the node must not try to connect to j in order to avoid a cycle.

Mutex

The Mutex cycle avoidance algorithm essentially ensures that the entire broadcast tree is in a consistent state at any point in time. To this end, a node j that decides to connect to a node i first locks its own sub-tree by notifying all its children C_j. In this process, all nodes in C_j also lock their respective sub-trees until all nodes below j are locked. The nodes that are included in a locked sub-tree are not allowed to change their parents or to accept new children until j unlocks them again. When j tries to connect to i and i is locked by j, j detects this situation and refrains from connecting to i to avoid giving rise to a cycle.

Ping-to-Source

The Ping-to-Source algorithm is a cycle detection algorithm. After connecting to parent i, node j sends a unicast Ping-to-Source packet to i, which, in turn, forwards the packet to its own parent Q_i. This process, in which nodes forward the Ping-to-Source packet to their parents, continues until one of three cases occurs. In the first case, the source node S receives the message. This occurs when there is no cycle, so S can drop the Ping-to-Source packet. In the second case, the Ping-to-Source packet arrives at an intermediate node that has no parent, in which case the packet is also dropped. In the third case, the Ping-to-Source packet eventually arrives at node j, which occurs if the broadcast tree has a cycle. In this case, j disconnects from i. Furthermore, if j was connected to a parent Q_j before trying to connect to i, it would attempt to reconnect to Q_j, which triggers the entire connection process discussed above.

4.2.2. Data Dissemination Phase

During the data dissemination phase, the source node S starts sending the data to its children $l \in C_S$ with a transmission power of $p_S(C_S)$ (see Equation (4)) using Application Data packets. All nodes l, in turn, relay the data to their respective children with a transmission power of $p_l(C_l)$. The Application Data packets contain a sequence number. Since the data size may exceed the size of a frame, e.g., 2304 bytes for Wi-Fi, the source node splits the data into chunks and increases the sequence number for every frame accordingly. Since node i transmits with a power of $p_i(C_i)$, other nodes $j \notin C_i$ may still receive the data because they may be reached by i when $p_{i,j} \leq p_i(C_i)$. In this case, j can still process and utilize such receptions as the packets are unambiguous due to their sequence number.

4.3. Protocol Packets

BTP uses several packet types for different purposes. Each BTP packet is sent with a transmission power of $p^{\max}$ during the broadcast tree construction phase, while Application Data packets are sent with a transmission power of $p_i(C_i)$ by each node $i \in V$ during the data dissemination phase. The following packet types exist in BTP:

Neighbor Discovery

Each node periodically broadcasts Neighbor Discovery packets to inform its neighbors of its presence. When a node receives a Neighbor Discovery packet, it checks whether the sender is a possible new parent node, as presented in Section 4.2.1. In particular, each Neighbor Discovery packet contains information about the transmission power levels required by its sender, as described in Section 4.2.1, allowing each receiver to decide whether to switch parents according to Equation (13).

Child Request

Once a node has identified a possible parent node, it requests a connection by sending a Child Request packet as a unicast message to that parent node.

Child Confirmation

A potential parent node that is able to accept a child request from another node sends a Child Confirmation packet in response as a unicast packet. When the child node receives the Child Confirmation packet, the child node and the parent node are considered to be connected.

Child Rejection

A potential parent that cannot accept a child request from another node sends a Child Rejection packet. A requesting child node that receives such a packet blocks this potential parent and tries to connect to another potential parent.

Child Revocation

Child nodes must inform their parent nodes when they want to disconnect, which is executed by sending a Child Revocation packet as a unicast message to the parent node. This packet type is used after successfully switching to a new parent or when detecting a cycle.

End of Construction

When a node ends the broadcast tree construction phase, it notifies its parent node by sending an End of Construction packet. This information is important for the parent node because the parent must wait for all its child nodes before it is allowed to finish the broadcast tree construction phase itself. As soon as a node finishes the broadcast tree construction phase, it sets a corresponding flag F in all its sent packets.

Application Data

In the data dissemination phase, application data are transmitted using the Application Data packet type. A node $i \in V$ sends all data packets with $p_i(C_i)$, i.e., with the transmission power that is required to reach its most distant child.

5. Matlab Simulation

This section presents Matlab simulations of BTP and other algorithms from the literature, allowing us to compare BTP with other algorithms. However, while simulations allow for control over various parameters, e.g., distances between nodes, transmission power, and channel gain, they do not always reflect the real world accurately.

5.1. Experimental Setup for the Matlab Simulation

Table 2 shows the parameters of our Matlab simulation. The nodes are randomly placed in a 500 m $\times$ 500 m square area, which allows us to assess BTP under various conditions while not taking advantage of an optimized node placement strategy in order to reflect practical constraints. Still, we require each node to have at least one neighbor according to Equation (5), while the maximum transmission power of a node is set to $p^{\max} = 20$ dBm. Further, the number of nodes varies between 10 and 90. The source node is chosen randomly for each simulation run. The channel is based on a path-loss model for which $|h_{i,j}|^2 = \frac{1}{d^{\alpha}}$, where d is the distance between the nodes i and j and α is the attenuation exponent, which is set to $\alpha = 3$. The SNR must exceed $\gamma^{min} = 10$ dB for correct reception, while the noise power σ^2 is set to -90 dBm. The finishing threshold is set to 10 unchanged iterations (see Section 4.2.1) since the broadcast tree did not improve further with a larger threshold in trial experiments. For each parameter combination, 1000 simulation runs were executed, i.e., a total of 63,000 simulations runs. We implemented the following algorithms for comparison with BTP:

Dijkstra

Dijkstra's algorithm [44] maintains a set of nodes whose shortest distance from the source is known and gradually expands this set until all nodes are included. It iteratively selects the node with the shortest distance, updates the distances to its neighbors, and continues this process until the shortest paths to all nodes have been established. It is important to note that Dijkstra's algorithm is used in our comparison to construct a spanning tree rather than to route packets to individual nodes.

BIP

Broadcast Incremental Power (BIP) [7] is an iterative algorithm that exploits the broadcast characteristics of wireless channels through centralized control. Starting from the source node, each iteration establishes a connection between the source and another node in the network either by using a direct single-hop connection or a multi-hop connection, thereby extending the sub-tree.

BIPSW

Broadcast Incremental Power with Sweep (BIPSW) [7] is a variation of the BIP algorithm. The efficiency of BIP is enhanced using a so-called sweep operation, which eliminates redundant transmissions in cases where a node can be served by multiple transmitters. This operation transforms the tree constructed using BIP into a spanning tree.

PCP

The Power Control Protocol (PCP) [14] employs an energy rank for each node in the network, which is defined as the energy that a node adds to a given path to the source. Nodes connect to a parent whose path to the source requires the least energy.

BPG

Broadcast trees with Potential Game (BGP) [5] is the original game-theoretical protocol from which BTP is derived. Section 4.1 provides a detailed description of BPG.

SBP

The Simple Broadcast Protocol (SBP) is a variant of BTP that always uses $p^{\max}$ to disseminate data. SBP is also employed for comparisons in the testbed experiments presented in Section 7 and provides an upper bound of the required transmission power.

Table 2. Parameters used for the Matlab simulation.

Parameter	Values
Protocols	BTP, BPG, BIP, BIPSW, PCP, SBP, Dijkstra
Nodes	10, 20, 30, 40, 50, 60, 70, 80, 90
Simulation Area	500 m × 500 m
$p^{\max}$	20 dBm
$\lvert h_{i,j}\rvert^2$	$\frac{1}{d^{\alpha}}$
α	3
γ^{min}	10
σ^2	−90 dBm
Finishing Threshold	10

5.2. Results of the Matlab Simulation

Figures 2 and 3 show the total transmission power as a function of the number of nodes in the network as introduced in Equation (6): $p = \sum_{i=1}^{|V|} p_i(C_i)$.

The x-axes show the number of nodes, while the y-axes show the total required transmission power (given in Watts on a logarithmic scale) for all nodes to disseminate data. The different colors denote different algorithms. Both Figures 2 and 3 show the same results, but Figure 3 does not include SBP to better distinguish between the other algorithms. The key indication of the results is that there are four performance categories. First, SBP shows the worst performance, and the more nodes that are added, the worse this performance becomes. While SBP constructs a spanning tree, it always uses the maximum transmission power p_{max}, resulting in this poor performance. Second, Dijkstra clearly uses less power than SBP but still more than the other algorithms since it does not consider the broadcast nature of the wireless channel. Instead, Dijkstra builds a spanning tree for unicast connections, resulting in suboptimal node connections in our broadcast scenario. Third, BIP and PCP are better than Dijkstra, but they still perform worse than BTP. On the one hand, BIP still has unnecessary and redundant connections between nodes. On the other hand, PCP builds a fixed action set at startup, which is not further optimized. In contrast, BTP allows each node to update its action in multiple iterations to respond to the actions of other nodes. Fourth, BTP, BPG, and BIPSW show the best performance. The performance of BIPSW is slightly better with fewer nodes, while that of BTP and BPG is slightly better with more nodes. BIPSW essentially outperforms BIP since it removes the redundant connections from BIP's solution. Furthermore, BIPSW is a centralized algorithm with global knowledge about the links and connections of a network, allowing it to potentially provide better decisions than BTP, which only uses local information. Besides BIPSW, all the algorithms except BTP and SBP also require global knowledge about the entire network. This is not realistic in a real-world scenario.

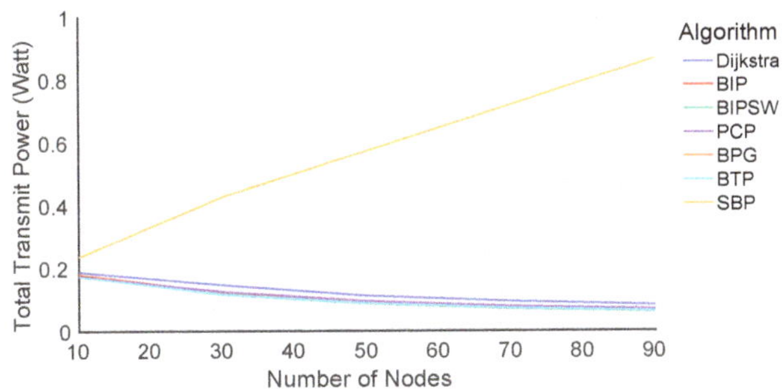

Figure 2. Total required transmission power for various algorithms.

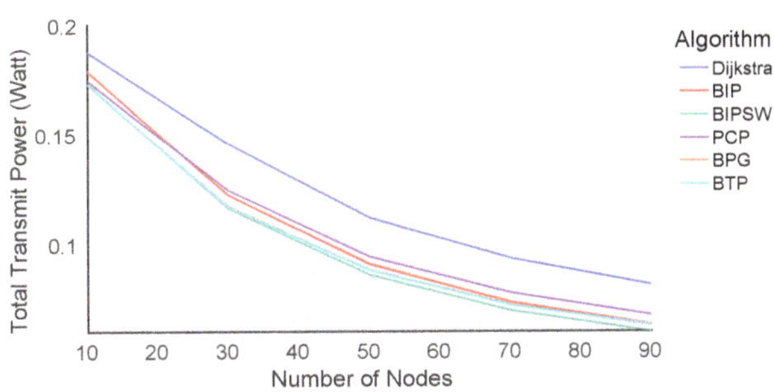

Figure 3. Total required transmission power for various algorithms (without SBP).

In summary, the results of the Matlab simulation show that BTP is on par with or better than the other algorithms. BTP is the only algorithm operating under realistic assumptions, using only local information from direct neighbors.

6. NS-3 Simulation

In this section, our NS-3 simulation of BTP and the corresponding results are presented. We investigate three aspects. First, while the original potential game has already been compared to alternative approaches in Section 5, in this section, we show that our approximation can minimize the total energy consumption. Second, using the NS-3 simulation, we can evaluate all three cycle avoidance and detection algorithms. Finally, the NS-3 simulation allows us to validate the scalability of BTP.

Although BTP is agnostic with respect to the underlying physical and link layer implementations, we selected Wi-Fi as the underlying wireless technology to allow for the comparison of the results of the NS-3 simulation with a real-world implementation evaluated using real hardware, namely, a Wi-Fi-based testbed (see Section 7). We assume that a data link layer ensures the reliable delivery of packets directly to other nodes. Access to the wireless transmission medium is coordinated through a Carrier Sense Multiple Access (CSMA) mechanism.

NS-3 is an open-source and event-driven simulator with several protocols ready to use, e.g., IP, TCP, or UDP. The individual components of the NS-3 simulator are divided into modules that can be used or extended to set up a self-defined environment.

By exploiting NS-3's energy framework, it is possible to simulate different energy sources. Based on the datasheets of the MAX28282 (https://datasheets.maximintegrated.com/en/ds/MAX2828-MAX2829.pdf (accessed on 4 July 2023)) and MAX28313 (https:

//datasheets.maximintegrated.com/en/ds/MAX2831-MAX2832.pdf (accessed on 4 July 2023)) Wi-Fi chips, we approximate the power consumption of a Wi-Fi chip as a function of the transmission power through a polynomial regression:

$$p_{mA}(p_{dBm}) = -0.000009708995023p^5 - 0.00089877372p^4 - 0.03112035853p^3 - 0.4798606017p^2 - 2.427503769p + 124.4196777$$

This function is essentially used to calculate the energy consumed over a period of time. The energy model of NS-3 tracks state-changes of the Wi-Fi module. If the state changes, for example, from receiving to transmitting, the energy consumed over the duration of the transmitting state is calculated and subtracted from the energy source. Switching losses, additional electronic components, or fluctuations due to heat are not simulated.

In our implementation, we used properties that model IEEE 802.11 characteristics, as follows. In wireless networks, preambles are used to announce the arrival of a packet and allow the receiver to synchronize with a received frame. During preamble detection, the signal-to-noise ratio (SNR) is determined. If the SNR is too low, the packet can be discarded because it may not be possible to decode it without an error. To reproduce the characteristics of Wi-Fi, the noise power of the channel can be modeled as follows:

$$\sigma^2 = k_B * 290 * w + \sigma^2_{int.} \tag{14}$$

where k_B is the Boltzmann constant, w is the channel width in Hz, and $\sigma^2_{int.}$ is the noise of interfering transmissions [45]. Furthermore, γ^{min}, i.e., the minimal SNR that still allows for the decoding of a packet, was set to 4 dB. For path loss $|h_{i,j}|^2$, we used the YANS model (https://www.nsnam.org/docs/release/3.24/doxygen/classns3_1_1_yans_wifi_channel.html#details (accessed on 4 July 2023)). Further, we set the maximum possible transmission power to $p^{\max} = 23$ dBm.

6.1. Experimental Setup for the NS-3 Simulation

Table 3 summarizes the parameters used for the evaluation of the NS-3 simulation.

We simulated six different node numbers from 50 to 300 in steps of 50 in five areas of different sizes, ranging from 100 m × 100 m to 500 m × 500 m, which we refer to as area configurations (1) to (5) in the remainder of this article. All nodes were placed randomly within the plane, while a pseudo-random number generator was initialized with a fixed seed and increased with every iteration of an experimental configuration. This allowed us to evaluate the protocols under various conditions, without relying on an optimized node placement strategy. Furthermore, we evaluated all three cycle-handling algorithms presented in Section 4.2.1. The finishing threshold for unchanged game rounds (see Section 4.2.1) was set to 10. Besides BTP, we also evaluated the SBP protocol. A data size of 1 KiB was used, i.e., the payload occupying a single Wi-Fi frame. Finally, each experimental run was executed until the payload was transmitted to all nodes or aborted after 20 s, and each configuration was repeated 30 times. In total, 5400 experimental runs were executed.

Table 3. Parameters used for the NS-3 simulation's evaluation.

Parameter	Values
Protocols	BTP, SBP
Cycle Handling	Ping-to-Source, Path-to-Source, Mutex
Nodes	50, 100, 150, 200, 250, 300
Simulation Area	(1) 100 m × 100 m, (2) 200 m × 200 m, (3) 300 m × 300 m, (4) 400 m × 400 m, (5) 500 m × 500 m
Data Size	1 KiB
Finishing Threshold	10
Runs	30

6.2. Results of the NS-3 Simulation

6.2.1. Total Energy Consumption

We first compare the total energy consumption values of BTP and SBP throughout the entirety of their experiments, i.e., incorporating both the broadcast tree construction phase and the data dissemination phase. The results are shown in Figure 4.

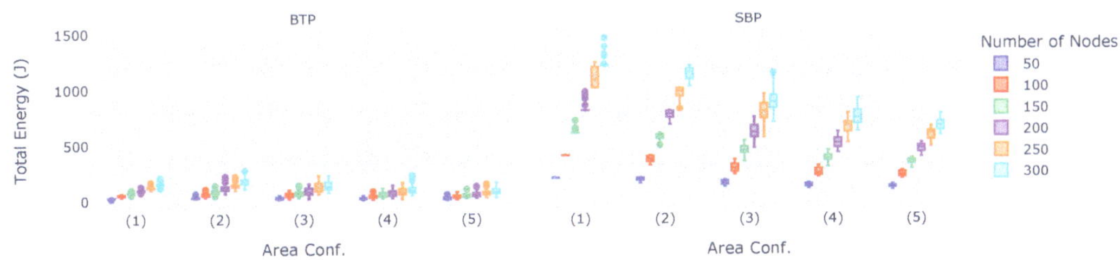

Figure 4. Total energy consumption values of BTP and SBP.

The x-axes denote the simulation area, and the y-axes show the total energy in Joules. The colors denote the different quantities of nodes, while the left and right sub-plots show the results regarding BTP and SBP, respectively. It is evident that BTP requires significantly less energy than SBP. In fact, depending on the experimental configuration, an energy reduction of between 83% and 92% can be achieved when using BTP. Moreover, in the worst case (i.e., 300 nodes, area configuration (1)), BTP requires only about 150 J, and even in the best case (50 nodes, area configuration (1)), with 203 J, it requires less energy, than SBP. This energy reduction primarily results from two effects. First, using a broadcast tree based approach reduced the total number of packets sent in the network. In fact, BTP required about 70% fewer data packets than SBP, resulting in 70% less energy consumption. Second, the remaining 13% to 22% reduction in energy consumption is a result of the optimal broadcast tree. It is also noticeable that the number of nodes is positively correlated with energy consumption, which is reasonable since more nodes need more energy, even if they send with minimal transmission power. The simulation area, however, is negatively correlated with energy consumption, i.e., the larger the area, the lower the total amount of energy consumed. This is due to the fact that an increasing area leads to the greater distances between nodes, thus not becoming part of the broadcast tree, which leads to their inability to disseminate data, thereby reducing overall energy consumption. This means it is likely that not all nodes are part of the network in these larger area configurations. This is further analyzed in Section 6.2.5.

6.2.2. Protocol Overhead

Figure 5 depicts the average energy overhead per node required to construct the broadcast tree in comparison to disseminating data.

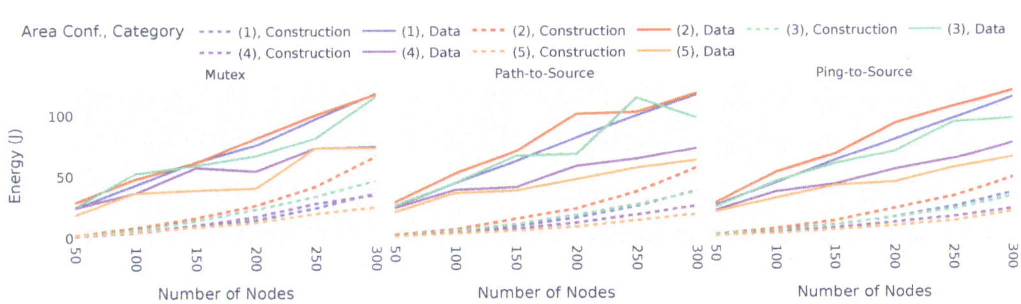

Figure 5. Energy usage of broadcast tree construction phase and data dissemination phase.

On the x-axes, the number of nodes is shown, while the y-axes denote the used average energy per node in Joules. The sub-plots depict the three cycle-handling algorithms, the colors denote different simulation areas, and the line style denotes the energy for broadcast tree construction (dotted) and data dissemination (solid). This plot shows that the energy used is only dependent on the number of nodes and the simulation area, while the energy per node is not significantly influenced by the cycle-handling algorithms. The key takeaway, however, is that disseminating data requires about three times more energy than constructing the broadcast tree in the configuration with many nodes. The fewer nodes that are involved, the greater the energy requirement for disseminating data compared to constructing the broadcast tree. For example, configurations with 50 nodes require up to 17 times more energy for data dissemination. The energy required for disseminating data grows linearly with the number of nodes, whereas the energy required for constructing the broadcast tree grows faster since the coordination and the resulting number of packets do not increase linearly. In summary, the experiments show the feasibility of BTP since the construction of the broadcast tree is only performed once, while the dissemination of data via the ready-to-use broadcast tree can be performed repeatedly afterwards.

6.2.3. Time for Broadcast Tree Construction Phase

Another important metric for the practical use of broadcast trees is the time required to construct a broadcast tree, i.e., the amount of time required until the data can be sent to the nodes. This is shown in Figure 6.

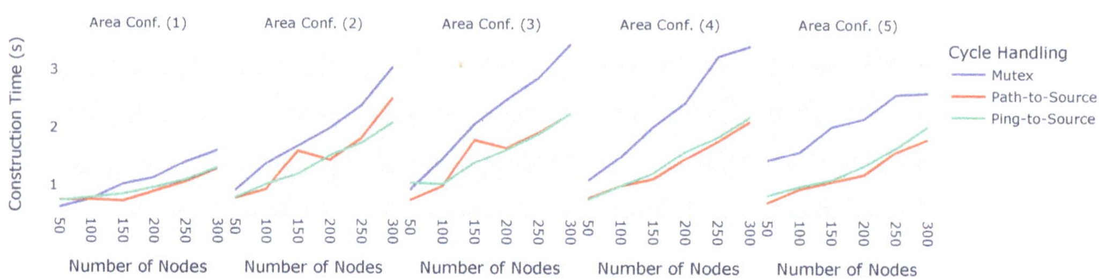

Figure 6. Time taken to construct the broadcast tree.

The x-axes show the number of nodes, while the y-axes denote the time taken to construct the broadcast tree in seconds. The simulation area is denoted in the sub-figures, and the cycle-handling algorithms are represented with different colors. The three main properties that significantly influence the time required to construct a broadcast tree are the number of nodes, the simulation area, and the cycle-handling algorithm employed. The more nodes that are in the network and the larger the area, the longer it takes to construct the broadcast tree. Regarding the cycle-handling algorithms, it is noticeable that Ping-to-Source and Path-to-Source show no significant differences. Mutex, however, can take up to 65% longer to construct a broadcast tree. This is because locking an entire sub-tree, switching the parent, and unlocking it again takes time that is not required using the other two algorithms. Due to this behavior exhibited by Mutex, we did not consider it in our real-world implementation.

6.2.4. Cycle Handling

Under certain circumstances, cycles may be either not avoided or detected or are not broken up before the experiment is finished. Figure 7 depicts how many cycles were still present when ending the experiment.

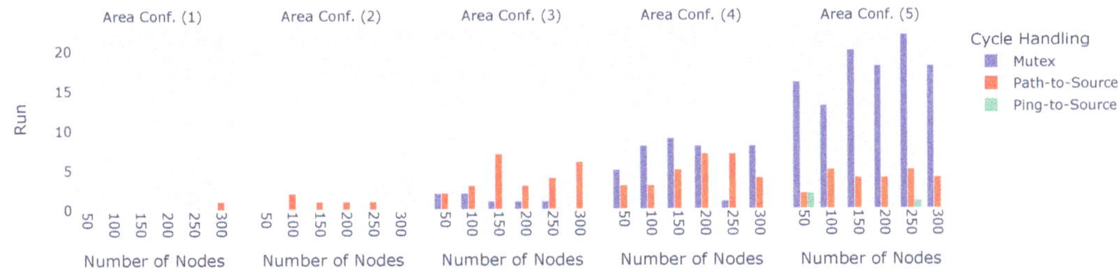

Figure 7. Number of cycles lasting until the end of an experiment.

The y-axes show the number of runs in which at least one cycle lasted until the end, while the x-axes show the different numbers of nodes. The sub-plots illustrate the results in the different areas, and the colors denote the three cycle-handling algorithms. It is evident that the number of nodes in the network does not significantly influence the number of cycles lasting until the end of the experiment. Area size, on the other hand, does have a significant influence, especially for the Mutex cycle avoidance algorithm, but also for Path-to-Source. Ping-to-Source is also affected by the area size but significantly less so since cycles could not be detected or broken up in only three out of 1800 Ping-to-Source experimental runs.The Mutex's behavior can be explained by the problem wherein locking an entire sub-tree leads to an almost complete halt in construction. Suppose the broadcast tree is in a somewhat advanced stage. When one node in the middle of the broadcast tree decides to change its parent, the entire sub-tree will be locked and stops any further optimization until the lock is eventually released. Finally, both cycle avoidance algorithms do not seem to avoid all cycles. This is because they both rely on transmitting information required for the respective algorithm. For Mutex, this transmission entails broadcasting the packet that locks and unlocks sub-trees. When the area is large, it may simply occur that lock or unlock packets are lost or not received by the children, leaving the entire broadcast tree in an unfinished state. The same applies to the Path-to-Source algorithm. This makes Mutex and Path-to-Source effectively unusable in any scenario; therefore, both were not considered in our real-world implementation.

6.2.5. Unconnected Nodes

Figure 8 shows the number of nodes that are not part of the broadcast tree at the end of the experiment.

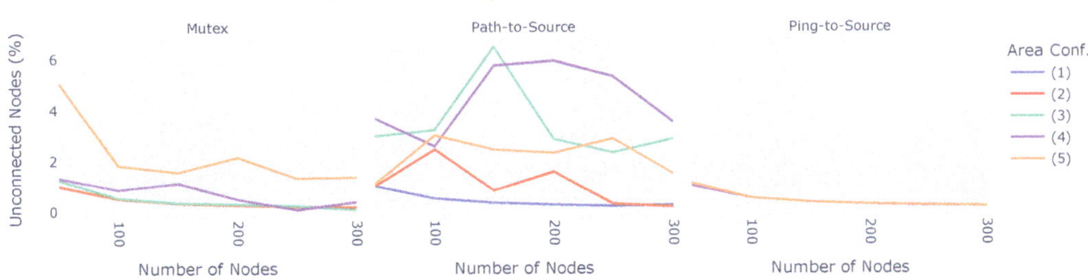

Figure 8. Percentage of nodes not part of the broadcast tree.

On the x-axes, the number of nodes is shown, while the three sub-plots depict the three different cycle-handling algorithms. The colors denote different area configurations. The overall ratio of unconnected nodes across all experiments is relatively small, amounting to about 6% in the worst case. However, the cycle-handling algorithms show quite large differences. Ping-to-Source shows good results when only a few nodes are spread out in the area, about 1%, i.e., one node not connected to the broadcast tree (on average). With more nodes in the network, the rate of unconnected nodes decreases to 0.1%. Mutex encounters difficulty when used in large areas, presenting 5% unconnected nodes in area configuration (5) and few nodes. However, while Ping-to-Source and Mutex show reasonable results, Path-to-Source behaves quite erratically. When using the Path-to-Source algorithm, while a node j connects to a parent i, another node k might attempt to connect to j, which is the parent of i. Since j and i are not connected yet, the Path-to-Source algorithm fails to detect and avoid this cycle. Therefore, Path-to-Source was not considered in our real-world implementation.

7. Real-World Implementation

Since the Matlab and NS-3 simulations in Sections 5 and 6 provided encouraging results, we also assessed the performance of BTP using real hardware to evaluate its practicability under real-world conditions.

We implemented BTP in userland C for Linux using standard libraries. Thus, we relied on Wi-Fi as the wireless technology since it is widely available on Linux systems. However, as shown in Section 4.1, BTP is not limited to Wi-Fi; it can be used with any radio technology. The SNR calculation of Equation (1) requires information about the wireless transmission characteristics of a received frame. This information, contained in the RadioTap header, is not available to userland C programs or even the Linux kernel since it becomes stripped off by the Wi-Fi chip's firmware. To bypass this limitation, we used the Nexmon framework [46] (https://nexmon.org (accessed on 4 July 2023)). We created a Nexmon patch that preserves the RadioTap header for MAC frames with our BTP EtherType. All other frames were handled normally to avoid interference with any other programs. Furthermore, we utilized Linux raw sockets for two reasons. First, their use further increases performance since the kernel's TCP/IP stack is bypassed and a packet is more or less directly passed to the Wi-Fi chip. Second, since BTP is located on the network layer of the ISO/OSI stack, we were able to define a custom EtherType (`0x88DF`) and enable the reception of BTP packets in userland. Addressing was handled on the Link Layer, with each node identified according to its Wi-Fi MAC address. Neighbour Discovery packets and Application Data packets were sent as MAC-broadcasts; all other packets were unicasts.

In Section 4.2, we discussed the tree construction phase in detail but left the data dissemination phase open to a specific implementation. To reduce the likelihood of collisions, we used an MTU of 1200 bytes, where 37 bytes were used for Ethernet and BTP headers and the remaining 1163 bytes were available for the payload. Furthermore, we added a delay between individual data frames to avoid sending all data frames at once, and the relaying nodes were given a rate limit per sequence number to further reduce the likelihood of collisions and increase the chance of successful data delivery. Additionally, the protocol did not specify what action to take if a node received the entire dataset. In our implementation, leaf nodes do not relay data frames. As soon as a leaf node has received the entire dataset, it disconnects from the parent. This leads to intermediate nodes eventually becoming leaf nodes that also do not relay data frames anymore and disconnect from their parent after receiving the entire data. In this way, the tree is eventually deconstructed.

7.1. Testbed

To evaluate BTP, we utilized a testbed deployed at our university. It consists of 75 Raspberry Pis spread over a university building across four floors. Besides Wi-Fi, all nodes also have an Ethernet uplink to a central management server and are configured for network booting to support easy, large-scale deployment. We created a network-bootable image for our experiments using the Pimod framework [47] (https://github.com/Nature40/pimod (accessed on 4 July 2023)). Although we used RAW sockets with a custom EtherType, the Wi-Fi chip did not receive any frames if it was not part of a Basic Service Set (BSS), i.e., associated with an access point or part of the same ad hoc network. Therefore, the Raspberry Pis were set to ad hoc mode with the same BSS ID. Furthermore, the Linux kernel will refuse to send frames to the Wi-Fi interface if an IP address has not been set.

Thus, we assigned a random IP address to all Raspberry Pis. We used channel 1 of the 802.11n mode and employed the RTS/CTS method of the 802.11 CSMA/CA mechanism.

7.2. Experimental Setup for the Real-World Implementation

Table 4 summarizes the parameters used in our evaluation.

Table 4. Parameters used in our evaluation.

Parameter	Values
Source Nodes	3
Data Sizes	1 KiB, 4 KiB, 16 KiB
Finishing Threshold	5, 15, 25
Protocols	BTP, SBP
Runs	5

Since we could not alter the position of the nodes (in contrast to our Matlab and NS-3 simulations), we used three source nodes located in the northern part of the building, in its center, and in the southern part. Furthermore, we used three data sizes, representing simple sensor values (1 KiB), network diagnosis (4 KiB), and device updates (16 KiB). Although both simulations showed that 10 iterations constitute a sweet spot between overhead and finding the optimal tree, we sought to evaluate BTP in greater depth for practical implementation. Therefore, we used three finishing thresholds for the counter of the number of iterations without topology changes. Finally, we again compared BTP to SBP. Each experimental configuration was repeated five times, resulting in 270 experimental runs.

7.3. Results of the Real-World Implementation

7.3.1. Total Energy Consumption

Due to the testbed's setup, we could not measure the power draw of all the nodes. Hence, we decided to employ a model to compute the energy consumption in mJ based on the used parameters of the physical and data link layers to estimate the energy used in each run. Figure 9 shows the total energy consumption for different parameter sets. The y-axis shows energy in mJ, while the x-axis denotes different data sizes, where different colors represent different finishing thresholds and SBP, respectively. The values include frames for tree construction as well as the data themselves. BTP requires between 68% (1 KiB and a finishing threshold of 5) and 90% (16 KiB and a finishing threshold of 25) less energy compared to SBP, depending on the data size and the values of the finishing thresholds. This is due to the fact that the broadcast tree resulting from BTP is optimal with respect to the energy requirements. Among the BTP parameters, energy use does not show significant differences with increasing finishing threshold values. While SBP requires about 70% more energy for 16 KiB compared to 1 KiB data, BTP only requires about 20% more energy. This is counter-intuitive since one would expect about the same increase in energy consumption. We discovered that this result was due to RF interference. Neither BTP nor SBP include any advanced MAC mechanism, which leads to congestion in the RF spectrum for larger data sizes. However, because BTP uses less power to transmit data, there is less interference causing retransmission, resulting in a moderate increase in overall energy consumption.

Figure 9. Total required energy.

7.3.2. Energy Consumption for Tree Construction and Data Dissemination

Figure 10 presents the energy required for broadcast tree construction and for data dissemination. The x-axis shows different data sizes, while the y-axis shows the energy used for Application Data packets (blue) and for the tree construction packets (red) for BTP and all the finishing thresholds. The energy required for tree construction does not depend on the size of the data but only on the number of nodes; thus, the red plots do not show significant differences. The energy required for data dissemination increases with the size of the data. For the 1 KiB data, about 30% of the energy is used for data dissemination, and about 70% is used for tree construction. However, even in this scenario, BTP performs significantly better than SBP in terms of total energy consumption, as shown in Figure 9. BTP requires about 1.5 MiB of data to construct a tree, regardless of the finishing threshold or data size, whereas data dissemination requires about 150 MiB of total data transfer for 1 KiB experiments, 300 MiB for 4 KiB experiments, and about 400 MiB for 16 KiB experiments. Note that there are outliers in experiments, since the environment can vary under real conditions. For example, in the 4 KiB experiments, three runs required between 740 mJ and 800 mJ of energy, which is outside the 1.5-fold interquartile range.

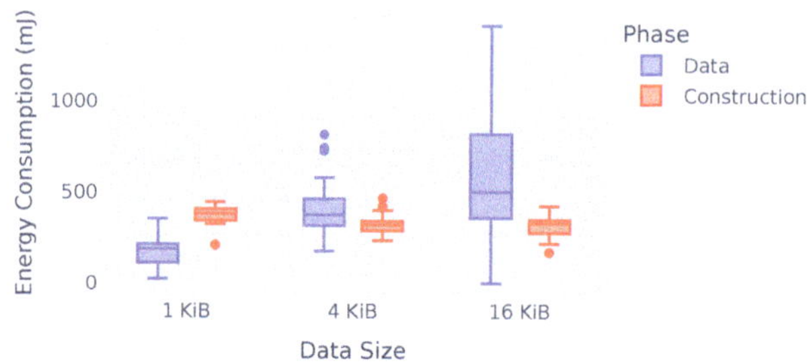

Figure 10. Energy required for both broadcast tree phases.

7.3.3. Successful Receptions

Figure 11 shows the results in terms of successful deliveries. The x-axis shows experiments with different data sizes, while the y-axis shows the percentage of nodes that received the entire dataset. The colors denote different finishing thresholds and SBP, respectively. BTP produces results that are at least on par with SBP; with larger data, SBP's performance becomes even worse. For 1 KiB and 4 KiB data sizes, BTP achieves a nearly 100% delivery ratio, with a few runs producing outliers. SBP only achieves this success ratio for 1 KiB, while for 4 KiB and 16 KiB data sizes, the average delivery rate falls below 85%. For a data size of 16 KiB, BTP still achieves an average delivery ratio of about 98%. This result (and the failed runs for 1 KiB and 4 KiB data sizes) is mainly due to the varying conditions of the network and the wireless medium. Even though the tests were conducted in a testbed, there were no lab conditions. The testbed was deployed in a building in the university, where there are offices and lecture rooms with employees and students and a number of other Wi-Fi networks; thus, some of the experiments were not completed. The poor performance of SBP, however, is due to the fact that BTP makes better use of the wireless medium by refraining from flooding the network with maximum transmission power, thus producing less interference between stations. Note that there are outliers in experiments, since the environment can vary under real conditions. For example, in the 1 KiB and 16 KiB experiments, there are runs that have no successful receptions at all, which is outside of the 1.5-fold interquartile range.

Figure 11. Ratio of nodes successfully receiving data.

7.3.4. Time for Broadcast Tree Construction Phase

Figure 12 shows the time taken to construct the broadcast tree. The x-axis shows the different data sizes, while the y-axis shows the construction time in seconds. The different colors represent different finishing thresholds. This figure has three key takeaways. First, the size of the data does not influence the time it takes to construct the broadcast tree. Second, the higher the finishing threshold, the longer it takes to construct the tree, which is the expected result. Third, the average time of seven seconds for the initial construction of the broadcast tree, which is only performed once, is reasonable since the constructed broadcast tree can subsequently be used for an arbitrarily long time period. Note that there are outliers in experiments, since the environment can vary under real conditions. For example, in the 1 KiB SBP experiments, there are runs that take up to 40 s to construct the tree, which is outside of the 1.5-fold interquartile range.

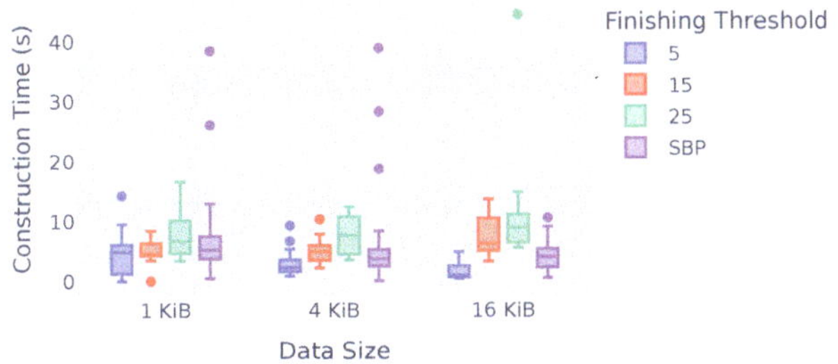

Figure 12. Time taken to construct the broadcast tree.

7.3.5. Contributions of Individual Nodes

Our last evaluation shows how much energy is contributed by individual nodes in the network. Figure 13 is a scatter plot where every dot represents the proportional energy used by a node in a particular experimental run.

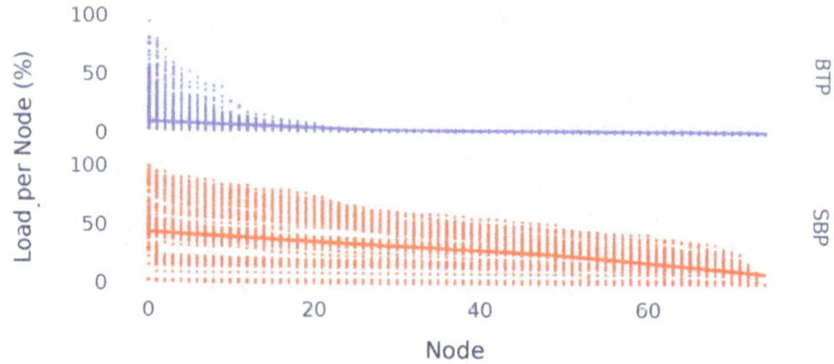

Figure 13. Scatter plot of the load distribution per node over different experimental runs.

The x-axis arranges nodes according to their loads (not their topological relationships) in relation to the total energy consumed over the entire experiment. This means that, for example, node 1 is not next to node 2, but node 1 contributes more to the total amount of energy required than node 2. The y-axis shows how much an individual node contributes to the amount of energy required, e.g., a dot at 50% at 0 means that node 0 contributes 50% of the required energy. The sub-plots and colors represent the two protocols with all experimental configurations. Using BTP, only a few nodes had a relatively high level of energy usage, while the nodes beyond node 20 barely contributed to the overall level of energy consumption. In SBP, on the other hand, all nodes contributed to overall energy utilization. This shows that although there are nodes in BTP that have a high load, the energy for most nodes is preserved, giving them a higher lifetime when battery-powered. Note that Figure 13 is a scatter plot covering all experimental runs; thus, it is not easy to see which values sum to 100%. There were experiments where, taking SBP as an example, a single node was responsible for 100% of the energy consumed. In this specific experiment, no other node sent any data, indicating that one node in the neighborhood of the source node received the data. This was the case for the unfinished experiments with a 1 KiB data size and a finishing threshold of 5, as indicated in Figure 11.

8. Conclusions

In this study, we presented BTP, a novel broadcasting protocol for wireless multi-hop networks based on a game-theoretical model and designed to function in practical implementations. BTP constructs a spanning tree in a decentralized manner to minimize the total energy consumption of an entire network by minimizing the transmission power at each node. In this section, we highlight our contributions as well as directions for future work.

8.1. Contributions

We made the following contributions. First, we adopted a game-theoretical model for the design of BTP while preserving the relevant convergence characteristics of the game-theoretical model. Second, we addressed the practical constraints in the implementation of BTP, i.e., (a) each node operates only on information that is locally available in its neighborhood, (b) all nodes operate asynchronously without requiring any form of time synchronization, and (c) ping-pong effects possibly occurring in the original game-theoretical model are avoided. Third, we integrated three algorithms capable of inhibiting the creation of graph cycles into the design of BTP. Finally, we evaluated BTP with respect to various aspects. We performed simulations to compare BTP to other algorithms from the literature and investigate the scalability of BTP. A practical implementation of BTP on a testbed with 75 Raspberry Pis allowed us to evaluate BTP under realistic conditions. BTP was able to achieve a total energy reduction of up to 90% compared to a simple broadcast protocol in our testbed.

8.2. Future Work

There are several areas for future work. For example, the current BTP implementation does not consider node mobility, which would require the spanning tree to be maintained continuously, including during the data dissemination phase. Furthermore, BTP can be enhanced with mechanisms that increase the reliability of data transfers, such as acknowledgments or checksums. In its current form, BTP also only allows data dissemination from the source node to the other nodes. However, as soon as a broadcast tree is constructed, bi-directional data transfer should be supported, e.g., to allow the source node to function as a data sink for sensor nodes. Moreover, once the broadcast tree has been built, routing protocols such as AODV [48] or DSR [49] may be used to transmit data to a particular node. Finally, BTP is based on a conventional medium access control mechanism with random back-off times, which works well in many scenarios. Future work might also incorporate coordinated medium access control mechanisms to provide real-time guarantees and further enhance the reliability of BTP [50].

Author Contributions: Conceptualization, A.S. and R.K.; Data curation, A.S., R.K. and J.H.; Formal analysis, A.S., R.K. and B.S.; Funding acquisition, A.K., M.H. and B.F.; Methodology, A.S., R.K., M.S. and J.H.; Software, A.S., R.K., M.S., J.H. and J.L.; Supervision, A.K., M.H. and B.F.; Validation, A.S. and R.K.; Visualization, A.S., R.K., M.S. and J.H.; Writing—original draft, A.S., R.K. and B.F.; Writing—review & editing, A.S., R.K., B.S., A.K., M.H. and B.F. All authors have read and agreed to the published version of the manuscript.

Funding: This work was funded by the Hessian State Ministry for Higher Education, Research and the Arts (HMWK) (LOEWE emergenCITY, LOEWE Natur 4.0), and the German Research Foundation (DFG, Project 210487104—Collaborative Research Center SFB 1053 MAKI, Projects A3, B3, C1, and C5).

Institutional Review Board Statement: Not applicable.

Informed Consent Statement: Not applicable.

Data Availability Statement: Publicly available datasets were analyzed in this study. This data can be found here: https://uni-marburg.de/nXNAEL (accessed on 4 July 2023). Furthermore, the code written for the protocol, the NS-3 simulation, and the experiments can be found in the following Github repositories: https://github.com/umr-ds/broadcast-tree (accessed on 4 July 2023), https://github.com/umr-ds/broadcast-tree-protocol (accessed on 4 July 2023), https://github.com/umr-ds/broadcast-tree-ns3 (accessed on 4 July 2023).

Acknowledgments: We would like to thank Mahdi Mousavi for making his original Matlab code [5] available to us.

Conflicts of Interest: The authors declare no conflict of interest.

References

1. Kadota, I.; Sinha, A.; Uysal-Biyikoglu, E.; Singh, R.; Modiano, E. Scheduling Policies for Minimizing Age of Information in Broadcast Wireless Networks. *Trans. Netw.* **2018**, *26*, 2637–2650. [CrossRef]
2. Malkhi, D.; Reiter, M.K.; Rodeh, O.; Sella, Y. Efficient Update Diffusion in Byzantine Environments. In Proceedings of the 20th Symposium on Reliable Distributed Systems, New Orleans, LA, USA, 28–31 October 2001; pp. 90–98.
3. Kuorilehto, M.; Hännikäinen, M.; Hämäläinen, T.D. A Survey of Application Distribution in Wireless Sensor Networks. *EURASIP J. Wirel. Commun. Netw.* **2005**, *2005*, 1–15. [CrossRef]
4. Li, D.; Jia, X.; Liu, H. Energy Efficient Broadcast Routing in Static Ad Hoc Wireless Networks. *Trans. Mob. Comput.* **2004**, *3*, 144–151.
5. Mousavi, M.; Al-Shatri, H.; Wichtlhuber, M.; Hausheer, D.; Klein, A. Energy-efficient Data Dissemination in Ad Hoc Networks: Mechanism Design with Potential Game. In Proceedings of the International Symposium on Wireless Communication Systems, Brussels, Belgium, 25–28 August 2015; pp. 616–620.
6. Wattenhofer, R.; Li, L.; Bahl, P.; Wang, Y.M. Distributed Topology Control for Power Efficient Operation in Multihop Wireless Ad Hoc Networks. In Proceedings of the 20th International Conference on Computer Communications, Anchorage, AK, USA, 22–23 April 2001; pp. 1388–1397.
7. Wieselthier, J.E.; Nguyen, G.D.; Ephremides, A. Energy-efficient Broadcast and Multicast Trees In Wireless Networks. *Mob. Netw. Appl.* **2002**, *7*, 481–492. [CrossRef]

8. Rahnavard, N.; Vellambi, B.N.; Fekri, F. Distributed Protocols for Finding Low-cost Broadcast and Multicast Trees in Wireless Networks. In Proceedings of the 5th Annual Conference on Sensor, Mesh and Ad Hoc Communications and Networks, San Francisco, CA, USA, 16–20 June 2008; pp. 551–559.
9. Khabbazian, M.; Saffar, K.G. The Gain of Energy Accumulation in Multi-Hop Wireless Network Broadcast. *Trans. Netw.* **2019**, *27*, 1830–1844. [CrossRef]
10. Cheng, L.; Niu, J.; Luo, C.; Shu, L.; Kong, L.; Zhao, Z.; Gu, Y. Towards Minimum-delay and Energy-efficient Flooding in Low-duty-cycle Wireless Sensor Networks. *Comput. Netw.* **2018**, *134*, 66–77. [CrossRef]
11. Moon, K.; Yoo, D.S.; Lee, W.; Oh, S.J. Receiver Cooperation in Topology Control for Wireless Ad-hoc Networks. *Trans. Wirel. Commun.* **2014**, *14*, 1858–1870. [CrossRef]
12. Han, R.; Yang, W.; Wang, Y.; You, K. DCE: A Distributed Energy-efficient Clustering Protocol For Wireless Sensor Network Based on Double-phase Cluster-head Election. *Sensors* **2017**, *17*, 998. [CrossRef]
13. Yang, L.; Lu, Y.; Xiong, L.; Tao, Y.; Zhong, Y. A Game Theoretic Approach for Balancing Energy Consumption in Clustered Wireless Sensor Networks. *Sensors* **2017**, *17*, 2654. [CrossRef]
14. Kuehne, A.; Le, H.Q.; Mousavi, M.; Wichtlhuber, M.; Hausheer, D.; Klein, A. Power Control in Wireless Broadcast Networks Using Game Theory. In Proceedings of the 10th International ITG Conference on Systems, Communications and Coding, Hamburg, Germany, 2–5 February 2015; pp. 1–5.
15. Kumar, S.; Goswami, A.; Gupta, R.; Singh, S.P.; Lay-Ekuakille, A. A Game-theoretic Approach for Cost-effective Multicast Routing in the Internet of Things. *Internet Things* **2022**, *9*, 18041–18053. [CrossRef]
16. Mousavi, M.; Al-Shatri, H.; Klein, A. Cost Sharing Games for Energy-efficient Multi-hop Broadcast in Wireless Networks. *Trans. Wirel. Commun.* **2019**, *19*, 310–324. [CrossRef]
17. Kim, B.S.; Suh, B.; Seo, I.J.; Lee, H.B.; Gong, J.S.; Kim, K.I. An Enhanced Tree Routing Based on Reinforcement Learning in Wireless Sensor Networks. *Sensors* **2022**, *23*, 223. [CrossRef] [PubMed]
18. Ahmed, A.; Abdullah, S.; Bukhsh, M.; Ahmad, I.; Mushtaq, Z. An Energy-Efficient Data Aggregation Mechanism for IoT Secured by Blockchain. *Access* **2022**, *10*, 11404–11419. [CrossRef]
19. Chen, F.W.; Kao, J.C. Game-based Broadcast over Reliable and Unreliable Wireless Links in Wireless Multihop Networks. *Trans. Mob. Comput.* **2012**, *12*, 1613–1624. [CrossRef]
20. Karimi-Bidhendi, S.; Guo, J.; Jafarkhani, H. Energy-efficient Deployment in Static and Mobile Heterogeneous Multi-hop Wireless Sensor Networks. *Trans. Wirel. Commun.* **2021**, *21*, 4973–4988. [CrossRef]
21. Khan, M.N.; Rahman, H.U.; Almaiah, M.A.; Khan, M.Z.; Khan, A.; Raza, M.; Al-Zahrani, M.; Almomani, O.; Khan, R. Improving Energy Efficiency with Content-based Adaptive Scheduling in Wireless Sensor Networks. *Access* **2020**, *8*, 176495–176520. [CrossRef]
22. Dusia, A.; Ramanathan, R.; Ramanathan, W.; Servaes, C.; Sethi, A.S. ECHO: Efficient Zero-control-packet Broadcasting for Mobile Ad Hoc Networks. *Trans. Mob. Comput.* **2021**, *21*, 3163–3175. [CrossRef]
23. Maric, I.; Yates, R.D. Cooperative Multihop Broadcast for Wireless Networks. *J. Sel. Areas Commun.* **2004**, *22*, 1080–1088. [CrossRef]
24. Marques, D.; Senna, C.; Luís, M. Forwarding in Energy-constrained Wireless Information Centric Networks. *Sensors* **2022**, *22*, 1438. [CrossRef] [PubMed]
25. Jiang, P.; Xu, Y.; Liu, J. A Distributed And Energy-efficient Algorithm for Event K-Coverage in Underwater Sensor Networks. *Sensors* **2017**, *17*, 186. [CrossRef]
26. Sinha, A.; Modiano, E. Throughput-optimal Broadcast in Wireless Networks with Point-to-multipoint Transmissions. *Trans. Mob. Comput.* **2021**, *20*, 232–246. [CrossRef]
27. Reina, D.; Günes, M.; Toral, S. Real Experimentation of Probabilistic Broadcasting Algorithms Based on Dissimilarity Metrics for Multi-hop Ad Hoc Networks. *Ad Hoc Netw.* **2016**, *47*, 1–15. [CrossRef]
28. Liu, W.; Nakauchi, K.; Shoji, Y. A Neighbor-based Probabilistic Broadcast Protocol for Data Dissemination in Mobile IoT Networks. *Access* **2018**, *6*, 12260–12268. [CrossRef]
29. Mager, F.; Baumann, D.; Trimpe, S.; Zimmerling, M. Toward Fast Closed-loop Control Over Multi-hop Low-power Wireless Networks. In Proceedings of the 17th International Conference on Information Processing in Sensor Networks, Porto, Portugal, 11–13 April 2018; pp. 158–159.
30. Park, H.; Kim, B.S.; Kim, K.H.; Shah, B.; Kim, K.I. A Tree Based Broadcast Scheme For (m, k)-Firm Real-time Stream in Wireless Sensor Networks. *Sensors* **2017**, *17*, 2578. [CrossRef]
31. Tai, K.Y.; Liu, B.C.; Hsiao, C.H.; Tsai, M.C.; Lin, F.Y.S. A Near-optimal Energy Management Mechanism Considering QoS and Fairness Requirements in Tree Structure Wireless Sensor Networks. *Sensors* **2023**, *23*, 763. [CrossRef]
32. Clausen, T.H.; Jacquet, P. *Optimized Link State Routing Protocol (OLSR)*; The Internet Society: Reston, VA, USA, 2003.
33. Ntlatlapa, N.; Johnson, C.A.D. Simple pragmatic approach to mesh routing using BATMAN. In Proceedings of the International Symposium on Wireless Communications and Information Technology in Developing Countries, Pretoria, South Africa, 6–8 October 2008; pp. 6–7.
34. Ibrahim, M.S.; Shanmugaraja, P.; Raj, A.A. Energy-efficient OLSR Routing Protocol for Flying Ad Hoc Networks. In *Advances in Information Communication Technology and Computing*; Springer: Singapore, 2022; pp. 75–88.

35. De Rango, F.; Fotino, M.; Marano, S. EE-OLSR: Energy Efficient OLSR routing protocol for Mobile Ad-hoc Networks. In Proceedings of the Military Communications Conference, San Diego, CA, USA, 16–19 November 2008; pp. 1–7.
36. Oda, H.; Kulla, E.; Katayama, K. A Modified Energy-aware BATMAN Routing Protocol. In Proceedings of the 20th International Conference on Network-Based Information Systems, Bratislava, Slovakia, 24–26 August 2018; pp. 949–958.
37. Lin, Z.; Lin, M.; De Cola, T.; Wang, J.B.; Zhu, W.P.; Cheng, J. Supporting IoT with rate-splitting multiple access in satellite and aerial-integrated networks. *IEEE Internet Things J.* **2021**, *8*, 11123–11134. [CrossRef]
38. Lin, Z.; Niu, H.; An, K.; Wang, Y.; Zheng, G.; Chatzinotas, S.; Hu, Y. Refracting RIS-aided hybrid satellite-terrestrial relay networks: Joint beamforming design and optimization. *IEEE Trans. Aerosp. Electron. Syst.* **2022**, *58*, 3717–3724. [CrossRef]
39. Huang, C.; Yang, Z.; Alexandropoulos, G.C.; Xiong, K.; Wei, L.; Yuen, C.; Zhang, Z.; Debbah, M. Multi-hop RIS-empowered terahertz communications: A DRL-based hybrid beamforming design. *IEEE J. Sel. Areas Commun.* **2021**, *39*, 1663–1677. [CrossRef]
40. Lin, Z.; An, K.; Niu, H.; Hu, Y.; Chatzinotas, S.; Zheng, G.; Wang, J. SLNR-based secure energy efficient beamforming in multibeam satellite systems. *IEEE Trans. Aerosp. Electron. Syst.* **2022**, *59*, 2085–2088. [CrossRef]
41. Castañeda, E.; Silva, A.; Gameiro, A.; Kountouris, M. An Overview on Resource Allocation Techniques for Multi-User MIMO Systems. *IEEE Commun. Surv. Tutor.* **2017**, *19*, 239–284. [CrossRef]
42. Monderer, D.; Shapley, L.S. Potential Games. *Games Econ. Behav.* **1996**, *14*, 124–143. [CrossRef]
43. MacKenzie, A.B.; DaSilva, L.A. *Game Theory for Wireless Engineers*; Springer Nature: Berlin/Heidelberg, Germany, 2022.
44. Mehlhorn, K.; Sanders, P. *Algorithms and Data Structures: The Basic Toolbox*, 1st ed.; Springer Publishing Company: Berlin/Heidelberg, Germany, 2008.
45. Rappaport, T.S. *Wireless Communications: Principles and Practice*; Prentice Hall PTR: Upper Saddle River, NJ, USA, 1996; Volume 2.
46. Schulz, M.; Wegemer, D.; Hollick, M. The Nexmon Firmware Analysis and Modification Framework: Empowering Researchers to Enhance Wi-Fi Devices. *Comput. Commun.* **2018**, *129*, 269–285. [CrossRef]
47. Höchst, J.; Penning, A.; Lampe, P.; Freisleben, B. PIMOD: A Tool for Configuring Single-board Computer Operating System Images. In Proceedings of the Global Humanitarian Technology Conference, Seattle, WA, USA, 29 October–1 November 2020; pp. 1–8.
48. Chakeres, I.D.; Belding-Royer, E.M. AODV Routing Protocol Implementation Design. In Proceedings of the 24th International Conference on Distributed Computing Systems Workshops, Tokyo, Japan, 23–24 March 2004; IEEE: Piscataway, NJ, USA, 2004; pp. 698–703.
49. Johnson, D.B.; Maltz, D.A.; Broch, J. DSR: The Dynamic Source Routing Protocol for Multi-Hop Wireless Ad Hoc Networks. *Ad Hoc Netw.* **2001**, *5*, 139–172.
50. Klose, R. Decentralized Ultra-Reliable Low-Latency Communications through Concurrent Cooperative Transmission. Ph.D. Thesis, Technische Universität Darmstadt, Darmstadt, Germany, 2023.

sensors

MDPI

Article

Efficient Sigma–Delta Sensor Array Beamforming

Sammy Johnatan Carbajal Ipenza [1] and Bruno Sanches Masiero [2,*]

1 NXP Semiconductors N.V., 5656 Eindhoven, The Netherlands
2 School of Electrical and Computer Engineering, Universidade Estadual de Campinas, Campinas 13083-852, Brazil
* Correspondence: masiero@unicamp.br

Abstract: Nowadays, sensors with built-in sigma–delta modulators (ΣΔMs) are widely used in consumer, industrial, automotive, and medical applications, as they have become a cost-effective and convenient way to deliver data to digital processors. This is the case for micro-electro-mechanical system (MEMS), digital microphones that convert analog audio to a pulse-density modulated (PDM) bitstream. However, as the ΣΔMs output a PDM signal, sensors require either built-in or external high-order decimation filters to demodulate the PDM signal to a baseband multi-bit pulse-code modulated (PCM) signal. Because of this extra circuit requirement, the implementation of sensor array algorithms, such as beamforming in embedded systems (where the processing resources are critical) or in very large-scale integration (VLSI) circuits (where the power and area are crucial) becomes especially expensive as a large number of parallel decimation filters are required. This article proposes a novel architecture for beamforming algorithm implementation that fuses delay and decimation operations based on maximally flat (MAXFLAT) filters to make array processing more affordable. As proof of concept, we present an implementation example of a delay-and-sum (DAS) beamformer at given spatial and frequency requirements using this novel approach. Under these specifications, the proposed architecture requires 52% lower storage resources and 19% lower computational resources than the most efficient state-of-the-art architecture.

Keywords: MAXFLAT; PDM; sigma–delta; microphone; sensor array; decimation; beamforming

Citation: Carbajal Ipenza, S.J.; Masiero, B.S. Efficient Sigma–Delta Sensor Array Beamforming. *Sensors* **2023**, *23*, 7577. https://doi.org/10.3390/s23177577

Academic Editor: Josip Lorincz

Received: 30 June 2023
Revised: 23 August 2023
Accepted: 25 August 2023
Published: 31 August 2023

1. Introduction

In the last decades, sensor array processing has emerged as an active area of research in estimating space-time parameters. Array-processing applications are applied to solve many real-world problems. In telecommunications, for example, antenna arrays are steered in one user direction to reduce user interference. Radar and sonar use arrays of antennas and hydrophones, respectively, to calculate parameters like direction of arrival (DoA), velocity, and range. In medicine, sensor arrays are used for medical imaging, and planar biomagnetic sensor arrays are used in electrocardiograms to localize brain activity. In industry, sensor arrays are used in automatic monitoring and fault detection [1].

More recently, microphone array processing has emerged to increase the audio quality in consumer devices like mobile phones, speakerphones, and smart speakers, which are broadly used in conference rooms, desktop devices, and intelligent virtual assistants (IVA), in both consumer and industrial devices. Most frequently, the signals from several microphones are combined via a beamforming algorithm to enhance the sound coming from a desired direction while attenuating ambient noise and interference [1].

However, microphone array implementations are still expensive due to the complex characteristics of speech signals (non-static source, intermittent, and broadband) and the usual environmental conditions (reverberation and non-stationary additive noise). Adding an extra microphone in the design requires new routing, new placement conditions, and more processing resources, increasing the system cost and power consumption, a critical factor for internet of things (IoT) and mobile applications.

Digital MEMS microphones (introduced in 2006 [2]) have emerged as an alternative to overcome the size and cost limitations. As these microphones have an analog-to-digital converter (ADC) incorporated as a pre-amplifier [3], they have a single line PDM output; because of that, they are also known as PDM microphones (PDM-mics). A decimation filter (also known as a PDM-to-PCM converter) demodulates this PDM bitstream output to a PCM signal. Unfortunately, implementing this decimation filter is still not cheap, as its cost (measured in die area and power) increases with the quality of the desired audio signal. Take, for example, the case of a microphone array using these PDM-mics. This architecture requires a decimation filter for each microphone input so that the implementation cost and power consumption will increase proportionally with the number of microphones, being even more expensive for practical applications.

This paper proposes a novel and economical method to implement beamforming algorithms with arrays of MEMS digital microphones. We apply the new architecture to a DAS beamformer as a proof of concept, but it can also be used with other beamforming strategies. This method merges a conventional beamformer's filtering and delays operations into a single structure dubbed as *delayed decimation filter*. We propose a J-stage decimation filter whose penultimate stage ($J-1$) is a Samadi filter, and its last stage (J) is an equiripple filter. The Samadi filter controls the overall filter delay by adjusting a single parameter, and the last equiripple stage compensates for the magnitude and phase distortion caused by the Samadi filter under a specific limit.

In the end, the proposed delayed decimation filter is an "all-in-one" filter that performs the same filtering and downsampling operations as any state-of-the-art decimation filter, has the capability of altering its group delay without any change in its structure or additional delay chain, and provides storage and computational resources savings in comparison to state-of-the-art architectures.

To explain the working principle of the proposal, we first recapitulate the implementation of a DAS beamformer in Section 2. We then present a novel beamformer based on delayed decimation filters in Section 3, where we introduce multirate and decimation filters, as well as how a Samadi filter can be used with these structures. To conclude, as a proof of concept, we present in Section 4 an implementation of this novel architecture, and in Section 5, we compare it to state-of-the-art DAS beamformer architectures.

2. DAS Beamformer

The DAS beamformer is the oldest and simplest array signal processing algorithm [1]. The underlying idea is to delay each microphone input by an appropriate time delay and then add all delayed microphone signals together. In this sense, the audio signal arriving from a particular direction at the array is reinforced in relation to signals coming from different directions and incoherent noise.

The traditional or discrete-time DAS beamformer (In the literature, the traditional DAS does not have the weights w_m in its temporal representation because these weights only show up if you use a "weighted DAS" or a frequency representation; however, in this work the "weighted DAS" is referred to as the traditional DAS, as w_m can implement the averaging process.) is the result of

$$z[k] = \sum_{m=0}^{M-1} w_m y_m[k-k_m], \tag{1}$$

where y_m is the mth microphone's output in PCM representation and k_m is the integer delay associated with the mth microphone, such that

$$k_m = [\Delta_m/T] = [\Delta_m f_o], \tag{2}$$

where Δ_m is the required delay in the mth microphone, $[x]$ means the nearest integer to x, and f_o and T are the sampling rate and period in y_m, respectively.

In case of PDM-mics, Equation (1) can be represented as shown in Figure 1, such that y_m is the decimation filter's output and x_m is the PDM bitstream incoming from the respective mth PDM-mic.

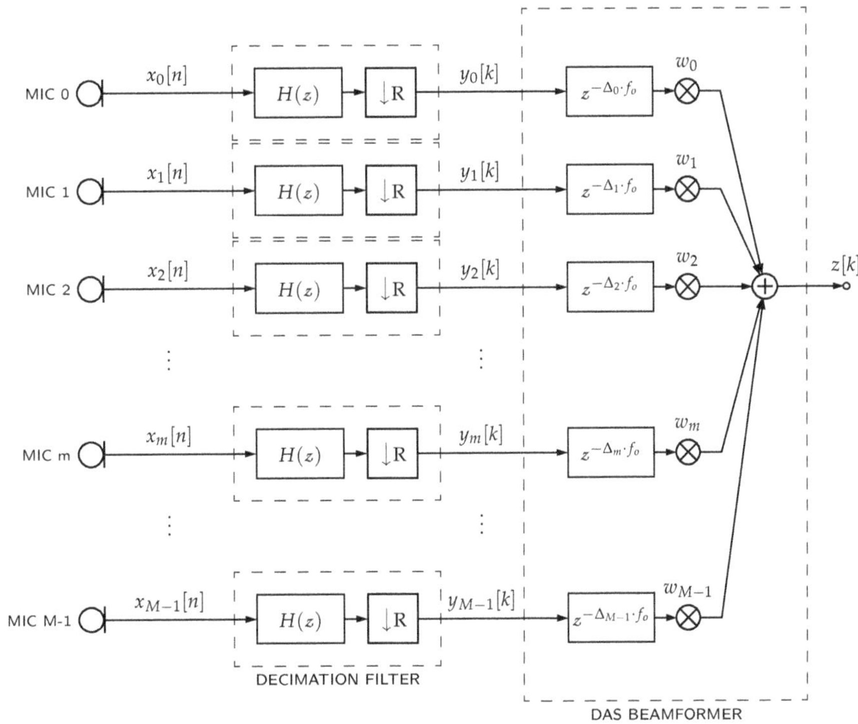

Figure 1. PDM microphones' DAS beamformers. Each PDM-mic requires a decimation filter with $H(z)$ frequency response and R downsampling. Then, each filter output $y_m[k]$ is delayed by a Δ_m factor. Finally, all delayed signals are weighted (factor w_m) and summed together.

Due to the integer nature of k, the DAS beamformer does not allow one to form sums that involve noninteger multiples of T. Consequently, beams cannot be steered in arbitrary directions, resulting in a directivity pattern with a stepped response due to the integer nature of the delay elements, which limits the beamformer resolution (as exemplified in Figure 2).

Also, if one assumes uncorrelated noise at the locations of the sensors and that the beamformer's delays are appropriately matched to the wave's DoA, it can be proven [4] that the beamformer gain (G) depends only on the weights w_m and the number of microphones:

$$G = \frac{\left(\sum_{m=0}^{M-1} w_m\right)^2}{\sum_{m=0}^{M-1} w_m^2}, \tag{3}$$

so that, for the beamformer in Figure 2, with $M = 40$ and $w_m = 1$, the white noise gain will be $G = 40$ or 32 dB. Furthermore, the dynamic range depends only on the number of elements in the array. The array used for the current example provides a dynamic range of 13 dB.

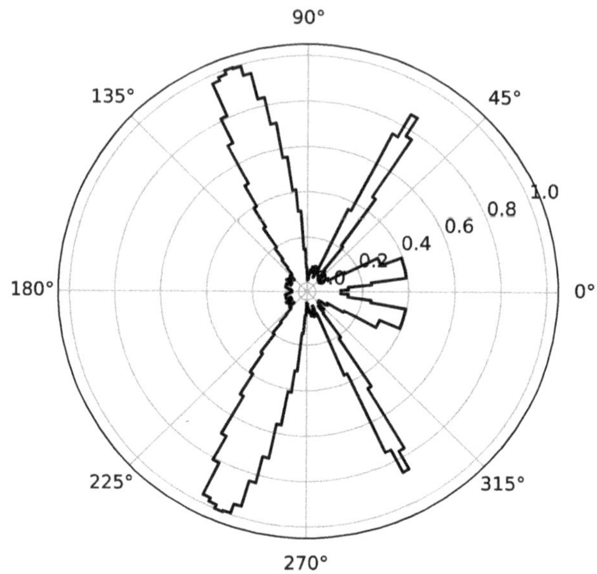

Figure 2. Normalized power (polar) of a uniform linear array of an $M = 40$ microphones DAS beamformer. Three audio sources of 1 kHz, 3 kHz, and 5 kHz are located at 20, 60, and 110 degrees, respectively, i.e., the three with equal strength. The beamformer is placed on the X-axis. Therefore, its directivity pattern is symmetric about this axis.

3. Beamformer Based on Delayed Decimation Filter

Figure 1 describes a typical architecture for implementing DAS beamformers with PDM-mics. For each PDM-mic, there is an associated decimation filter to convert the PDM bitstream into a PCM bitstream and a delay line to steer the beamformer. To devise a more economical implementation of this architecture, we propose to merge the decimation filtering and the delaying operations into a single structure. To explain how a Samadi filter can be used for this purpose, we first review the concept of multirate and decimation filters, present the Samadi filter structure, show how it can be used as a multirate filter, and finally propose a new beamforming architecture based on this multirate filter (delayed decimation filter).

3.1. Multirate and Decimation Filters

Multirate filters are digital filters whose different parts operate at different rates. The most obvious application of such a filter is when the input and output sample rates must differ (decimation or interpolation). A decimation filter is a class of multirate filters [5] that decreases a signal sampling rate by an integer or fractional factor. Figure 3 shows a generic decimation filter structure, where the input signal at f_i sampling rate passes through a low-pass filter (LPF) with impulse response $H(z)$, and then it is downsampled by a factor R to an output sampling rate $f_o = f_i/R$. In the case of a PDM-mic, usually, $x[n]$ has a one-bit width only while $y[k]$ is a multi-bit output.

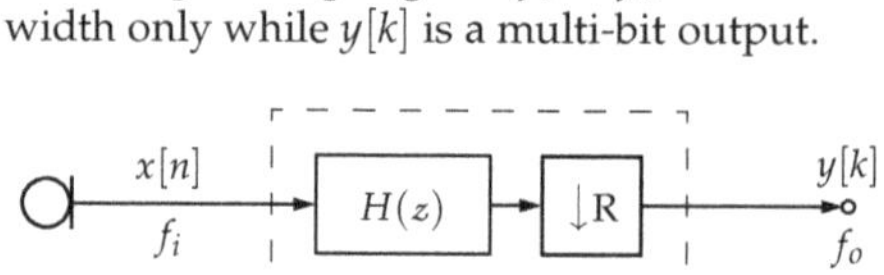

Figure 3. Generic decimation filter structure. In order to avoid aliasing, the input data $x[n]$ at f_i sampling rate is low-pass filtered and then downsampled by R. If correctly filtered, the output data $y[n]$ at f_o sampling rate contain the same information as $x[n]$ decimated by R.

For a given application, there are many design parameters to be taken into account for the LPF design, such as filter passband frequency F_p, stopband frequency F_s, passband ripple δ_p, and stopband ripple δ_s, as exemplified in Figure 4. Those LPF design parameters are related as follows:

$$U_p = \{f : f \in [0, F_p]\} \tag{4a}$$
$$U_s = \{f : f \in [F_s, f_i]\} \tag{4b}$$

$$\delta_p = \max(||H(e^{2\pi i f/f_i})| - 1|) \quad \forall f \in U_p, \tag{4c}$$
$$\delta_s = \max(|H(e^{2\pi i f/f_i})|) \quad \forall f \in U_s, \tag{4d}$$

where U_p and U_s are the passband and stopband frequency ranges, respectively. Also, the angular passband and stopband frequencies can be expressed as

$$\omega_p = \frac{2\pi F_p}{f_i}, \tag{5a}$$
$$\omega_s = \frac{2\pi F_s}{f_i}, \tag{5b}$$

and U_p and U_s intervals can be scaled to angular frequency domain as

$$V_p = \frac{2\pi U_p}{f_i}, \tag{6a}$$
$$V_s = \frac{2\pi U_s}{f_i}. \tag{6b}$$

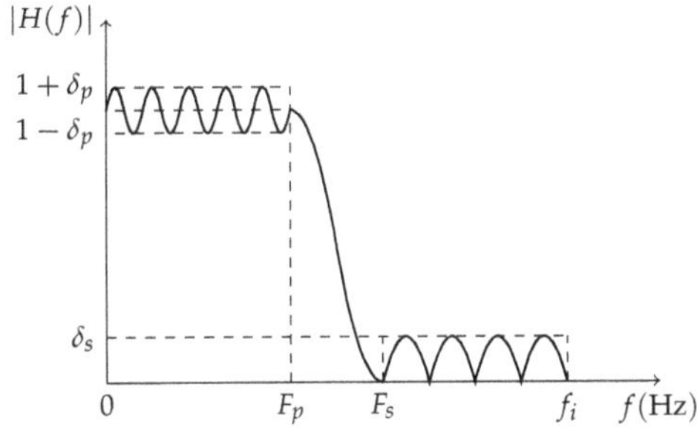

Figure 4. Low-pass filter design parameters. The passband and stopband regions are defined by F_p and F_s, respectively, and their respectives ripples are defined by δ_p and δ_s. The whole filter frequency response is constrained to the input sampling rate (f_i).

In the case of audio sensors such as MEMS microphones, a decimation filter is required to convert the oversampled output from the internal ADC to a standard audio PCM output. Baseband signal quality parameters such as linearity, signal-to-noise ratio (SNR), total harmonic distortion (THD), and total harmonic distortion plus noise (THD+N) can be worsened at the filter output if the LPF is not properly designed [6]. Also, the LPF structure should be carefully chosen to obtain a proper phase response. A Finite Impulse Response (FIR) structure, for example, can be used if a linear phase is required; otherwise, Infinite Impulse Response (IIR) filters are preferred, as, usually, IIR filters are smaller than their equivalent FIR implementations. Moreover, some applications tolerate some degree of non-linearity in phase; in this case, quasi-linear filters, a mixture of FIR and IIR filters, can be used.

3.2. Universal Maximally Flat Samadi Filter

As derived in [7], the transfer function in Samadi filters is defined by

$$H_{N,K,d}(z) = \sum_{j=0}^{N-K} c_j \left(\frac{1-z^{-1}}{2}\right)^j \left(\frac{1+z^{-1}}{2}\right)^{N-j}, \tag{7}$$

where

$$c_j = \sum_{i=0}^{j} (-1)^{j-i} \binom{\frac{N}{2}-d}{i} \binom{\frac{N}{2}+d}{j-i}, \tag{8}$$

K is the number of zeros at $z = -1$, N is the filter order, and the delay parameter d is a real number defined as

$$d = \alpha - \frac{N}{2}. \tag{9}$$

For a given group delay α, such that $0 \leq \alpha \leq N$, from (9), one can verify that

$$-\frac{N}{2} \leq d \leq \frac{N}{2} \tag{10}$$

or

$$|d| \leq d_{\max}, \tag{11}$$

where $d_{\max} = N/2$ is the maximum allowed delay parameter and the binomial coefficients in (8) are defined as

$$\binom{r}{s} = \begin{cases} \prod_{q=0}^{s-1} \frac{r-q}{q+1}, & s \geq 1 \\ 1, & s = 0 \\ 0. & s < 0 \end{cases} \tag{12}$$

This filter becomes a maximally flat (MAXFLAT) linear phase FIR when $d = 0$. As shown in [8,9], the angular passband frequency (ω_p) of these linear phase filters is related with N as

$$L \simeq \lceil N\omega_p/\pi + 0.5 \rceil \tag{13}$$

where L is defined for convenience as

$$L = N - K. \tag{14}$$

The cutoff frequency of these linear phase filters increases almost linearly with L, as shown in Figure 5 for different values of N. Also, as demonstrated in [7] and shown in Figure 5a, for linear phase filters ($d = 0$), the coefficient of (7) is

$$c_j|_{d=0} = 0, \quad j \text{ odd}. \tag{15}$$

Then, the magnitude frequency spectra of $L = 2j$ and $L = 2j + 1$ are the same for $j \in \{0, \ldots, \lfloor N/2 - 1 \rfloor\}$. Figure 5 also shows that the filter has a linear phase and that the group delay for $d = 0$ is $\alpha = N/2$, as expected by (9).

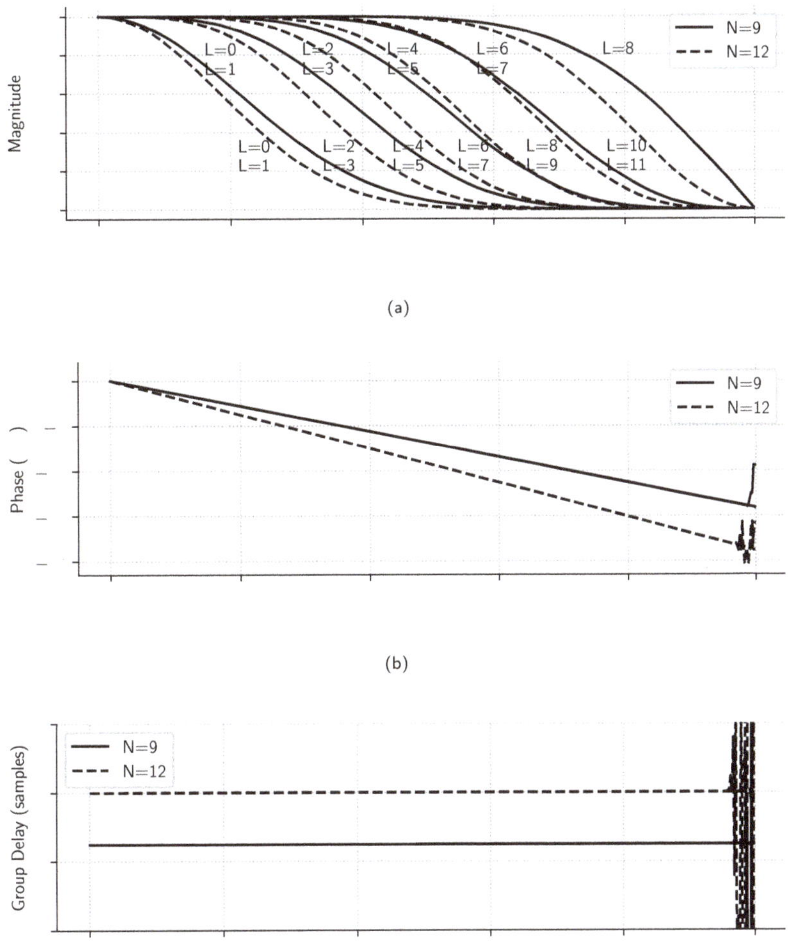

Figure 5. Normalized frequency spectra of linear-phase Samadi filters ($d = 0$) with $N = 9$ and $N = 12$: (**a**) magnitude, (**b**) phase, and (**c**) group delay. It is observed that, in $d = 0$ case, ω_p changes linearly with L, that the phase is linear for both N values and that the group delay is proportional to N.

On the other hand, when $d \neq 0$, the Samadi filter becomes a MAXFLAT nonlinear phase filter. The most interesting characteristic of this filter class is the ability to modify its group delay with the filter delay parameter (d), as given by (9). Figure 6 shows how the flatness of the magnitude and phase of the filter's frequency response is affected when d increases—we see that passband δ_p's ripples worsen as d increases. However, it is also shown that the phase is still linear inside the passband region for $\omega < 0.15\pi$ and that the decimation filter continues under the same specification for all values $|d| \leq 5$. This suggests that this filter can be used as an intermediary stage in a multirate filter chain to adjust the overall filter delay (Δ) and perform low-pass filtering at the same time, as discussed in the following sections.

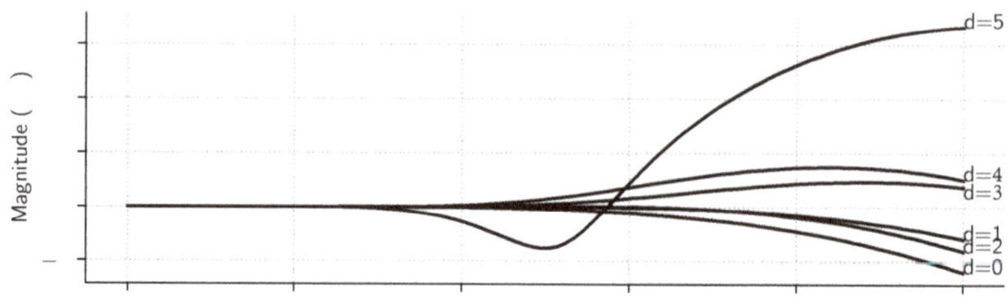

(a)

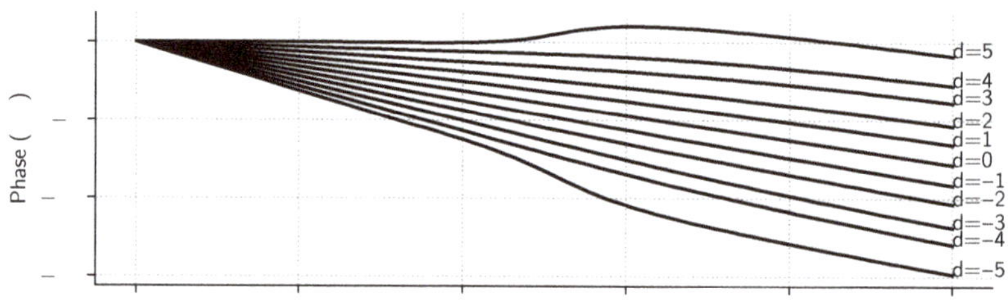

(b)

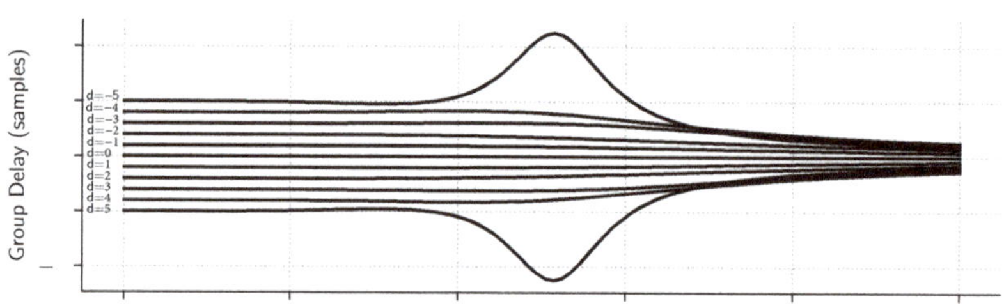

(c)

Figure 6. Normalized frequency spectra of Samadi filters with $N = 10$ and $d \in \{-5, \ldots, 5\}$: (**a**) magnitude, (**b**) phase, and (**c**) group delay. It is observed that, approximately until $\omega/\pi < 0.15$, the magnitude is flat, the phase is linear, and the group delay is proportional to d. For $\omega/\pi \geq 0.15$, the frequency response is nonlinear in magnitude, phase, and group delay.

Finally, we propose Algorithm 1 to calculate the minimum K and N Samadi filter values for a given d, matching a given filter specification with the following parameters: V_p, V_s, δ_p, and δ_s. In lines 2–4, the algorithm initializes ω_p, L, and N values to the minimum possible ones. Then, in line 5, it starts to iterate to calculate the minimum K and N values. In line 6, K is updated. In lines 7–8, δ_p and δ_s are calculated from the filter frequency response for V_p and V_s ranges, respectively, and for the current K and N values. If δ_p and δ_s meet the specification, it returns the parameter values in line 10. Else, in lines 12–26, it increases the N or L value, depending on the d weight or if the filter parameters are inside ranges defined in (13)–(15).

Algorithm 1 Samadi Filter minimum N and K calculation algorithm

```
procedure SamadiMinN(d, δ_p, δ_s, V_p, V_s)
    L ← 0
    N ← 2⌈d⌉
    ω_p ← max(V_p)
    loop
        K = N − L
        δ'_p ← max(|H_{N,K,d}(e^{iω}) − 1|)   ∀ω ∈ V_p
        δ'_s ← max(|H_{N,K,d}(e^{iω})|)   ∀ω ∈ V_s
        if δ'_p ≤ δ_p and δ'_s ≤ δ_s then
            return N, K
        else
            if d = 0 then                                   ▷ Linear-phase filter
                if L > ⌈Nω_p/π + 0.5⌉ then
                    L ← 0
                    N ← N + 1
                else
                    L ← L + 2
                end if
            else                                            ▷ Nonlinear-phase filter
                if δ'_s ≥ 1 or L ≥ N then
                    L ← 0
                    N ← N + 1
                else
                    L ← L + 1
                end if
            end if
        end if
    end loop
end procedure
```

Figure 7 shows minimum N and K values, calculated using Algorithm 1 for $d \in \{0, \ldots, 26\}$ and different values of ω_p. It is shown that the minimum N, required for any d, decreases with ω_p increments, and it is almost three times d when $\omega_p/\pi = 0.28$.

Also, it is essential to remark that, if the Samadi filter is designed for $d_{\max}$, the decimation filter continues under the same specification for values $|d| \leq d_{\max}$. This effect can be observed in Figure 6a, where δ_p decreases for lower values of d, and, in Figure 7, where, for $d \geq 3$, if N is kept constant and d is decreased, ω_p tends to increase so that the flatness is improved.

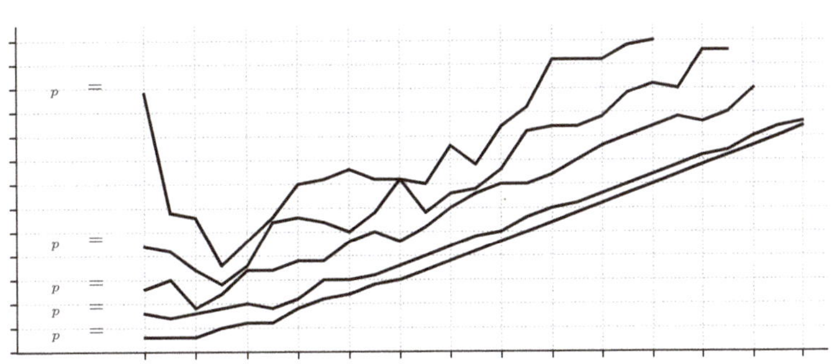

(a)

Figure 7. *Cont.*

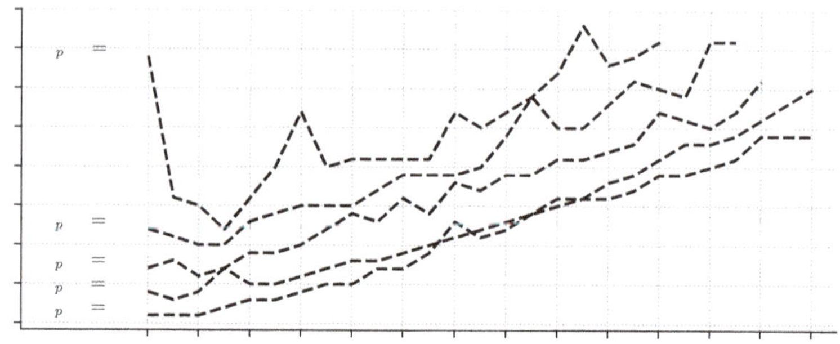

(b)

Figure 7. Minimum (**a**) N and (**b**) K values calculated using Algorithm 1 for $\delta_s = -80\,\text{dB}$ and different values of d and ω_p. It is observed that ω_p and d have a negative correlation for a given N value i.e., when ω_p increases, d decreases.

3.3. Delayed Decimation Filter

Because of its configurable group delay property, a single Samadi filter could be used as the LPF of a multirate filter with adjustable overall filter delay, as shown in Figure 8a—this structure is dubbed in this paper as delayed decimation filter. However, as a Samadi filter does not have the flexibility to be designed for specific F_p and F_s values without changing other filtering parameters, its frequency response needs to be compensated to keep the overall decimation filter's parameters under specification for different delay values (d). For this reason, we propose a J-stages decimation filter architecture whose penultimate stage ($J-1$) is a Samadi filter and its last stage (J) is an equiripple filter, as shown in Figure 8b. The Samadi filter can then be decomposed iton its binomial components, as shown in Figure 8c.

The Samadi filter controls the overall filter delay (Δ) by setting its respective d parameter, and the last equiripple stage compensates for the magnitude and phase distortion caused by the Samadi filter under a specified limit. Also, as this is a multi-stage filter, other filtering stages (1 to $J-2$) can be optionally added to help with decimation and filtering.

The overall filter delay Δ depends on the d, R_{J-1}, and R_J parameters in such a way:

$$\Delta = \frac{d}{R_J R_{J-1} f_o}. \tag{16}$$

If we replace (16) in (11), it is observed that the maximum required delay ($\Delta_{\max}$) is limited by the $d_{\max}$ parameter as follows:

$$|\Delta| \leq \frac{d_{\max}}{R_J R_{J-1} f_o} \quad . \tag{17}$$

Therefore, since $d_{\max} = \Delta_{\max} R_J R_{J-1} f_o$, the minimum K and N parameters can be calculated using Algorithm 1 for $d = d_{\max}$ and the desired filter specification parameters: $\delta_p = \delta_p^j$, $\delta_s = \delta_s^j$, $V_p = V_p^j$, and $V_s = V_s^j$ for $j = J-1$.

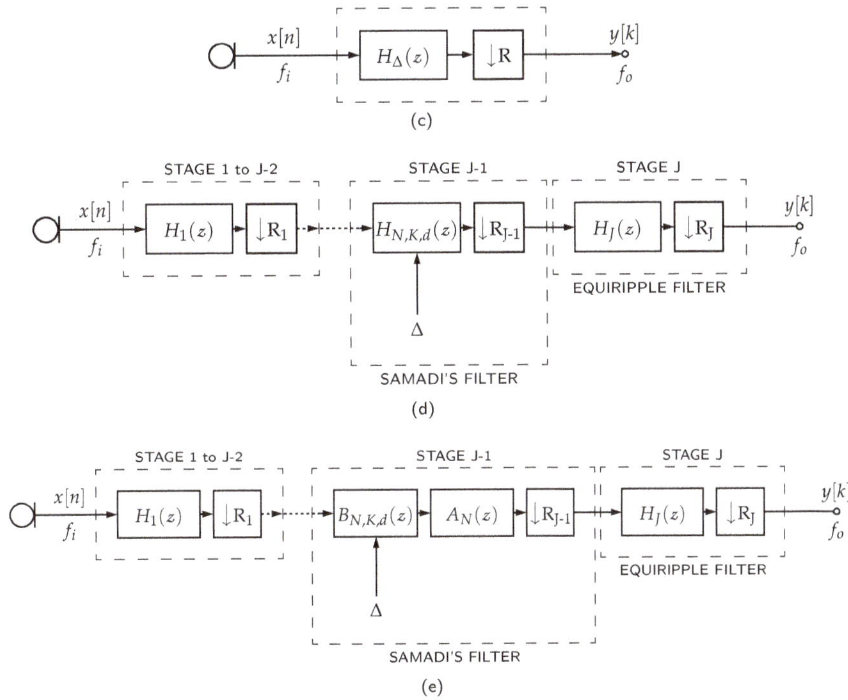

Figure 8. (**a**) Delayed decimation filter, (**b**) its version as a multi-stage decimation filter with the $J-1$ stage being a Samadi filter, and (**c**) its version with Samadi filter decomposed into its binomial components. Samadi filter stage is meant to control the overall filter delay (Δ) and the equiripple filter to compensate the non-linear response of the Samadi filter in its non-flat band. The optional Stages 1 to $J-2$ are meant to compensate and downsample the overall frequency response.

3.4. Optimized Beamformer Structure

Since the Samadi filter is a binomial filter sequence (as first proposed by Haddad in [10]), (7) can be rearranged to allow the filter to be expressed as

$$H_{N,K,d}(z) = \left(\frac{1+z^{-1}}{2}\right)^{N} \sum_{j=0}^{N-K} c_j \left(\frac{1-z^{-1}}{1+z^{-1}}\right)^{j}. \tag{18}$$

The binomial filter in Equation (18) can be realized as a cascade of two filters:

$$H_{N,K,d}(z) = A_N(z) B_{N,K,d}(z), \tag{19}$$

where

$$A_N(z) = \left(\frac{1+z^{-1}}{2}\right)^{N}, \quad B_{N,K,d}(z) = \sum_{j=0}^{N-K} c_j \left(\frac{1-z^{-1}}{1+z^{-1}}\right)^{j}. \tag{20}$$

The Samadi filter stage in a delayed decimation filter in Figure 8c can be expressed in its binomial representation in such a way that the latter part of the filter chain does not depend on Δ, as d is used only for the calculation of c_j. Therefore, if M delayed decimation filters are placed in parallel, the weightings by w_m are placed just before the $A_N(z)$ filter and their outputs are added to form a beamformer. Note that the latter part after $B_{N,K,d}(z)$ can be shared between all microphone channels, as shown in Figure 9.

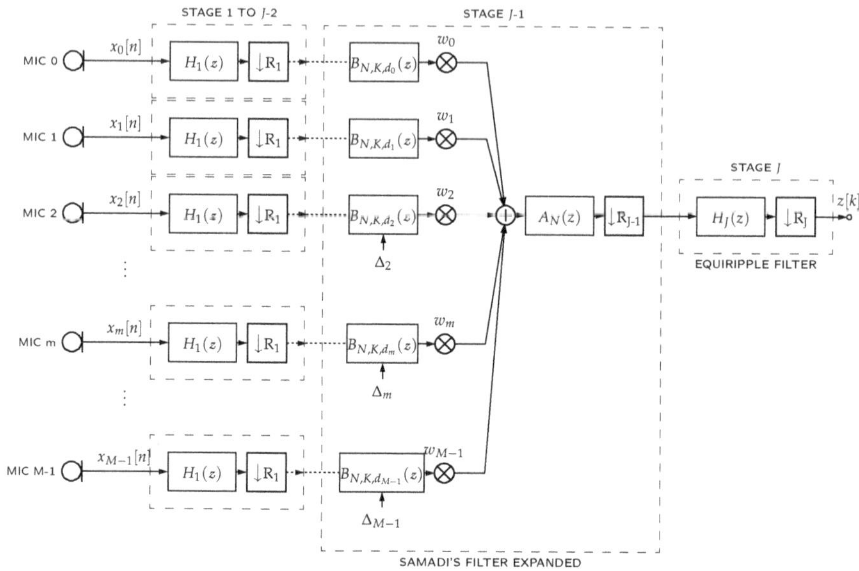

Figure 9. PDM-mic array DAS beamformer using delayed decimation filters.

4. Proof of Concept

We now evaluate the proposed architecture. We determine the delayed decimation filter parameters for a given specification and compare the proposed architecture to state-of-the-art DAS beamformer architectures.

4.1. Decimation Filter Specifications

Filter specifications and array geometries change depending on the beamformer application. Therefore, to compare the efficiency between the proposed method and the straightforward DAS beamformer implementation, we use the specification shown in Table 1 as the basis of all our decimation filter designs, as it is considered enough for most PDM-mic types and speech-processing applications.

Table 1. Decimation filter specifications.

Parameter	Value
input sampling rate (f_i)	3072.0 kHz
output sampling rate (f_o)	16.0 kHz
passband frequency (F_p)	7.5 kHz
stopband frequency (F_s)	8.0 kHz
passband ripple (δ_p)	$\leq$0.0116 ($\leq$0.1 dB)
stopband ripple (δ_s)	$\leq$0.0001 ($\leq -$80.0 dB)
decimation factor (R)	192
filter input length (L_{in})	1
filter output length (L_{out})	24
phase response	linear or almost linear

4.2. Beamformer Specification

The delay from the array center to the mth microphone (Δ_m) in an array is constrained to

$$|\Delta_m| \leq \Delta_{\max} \quad \text{for } m = 0, 1, \ldots, M-1 \tag{21}$$

such that

$$\Delta_{\max} = \frac{|\bar{x}_{\max} - \bar{x}_c|}{c}, \tag{22}$$

where $\bar{x}_{\max}$ is the furthest sensor location in relation to $\bar{x}_c$ (which is the array's center reference), M is the number of microphones, and c is the sound speed (typically 343.0 m/s).

Assume that we require a microphone array for hands-free applications that, when placed 80 cm from the voice source, would attain the same SNR as the SNR obtained by a single microphone placed 2 cm from the same source [11]. Then, by (3), the desired microphone array requires $M = 40$ microphones.

Also, as the minimum distance between microphones should be $D_{\min} \leq c/2F_p$ to avoid spatial aliasing, if the frequency range is limited to $F_p = 7.5$ kHz, then the desired microphone array will require $D_{\min} \leq 2$ cm. Finally, as $M = 40$, if a 5×8 microphone array is assumed, then the $\Delta_{\max}$ can be calculated using (22), with the resulting value shown in Table 2.

Table 2. Microphone array specifications.

Parameter	Value
number of microphones (M)	40 (5×8)
minimum distance between microphones ($D_{\min}$)	22.0 mm
array dimensions	110.0 mm × 176.0 mm
maximum required delay ($\Delta_{\max}$)	314.47 µs
mth-filter channel gain (w_m)	1
frame length (for frequency domain implementations) (L_{frame})	4.0 ms

4.3. Filter Design

A delayed decimation filter was designed according to specifications listed in Table 1. The filter has a three-stage architecture ([*lthband, maxflat, equir*]) with respective decimation rates [48, 2, 2]. The *lthband* stage is an LPF whose cutoff frequency is π/L, and the impulse response is zero for every L-th sample [5]. The second stage is a *maxflat* Samadi filter, and the last is an equiripple filter [12]. As $R_J = R_{J-1} = 2$, by (17), $d_{\max} = 20.13$; the parameters N and K of the *maxflat* stage are calculated using Algorithm 1 so that the overall filter specification is kept for all $|d| \leq 20.13$.

Figure 10a shows the individual frequency spectrum of each internal stage for $d_{\max} = 20.13$, and Figure 10b zooms in the passband frequency region. Note that even though the *maxflat* stage has a bumpy frequency spectrum above the passband frequency (F_p), this is compensated by the last stage equiripple filter (*equir*). Figure 11a also shows that the magnitude in the overall frequency spectrum of the delayed decimation filter is inside the required passband and stopband filter specifications, while Figure 11b,c show that the filter phase and magnitude response is almost linear in the passband range.

The advantage of using a Samadi filter is that it allows one to change its group delay by changing some coefficients, i.e., without changing the whole filter structure. Figure 12 shows the group delay of this multi-stage filter for many values of its d parameter. It is easy to see how the group delay is directly proportional to the d parameter.

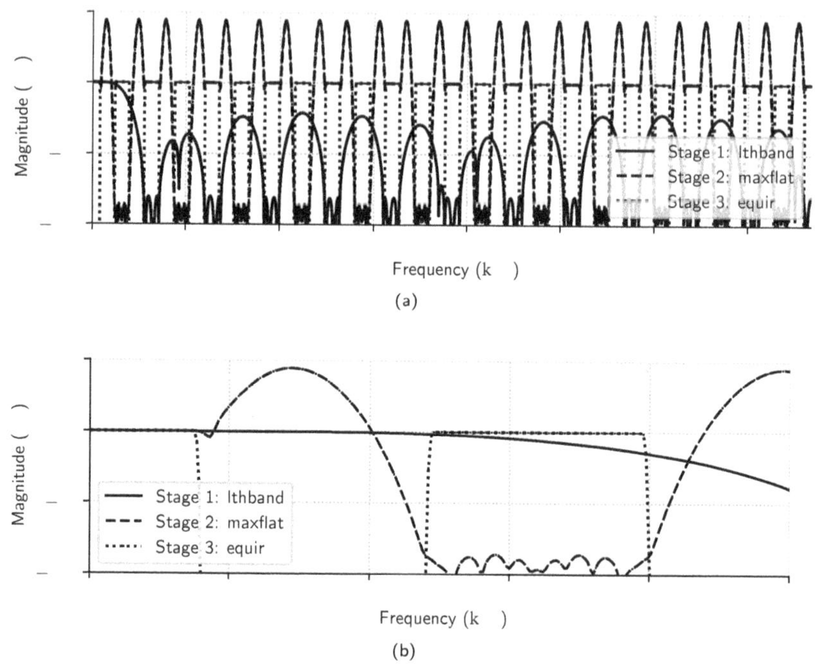

Figure 10. (**a**) Magnitude frequency spectrum of internal stages of the delayed decimation filter in the whole input range, and (**b**) the same frequency spectrum in the 0 kHz to 50 kHz range.

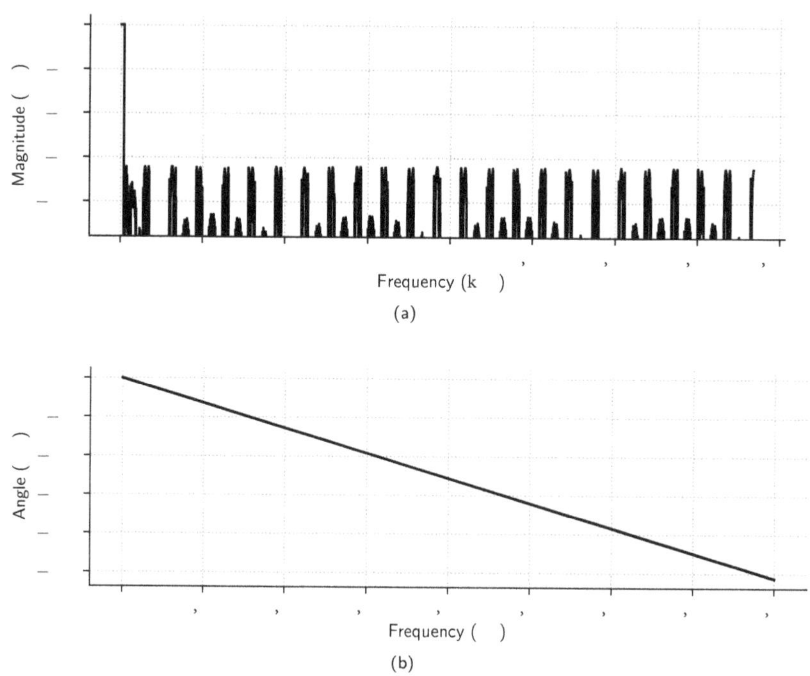

Figure 11. *Cont.*

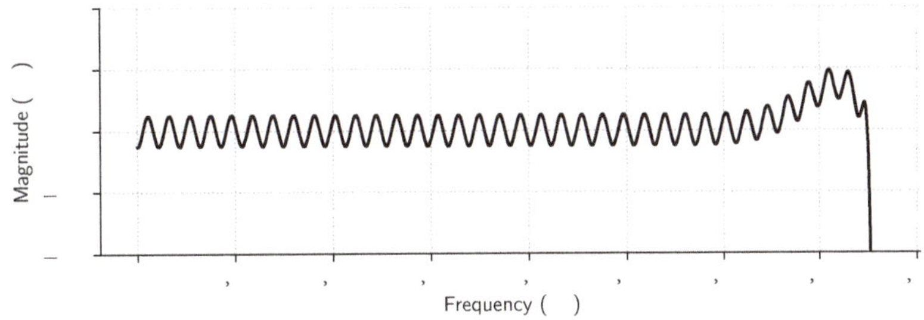

(c)

Figure 11. (**a**) Magnitude and (**b**) phase frequency spectrum of the delayed decimation filter. (**c**) Passband ripple frequency spectrum.

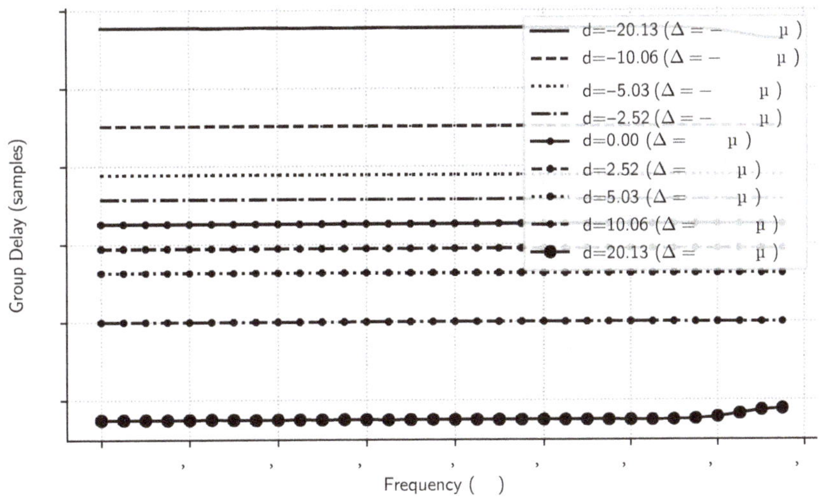

Figure 12. Delayed decimation filter group delay.

Table 3 shows the resources required to implement a DAS beamformer based on this three-stage delayed decimation filter designed for array specifications listed in Table 2, and Table 4 shows the breakdown of resources required per filter stage.

Table 3. Required resources to implement a beamformer using 40 shared delayed decimation filters.

	Value	**Unit**
beamformer's storage requirement (S^z_{bf})	39,478	bit
beamformer's number of multiplications per second (S^*_{bf})	6.9624×10^8	MPS
beamformer's number of additions per second (S^+_{bf})	8.81696×10^8	APS
beamformer's total number of additions per second (S^o_{bf})	2.45858×10^9	APS
estimated minimum frequency in a single-core/single-adder processor (f_{cpu})	2458.58	MHz
estimated number of adders in an FPGA running at 64 MHz (T^+_{FPGA})	39	-
estimated number of adders in a VLSI circuit running at 10 MHz (T^+_{lp})	246	-

Table 4. Delayed decimation filter resource requirements breakdown. The first row corresponds to the Lth-band filter stage, the second and third ones are to the $B_{N,K,d}(z)$ and $A_n(z)$ parts of the Samadi filter, respectively, and the last one to the equiripple filter.

Stage	S^z_{bf} (bit)	S^*_{bf} (MPS)	S^+_{bf} (APS)	S^o_{bf} (APS)	f_{cpu} (MHz)	T^+_{FPGA}	T^+_{lp}
lthband	138	15,680,000	15,616,000	15,616,000	15.62	1	2
maxflat — $B_{N,K,d}(z)$	552	1,536,000	6,144,000	39,936,000	39.94	1	4
maxflat — $A_n(z)$	2714	0	3,776,000	3,776,000	3.78	1	1
equir	9164	5,040,000	5,024,000	171,344,000	171.34	3	18

5. Results

Results from Table 3 are compared to other state-of-the-art DAS beamformer architectures (more details in [13]) in Table 5.

The *pcm_multi* architecture is the same as shown in Figure 1 but uses a multi-stage decimator filter structure for each channel. It has more beamformer's storage requirement and additions per second because of the parallel architecture for delaying and filtering.

The *pcm_single_memsav* architecture is also the same as shown in Figure 1 but uses a single-stage decimation filter with a memory-saving polyphase implementation [14] for each channel. This architecture has the lowest beamformer's storage requirement because of the polyphase implementation. Still, conversely, it also has the most additions per second because more operations are performed at higher sampling rates before downsampling.

The *pdm_multi* architecture is the same as shown in Figure 13. Still, using a multi-stage decimator filter structure in the output is the most efficient state-of-the-art architecture because only a single decimation filter is required, and the delaying operations require only a few bits per channel.

The *pdm_single_memsav* architecture is also the same as shown in Figure 13 but using a single-stage decimation filter with a memory-saving polyphase implementation [14] in the output. It has lower beamformer's storage requirement because of the polyphase implementation, but, conversely, it also requires more additions per second because more operations are performed at higher sampling rates before downsampling.

Table 5 shows that, for the given specification and because of the shared resources for delaying and filtering, the proposed architecture (*delayed_bf*) requires about 19% lower computational resources (additions per second) and 52% lower storage (beamformer's storage requirement) than the most efficient state-of-the-art architecture (*pdm_multi*).

It is also observed that the proposed architecture's storage efficiency is ranked just after the *pcm_single_memsav* architecture. However, as the *pcm_single_memsav* architecture also requires a prohibitive quantity of computational resources (about 697% more), it can be concluded that the proposed beamformer based on delayed decimation filters is the most resource-efficient beamformer architecture for the given specification.

Finally, we see that, because of the lowest computational resources requirement, in practical cases such as implementing the beamformer either in a single-core/single-adder CPU, in a Field-Programmable Gate Array (FPGA) running at 64 MHz, or in an integrated circuit (VLSI) running at 10 MHz, the proposed architecture will be, in all cases, about 19% more efficient.

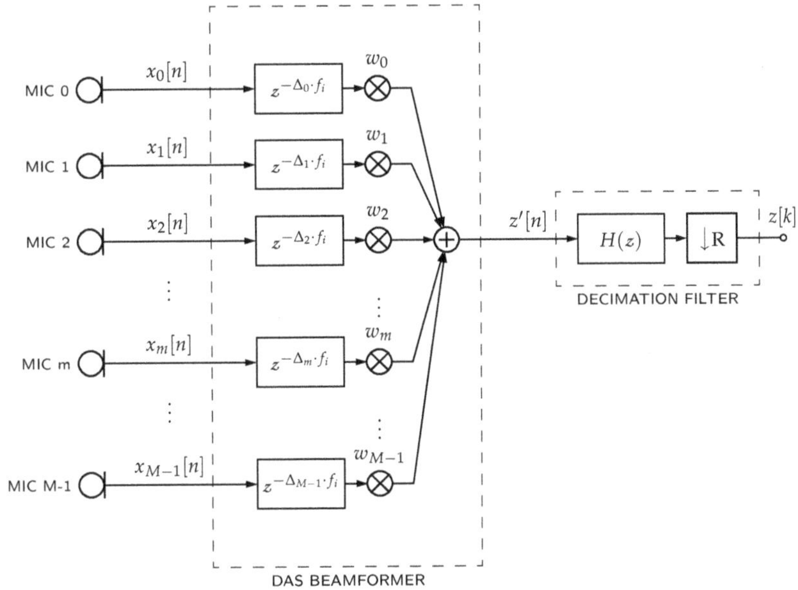

Figure 13. PDM microphones' DAS beamformer at PDM domain. Each PDM-mic output $x_m[n]$ is delayed by a Δ_m factor, then all delayed signals are weighted (factor w_m) and summed together. Finally, the resulting sum is filtered and downsampled.

Table 5. Comparison of the proposed beamformer architecture based on delayed decimation filter (*delayed_bf*) and other state-of-the-art beamformer architectures implementing a DAS beamformer, as specified in Tables 1 and 2. All percentages are related to the respective value for the *pdm_multi* architecture, the most efficient state-of-the-art architecture found for the given specification [13].

DAS Beamformer Architecture	Beamformer's Storage Requirement (S^z_{bf}) in Bit $\times 10^3$	Beamformer's Total Number of Additions per Second (S^o_{bf}) in APS $\times 10^8$	Estimated Minimum Frequency in a Single-Core/Single-Adder Processor (f_{cpu}) in MHz	Estimated Number of Adders in an FPGA Running at 64 MHz (T^+_{FPGA})	Estimated Number of Adders in a VLSI Circuit Running at 10 MHz (T^+_{lp})
Using delayed decimation filter (*delayed_bf*)	39.5 (−52%)	24.6 (−19%)	2458.58 (−19%)	39 (−19%)	246 (−20%)
Using a multi-stage decimation filter (*pcm_multi*)	210.0 (+155%)	41.8 (+37%)	4184.94 (+37%)	66 (+37%)	419 (+37%)
Using a single-stage memory saving decimation filter (*pcm_single_memsav*)	27.4 (−67%)	243.0 (+697%)	24,303.98 (+697%)	380 (+692%)	2431 (+694%)
Using a multi-stage decimation filter at PDM domain (*pdm_multi*)	82.3 (0%)	30.5 (0%)	3050.29 (0%)	48 (0%)	306 (0%)
Using a single-stage memory saving decimation filter at PDM domain (*pdm_single_memsav*)	77.8 (−6%)	35.5 (+16%)	3553.26 (+16%)	56 (+17%)	356 (+16%)

6. Conclusions

In this study, we proposed combining the decimation filters found in PDM-mics with the delay line required in the traditional DAS beamformer. This was achieved by designing a decimation filter that includes a stage realized with the Samadi filter structure, which easily allows its group delay to be altered by the varying a single parameter.

We evaluated the proposed architecture by comparing it to other state-of-the-art DAS beamformer architectures. To facilitate the comparison, we established a set of filter specifications as a baseline for all decimation filter designs. These specifications were sufficient for various PDM-mics and speech-processing applications.

The designed filter demonstrated satisfactory performance, as exemplified in the frequency response and group delay plots. Furthermore, using a Samadi filter provided flexibility in adjusting the group delay without altering the overall filter structure.

Overall, the proposed architecture showed promising filter design and resource requirements results, providing the best trade-off between storage and computational resources. The presented specification requires 52% lower storage resources and 19% lower computational resources than the most efficient state-of-the-art architecture. The findings support the feasibility and effectiveness of the proposed approach for beamforming applications applied, but not limited, to DAS beamformers.

Author Contributions: Conceptualization, S.J.C.I. and B.S.M.; methodology, S.J.C.I. and B.S.M.; software, S.J.C.I.; validation, S.J.C.I.; formal analysis, S.J.C.I. and B.S.M.; investigation, S.J.C.I.; resources, B.S.M.; data curation, S.J.C.I. and B.M; writing—original draft preparation, S.J.C.I.; writing—review and editing, S.J.C.I. and B.S.M.; visualization, S.J.C.I.; supervision, B.S.M.; project administration, B.S.M.; funding acquisition, B.S.M. All authors have read and agreed to the published version of the manuscript.

Funding: This research was funded by São Paulo Research Foundation (FAPESP), grant #2017/08120-6.

Conflicts of Interest: The authors declare no conflict of interest.

Abbreviations

The following abbreviations are used in this manuscript:

Symbols

$B_{N,K,d}(z)$	Samadi filter binomial component.
c	sound speed.
$D_{\min}$	minimum distance between microphones.
δ_p^j	jth-stage passband ripple.
Δ_m	delay from the array center to the mth microphone.
$d_{\max}$	maximum allowed delay parameter.
$\Delta_{\max}$	maximum required delay.
Δ	overall filter delay.
δ_p	passband ripple.
δ_s	stopband ripple.
d	Samadi filter delay parameter.
δ_s^j	jth-stage stopband ripple.
f_{cpu}	estimated minimum frequency in a single-core/single-adder processor.
f_i	input sampling rate.
f_o	output sampling rate.
F_p	passband frequency.
F_s	stopband frequency.
G	beamformer gain.
α	group delay.
$H(e^{2\pi i f/f_i})$	overall low-pass filter impulse response.
$H_{N,K,d}(z)$	Samadi filter impulse response.

$H(z)$	low-pass filter impulse response.
S^{z}_{bf}	beamformer's storage requirement.
K	number of zeros at $z = -1$ in a Samadi filter.
L_{frame}	frame length (for frequency domain implementations).
L_{in}	filter input length.
L_{out}	filter output length.
M	number of microphones.
N	Samadi filter order.
R	decimation factor.
T^{+}_{FPGA}	estimated number of adders in an FPGA running at 64 MHz.
T^{+}_{lp}	estimated number of adders in a VLSI circuit running at 10 MHz.
S^{+}_{bf}	beamformer's number of additions per second.
S^{*}_{bf}	beamformer's number of multiplications per second.
S^{o}_{bf}	beamformer's total number of additions per second.
U_p	passband frequency range.
U_s	stopband frequency range.
V_p	passband angular frequency range.
V^{j}_{p}	jth-stage passband angular frequency range.
V_s	stopband angular frequency range.
V^{j}_{s}	jth-stage stopband angular frequency range.
w_m	mth-filter channel gain.
ω_p	angular passband frequency.
ω_s	angular stopband frequency.

Abbreviations

ΣΔM	sigma–delta modulator.
ADC	analog-to-digital converter.
APS	additions per second.
DAS	delay-and-sum.
DoA	direction of arrival.
FIR	Finite Impulse Response.
FPGA	Field-Programmable Gate Array.
IIR	Infinite Impulse Response.
IoT	internet of things.
IVA	intelligent virtual assistants.
LPF	low-pass filter.
MAXFLAT	maximally flat.
MEMS	micro-electro-mechanical system.
PCM	pulse-code modulated.
PDM	pulse-density modulated.
PDM-mic	PDM microphone.
SNR	signal-to-noise ratio.
THD	total harmonic distortion.
THD+N	total harmonic distortion plus noise.
VLSI	very large-scale integration.

References

1. Krim, H.; Viberg, M. Two decades of array signal processing research: The parametric approach. *IEEE Signal Process. Mag.* **1996**, *13*, 67–94. [CrossRef]
2. Lawes, R. *MEMS Cost Analysis: From Laboratory to Industry*; Pan Stanford: Stanford, CA, USA, 2014.
3. Vardhini, P.H.; Makkena, M.L. Design and comparative analysis of on-chip sigma delta ADC for signal processing applications. *Int. J. Speech Technol.* **2021**, *24*, 401–407. [CrossRef]
4. Johnson, D.H.; Dudgeon, D.E. *Array Signal Processing: Concepts and Techniques*; Simon & Schuster, Inc.: New York, NY, USA, 1992.
5. Milic, L. *Multirate Filtering for Digital Signal Processing: MATLAB Applications*; Premier Reference Source, Information Science Reference; IGI Global: Hershey, PA, USA, 2009.
6. Metzler, B. *Audio Measurement Handbook*; Audio Precision: Raleigh, NC, USA, 1993.

7. Samadi, S.; Nishihara, A.; Iwakura, H. Universal maximally flat lowpass FIR systems. *IEEE Trans. Signal Process.* **2000**, *48*, 1956–1964. [CrossRef]
8. Herrmann, O. On the approximation problem in nonrecursive digital filter design. *IEEE Trans. Circuit Theory* **1971**, *18*, 411–413. [CrossRef]
9. Rajagpoal, L.; Roy, S.D. Design of maximally-flat FIR filters using the Bernstein polynomial. *IEEE Trans. Circuits Syst.* **1987**, *34*, 1587–1590. [CrossRef]
10. Haddad, R. A class of orthogonal nonrecursive binomial filters. *IEEE Trans. Audio Electroacoust.* **1971**, *19*, 296–304. [CrossRef]
11. Van Compernolle, D. Future Directions in Microphone Array Processing. In *Microphone Arrays: Signal Processing Techniques and Applications*; Springer: Berlin/Heidelberg, Germany, 2001; pp. 389–394. [CrossRef]
12. McClellan, J.; Parks, T. A unified approach to the design of optimum FIR linear-phase digital filters. *IEEE Trans. Circuit Theory* **1973**, *20*, 697–701. [CrossRef]
13. Carbajal Ipenza, S.J. Efficient Pulse-Density Modulated Microphone Array Processing. Master's Thesis, Universidade Estadual de Campinas, Campinas, Brazil, 2020.
14. Fliege, N. *Multirate Digital Signal Processing: Multirate Systems, Filter Banks, Wavelets*; Wiley: New York, NY, USA, 1994.

MDPI
St. Alban-Anlage 66
4052 Basel
Switzerland
www.mdpi.com

Sensors Editorial Office
E-mail: sensors@mdpi.com
www.mdpi.com/journal/sensors

www.ingramcontent.com/pod-product-compliance
Lightning Source LLC
Chambersburg PA
CBHW041941260926
51850CB00018B/368
* 9 7 8 3 0 3 6 5 8 9 6 6 4 *

Energy-Efficient Communication Networks and Systems